Ecohydrology: Future Perspectives

Ecohydrology: Future Perspectives

Editor: Jonah Lyons

MURPHY & MOORE
www.murphy-moorepublishing.com

Murphy & Moore Publishing,
1 Rockefeller Plaza,
New York City, NY 10020, USA

Visit us on the World Wide Web at:
www.murphy-moorepublishing.com

ISBN: 978-1-63987-175-9 (Hardback)

Cataloging-in-Publication Data

Ecohydrology : future perspectives / edited by Jonah Lyons.
p. cm.
Includes bibliographical references and index.
ISBN 978-1-63987-175-9
1. Ecohydrology. 2. Hydrology. 3. Aquatic ecology. I. Lyons, Jonah.
QH541.15.E19 E26 2022
577.6--dc23

Table of Contents

Preface

The sub-discipline of hydrology that deals with the study of interactions between water and ecological systems is termed as ecohydrology. Terrestrial and aquatic systems are studied under its domain. Terrestrial ecosystems include forests and deserts. It involves the interactions among vegetation, the land surface, the vadose zone and the groundwater. Aquatic ecosystems include streams, lakes, rivers and wetlands. It studies how their structures and functions are affected by water chemistry, geomorphology and hydrology. Transpiration and plant water use, the adaption of organisms to their water environment, the influence of vegetation and benthic plants on streamflow, are some of the main areas of research of this discipline. This book contains some path-breaking studies in the field of ecohydrology. The topics included herein are of utmost significance and bound to provide incredible insights to readers. Those with an interest in ecohydrology would find this book helpful.

Various studies have approached the subject by analyzing it with a single perspective, but the present book provides diverse methodologies and techniques to address this field. This book contains theories and applications needed for understanding the subject from different perspectives. The aim is to keep the readers informed about the progresses in the field; therefore, the contributions were carefully examined to compile novel researches by specialists from across the globe.

Indeed, the job of the editor is the most crucial and challenging in compiling all chapters into a single book. In the end, I would extend my sincere thanks to the chapter authors for their profound work. I am also thankful for the support provided by my family and colleagues during the compilation of this book.

Editor

1

Portfolio optimisation for hydropower producers that balances riverine ecosystem protection and producer needs

X. A. Yin, Z. F. Yang, and C. L. Liu

State Key Laboratory of Water Environmental Simulation, School of Environment, Beijing Normal University, Beijing, China

Correspondence to: Z. F. Yang (zfyang@bnu.edu.cn)

Abstract. In deregulated electricity markets, hydropower portfolio design has become an essential task for producers. The previous research on hydropower portfolio optimisation focused mainly on the maximisation of profits but did not take into account riverine ecosystem protection. Although profit maximisation is the major objective for producers in deregulated markets, protection of riverine ecosystems must be incorporated into the process of hydropower portfolio optimisation, especially against a background of increasing attention to environmental protection and stronger opposition to hydropower generation. This research seeks mainly to remind hydropower producers of the requirement of river protection when they design portfolios and help shift portfolio optimisation from economically oriented to ecologically friendly. We establish a framework to determine the optimal portfolio for a hydropower reservoir, accounting for both economic benefits and ecological needs. In this framework, the degree of natural flow regime alteration is adopted as a constraint on hydropower generation to protect riverine ecosystems, and the maximisation of mean annual revenue is set as the optimisation objective. The electricity volumes assigned in different electricity submarkets are optimised by the noisy genetic algorithm. The proposed framework is applied to China's Wangkuai Reservoir to test its effectiveness. The results show that the new framework could help to design eco-friendly portfolios that can ensure a planned profit and reduce alteration of the natural flow regime.

1 Introduction

Since the global electricity reform process began in the 1980s (Zelner et al., 2009; Wang and Chen, 2012), and especially after the 1990s, market-oriented reforms in the electric power industry were implemented in many countries to allocate electricity more efficiently through market mechanisms (Cai et al., 2009, 2011; Williams and Dubash, 2004; Tsai, 2011; Wu, 2012). In the process of reform, vertically integrated electricity utilities were restructured and unbundled, and competition has been introduced into generation as well as the wholesale and retail segments of the industry (Pollitt, 2009; Holmberg, 2011; Mulder, 2011). In deregulated markets, hydropower producers usually own generation resources and are allowed to participate in any submarkets such as bilateral contract and spot markets (Karandikar et al., 2010; Ramos et al., 2010). Maximising profits is usually their sole objective for participating in the market (Liu et al., 2009). To maximise profits, hydropower producers need to devise their own strategies for portfolio design (Shen and Yang, 2012).

Extensive research has been performed to optimise hydropower portfolios. Bjørgan et al. (1999) integrated the optimisation of future contract and power scheduling based on risk management in a static mean-variance framework, and the efficient frontier was used as a tool to identify a preferred contract portfolio. Using a continuous-time framework, Keppo (2002) proposed a model for optimal long-term electricity trading strategies and the associated production process by maximising production and terminal water reservoir level in the case of multi-reservoir hydropower systems. Fleten et al. (2002) used a four-stage stochastic programming model with 256 scenarios for simultaneous risk management via contracts and hydropower generation planning on a 1.5-year horizon. Shrestha et al. (2005) presented a portfolio management technique to optimise expected revenue for a hydropower producer, a scenario that utilises tree analysis with corrective recourse actions for probable scenarios. The effect of contract position adjustment is also analysed to minimise revenue variation from the expected values for risk-averse producers. Liu et al. (2009) present a stochastic linear programming framework for hydropower portfolio management with uncertainty in market prices and inflows on medium term, in which the uncertainty was modelled as a scenario tree using the Monte Carlo simulation method, to maximise the expected revenue over the entire scenario tree. These approaches could effectively optimise hydropower portfolios and maximise the total profit of hydropower producers. However, none of the previous research on hydropower portfolio optimisation considered the need to protect riverine ecosystems.

The determined portfolios could give two important results: (1) the optimal electricity volume that should be generated for each period (day, week, month, etc.); (2) the optimal allocation of the generated hydropower among different electricity submarkets. The electricity volume determines the water volume that should be released to the downstream rivers, and in turn determines the degree of flow regime alteration. Flow regime alteration is the major cause of riverine ecosystem degradation (Bunn and Arthington, 2002; Babel et al., 2011; Kern et al., 2012; Bhatt and Khanal, 2012). Due to the difference of electricity price among different submarkets, the electricity allocation among different submarkets as well as the total generated electricity volume determines the revenue of hydropower producers. Thus, the determination of hydropower portfolio before hydropower generation is essential for both riverine ecosystem protection and producer needs. Although profit maximisation is the major objective for producers in deregulated markets, the need to reduce flow regime alteration must be incorporated into the process of hydropower portfolio optimisation for riverine ecosystem protection, especially with the background of increasing attention to environmental protection and stronger opposition to hydropower generation (Jager and Smith, 2008; Chen et al., 2012).

Many new reservoir operating methods have been proposed to better sustain environmental flows (e-flows, the volume of water that should remain in a river and the variation of this provision over time to maintain specific indicators of ecosystem health) in rivers (Richter, 2010; Shiau and Wu, 2010; Brown and King, 2012). These methods are river protection approaches at the hydropower generation stage. Portfolio determination is a task that occurs before hydropower generation. The designed hydropower portfolio can significantly influence reservoir operation through the influence of planned electricity volume and in turn the water volume released to the downstream rivers (Chen et al., 2012; Yin et al., 2011, 2012). Even if the reservoir operating rules are refined, their ecological protection effects may not be as effective as expected under the conditions of improperly designed portfolios. Meanwhile, the effects of portfolio optimisation on river protection need to be displayed by the influence on hydropower operation. Research on the optimal portfolios must be performed with the consideration of both economic benefits and ecological needs, which can provide the basis for developing eco-friendly reservoir operating rules.

This research seeks to remind hydropower producers of the requirement of river protection when they design the portfolios and to help the producers design eco-friendly portfolios. In this work, we extend previous research on hydropower portfolio optimisation and establish a framework to determine the optimal portfolio of a hydropower reservoir, accounting for both economic benefits and ecological needs. This framework may help mitigate the impact of hydropower generation on riverine ecosystems, especially for the situation where the tension between river protection and power generation is very severe.

2 Methods

2.1 Framework for determining optimal hydropower portfolios for both riverine ecosystem and producer needs

2.1.1 Determining the submarkets for participation

Submarkets available for hydropower are not the same for each country in the world. Three common markets are bilateral contract (future), day-ahead, and real-time balancing (Alaywan and Wu, 2002; Kranz et al., 2002). Day-ahead and real-time balancing markets are also called spot markets. Participants in the electricity market face the risks of unknown demand and price (Aggarwal et al., 2009; Eichhorn et al., 2009). To avoid revenue risks, hydropower producers and grid companies like to make long-term or mid-term electricity supply contracts, forming the bilateral contract market (Lin and Wu, 2008). In a bilateral contract, the trading power volume and power price are designed by the power producers and grid companies, and will not change during the contract period. In the day-ahead market, participants submit sell/bid offers for electricity for the following day. These offers consist of a quantity of energy to be sold or purchased and a desired price, where sell offers correspond roughly to each producer's marginal cost of energy production. The system operators then rank sell offers from least to most expensive; the last sell offer required to satisfy day-ahead forecast demand clears the market, and the marginal cost of increasing power supply by one additional megawatt determines the market-clearing energy price. Sellers with offers equal to or below this price then generate revenue equal to their respective bid quantities multiplied by the market-clearing price (Rothwell and Gomez, 2003). In the real-time balancing market, system managers coordinate an hourly real-time energy market, which is used to meet real-time electricity demand when it varies relative to day-ahead forecasts. An hourly real-time market clearing price is determined in a manner similar to that of the day-ahead market (i.e. via the ranking of bid/sell offers), and transactions are consummated as necessary to meet real-time demand (Lambert, 2001). The price in the spot market is uncertain, and it could be higher than, lower than or equal to the price in the contract market.

The hydropower producers choose the submarkets in which they want to participate. The selection depends to some extent on the risk preferences of the producers (Shrestha et al., 2005; Botterud et al., 2010). Risk-taking producers tend to participate in the spot market for possibly higher profit despite higher risk. Risk-averse producers participate in the bilateral contract market that is more reliable, although extensive research supports short-term bidding for potentially higher revenue. Risk-neutral producers tend to participate in both the bilateral contract and spot markets.

2.1.2 Determining the rules for reducing flow regime alteration

Although sustaining the natural flow regime is a basic principle for river protection and e-flow management (Poff et al., 1997, 2010), hydropower generation will inevitably lead to changes. A possible method to reduce flow regime alteration would be to apply a type of e-flow provision strategy (such as sustaining the minimum e-flows and ensuring several high flow pulses, etc.) and optimise the parameters related to the hydropower portfolio to minimise alteration of the natural flow regime or restrict the degree of alteration to a specified threshold. This method has been applied extensively in research (Black et al., 2005; Richter and Thomas, 2007; Jager and Smith, 2008; Yin et al., 2011, 2012).

Reservoir operators can use different e-flow provision strategies depending on their attitude toward riverine ecosystems. Sustaining the minimum e-flow is a commonly used strategy for real-world e-flow provisions. This strategy can provide basic protection of the riverine ecosystem, avoiding severe degradation. To sustain ecological functions related to high flows, some research proposes occasional high-flow releases for habitat improvement (Ligon et al., 1995; Gore et al., 2001; Renofalt et al., 2009). To better protect riverine ecosystems, hydropower operators can develop and apply more sophisticated e-flow strategies.

2.1.3 Setting the portfolio optimisation objectives and constraints

Maximising the total profit for a given planning period and reducing the degree of flow regime alteration are two objectives for hydropower portfolio design. However, the two objectives are in conflict and cannot be achieved simultaneously, which is a typical multi-objective problem. To address this problem, the optimisation objectives need to be set first. An optimisation objective can be set as one of two conflicting objectives, i.e. maximising the overall profit for a given planning period or reducing the degree of alteration of the flow regime. The other objective can be set as one constraint by assigning a threshold accepted by the hydropower producers or river protectors. Alternatively, the two objectives could be integrated into one by some mathematical method such as compromise programming or weighted average.

Constraints for portfolio optimisation include maximum power generation capacity, water release capacity, reservoir maximum storage capacity, etc. If either of the two objectives is chosen as a single optimisation objective, the constraints also need to include the threshold for the other objective, i.e. the minimum acceptable overall profit or the maximum acceptable degree of flow regime alteration. The water mass balance also needs to be considered (Liu et al., 2009).

2.1.4 Choosing the solution method under uncertainty

In a deregulated market, portfolio optimisation is faced with the uncertainty of reservoir inflows and spot electricity prices (Fleten et al., 2002). A series of methods has been developed to obtain optimal solutions under uncertainty, such as stochastic dynamic programming, stochastic dual dynamic programming, stochastic programming combined with scenario trees, and noisy genetic algorithms (Chang et al., 2005; Chen, 2003; Chen et al., 2007). The optimisation method needs to address these uncertainties.

2.2 Methods used in the case study

2.2.1 Range of variability approach

The range of variability approach (RVA) (Richter et al., 1996, 1997, 1998) has been widely used for assessing flow regime alteration and directing hydraulic facility operations (Galat and Lipkin, 2000; Shiau and Wu, 2004, 2006, 2007; Zhang et al., 2009). According to the RVA, a range of variation for each hydrological indicator was derived from the natural hydrological time series and was set as the flow management target. A range defined by the 75th and 25th percentile flows has been recommended as the management target (Richter et al., 1998). The degree of alteration, D_m, was used to measure the deviation of the impacted flow regime from the natural one for the mth hydrologic indicator, which was defined by

$$D_m = \left| \frac{N_{o,m} - N_{e,m}}{N_{e,m}} \right| \times 100\,\%, \tag{1}$$

where $N_{o,m}$ was the observed number of post-impact years in which the value of the mth hydrologic indicator fell within its RVA target range, and $N_{e,m}$ was the expected number of post-impact years in which the indicator value fell within the RVA target range. The average degree of alteration of these hydrologic indicators was applied to quantify the river's overall impact, which can be expressed as follows:

$$D = \frac{1}{G}\sum_{m=1}^{G} D_m, \tag{2}$$

where D was the overall degree of flow regime alteration, and G was the number of hydrological indicators (G is equal to 32 in this research). The degree of flow regime alteration can be categorised further into three levels: low alterations (values of D between 0 and 0.33), moderate alterations (values of D between 0.33 and 0.67), and high alterations (values of D between 0.67 and 1.0) (Richter et al., 1998).

2.2.2 Tennant method

To illustrate the applicability of the proposed approach, we use the Tennant method (Tennant, 1976), a simple and widely used method, to determine seasonal minimum e-flows. Tennant method was recommended for e-flow assessment in the "technical guidelines for environmental impact assessment for ecological water usage, low temperature water and fish habitat facilities in the hydraulics projects" by the State Environment Protection Administration of China (2006), and was applied by the Haihe Water Conservation Commission (2008) for the Hai River basin. Accordingly, the wet season e-flow was set at 30 % of average daily flow (ADF), and the dry season e-flow was set at 10 % ADF. More sophisticated methods could be used to replace the Tennant method if enough hydrological, biological, and geomorphological data are available.

2.2.3 Optimisation objectives and constraints

In the following case study, the hydropower producer participates in the contract and day-ahead markets according to the hydropower generation planning. The goal for hydropower portfolio optimisation is to maximise the mean annual revenue subject to specified e-flow management requirements. These requirements include the e-flow provision strategy and the specified threshold for the degree of flow regime alteration. The optimisation problem can be expressed by the following equation:

$$L = \max \frac{1}{T}\sum_{j=1}^{T}\sum_{k=1}^{365}\left(\mathrm{PC}_{kj}\cdot \mathrm{CL}_{kj} + \mathrm{PD}_{kj}\cdot \mathrm{DL}_{kj}\right), \tag{3}$$

which is subject to

$$R_{kj} \geq \mathrm{EF}_{kj}, \tag{4}$$

$$D \leq D_0, \tag{5}$$

where L denotes the overall optimisation objective, and PC_{kj} is the designed hydropower price in the hydropower supply contract for day k of year j (constant within 1 month, RMB kwh^{-1}); CL_{kj} is the designed hydropower volume in the hydropower supply contract for day k of year j (constant within 1 month, kwh); PD_{kj} is the hydropower price in the day-ahead market for day k of year j (RMB kwh^{-1}); DL_{kj} is the bidding volume for power in the day-ahead market for day k of year j (kwh); R_{kj} is the actual reservoir water release for day k of year j; EF_{kj} is the minimum e-flow for day k of year j; D is the degree of actual flow regime alteration under a certain portfolio; and D_0 is the specified threshold of degree of flow regime alteration.

In the contract market, the producer and electricity grid make an agreement on the contract load for each month and the associated power price. In the day-ahead market, the producers need to determine the bidding volume for power. In this case, we assume that the producers first use inflow and the water in the reservoir to produce electricity to satisfy the contract load, and the producer will buy electricity from the market to satisfy the contract load only when the available electricity is not sufficient. Because the electricity cannot be stored, the bought electricity is set equal to the difference between the contract volume and the available power volume. We use the following equations to determine the bidding volume of power in the day-ahead market:

$$\text{if } \mathrm{AE}_{kj} - \mathrm{CL}_{kj} > 0,\ \mathrm{DL}_{kj} = \min\left[k_{kj}\left(\mathrm{AE}_{kj} - \mathrm{CL}_{kj}\right)\mathrm{PD}_{kj},\ \mathrm{ME} - \mathrm{CL}_{kj}\right]; \tag{6}$$

$$\text{if } \mathrm{AE}_{kj} - \mathrm{CL}_{kj} \leq 0,\ \mathrm{DL}_{kj} = \mathrm{AE}_{kj} - \mathrm{CL}_{kj}, \tag{7}$$

where AE_{kj} is the available electricity volume that the hydropower plant can generate with the water in the reservoir and inflow for day k of year j (kwh); ME is the maximum electricity production capacity for 1 day (kwh); and k_{kj} is the parameter for day k of year j (kwh RMB^{-1}).

In Eq. (6), $k_{kj}(\mathrm{AE}_{kj} - \mathrm{CL}_{kj})\mathrm{PD}_{kj}$ means that the higher the available electricity volume, the higher the day-ahead power price, and the higher the bidding volume for power. There may be some alternative and more sophisticated equations to replace $k_{kj}(\mathrm{AE}_{kj} - \mathrm{CL}_{kj})P_{kj}$. The equation forms may also influence the revenue of hydropower producers and the effects of riverine ecosystem protection. Further research would be valuable to analyse the influence of equation forms and the optimal forms. The parameters k_{kj} and CL_{kj} are two variables that need to be optimised. We assume k_{kj} and CL_{kj} do not change over the period of 1 month and are the same for each year. Thus, k_{kj} and CL_{kj} both have 12 values.

The real electricity produced by a hydropower reservoir is related to many factors such as the turbine release water discharge for power generation, the water head, and the coefficient of hydropower station power generation. The equations for hydropower generation have been presented extensively in the literature (e.g. Cheng et al., 2008; Li et al., 2009; Liu et al., 2011), and thus are not listed in the present paper.

2.2.4 Noisy genetic algorithm

In hydropower portfolio optimisation, the future inflow and spot price are uncertain. The noisy genetic algorithm (NGA) is an effective method to determine the optimal values of parameters under uncertainty (Miller and Goldberg, 1996). The NGA has been applied for stochastic reservoir operation (Yun et al., 2010), ground water remediation (Aly and Peralta, 1999) and groundwater sampling network design (Wu et al., 2005) under uncertainty. In this research, NGA is applied to optimise the hydropower portfolio under uncertainty of flows and spot price.

The term *noise* can be defined as any factor that hinders the accurate evaluation of the fitness of a given trial solution. In this study, noise refers to the stochastic nature of the inflows and the power price in the day-ahead market. Most components in the NGA are the same as in a simple genetic algorithm (GA). The main difference between the NGA and GA is in the fitness function. In the NGA, the fitness value cannot be evaluated accurately because of the variability of monthly inflows. To overcome this difficulty, the fitness value is substituted by the expected fitness value. The details of the NGA can be found in Miller and Goldberg (1996) and Yun et al. (2010).

3 Study site

The Wangkuai Reservoir is a key hydraulic facility in the Hai River basin of China. The present effective storage capacity of the Wangkuai Reservoir is $6.52 \times 10^8\,m^3$, and the dead storage capacity is $0.88 \times 10^8\,m^3$. The catchment area of the reservoir is $3770\,km^2$. The installed hydropower generation capacity of the Wangkuai Reservoir is 21.5 MW. In this research, we focus on the hydropower generation function of the Wangkuai Reservoir. The inflow data from 1971 to 1993 and the physical characteristics are used to simulate hydropower generation and optimise the hydropower portfolio. The producer considered in this paper is a price taker. The hydropower price in the day-ahead market is shown in Table 1 (Liu et al., 2009; Liu, 2009). The bilateral contract price between the Wangkuai Reservoir and the grid company is $0.36\,RMB\,kwh^{-1}$ (Hebei Province Municipal Price Bureau, 2009).

In this research, we consider two e-flow provision strategies. In the first e-flow strategy, only the minimum e-flows are sustained. This strategy is most commonly used in real-world e-flow provisions. In the second e-flow strategy, in addition to the minimum e-flows, occasional high-flow (flows falling above the 75th percentile of all flows) releases are required to sustain the ecological functions related to high flows. In this research, we also assume for demonstration that after three high-flow events have occurred in a season, no further high flows are released, following the research by Vogel et al. (2007).

Table 1. Details of price variations in the day-ahead market.

Month	Mean ($RMB\,kwh^{-1}$)	Maximum ($RMB\,kwh^{-1}$)	Minimum ($RMB\,kwh^{-1}$)	Standard deviation ($RMB\,kwh^{-1}$)
1	0.29	0.45	0.20	0.03
2	0.28	0.45	0.20	0.03
3	0.27	0.45	0.20	0.03
4	0.29	0.45	0.20	0.04
5	0.33	0.45	0.20	0.03
6	0.37	0.45	0.20	0.04
7	0.43	0.45	0.20	0.02
8	0.40	0.45	0.20	0.04
9	0.41	0.45	0.20	0.03
10	0.38	0.45	0.20	0.05
11	0.36	0.45	0.20	0.05
12	0.32	0.45	0.20	0.04

4 Results

Matlab 6.5 was used to apply NGA to determine the optimal hydropower portfolio. The generation size and evolution times were set at 600 and 1000, respectively. According to the Tennant method (Tennant, 1976), for the dry season (November to April), the seasonal minimum e-flow (10 % average daily flow) was $1.8\,m^3\,s^{-1}$; for the wet season (May to October), the seasonal minimum e-flow (30 % average daily flow) was $5.4\,m^3\,s^{-1}$. The threshold for the degree of flow regime alteration is set at 0.67, the upper value for moderate alteration of the flow regime (Richter et al., 1996, 1997, 1998).

The optimised parameters are listed in Tables 2 and 3. These tables show that, during 5 of the 6 months in the wet season, the contract load under the second e-flow provision strategy is higher than the contract load under the first strategy because, under the second e-flow strategy, the reservoir is required to maintain several high flow pulses. The releases of greater flows required by the second e-flow strategy make higher contract loads reasonable. In addition, most (9 of 12) of the values for parameter k under the first e-flow strategy are higher than the values of parameter k under the second e-flow strategy, indicating that more water will be released to produce hydropower for the day-ahead market under the first e-flow strategy than under the second strategy.

The optimised mean annual revenues are 8.72×10^6 RMB and 7.55×10^6 RMB under the two e-flow provision strategies. The mean annual revenue under the second strategy is lower than the mean annual revenue under the first strategy, possibly because the extra water releases to maintain the high flows under the second e-flow strategy increase the proportion of electricity assigned in the contract market, reducing the proportion of electricity assigned in the spot market, which sometimes has higher prices than the contract price. Thus, the extra requirement to sustain high flows for river ecosystem protection potentially reduces the profits from hydropower generation.

Table 2. Optimised monthly volume of hydropower in the contract for the Wangkuai Reservoir ($\times 10^5$ kwh).

	Jan	Feb	Mar	Apr	May	Jun	Jul	Aug	Sep	Oct	Nov	Dec
E-flow strategy 1	8.86	7.27	10.35	10.18	14.18	14.91	16.39	18.65	18.21	17.85	7.50	7.78
E-flow strategy 2	9.21	7.86	9.47	9.26	17.69	18.35	19.47	21.11	20.81	17.36	8.75	7.70

Table 3. Optimised value of k for each month for the Wangkuai Reservoir (kwh RMB^{-1}).

	Jan	Feb	Mar	Apr	May	Jun	Jul	Aug	Sep	Oct	Nov	Dec
E-flow strategy 1	1.13	0.86	0.78	0.72	0.81	0.92	1.03	1.35	1.48	0.78	1.31	0.98
E-flow strategy 2	0.96	0.76	1.32	0.64	0.84	0.68	0.64	1.08	0.72	0.92	0.96	0.88

5 Discussion

5.1 The influence and significance of incorporating environmental policies in portfolio management

In previous research on portfolio optimisation, no specific rules were used for e-flow provision. The e-flows are supplied only by the water released for hydropower generation. This strategy is called non-e-flow for short. In the following section, we explore the influence of incorporating environmental policies in portfolio management by comparing revenue and flow regime alteration under the three e-flow strategies (i.e. the non-e-flow strategy and the two strategies established in the study site section), possibly helping to test the significance of incorporating environmental policies in portfolio optimisation.

We first determine the maximum mean annual revenue (without the constraint of the threshold for flow regime alteration degree) and corresponding degree of flow regime alteration under the three e-flow strategies. Under the three strategies, the maximum revenue and the corresponding degree of alteration are 12.38×10^6 RMB and 0.82 (non-e-flow strategy), 10.27×10^6 RMB and 0.75 (strategy 1), and 9.37×10^6 RMB and 0.68 (strategy 2). In comparison with the maximum annual revenue under the e-flow strategies 1 and 2 (10.27×10^6 RMB and 9.37×10^6 RMB), the non-e-flow portfolio optimisation method achieves higher revenue. However, the degree of flow regime alteration corresponding to this high revenue is 0.82, obviously greater than the degree of alteration under the first and second e-flow provision strategies (0.75 and 0.68). Thus, although the non-e-flow portfolio optimisation method could yield higher revenue, it would come at the cost of degradation of the river ecosystem. To avoid severe degradation of riverine ecosystems, incorporation of an e-flow provision strategy into the hydropower portfolio optimisation process is necessary, at least with regard to sustaining the minimum e-flows.

The minimum degrees of flow regime alteration are also determined under the three e-flow strategies. The lowest degrees of alteration under the non-e-flow and the first e-flow strategies are the same (0.31) because, under the non-e-flow strategy, the releases are also greater than the minimum e-flows in each month to achieve the minimum degree of flow regime alteration, and the contract load and k are the same under the two e-flow strategies. Thus, if reducing the degree of flow regime alteration is taken as the key objective, it is not necessary to incorporate the minimum e-flow requirement as a constraint. The degree of flow regime alteration (0.21, corresponding to the revenue of 3.59×10^6 RMB) under the second e-flow strategy is obviously lower than the degree of flow regime alteration under the other two strategies, demonstrating the significance of assigning some high flows in the e-flow provision rules.

Table 4. Minimum degree of flow regime alteration under different planned revenues and different e-flow provision strategies.

Revenue	9.37×10^6 RMB	6.63×10^6 RMB	3.89×10^6 RMB
Non-e-flow strategy	0.78	0.69	0.36
E-flow strategy 1	0.72	0.61	0.31
E-flow strategy 2	0.68	0.57	0.28

We further determine the minimum degree of flow regime alteration corresponding to the mean annual revenues of 9.37×10^6 RMB (the maximum mean annual revenue that all the three strategies can achieve), 3.89×10^6 RMB (the minimum mean annual revenue that all three strategies can achieve), and 6.63×10^6 RMB (the median revenue that all the three strategies can achieve) for the three e-flow provision strategies. The results are listed in Table 4. Table 4 shows that the degree of flow regime alteration under the non-e-flow strategy is always greater than the degree of flow regime alteration under the other two strategies, and the degree of alteration under the second strategy is always less than the degree of alteration under the other two strategies. Thus, the incorporation of a specific e-flow strategy can result in a lower degree of flow regime alteration with the same annual revenue. It further demonstrates the importance of

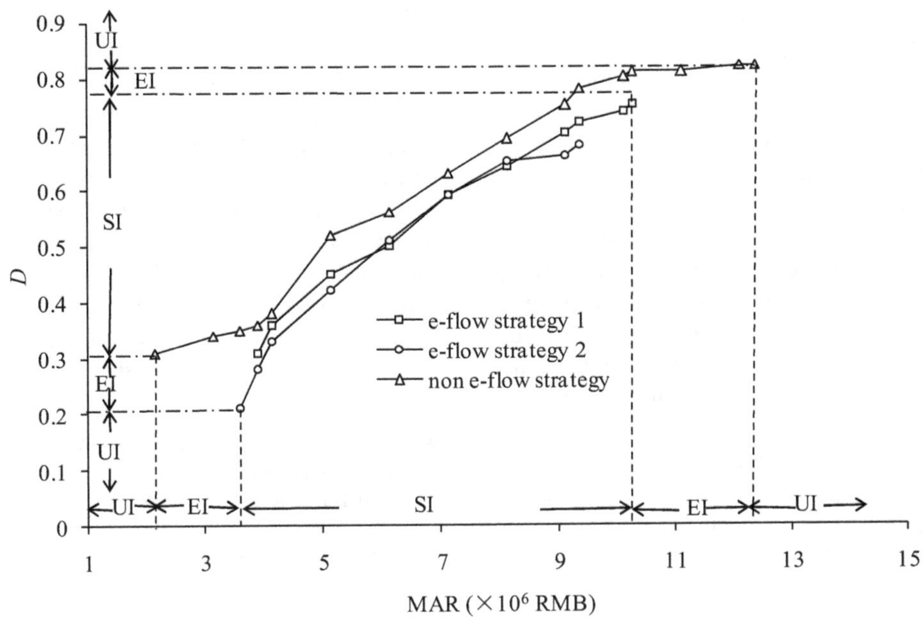

Fig. 1. Minimum degree of flow regime alteration for different planned mean revenues under three e-flow provision strategies. *D* is the degree of flow regime alteration; MAR is the mean annual revenue; UI is the unachievable interval; EI is the exclusive interval; SI is the shared interval.

incorporatinge-flow strategy into portfolio optimisation process for a specified revenue.

5.2 Determining the optimal e-flow provision strategy

Different e-flow provision strategies will result in different mean annual revenues and different degrees of flow regime alteration. The basic principles for e-flow strategy determination can be stated as follows: if the planned revenue can be achieved by several e-flow strategies, the strategy that results in the lowest flow regime alteration is chosen. If the planned degree of flow regime alteration can be achieved by several e-flow strategies, the strategy that results in the highest revenue should be chosen. On the basis of these principles, we have drawn the curves for mean annual revenue and minimum degree of alteration under the three e-flow provision strategies. The results are shown in Fig. 1.

On the basis of Fig. 1, the mean annual revenue can be divided into three types of intervals, i.e. exclusive intervals (the intervals that only one e-flow strategy can achieve), shared intervals (the intervals that more than one type of e-flow strategy can achieve), and unachievable intervals (the intervals that no available e-flow strategy can achieve). From the planned revenue, we can easily know which intervals the planned revenue is within. If the planned revenue is within an exclusive interval, the e-flow strategy corresponding to that interval can be applied in the portfolio optimisation process. If the planned revenue is within a shared interval, the e-flow strategy that results in the lowest degree of flow regime alteration should be applied to maintain the riverine ecosystem. For planned revenue within an unachievable interval, if the revenue is above the highest value for all e-flow strategies, no strategy can achieve a low degree of flow regime alteration in the contract and day-ahead market. The revenue can possibly be achieved by participating in other electricity submarkets, such as the real-time balancing market (the price in this submarket can be higher, but the price is more uncertain). If the revenue is below the lowest value for all e-flow strategies, which indicates a very favourable attitude toward river protection, the hydropower producers can apply a more favourable e-flow strategy that could result in a lower degree of flow regime alteration.

On the basis of Fig. 1, the degree of flow regime alteration can also be divided into three categories, i.e. exclusive intervals (an interval that only one e-flow strategy can achieve), shared intervals (intervals that more than one type of e-flow strategy can achieve), and unachievable intervals (intervals that no e-flow strategy can achieve). Like the procedure for the mean annual revenue discussed above, the most suitable e-flow provision strategy for a planned degree of flow regime alteration can be determined. If the planned degree of flow regime alteration is within an exclusive interval, the e-flow strategy corresponding to that interval can be applied in the portfolio optimisation process. If the planned degree of alteration is within a shared interval, the e-flow strategy that results in the highest revenue should be adopted. If the planned degree of alteration is within an unachievable interval and is below the lowest value for all strategies, the hydropower producers should develop and apply more favourable e-flow strategies.

Choosing an e-flow provision strategy is a process of compromising between hydropower producers and river protectors. The river protectors usually have the authority to ask the hydropower producers to obey some basic rules for riverine ecosystem protection, such as a minimum e-flow release. If the river protectors would like to improve the health of the riverine ecosystem by changing the hydropower generation scheme, a more ecologically favourable e-flow provision strategy should become a legal requirement, or ecological compensation should be given to the hydropower producers. On the basis of Fig. 1, we can make a preliminary assessment of the compensation criteria. For example, if the present legally required e-flow rules are to sustain the minimum e-flows and the river protectors want hydropower producers to use the second e-flow strategy, the compensation criteria should be approximately 0.9×10^6 RMB, i.e. the difference between the maximum possible mean annual revenue under the first (10.27×10^6 RMB) and second e-flow strategies (9.37×10^6 RMB).

6 Conclusions

Previous research on hydropower portfolio optimisation focused mainly on the maximisation of profits but neglected the requirement of riverine ecosystem protection. This research seeks mainly to remind hydropower producers of the requirement for river protection when they design portfolios and help make a shift of portfolio optimisation from economically oriented to ecologically friendly. In this study, a new framework has been developed to determine optimal hydropower portfolios considering both economic benefits and ecological needs. Within this framework, the degree of flow regime alteration is adopted as a constraint for e-flow provision and riverine ecosystem protection, and the maximisation of mean annual revenue is set as the optimisation objective. The following objectives are achieved.

- For the same planned mean annual revenue, the alteration in flow regime is lower for portfolios with e-flow provision strategies than for the portfolio without an e-flow strategy. A lower degree of alteration indicates a lower degree of river degradation. If a planned mean annual revenue can be achieved under either an e-flow strategy or a non-e-flow strategy, incorporating the e-flow strategy into the portfolio optimisation process is both economically and ecologically beneficial.
- The proper e-flow provision strategy depends on the planned revenue and the planned degree of flow regime alteration. If the planned revenue can be achieved by several e-flow strategies, the strategy that results in the lowest flow regime alteration should be chosen. If the planned degree of flow regime alteration can be achieved by several e-flow strategies, the strategy that results in the highest revenue should be chosen.

In the case study of Wangkuai Reservoir, the contract and day-ahead markets participated according to the planned needs of the reservoir. In future research on portfolio optimisation in other cases that account for both riverine ecosystem and producer needs, the real-time balancing market may also participate according to the planning needs of these specific cases and the risk preferences of the hydropower producers. In these cases, the proposed framework for portfolio optimisation can also be used. The difficulties are the precise prediction of hydropower price in the real-time balancing market and the assessment of hydrological alteration at hourly time steps, which require further research.

Acknowledgements. We thank the International Science & Technology Cooperation Program of China (no. 2011DFA72420), the National Basic Research Program of China (no. 2013CB430402), the National Science Foundation for Innovative Research Group (no. 51121003), and the National Natural Science Foundation of China (no. 51309012) for their financial support.

Edited by: Y. Cai

References

Aggarwal, S. K., Saini, L. M., and Kumarv, A.: Electricity price forecasting in deregulated markets: a review and evaluation, Int. J. Electr. Power Energ. Syst. Eng., 31, 13–22, doi:10.1016/j.ijepes.2008.09.003, 2009.

Alaywan, Z. and Wu, T.: Effects of Firm Transmission Rights on Transmission Revenue Allocation in California, available at: http://citeseerx.ist.psu.edu/viewdoc/download?doi=10.1.1.200.1503&rep=rep1&type=pdf (last access: 10 December 2013), 2002.

Aly, A. A. and Peralta, R. C.: Optimal design of aquifer cleanup systems under uncertainty using a neural network and a genetic algorithm, Water Resour. Res., 35, 2523–2532, doi:10.1029/98WR02368, 1999.

Babel, M. S., Chien, N. D., Mullick, M. A., and Nanduri, U. V.: Operation of a hydropower system considering environmental flow requirements: a case study in La Nga river basin, Vietnam, J. Hydro-Environ. Res., 6, 63–73, doi:10.1016/j.jher.2011.05.006, 2011.

Bhatt, R. P. and Khanal, S. N.: A study on change in flow regime and discharge impacts on water quality of hydropower operation, Int. J. Ecol. Dev., 21, 76–88, 2012.

Bjørgan, R., Liu, C. C., and Lawarree, J.: Financial risk management in a competitive electricity market, IEEET Power Syst., 14, 1285–1291, doi:10.1109/59.801886, 1999.

Black, A. R., Rowan, J. S., Duck, R. W., Bragg, O. M., and Clelland, B. E.: DHRAM: a method for classifying river flow regime alterations for the EC Water Framework Directive, Aquat. Conserv., 15, 427–446, doi:10.1002/aqc.707, 2005.

Botterud, A., Kristiansen, T., and Ilic, M. D.: The relationship between spot and futures prices in the Nord Pool electricity market, Energ. Econ., 32, 967–978, doi:10.1016/j.eneco.2009.11.009, 2010.

Brown, C. and King, J.: Modifying dam operating rules to deliver environmental flows: experiences from southern Africa, Int. J. River Basin Manage., 10, 13–28, doi:10.1080/15715124.2011.639304, 2012.

Bunn, S. E. and Arthington, A. H.: Basic principles and ecological consequences of altered flow regimes for aquatic biodiversity, Environ. Manage., 30, 492–507, doi:10.1007/s00267-002-2737-0, 2002.

Cai, Y. P., Huang, G. H., Yang, Z. F., Lin, Q. G., and Tan, Q.: Community-scale energy management systems design under uncertainty – An interval chance-constrained programming approach, Renew. Sustain. Energ. Rev., 13, 721–735, 2009.

Cai, Y. P., Huang, G. H., Tan, Q., and Yang, Z. F.: An integrated approach for climate-change impact analysis and adaptation planning under multi-level uncertainties, Part I: Methodology, Renew. Sustain. Energ. Rev., 15, 2779–2790, 2011.

Chang, F. J., Chen, L., and Chang, L. C.: Optimizing the reservoir operating rule curves by genetic algorithms, Hydrol. Process., 19, 2277–2289, 2005.

Chen, L.: Real-coded genetic algorithm optimization of long-term reservoir operation, J. Am. Water Resour. As., 39, 1157–1165, 2003.

Chen, L., McPhee, J., and Yeh, W. W.: A diversified multiobjective GA for optimizing reservoir rule curves, Adv. Water Resour., 30, 1082–1093, 2007.

Chen, Q. W., Chen, D., Li, R. N., Ma, J. F., and Blanckaert, K.: Adapting the operation of two cascaded reservoirs for ecological flow requirement of a de-watered river channel due to diversion-type hydropower stations, Ecol. Model., 252, 266–272, doi:10.1016/j.ecolmodel.2012.03.008, 2012.

Cheng, C. T., Wang, W. C., Xu, D. M., and Chau, K. W.: Optimizing hydropower reservoir operation using hybrid genetic algorithm and chaos, Water Resour. Manage., 22, 895–909, doi:10.1007/s11269-007-9200-1, 2008.

Eichhorn, A., Heitsch, H., and Römisch, W.: Scenario tree approximation and risk aversion strategies for stochastic optimization of electricity production and trading, in: Optimization in the Energy Industry, Springer, Berlin, Heidelberg, Berlin, Germany, 321–346, 2009.

Fleten, S. E., Wallace, S. W., and Ziemba, W. T.: Hedging electricity portfolios via stochastic programming, IMA V. Math., 128, 71–94, 2002.

Galat, D. L. and Lipkin, R.: Restoring ecological integrity of great rivers: historical hydrographs aid in defining reference conditions for the Missouri River, Hydrobiologia, 422–423, 29–48, doi:10.1023/A:1017052319056, 2000.

Gore, J. A., Layzer, J. B., and Mead, J.: Macroinvertebrate instream flow studies after 20 year: a role in stream management and restoration, Regul. River., 17, 527–542, doi:10.1002/rrr.650, 2001.

Haihe Water Conservation Commission: Comprehensive water resources planning for Hai River basin, Ministry of Water Resources of China, Tianjin, 2008.

Hebei Province Municipal Price Bureau – HPMPB: Announce for the adjustment of electricity price in Hebei Province, HPMPB, Shijiazhuang, 2009.

Holmberg, P.: Strategic forward contracting in the wholesale electricity market, Energy J., 32, 169–202, 2011.

Jager, H. I. and Smith, B. T.: Sustainable reservoir operation: can we generate hydropower and preserve ecosystem values?, River Res. Appl., 24, 340–352, doi:10.1002/rra.1069, 2008.

Karandikar, R. G., Khaparde, S. A., and Kulkarni, S. V.: Strategic evaluation of bilateral contract for electricity retailer in restructured power market, Int. J. Elec. Power, 32, 457–463, doi:10.1016/j.ijepes.2009.09.018, 2010.

Keppo, J.: Optimality with hydropower system, IEEE T. Power Syst., 17, 583–589, doi:10.1109/TPWRS.2002.800898, 2002.

Kern, J. D., Characklis, G. W., Doyle, M. W., Blumsack, S., and Whisnant, R. B.: Influence of deregulated electricity markets on hydropower generation and downstream flow regime, J. Water Res. Pl.-ASCE, 138, 342–355, doi:10.1061/(ASCE)WR.1943-5452.0000183, 2012.

Kranz, B., Pike, R., and Hirst, E.: Integrated electricity markets in New York, Electricity J., 16, 54–65, doi:10.1016/S1040-6190(03)00014-9, 2002.

Lambert, R. A.: Contracting theory and accounting, J. Account. Econ., 32, 3–87, doi:10.1016/S0165-4101(01)00037-4, 2001.

Li, Y. P., Huang, G. H., and Chen, X.: Multistage scenario-based interval-stochastic programming for planning water resources allocation, Stoch. Environ. Res. Risk. A., 23, 781–792, doi:10.1007/s00477-008-0258-y, 2009.

Ligon, F. K., Dietrich, W. E., and Trush, W. J.: Downstream ecological effects of dams, BioScience, 45, 183–192, doi:10.2307/1312557, 1995.

Lin, M. and Wu, F. L.: Trading strategy of generation companies in electricity market, Proc. CSEE, 28, 111–117, 2008.

Liu, H. L.: Research on Optimal Hydroelectric Scheduling and Risk Management in Electricity Market, Ph.D. thesis, Shanghai Jiaotong University, Shanghai, 138 pp., 2009.

Liu, H. L., Jiang, C. W., and Zhang, Y.: Portfolio management of hydropower producer via stochastic programming, Energ. Convers. Manage., 50, 2593–2599, doi:10.1016/j.enconman.2009.06.010, 2009.

Liu, X. Y., Guo, S. L., Liu, P., Chen, L., and Li, X.: Deriving optimal refill rules for multi-purpose reservoir operation, Water Resour. Manage., 25, 431–448, doi:10.1007/s11269-010-9707-8, 2011.

Miller, B. L. and Goldberg, D. E.: Optimal sampling for genetic algorithms, in: Proceedings of the Artificial Neural Networks in Engineering (ANNIE'96) Conference, 10–13 November 1996, St. Louis, USA, 291–296, 1996.

Mulder, M.: Competition and Long-Term Contracts in the Dutch Electricity Market, Competition, Contracts and Electricity Markets: a New Perspective, Edward Elgar Publishing, Northampton, 2011.

Poff, N. L., Allan, J. D., Bain, M. D., Karr, J. R., Prestegaard, K. L., Richter, B. D., Sparks, R. E., and Stromberg, J. C.: The natural flow regime: a paradigm for river conservation and restoration, BioScience, 47, 769–784, doi:10.2307/1313099, 1997.

Poff, N. L., Richter, B., Arthington, A., Bunn, S. E., Naiman, R. J., Kendy, E., Acreman, M., Apse, C., Bledsoe, B. P., Freeman, M., Henriksen, J., Jacobsen, R. B., Kennen, J., Merritt, D. M., O'Keefe, J., Olden, J., Rogers, K., Tharme, R. E., and Warner, A.: The ecological limits of hydrologic alteration (ELOHA): a new framework for developing regional environmental flow standards, Freshwater Biol., 55, 147–170, doi:10.1111/j.1365-2427.2009.02204.x, 2010.

Pollitt, M.: Evaluating the evidence on electricity reform: lessons for the South East Europe (SEE) market, Util. Pol., 17, 13–23, doi:10.1016/j.jup.2008.02.006, 2009.

Ramos, L., de Sousa, J., Silva, T., Jeroìnimo, R., and Lima, J. A.: A risk management model for trading electricity in the spot market and through bilateral contracts, in: Energy Market (EEM)/2010 7th International Conference on the European, 23–25 June 2010, Madrid, Spain, 1–7, doi:10.1109/EEM.2010.5558701, 2010.

Renofalt, B., Jansson, R., and Nilsson, C.: Effects of hydropower generation and opportunities for environmental flow management in Swedish riverine ecosystems, Freshwater Biol., 55, 49–67, doi:10.1111/j.1365-2427.2009.02241.x, 2009.

Richter, B. D.: Re-thinking environmental flows: from allocations and reserves to sustainability boundaries, River Res. Appl., 26, 1052–1063, doi:10.1002/rra.1320, 2010.

Richter, B. D. and Thomas, G. A.: Restoring environmental flows by modifying dam operations, Ecol. Soc., 12, 12, 2007.

Richter, B. D., Baumgartner, J. V., Powell, J., and Braun, D. P.: A method for assessing hydrologic alteration within ecosystems, Conserv. Biol., 10, 1163–1174, doi:10.1046/j.1523-1739.1996.10041163.x, 1996.

Richter, B. D., Baumgartner, J. V., Wigington, R., and Braun, D. P.: How much water does a river need?, Freshwater Biol., 37, 231–249, doi:10.1046/j.1365-2427.1997.00153.x, 1997.

Richter, B. D., Baumgartner, J. V., Braun, D. P., and Powell, J.: A spatial assessment of hydrologic alteration within a river network, Regul. River, 14, 329–340, 1998.

Rothwell, G. S. and Gomez, T.: Electricity Economics: Regulation and Deregulation, IEEE Press, New York, 2003.

Shen, D. and Yang, Q.: Electricity market regulatory reform and competition-case study of the New Zealand electricity market, in: Energy Market Integration in East Asia: Theories, Electricity Sector and Subsidies, ERIA, Jakarta, Indonesia, 103–139, 2012.

Shiau, J. T. and Wu, F. C.: Assessment of hydrologic alterations caused by Chi-Chi diversion weir in Chou-Shui Creek, Taiwan: opportunities for restoring natural flow conditions, River Res. Appl., 20, 401–412, doi:10.1002/rra.762, 2004.

Shiau, J. T. and Wu, F. C.: Compromise programming methodology for determining instream flow under multiobjective water allocation criteria 1, J. Am. Water Resour. As., 42, 1179–1191, doi:10.1111/j.1752-1688.2006.tb05293.x, 2006.

Shiau, J. T. and Wu, F. C.: Pareto-optimal solutions for environmental flow schemes incorporating the intra-annual and interannual variability of the natural flow regime, Water Resour. Res., 43, W06433, doi:10.1029/2006WR005523, 2007.

Shiau, J. T. and Wu, F. C.: A dual active-restrictive approach to incorporating environmental flow targets into existing reservoir operation rules, Water Resour. Res., 46, W08515, doi:10.1029/2009WR008765, 2010.

Shrestha, G. B., Pokharel, B. K., Lie, T. T., and Fleten, S. E.: Medium term power planning with bilateral contracts, IEEE T. Power Syst., 20, 627–633, doi:10.1109/TPWRS.2005.846239, 2005.

State Environment Protection Administration of China: Technical guidelines for environmental impact assessment for ecological water usage, low temperature water and fish habitat facilities in hydraulics projects, report, Beijing, 2006.

Tennant, D. L.: Instream flow regimens for fish, wildlife, recreation and related environmental resources, Fisheries, 1, 6–10, doi:10.1577/1548-8446(1976)001<0006:IFRFFW>2.0.CO;2, 1976.

Tsai, C. M.: The reform paradox and regulatory dilemma in China's electricity industry, Asian Surv., 51, 520–539, 2011.

Vogel, R. M., Sieber, J., Archfield, S. A., Smith, M. P., Apse, C. D., and Huber-Lee, A.: Relations among storage, yield, and instream flow, Water Resour. Res., 43, W05403, doi:10.1029/2006WR005226, 2007.

Wang, Q. and Chen, X.: China's electricity market-oriented reform: from an absolute to a relative monopoly, Energ. Pol., 51, 143–148, doi:10.1016/j.enpol.2012.08.039, 2012.

Williams, J. H. and Dubash, N. K.: Asian electricity reform in historical perspective, Pac. Aff., 77, 411–436, 2004.

Wu, J., Zheng, C., and Chien, C. C.: Cost-effective sampling network design for contaminant plume monitoring under general hydrogeological conditions, J. Contam. Hydrol., 77, 41–65, doi:10.1016/j.jconhyd.2004.11.006, 2005.

Wu, Y.: Electricity market integration: global trends and implications for the EAS region, in: Energy Market Integration in East Asia: Theories, Electricity Sector and Subsidies, ERIA, Jakarta, Indonesia, 59–81, 2012.

Yin, X. A., Yang, Z. F., and Petts, G. E.: Reservoir operating rules to sustain environmental flows in regulated rivers, Water Resour. Res., 47, W08509, doi:10.1029/2010WR009991, 2011.

Yin, X. A., Yang, Z. F., and Petts, G. E.: Optimizing environmental flows below dams, River Res. Appl., 28, 703–716, doi:10.1002/rra.1477, 2012.

Yun, R., Singh, V. P., and Dong, Z. C.: Long-term stochastic reservoir operation using a noisy genetic algorithm, Water Resour. Manage., 24, 3159–3172, doi:10.1007/s11269-010-9600-5, 2010.

Zelner, B. A., Henisz, W. J., and Holburn, G. L. F.: Contentious implementation and retrenchment in neoliberal policy reform: the global electric power industry, 1989–2001, Admin. Sci. Quart., 54, 379–412, doi:10.2189/asqu.2009.54.3.379, 2009.

Zhang, Q., Xu, C. Y., Chen, Y. D., and Yang, T.: Spatial assessment of hydrologic alteration across the Pearl River Delta, China, and possible underlying causes, Hydrol. Process., 23, 1565–1574, doi:10.1002/hyp.7268, 2009.

2

Hydrological threats to riparian wetlands of international importance – a global quantitative and qualitative analysis

Christof Schneider[1], **Martina Flörke**[1], **Lucia De Stefano**[2], **and Jacob D. Petersen-Perlman**[3]

[1]Center for Environmental Systems Research, University of Kassel, Kassel, Germany
[2]Department of Geodynamics, Complutense University of Madrid, Madrid, Spain
[3]Department of Geography, Western Oregon University, Monmouth, Oregon, USA

Correspondence to: Christof Schneider (schneider@usf.uni-kassel.de)

Abstract. Riparian wetlands have been disappearing at an accelerating rate. Their ecological integrity as well as their vital ecosystem services for humankind depend on regular patterns of inundation and drying provided by natural flow regimes. However, river hydrology has been altered worldwide. Dams cause less variable flow regimes and water abstractions decrease the amount of flow so that ecologically important flood pulses are often reduced. Given growing population pressure and projected climate change, immediate action is required. However, the implementation of counteractive measures is often a complex task. This study develops a screening tool for assessing hydrological threats to riparian wetlands on global scales. The approach is exemplified on 93 Ramsar sites, many of which are located in transboundary basins. First, the WaterGAP3 hydrological modeling framework is used to quantitatively compare current and future modified flow regimes to reference flow conditions. In our simulations current water resource management seriously impairs riparian wetland inundation at 29 % of the analyzed sites. A further 8 % experience significantly reduced flood pulses. In the future, eastern Europe, western Asia, as well as central South America could be hotspots of further flow modifications due to climate change. Second, a qualitative analysis of the 93 sites determined potential impact on overbank flows resulting from planned or proposed dam construction projects. They take place in one-third of the upstream areas and are likely to impair especially wetlands located in South America, Asia, and the Balkan Peninsula. Third, based on the existing legal/institutional framework and water resource availability upstream, further qualitative analysis evaluated the capacity to preserve overbank flows given future streamflow changes due to dam construction and climate change. Results indicate hotspots of vulnerability exist, especially in northern Africa and the Persian Gulf.

1 Introduction

Natural wetland areas have declined at the global scale by 31 % between 1970 and 2008 (Dixon et al., 2016) and even higher numbers are likely for floodplain wetlands specifically. In Europe and North America up to 90 % of all natural floodplains are functionally extinct and in developing countries they are disappearing at an accelerating rate (Tockner and Stanford, 2002). Today, river systems belong to the most threatened ecosystems on the planet, and the global freshwater Living Planet Index, indicating changes in fish, bird, reptile, amphibian, and mammal populations, has declined by 76 % since 1970 (WWF, 2014). One of the main reasons for this situation is the alteration of natural flow regimes due to water resource development (Dynesius and Nilsson, 1994; Kingsford, 2000; Tockner and Stanford, 2002).

Dams are built for different purposes. On the one hand, they offer important benefits and contribute 12–16 % of global food production and 19 % of global electricity generation (WCD, 2000; Richter and Thomas, 2007). On the other hand, dams have been identified as the largest anthropogenic impact on the natural environment (Petts, 1984; Dynesius and Nilsson, 1994; Poff et al., 1997). A study by Nilsson et al. (2005) showed that dams affect 59 % of all large (i.e., natural annual discharge $\geq 350\,\mathrm{m^3\,s^{-1}}$) river systems globally. In the year 2000, the total cumulative storage capacity of large dams accounted for approximately 8300 km^3 (Chao et al., 2008; ICOLD, 2007), meaning that more than 20 % of global annual river discharge can be retained in reservoirs (Vörösmarty et al., 1997). In general, dams cause fewer variable flow regimes by considerably dampening flood peaks and elevating low flows. The downstream effects of individual dams reach up to tens or hundreds of kilometers, reducing the extent and frequency of floodplain wetland inundation (Collier et al., 1996; McCully, 1996; Poff et al., 2007). Further decreases in flow are caused by water abstractions of an exponentially growing world population. In the year 2014, 3986 km^3 of freshwater were withdrawn globally according to AQUASTAT statistics (FAO, 2016). The main fraction was used by agriculture (69 %), followed by the industrial (19 %) and domestic water supply sectors (12 %).

While floods are known as one of the most damaging natural disasters worldwide, affecting human lives and property (Jonkman, 2005; Doocy et al., 2013; Swiss Re, 2014), they are essential at pristine and not heavily altered floodplains, benefiting river-floodplain ecosystems. A natural river floodplain falls into the wetland category and represents an ecotone at the interface of aquatic and terrestrial realms, which is periodically flooded and dried (Gregory et al., 1991; Bayley, 1995). Here, as described by the flood pulse concept (Junk et al., 1989; Bayley, 1991; Tockner et al., 2000; Junk and Wantzen, 2004), the periodic occurrence of overbank flows is by far the single most important driving force (Welcomme, 1979; Tockner and Stanford, 2002) and engenders one of the most dynamic, diverse, and productive systems in the world (Naiman et al., 1993; Nilsson and Berggren, 2000; Allan et al., 2005). Costanza et al. (1997) estimated the monetary value of ecosystem services from floodplains and swamps at USD 3.2 trillion per year worldwide.

Due to population growth, climate change, and new dam initiatives, impacts on riparian wetlands are very likely to further increase in the next decades. Currently, major initiatives in hydropower development are taking place as a new source of renewable energy. At least 3700 major dams are either planned or under construction, which is supposed to further reduce the number of remaining free-flowing rivers by 21 % (Zarfl et al., 2014). These dams offer economic opportunities, but have the potential to negatively impact river ecosystem health (Lloyd et al., 2004; WWF, 2004; Poff and Zimmermann, 2010) and cause conflicts among upstream and downstream water users. Climate change may severely alter flow regimes over large regional scales as well (Nohara et al., 2006; Laize et al., 2014). Hydrological projections indicate that future flow regimes are likely to be different under climate change due to regionally and seasonally changing precipitation patterns and amounts (Schneider et al., 2013). The higher temperatures will influence timing and quantities of snowmelt (Verzano and Menzel, 2009) as well as frequency and intensity of extreme weather events such as floods (Milly et al., 2008). Okruszko et al. (2011) showed that, depending on the applied scenario, European wetlands could lose 26 to 46 % of their ecosystem services by 2050 due to climatic and socioeconomic impacts on hydrology.

In concept, there are different measures to counteract flow alteration threats to riparian wetlands. However, implementing such measures is a complex task and faces challenges such as setting strategic goals, identifying operation targets, having conflict resolution mechanisms in place, involving stakeholders, and monitoring the entire development (Pahl-Wostl et al., 2013). International reviews (Moore, 2004; Le Quesne et al., 2010) revealed that the main obstacles to environmental flow (eFlow) implementations around the world include insufficient legal and institutional capacities, as well as conflicts of interests regarding available water resources. This is especially the case in transboundary river basins. The more countries affect the water management upstream of a riparian wetland, the more groups of stakeholders with different interests are present. More interdependencies are created at different administrative levels both within and between the countries, and the potential for conflicts is higher (GWP, 2014). Hence, international water treaties and institutions are required to agree on common goals, coordinate basin-wide water management, and allocate water to different users (Le Quesne et al., 2010). In the past, ineffective governance systems have often led to overexploitation of water resources with detrimental effects for river ecosystems and, in the long term, for human well-being (Pahl-Wostl et al., 2013).

Despite the political and legal progress in the last decades (Naiman et al., 2002; Postel and Richter, 2003; Arthington et al., 2006; Poff and Matthews, 2013), most river reaches and wetlands remain vulnerable to overexploitation worldwide (Poff et al., 2009; Richter, 2009; Richter et al., 2012). Regional studies show that floodplain wetlands have been downsized and transformed into terrestrial ecosystems due to reduced flooding caused by water resource management (Hughes, 1988; Maheshwari et al., 1995; Barbier and Thompson, 1998; Kingsford, 2000; Nislow et al., 2002; Middelkoop et al., 2015). Today, the speed of river ecosystem destruction and biodiversity loss is exceeding the ability of scientists to review applied water management practices and ecological consequences for each river. Thus, there is an urgent need to complement more accurate but time-consuming case studies with global water assessments that cover large-scale developments (Poff and Matthews, 2013). Accordingly, different authors have assessed ecologically relevant flow regime alterations on larger scales in recent years (e.g., Smakhtin et al., 2004; Smakhtin and Eriyagama, 2008; Döll et al., 2009; Döll and Zhang, 2010; Vörösmarty et al., 2010; Schneider et al., 2013; Laize et al., 2014; Pastor et al., 2014; Grill et al., 2015).

Building on the work from these valuable papers, this study aims at establishing a screening tool to systematically identify riparian wetlands that are threatened due to river flow regime modifications. While most large-scale eFlow assessments focused on in-channel river flows, our assessment is the first that applies the flood pulse concept on a global scale. Complex flow-dependent ecosystem habitats such as floodplain wetlands are provided by specific flow events. Consequently, rather than changes in average flow conditions, our modeling approach focuses on overbank flows leading to inundation of adjacent riparian wetlands. Additionally, it considers different drivers of global change such as dam operation, water use, and climate change. As many ecological functions and habitats are facilitated by hydrological events that last only up to a few days (e.g., strong precipitation events, flood formation, and overbank flows), discharge simulations are carried out on a daily time step. The modeling is performed on a detailed river network with a very high spatial resolution for a global model and can be applied for single reaches of larger rivers with a global coverage.

Next to flow regime modifications, the threat to riparian wetlands also depends on the society's capacity to act, which is required to respond to hydrological changes and implement counteractive measures supporting the conservation of riparian wetlands. Capacity to act might be limited due to high water resource competition or deficits in the legal and institutional framework in place, and this kind of threat has not yet been taken into account in large-scale studies. In order to fill this gap, we combined quantitative with qualitative indicators which address upstream water resource availability as well as the presence of institutional arrangements facilitating the establishment of eFlows.

In this study, the proposed screening tool is exemplarily applied to 93 selected riparian wetlands of international importance to address the following research questions.

1. What is the impact of current water resource management on riparian wetland flooding?

2. At which sites is inundation likely to be further modified due to climate change and new dam construction?

3. At which sites could the implementation of conservation measures be hindered by a low capacity to act?

2 Methodology

In order to exemplify the proposed screening tool, we selected wetlands based on two criteria. First, we chose wetlands listed under the Ramsar Convention, which is a global framework for intergovernmental cooperation aiming for the conservation and sustainable use of wetlands. This criterion ensured international importance and the significant value of the selected wetlands for humanity as a whole. Second, the wetlands have to be dependent on lateral overspill of adjacent rivers (i.e., fluviogenic). The Ramsar Classification System describes different wetland types, but does not categorize riparian wetlands. However, riparian wetlands were selected from the Ramsar list on the basis of information provided by the Ramsar information sheets (RSIS, 2015), indicating a wetland's dependence on flooding. For Europe, a higher number of sites were chosen as the European wetland geodatabase (Okruszko et al., 2011) clearly defines wetland type and main source of water for each European Ramsar wetland. In total, 93 sites were selected, ranging from 5 to 55 374 km^2 in size and located in 48 countries and 47 river basins, respectively. The Danube basin had the most selected wetlands of all river basins, with 19 riparian Ramsar wetlands. A detailed list of all wetlands is provided in Sect. S1 in the Supplement.

Our wetland assessment combines a quantitative and a qualitative analysis. The quantitative analysis is based on the flood pulse concept, which describes the flood pulse as a major driver determining the extent of the river floodplain and the biota living within it (Junk et al., 1989; Tockner et al., 2000). For each site we determined the percentage change in flood volume caused by (i) current water resource management and (ii) future climate change. In each case, we compared the modified river flow regimes to reference conditions which reflect near-natural flow regimes.

The qualitative analysis addresses vulnerability due to new dam initiatives as well as a deficient capacity to act. New dam initiatives have the potential to further reduce wetland inundation in the near future. Capacity to act is required to implement complex counteractive measures at threatened sites and equitably allocate water resources to different water use sectors. However, capacity to act is often restricted by deficits in legal and institutional arrangements as well as water resource competition (Moore, 2004; Le Quesne et al., 2010). We address the third research question by identifying riparian wetlands where implementation of conservation measures will likely be hindered by a low capacity to act.

2.1 The quantitative assessment of threats

In order to quantitatively assess anthropogenic alterations of flood pulses, we applied WaterGAP3 (Eisner, 2016). WaterGAP3 is an integrated global modeling framework to assess impacts of global change on renewable freshwater resources. The model has been further improved to represent specific flow events (Verzano and Menzel, 2009; Verzano et al., 2012) and identify river ecosystems at risk (Schneider et al., 2013). WaterGAP3 was selected for the study because of the global coverage, the high spatial resolution of 5 by 5 arcmin ($\sim 9 \times 9$ km at the Equator) to represent hydrological processes, the temporal resolution of daily time steps which is important for modeling flood formation, the operation of currently >6000 dams with optimization schemes for different dam types, and the calculation of water withdrawals and consumption of five different water-related sectors (domestic, manufacturing industries, thermal electricity production, agricultural crop irrigation, and livestock).

Forced by climatic time series, the hydrology model of WaterGAP3 computes the macro-scale behavior of the terrestrial water cycle. The daily water balances for each grid

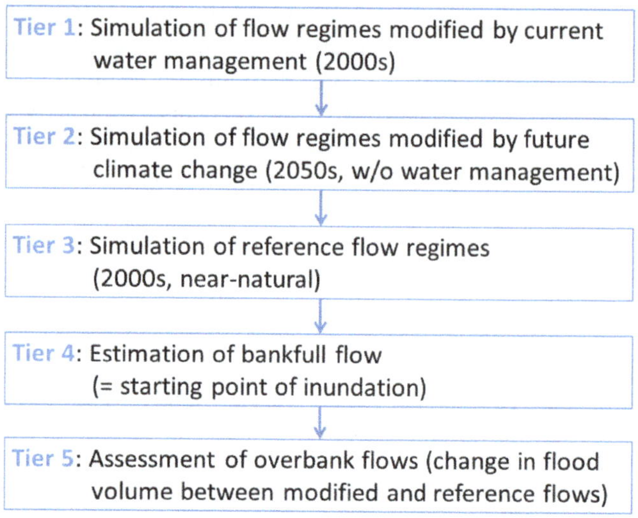

Figure 1. Schematic illustration of the steps taken in the quantitative analysis based on WaterGAP3 modeling.

cell take into account distributed physiographic characteristics from high spatial resolution maps describing slope, soil type, land cover, aquifer type, permafrost, and glaciers, as well as extent and location of lakes and wetlands. The total runoff in each grid cell, derived from the water balances of land and freshwater areas, is routed along a predefined drainage direction map (DDM5; Lehner et al., 2008) to the catchment outlet.

Simulated river flows are calibrated against observed annual discharge data from the Global Runoff Data Centre (GRDC, 2004) at about 1600 gauging stations globally. The calibration process adjusts only one model parameter (i.e., runoff coefficient γ), which has an effect on cell surface runoff generation at gauging stations (Eisner, 2016). In recent studies the model's ability to represent specific flow events has been proven for different maximum flow magnitudes (Schneider et al., 2011a; Schneider, 2015; Eisner, 2016).

In order to assess quantitative changes in floodplain inundation, we conducted different model experiments. Single steps of the entire approach are illustrated in Fig. 1 and described in detail in the following subchapters. In brief, we proceeded as follows: first, we simulated modified river flow regimes under current water resource management (tier 1; Sect. 2.1.1) and climate change (tier 2; Sect. 2.1.2). As assessment of river ecosystem health implies comparison of modified flows to natural flow conditions (Norris and Thoms, 1999), we simulated the reference flow regimes in tier 3 (Sect. 2.1.3) by not accounting for anthropogenic impacts (i.e., dam management and water use) except current climate and land-cover conditions existing in the 2000s. In tier 4 (Sect. 2.1.4), we estimated bankfull flow, which constitutes an important parameter in our analysis. It describes the flow where water just begins to enter the active floodplain and, thus, marks the starting point of inundation. As floodplain inundation requires overtopping of the banks, each daily flow above bankfull was a critical flow to investigate in tier 5 (Sect. 2.1.5). Here we compared modified (tiers 1 and 2) to reference flow regimes (tier 3) and quantified the changes in overbank flows (flood volume) as a proxy for changes in wetland inundation.

2.1.1 Simulation of modified flow regimes under current water resource management

For the simulation of flow regimes under current water resource management (i.e., 1981–2010), we took anthropogenic flow alterations due to water use and dam operation into account (tier 1, Fig. 1). Regarding water use, river discharge is reduced in each grid cell by water consumption as calculated by the global water use models of WaterGAP3. These models simulate spatially distributed water uses for the five most important water use sectors (aus der Beek et al., 2010; Flörke et al., 2013).

Net irrigation requirements are simulated for each grid cell based on climatic conditions, dominant crop type, and irrigated area around the year 2005 (GMIAv5; Siebert et al., 2013) assuming an optimal water supply to irrigated crops. Livestock water demands are determined by multiplying the number of animals per grid cell by the livestock-specific water use intensity (Alcamo et al., 2003). For the electricity production sector, the amount of cooling water consumed is calculated by multiplying the water use intensity of each power station by the equivalent annual thermal electricity production. The water use intensity is affected by the cooling system (once-through flow cooling, tower cooling, or ponds) and the type of fuel (coal and petroleum, natural gas and oil, nuclear, or biomass and waste) used at each power station (Flörke et al., 2012). Power station characteristics such as type, size, and location are derived from the World Electric Power Plants Data Set (UDI, 2004).

Consumptive water uses of the manufacturing and domestic sectors are computed on a country scale following data from national statistics and reports, which are subsequently allocated to the grid cells of the associated country by means of urban population and population density maps, respectively (Flörke et al., 2013). For the domestic sector, WaterGAP3 also considers water transfers of 480 larger cities, including their 1642 withdrawal points (City Water Map; McDonald et al., 2014).

In order to assess flow alterations due to dam operation, the number of dams implemented in the model has been further increased based on information provided by the Global Reservoir and Dam (GRanD) database (Lehner et al., 2011). From this dataset, WaterGAP3 now considers all large dams (i.e., dams with a height of ≥ 15 m) plus smaller dams exceeding a reservoir storage volume of 0.5 km^2; 6025 dams are currently allocated to the global WaterGAP3 stream net accounting for a total accumulative storage volume of

$6200\,km^3$. This is state-of-the-art in comparison to other global models (Haddeland et al., 2014).

The operation of dams is performed in WaterGAP3 as a function of dam type. Dams with the main purpose of irrigation are operated according to the algorithm of Hanasaki et al. (2006) with minor modifications by Döll et al. (2009). The annual reservoir release is a function of long-term average annual reservoir inflow, the relative reservoir storage at the beginning of the operational year, and the difference between precipitation and evaporation over the reservoir surface. Subsequently, monthly reservoir releases are calculated depending on the downstream consumptive water use in each month.

Other dam types are operated based on an optimization scheme provided by van Beek et al. (2011). Depending on the dam type, an objective function is applied that maximizes electricity production by maximizing the hydrostatic pressure head to the turbines (hydropower dams), minimizes flood damages by minimizing overbank flows (flood control dams), and aims for a constant outflow by minimizing deviations from the annual mean (water supply and navigation dams). Furthermore, we considered different constraints that reserve sufficient storage capacity to accommodate larger floods for 7 days (flood protection) and to keep sufficient water in the reservoir to safeguard a minimum flow for at least 30 days (minimum flow provisions).

Given current reservoir storage and monthly inflow data of the upcoming year, the overall modeling strategy is to find the monthly target storages (and corresponding monthly reservoir releases) that ensure optimal functioning of the dam. This strategy was realized in WaterGAP3 by evaluating objective functions and constraints through deterministic dynamic optimization (Bellman, 1957) and discretizing reservoir storage by the Savarenskiy scheme (Savarenskiy, 1940) considering a discretization width of 2 %. At the beginning of each month, the accumulated objective function value is computed for the upcoming 12 months taking into account every possible combination of the discrete reservoir storage classes. The combination, which provides the most suitable value for the objective function without harming any constraint, determines the monthly target storages. As inflow data, forecasted monthly values are used derived from average simulated flows of the last 5 years (rather than simulated values for the future year). This prospective scheme reflects more realistically the hydrological situation, where water managers have to deal with uncertain forecast as well (van Beek et al., 2011). The monthly target storages together with the actual incoming flow are subsequently used to calculate the daily reservoir releases.

In this modeling study we used WATCH-Forcing-Data-ERA-Interim (WFDEI; Weedon et al., 2014) for climate input representing current conditions. The time series consists of a set of daily, 30×30 arcmin ($\sim 50 \times 50$ km at the Equator) gridded meteorological forcing data, which were simply disaggregated to the 5 arcmin resolution as required by the model.

2.1.2 Simulation of modified flow regimes under climate change

To simulate future flow regimes modified only by climate change, additional model runs were conducted (tier 2, Fig. 1) for the 2050s (represented by the time period 2041–2070). Here, WaterGAP3 was driven with bias-corrected, daily climate data from five different general circulation models (GCMs), namely GFDL-ESM2M, HadGEM2-ES, IPSL-CM5A-LR, MIROC-ESM-CHEM, and NorESM1-M, provided by ISI-MIP (Hempel et al., 2013). We assumed climate drivers to follow the Representative Concentration Pathway leading to a radiative forcing (cumulative measure of human emissions of greenhouse gases from all sources) of $6.0\,W\,m^2$ (RCP6.0). Current CO_2 emissions are close to the upper end of the scenario range and RCP6.0 is a medium–high emission scenario with a global mean temperature increase of 2.2 °C until the end of the century compared to 1986–2005 (Riahi et al., 2011). Within the future time frame the differences between the emission scenarios (as represented by the radiative forcing) are smaller than between scenarios based on different GCMs. Thus we considered climate forcing of five different GCMs but only one emission scenario in order to address the uncertainty of projected climatic conditions. Although model outcomes of tier 2 will not reflect future conditions due to not taking into account future water management, this model experiment supports the identification of the sole effect of climate change on riparian wetland inundation. Therefore, we disabled dam operation and water use in these model runs.

2.1.3 Simulation of reference flow regimes

The aim of tier 3 was to simulate daily reference flow regimes reflecting near-natural conditions (Fig. 1). Hence, anthropogenic impacts such as dam operation and water use were disabled in these model runs. In order to be able to make comparisons with modified conditions, we conducted six different model runs for the reference period 1981–2010. We forced WaterGAP3 with WFDEI climate data to simulate reference flow regimes for the comparison with modified flows of tier 1 and with GCM data for the comparison with flow regimes of tier 2. Land cover data were derived from the Global Land Cover Characterization map (GLCC; USGS, 2008) and for EU countries from the CORINE Land Cover map (CLC2000; EEA, 2004) and kept constant over the entire model simulations.

2.1.4 Estimation of bankfull flow

Bankfull flow was estimated in our approach for each grid cell by flood frequency analysis (tier 4, Fig. 1). We applied the partial duration series (PDS) approach taking into ac-

Table 1. Thresholds for different levels of mean annual flood volume deviation (Δ) between modified and natural flow regimes (as suggested for global assessments by Hoekstra et al., 2011).

River status	Level of modification	Thresholds for reduction in flood volume
A	none/slightly	$\Delta \leq 20\,\%$
B	moderately	$20\,\% < \Delta \leq 30\,\%$
C	significantly	$30\,\% < \Delta \leq 40\,\%$
D	seriously	$\Delta > 40\,\%$

Table 2. Defined impact on a riparian wetland due to new dam initiatives in the upstream area.

Number of major dam initiatives	Potential impact	Number of affected wetlands
0	NONE	63
1–12	MED	16
28–276	HIGH	14

count 30-year time series of daily discharge data modeled by WaterGAP3, an increasing threshold censoring procedure, a declustering scheme, and the generalized Pareto distribution. In the PDS, bankfull flow is determined by a return period of 0.92 years. The approach including a validation of bankfull flow estimates against observed conditions is described in detail by Schneider et al. (2011a). Results of this study show that bankfull flow can be reasonably simulated by WaterGAP3 with a high model efficiency ($E_1 = 0.71$) and weighted correlation ($\omega r^2 = 0.90$) as well as a systematic overestimation of 22.8 %.

2.1.5 Assessment of overbank flow modifications

We used the flood volume (i.e., the cumulative amount of daily discharge above bankfull) as a measure of the extent of flooding in tier 5 (Fig. 1) which was determined as the long-term annual mean over the 30-year time period. The percentage change in flood volume between the modified and reference flow regimes describes the anthropogenic impact on floodplain inundation. Climate change impacts on flow regimes are presented as the ensemble median, which reflects the direction of change of at least three out of the five selected GCMs. The entire approach was carried out for each single grid cell of the global 5 arcmin raster, but only grid cells associated with riparian wetlands were examined further.

In order to evaluate the ecological consequences of flood volume alterations, thresholds needed to be defined. So far no generalizable relationships between flow alteration and ecological impact are available for large-scale assessments. Therefore we applied "thresholds for potential concern" (Hoekstra et al., 2011) for the deviation (Δ) in flood volume between the modified and reference flow regimes in order to distinguish distinct levels of modification (Table 1). These thresholds are based on the "presumptive standard" suggested by Richter et al. (2012) for daily flow alterations and likely indicating moderate to major changes in ecosystem structure and functions as well as initial thoughts from some water resource experts to set a global standard on eFlow requirements, though it has to be considered in our assessment that small reductions in flood volume can already result in large decreases in the extent of area flooded (Taylor et al., 1996; Kingsford, 2000; Tockner and Stanford, 2002).

In general it can be expected that the greater the deviation from natural conditions, the greater the expected ecological impact (Poff and Hart, 2002; Magilligan and Nislow, 2005). Quantitative relationships between peak flows and ecosystems are provided, e.g., by Wilding and Poff (2008) for rivers in the US state of Colorado. In their study, riparian vegetation responds with a maximum change of 12 % in community composition for each 10 % reduction in peak flows. Consequently, a reduction of 40 % in flood volume, which indicates a serious modification in our analysis, could lead to a 48 % change in riparian vegetation. Stream invertebrates, in turn, respond exponentially. A 40 % change in peak flow may cause a maximum response of 54 % change in invertebrates.

2.2 The qualitative assessment of threats

In order to evaluate further impairments in riparian wetland flooding in the coming decades, we conducted a qualitative assessment. The qualitative assessment addresses (i) future dam construction and (ii) the capacity to act, which is required to respond to ecological threats caused by flow regime alterations.

2.2.1 Future dam construction

Besides climate change, the construction of new dams will further modify flood pulses and, thus, put additional pressure on riparian wetlands. Therefore, for each selected site we determined the number of all upstream dam projects which are over 10 megawatts in capacity and were planned, proposed or under construction as of July 2014 (Petersen-Perlman, 2014). A number of sources were used to build this dataset: the United Nations Framework Convention on Climate Change's Clean Development Mechanisms (http://cdm.unfccc.int), International Rivers, and other organizations' websites known to fund dam construction (e.g., World Bank). If no dam initiatives were found in the upstream area, we assigned no impact. The remaining sites were divided into two groups to define a medium (1–12 dam initiatives) and high (28–276 dam initiatives) impact (Table 2). The cutoff threshold between medium and high impact was chosen with the intention to get two almost equally sized groups of affected wetlands and to respect the large gap between 12 and 28 upstream dams.

Table 3. Water availability for ecological allocations defined by means of the number of months with water scarcity upstream of the Ramsar site.

Number of months with water scarcity	Water availability for ecological allocation
6–12	LOW
2–5	MED
0–1	HIGH

2.2.2 Capacity to act

The implementation of counteractive measures is a complex task and depends on the local capacity to act. In order to assess that capacity for each site, we calculated two sub-indicators.

The first sub-indicator addresses the availability of water for ecological allocations. Flood pulse provisions especially require a relatively large amount of water at a specific time of the year. However, in some regions, water use alone can have a strong impact on the river flow regime. For example, the outflow of the Colorado and Murray–Darling rivers is reduced by water use to < 1 and 36 %, respectively, of its natural flow (Jolly, 1996; Cushing and Allan, 2001). A high level of water scarcity in the upstream area indicates high water resource competition between different water use sectors and reduces the potential to allocate adequate amounts of water for ecological requirements. Water scarcity was defined following the approach of Hoekstra et al. (2012), who suggested that no more than 20 % of monthly river discharge should be depleted by consumptive water use to maintain river ecosystem integrity. Depending on the average number of months per year with water scarcity (i.e., a consumption-to-availability ratio > 0.2) in the upstream area, water availability for ecological purposes was determined (Table 3). The cutoff thresholds for low (6–12 months), medium (2–5 months), and high (0–1 month) water availability were arbitrarily chosen.

The second sub-indicator addresses the legal and institutional framework in place, and distinguishes between transboundary and non-transboundary upstream areas. For the latter, the sub-indicator depicts whether the country where the riparian wetland is located has legal provisions or official recommendations for the establishment of eFlows (yes) or not (no). Having a legal provision is an important first step for setting strategic goals, advocating ecological water requirements with stakeholders, securing planning resources, and promoting eFlow implementation (Le Quesne et al., 2010). However, it is no guarantee that eFlows will actually be established in practice, enforced, or adequate. As most of this information on legal eFlow provisions was available in qualitative terms, we introduce a simple yes–no query to our capacity to act indicator. In particular, no quantitative information on eFlow provisions was found for the management of dams. The main sources of information for this sub-indicator were OECD (2015), Benítez Sanz and Schmidt (2012), Le Quesne et al. (2010), and the FAO Water Lex Legal Database (FAO, 2015).

Table 4. Institutional capacity in place in transboundary upstream areas of riparian wetlands based on formal arrangements such as international water treaties, river basin organizations, legal eFlow provisions, and specific treaty provisions.

Score	Institutional capacity
0–2	LOW
> 2–4	MED
> 4–6	HIGH

In transboundary upstream areas the sub-indicator takes into account five parameters as the complexity of water management increases. Here, we measured formal institutional capacity by (i) the presence of river basin organizations (RBOs), (ii) at least one relevant treaty, and specific treaty provisions such as (iii) a water allocation mechanism, (iv) a conflict resolution mechanism, and (v) flow variability management. Formal arrangements governing transboundary river basins, in the form of international water treaties and RBOs, can be particularly instrumental in managing disputes among different stakeholders involved in water resource management. The greater the institutional capacity, the higher the potential for eFlow allocations. Institutional frameworks can determine targets, responsible authorities, reoperation strategies, reallocation of water shares, monitoring efforts, and consequences of assessment outcomes (Le Quesne et al., 2010; Pahl-Wostl et al., 2013). For the calculation of this sub-indicator, we divided the upstream areas into basin-country units (BCUs, i.e., the portion of a country within a river basin shared by two or more countries). For each of the five parameters present at BCU level, one point was given, allowing for a score ranging from zero to five. In order to assign a score to each wetland reflecting upstream transboundary institutional capacity, we aggregated and weighted the scores of all upstream BCUs based on the contribution of each BCU to the runoff of the total upstream area.

We gave an additional point in case the country where the wetland is located has legal provisions or official recommendations for the establishment of eFlows. The scores were then grouped into three classes describing a low, mid, and high institutional capacity (Table 4).

All underly-ing data were obtained from De Stefano et al. (2012) and complemented with data embedded in international RBOs (Schmeier, 2014).

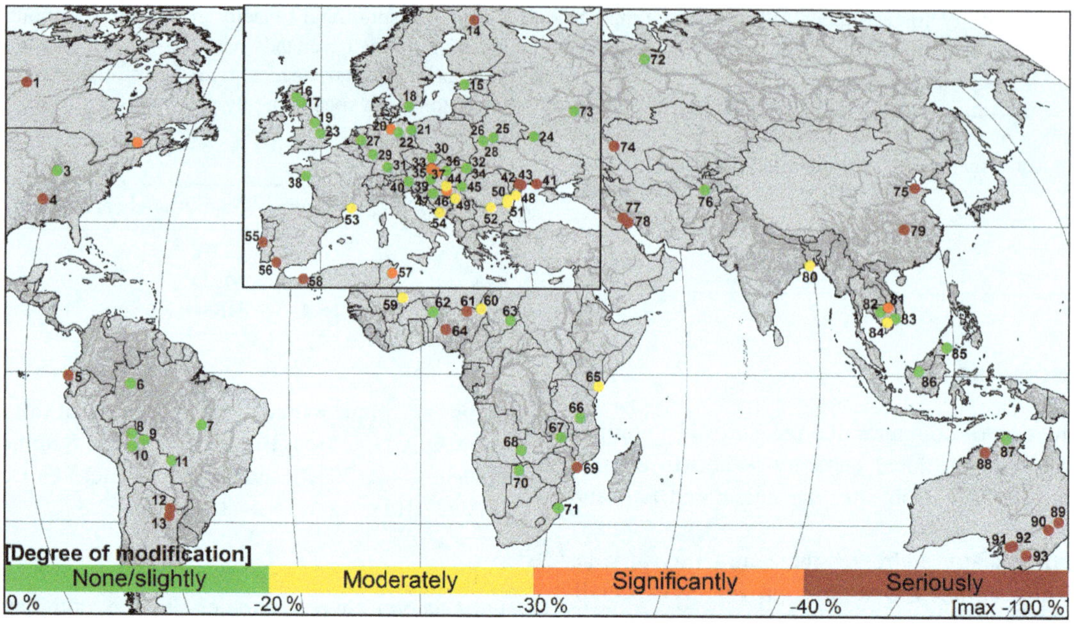

Figure 2. Global map of overbank flow alterations for selected riparian wetlands of international importance (nos. 1–93) as a consequence of current water resource management.

3 Results

3.1 Quantitative analysis

3.1.1 Overbank flow alterations caused by current water resource management

Figure 2 shows the degree of alteration in flood volume at the 93 wetland study sites caused by current (1981–2010) water management practices. When comparing modified to reference flow regimes, every second site (51 %) is impaired by at least moderately reduced flood volumes in our simulations. Almost every third site (29 %) is seriously and a further 8 % are significantly affected by the flow regime modifications. Seriously affected sites occur on all continents, but particularly in Australia, China, North America, and the Iberian Peninsula, as well as at rivers that drain into the Black Sea (e.g., Dnieper and Dniester rivers) or the Persian Gulf (e.g., Tigris and Karun rivers). We found that dams for hydropower generation are the most frequent dam type in almost one-third of the selected upstream areas, followed by irrigation dams in one-quarter of the upstream areas. However, irrigation dams are the most frequent dam type in almost half of the cases (48 %), when only wetlands with seriously modified inundation patterns are regarded.

In Australia, five (nos. 89–93) of six vulnerable sites are located in the Murray–Darling basin. In their upstream areas more than 100 % of the annual flow can be stored in reservoirs, indicating a high impact on flow regulation, which was also found by Grill et al. (2015). Intense agricultural irrigation is responsible for the highest water withdrawals and irrigation dams are the most frequent dam type in almost all upstream areas. This is underlined by Kingsford (2000), who reported that many floodplains in the Murray–Darling basin have turned into terrestrial ecosystems.

At the Volga River, the construction of dams for hydropower and navigation during the Soviet Union era substantially altered the flow regime, which seriously influences the dynamics of the Volga Delta (no. 74) in our analysis. This finding is in line with other studies. Khublaryan (2000) reported that mean high water flow decreased from 67 to 42 % of the annual flow in the Lower Volga River due to river regulation. Middelkoop et al. (2015) found that dam operation caused a decrease in magnitude and duration of spring peak flow in the Lower Volga.

Nine of the analyzed sites are located along the Danube River, for which we identified slightly (no. 31), moderately (nos. 44, 48, 50, 51, and 52) and significantly (nos. 35, 37, and 46) reduced flood volumes. Despite numerous dams, the lower storage capacities cause the Danube River to be more affected by fragmentation than flow regulation, as also shown by Grill et al. (2015).

We found the lowest number of vulnerable wetlands in South America and Africa. In South America many riparian wetlands possess only slightly modified inundation patterns. Only a few large dams are located in upstream areas, and many river reaches, especially in the Amazon basin, are still in pristine condition (see also Tockner and Stanford, 2002). Nevertheless, seriously modified inundation patterns exist as well at three (nos. 5, 12, and 13) of nine selected study sites located in Ecuador and Argentina. In Africa, about half of the sites are not or only slightly affected under current condi-

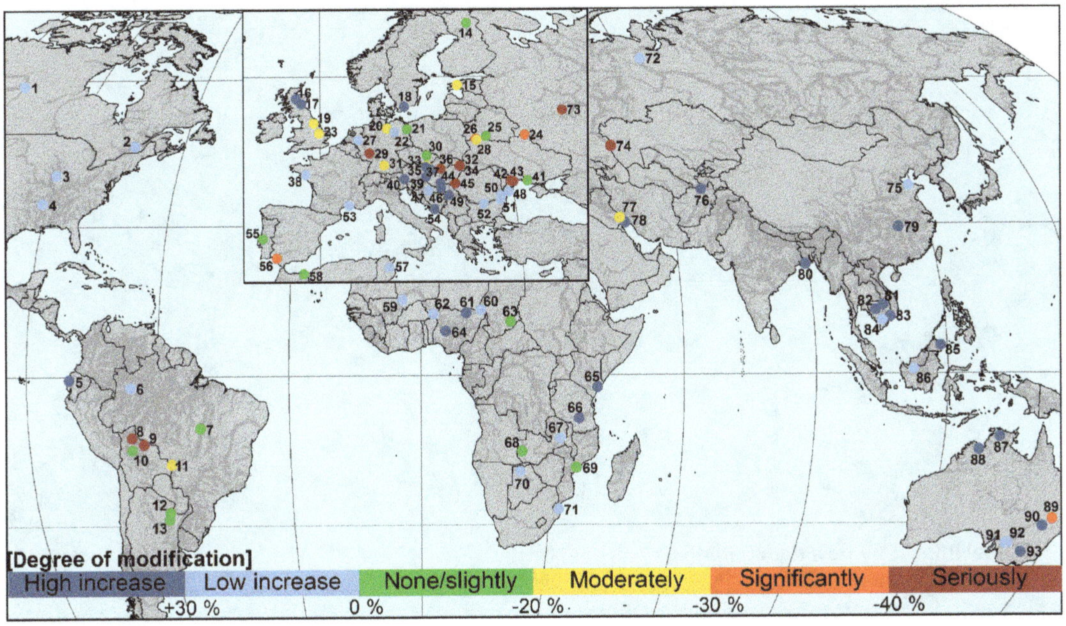

Figure 3. Global map of overbank flow alterations for selected riparian wetlands of international importance (nos. 1–93) as a consequence of the exclusive effect of climate change in the 2050s.

tions. However, one-third of the African sites are impaired by seriously or significantly altered overbank flow events. These sites are located in Morocco, Tunisia, Mozambique, and Nigeria. The threat to Nigeria's wetlands is also reported by Uluocha and Okeke (2004) inter alia due to population pressure and dam construction.

3.1.2 Overbank flow alteration due to climate change

In the future climate change is likely to further modify river flow regimes, as indicated by the model results driven by five GCM projections. According to the ensemble median, the average flood volume is expected to decrease at 41 % of the sites in the 2050s due to the exclusive effect of climate change (Fig. 3). At 16 % of the sites, reductions are significant or even serious (i.e., > 30 %). Overall two spatial hotspots could be identified where flood pulses are likely to be reduced under climate change, namely eastern Europe/western Asia as well as South America below the Amazon River. These WaterGAP3 results are in line with Dankers et al. (2013), who modeled changes in peak flows at the end of this century by nine global hydrology models.

In Europe, most sites of concern are located in eastern Europe, i.e., in the Ukraine (nos. 24, 26, and 43), Hungary (nos. 32 and 36), Slovakia (no. 34), Moldova (no. 42), and Romania (no. 45), but also in Spain (no. 56) and Germany (no. 29). Climate change will induce an additional threat for three of the sites (nos. 42, 43, and 56) which already experience seriously or significantly reduced flood volumes under current water management practices. In Asia, wetlands affected by reduced flooding under climate change are located in Russia (nos. 73 and 74) and Iraq (no. 77). Flood pulses are already seriously reduced under current water management at two of them (nos. 74 and 77). The expected reduction in wetland inundation in eastern Europe and western Asia in the future can be explained by changes in snowmelt. In these two regions characterized by continental climate, global warming is likely to cause a reduction in snow cover resulting in lower and earlier snowmelt-induced flood peaks in spring as found by Schneider et al. (2011b, 2013). Moreover, analyses of streamflow trends in European Russia indicate that spring flows have been decreasing since the mid-1970s (Georgiyevsky et al., 1995, 1996, 1997).

Increasing flood volumes, in contrast, can be found at 51 % of the selected riparian wetlands under climate change conditions. The rise in flood volume is expected to be higher than 30 % in the 2050s at almost every third (30 %) site. Those wetlands tend to be located closer to the coast and especially in Southeast Asia, southeastern Europe, Scotland, western Africa, Tanzania, and Kenya. In the analysis of Dankers et al. (2013), increases in flood hazard were projected consistently for Southeast Asia.

3.2 Qualitative analysis

3.2.1 Future dam construction

New dam initiatives have the potential to further impair riparian wetland flooding. New dams are currently planned or under construction in the upstream areas of one-third of the selected riparian wetlands (Fig. 4). In agreement with results of Zarfl et al. (2014), extensive dam construction is on

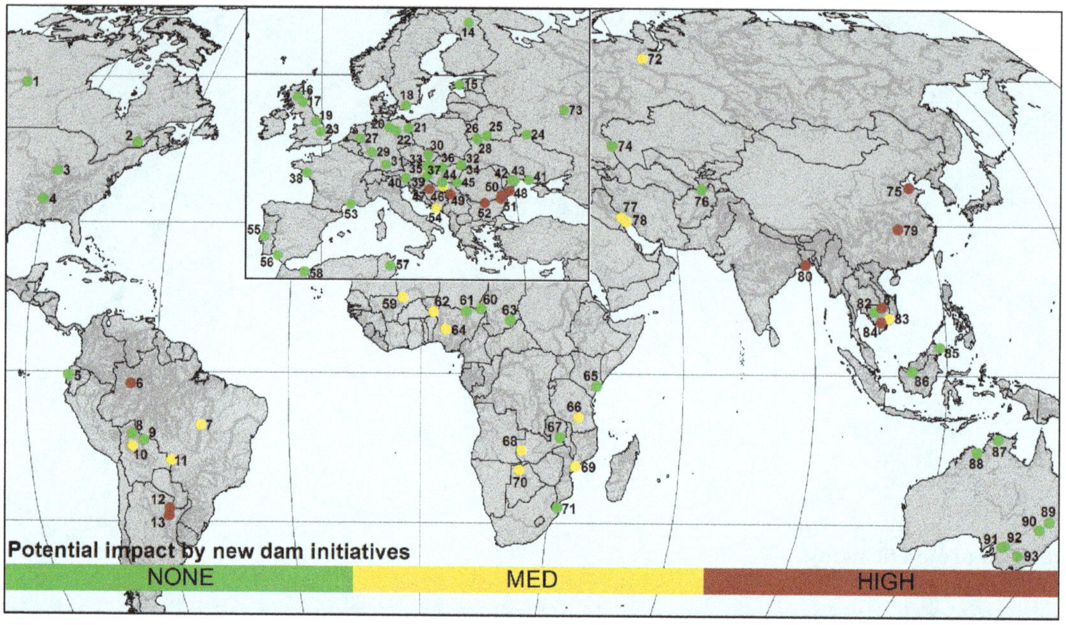

Figure 4. Potential impact of new dam initiatives taking into account dams currently planned or under construction in the upstream area of each riparian wetland.

the way, particularly in areas upstream of South American (67 %) and Asian (60 %) wetlands. We found that a large impact is likely in the upstream areas of wetlands located in the basins of Amazon, Parana, and Paraguay, as well as Yangtze, Yellow, Mekong, and Ganges–Brahmaputra. Riparian wetlands in China (nos. 75 and 79) and Argentina (nos. 12 and 13) are already characterized by seriously reduced flood volumes under current water management conditions. Dams are also planned or under construction upstream at about half (47 %) of the selected African sites, although the number of dams is relatively small in most upstream areas. Analyzing future trends for riverine floodplains, Tockner and Stanford (2002) also concluded in their assessment that in South America, Asia, and Africa, many floodplains will become reduced in size or even disappear in the future.

While a high number of dams have been constructed in North America and Australia in the last century, no further dams are planned or under construction upstream of the selected Ramsar sites. This is also the case for most parts of Europe, but a high number of new dams could be constructed upstream of riparian wetlands located in the Balkan Peninsula (i.e., in Croatia, Serbia, Bulgaria, and Romania), further threatening riparian wetlands in the lower Danube basin.

3.2.2 Capacity to act

Implementing counteractive measures requires that (i) sufficient water is available to satisfy water demands of different water use sectors and (ii) institutional arrangements are in place, enabling the establishment of eFlows. Considering these two factors, Fig. 5 displays the capacity to act in the upstream area for each riparian wetland.

Our analysis shows that the highest competition for water exists in the upstream area of the Lake Chad Wetlands (Nigeria) followed by wetlands of the Murray–Darling (Australia), Schatt al-Arab (Persian Gulf), Tana (Kenya), Moulouya (Morocco), and Yellow (China) River basins, where water scarcity occurs upstream in 6 to 10 months of the year on average. Lake Chad lost one-tenth of its size in the last 40 years (Uluocha and Okeke, 2004) and the United Nations Environment Programme stated that human water use is responsible for about half of the decrease (UNEP, 2008), which supports our finding. eFlow applications might also be challenging in the Iberian Peninsula as water availability for ecological purposes is rated medium, with water scarcity occurring on average in 5 months of the year due to high water requirements for agricultural irrigation.

Globally, normative eFlow provisions are considered in the national or State Water Act at about 50 % of the selected Ramsar sites. The highest percentage of sites without normative eFlow provisions occurs in Asia (87 %) and Africa (80 %). The lowest values for formal institutional capacity became obvious in the transboundary upstream areas of wetlands located in the Ukraine (nos. 24 and 41), Belarus (no. 25), and Russia (no. 72). In this study, eastern Europe and western Asia were identified as hotspot regions where climate change is likely to reduce flood pulses in the future. Thus, a high formal institutional capacity would be of importance here to conserve riparian wetlands and allocate water to different water users.

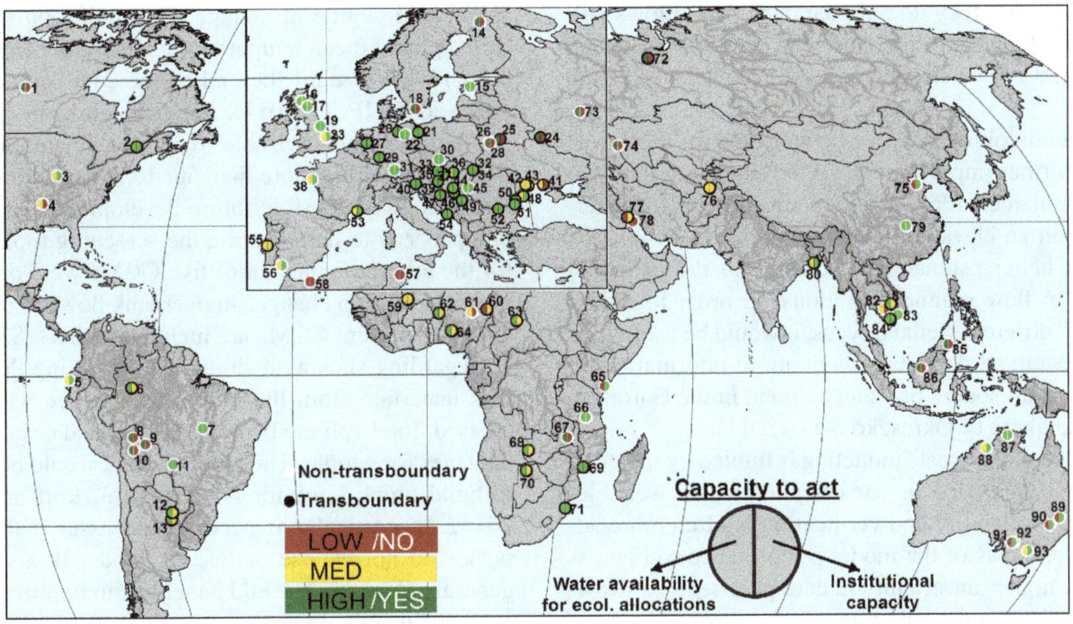

Figure 5. Current capacity to act with regard to anthropogenic flow regime modifications for selected riparian wetlands. The left semicircle represents the water availability for ecological allocations, while the right semicircle characterizes the institutional capacity in the upstream area. For wetlands with a non-transboundary upstream area (white border), the right semicircle represents the presence or absence of legal provisions or official recommendation to establish eFlows.

Considering both sub-indicators, the lowest values for capacity to act were found for riparian wetlands located in northern Africa (nos. 57 and 58), northeastern Nigeria (nos. 60 and 61), as well as at the Dnipro River Delta (no. 41) and the Persian Gulf (nos. 77 and 78). Detailed results for all indicators and wetlands are listed in Sect. S2.

4 Discussion

Currently, the concept of eFlows is transitioning from an era of ecosystem integrity and conservation at single river reaches to a period of globalization, where regional studies are complemented by global water assessments that cover large-scale developments. The main reasons are increasing threats at global scales (e.g., global warming) and the associated pace of ecosystem destruction, but more sophisticated global hydrology models are also available now (Poff and Matthews, 2013). In this study, we applied the WaterGAP3 global modeling framework, which has been further improved in recent years to model specific flow events such as floods. WaterGAP3 is a state-of-the-art global water model that performs well compared to other global and regional models (Beck et al., 2016; Eisner et al., 2017).

Despite the high number of dams (> 6000) operated in WaterGAP3, our results for the impacts related to water resource management should be regarded as an underestimation as only larger dams from a global dataset (GRanD, Lehner et al., 2011) are taken into account. The aggregated effect of remaining smaller dams has an impact on floodplain inundation as well (Rosenberg et al., 2000), so it can be assumed that the impacts are even higher for some wetlands.

Our analysis is based on dam operation rather than reservoir capacity and river fragmentation. WaterGAP3 operates dams by dynamic optimization schemes taking into account various objective functions and constraints. Since no global dataset exists that describes specific operation rules or management strategies of individual dams, the dam operation module as part of WaterGAP3 considers generic operation schemes reflecting the main purpose of each dam. Thus, the performance of our dam module can be regarded as lower compared to detailed reservoir models using site-specific information. Accordingly, eFlow provisions that are already enforced in reality are also not acknowledged in the model. Therefore our screening tool could flag vulnerable wetlands that are, at least to some degree, protected by eFlow provisions in practice. For example, eFlow provisions are part of Australian law and have also been defined for floodplain wetlands of the Murray–Darling basin (Poff and Matthews, 2013). Yet eFlows are defined at only a tiny fraction of rivers worldwide and in most cases are restricted to low flows (Poff et al., 2009; Richter et al., 2012), so that most wetlands remain vulnerable to flow regime modifications. Our study benefits from the qualitative assessment where we collected information on legal eFlow provisions from the related national or State Water Act, which we combined with our quantitative model outcomes. Legal eFlow provisions are regarded as a first important step for promoting eFlow implementation. However, they do not guarantee that eFlows will actually be established in practice, enforced, or adequate, and hence could not be considered in the model's operation scheme.

We used flood volume as a proxy indicator for the extent of flooding. Further improvements of the screening tool will address the implementation of floodplain storages in WaterGAP3 based on an elevation model on sub-grid scale. This will enable a better estimation of change in the extent of flooding due to flow regime alterations. In order to distinguish between different wetland types, it would be useful that future global wetland datasets provide more information on the wetland's main source of water as done in the European wetland geodatabase of Okruszko et al. (2011).

We stress that global-scale modeling is limited by the quality of all input data used in our calculations. In wealthier countries data availability and verification is often more advanced. Hence, a bias of the model performance can be expected, with a higher uncertainty in data-poor regions. However, our global-scale modeling approach allows transfer of knowledge to these regions by identifying hotspots of risk where further hydro-ecological research can be directed.

The implementation of appropriate counteractive measures is likely to be most urgent for the identified hotspots of current and future threats. Those measures encompass adaptive integrated dam management that reconciles interests of different water use sectors, improved flood management plans, water-use-efficiency enhancement, and sophisticated eFlow provisions, e.g., according to the Block Building Methodology (BBM; Tharme and King, 1998) or the Basic Flow Methodology (BFM; Palau and Alcazar, 2012). These two methodologies take account of ecologically relevant flow elements such as flood pulses for riparian wetlands. Dam reoperation strategies aiming at ecosystem restoration depend on the dam's main operating purpose (Watts et al., 2011). In our assessment hydropower dams were the most frequent dam type in the upstream areas of riparian wetlands. However, irrigation dams dominate in the upstream areas of seriously affected sites. Consequently, notably for irrigation and hydropower dams, innovative and integrative operating rules need to be developed, maintaining global food security and economic benefits while at the same time releasing eFlows for ecosystem health and biodiversity.

Depending on the location, climate change will increase or decrease floodplain inundation in the future. In our simulations, two hotspots (eastern Europe/western Asia as well as central South America) were identified with reduced floodplain wetland inundation under climate change. Especially these sites could benefit from achieving climate targets set in international agreements. Further application of the screening tool presented in this study could take into account a higher number of GCM projections as well as scenarios describing future socio-economic developments. Outcomes could be used for a comprehensive uncertainty analysis in order to make statements for each wetland about the probability and degree of change. Depending on the RCP, projected global mean temperature is likely to increase between 0.3 and 4.8 °C until the end of the 21st century relative to 1986–2005 (IPCC, 2014). For time horizons beyond 2050, it would also be advisable to select climate change projections representing more than one RCP to provide insight into a full range of possible future developments. As the goal of this paper is to demonstrate the screening tool, only results for the ensemble median of five GCMs were presented. The model results on changes in overbank flows as obtained from the five different GCMs are included in Sect. S3.

Regarding sites with simulated increasing flood volumes, it is uncertain from the global perspective whether the increased flood volume benefits the wetland or generates flood damages for people. The latter, in turn, would be an incentive to build more dams for flood control (Poff and Matthews, 2013). In particular it needs to be assured that high-flow pulses do not expose people to flood risk and damage. In general, all wetlands could benefit from improved flood management plans taking non-structural measures into account (Sparks, 1995). For example, restoring river floodplains and dead stream branches minimizes flood damages and reduces flood-control storages in reservoirs. This measure would increase the potential to allocate more water for hydropower generation, water supply, or eFlow provisions (Watts et al., 2011). Further measures encompass dyke relocation, buying land from farmers, defining maximum admissible dam releases for flood provisions, or establishing floodways that direct floodwater around human settlements.

Riparian wetlands are threatened by human-induced flow regime modifications, but also depend on society's capacity to act to prevent or reduce these modifications through the implementation of counteractive measures. These two dimensions together have not yet been considered in global assessments of hydrological threats to riparian wetlands. In order to fill this gap, our approach addresses both dimensions and combines qualitative information on the capacity to act with quantitative hydrological model results. In riparian wetlands that are located in countries where the capacity to act is limited by a low institutional capacity, the acknowledgement of ecological water requirements in its legislation could be assumed to be an important first step in eFlow implementation. This would show that ecological water requirements are on the agenda of legislators and water practitioners, and hence helps in advocating acceptance of ecological water requirements. In general, the more countries depend on the available water resources within a river basin, the more challenging the implementation of eFlows is. Therefore, for all transboundary upstream areas, we aimed at investigating whether RBOs, international water treaties, and specific treaty provisions have already been put in place to manage disputes and water resource allocation among different water users. At riparian wetlands in transboundary river basins where institutional capacity is low, the establishment of such formal arrangements could be supportive of wetland conservation. While the institutional capacity indicator considers national laws, international treaties, and RBO agreements, it is important to stress that the presence of formal arrangements is no guarantee of effective enforcement in practice.

Our screening tool helps to identify riparian wetlands where the capacity to act is limited by, e.g., high water resource competition. Especially at sites suffering from water shortage in the upstream area, measures are likely to be required that increase water use efficiency (e.g., water recycling, technological innovations, dripping irrigation, changing crop mix, importing agricultural products, water metering or other incentives to save water) in order to reduce water abstractions and raise the amount of water that can be allocated for ecological requirements.

We assessed the threat of future dam construction to riparian wetlands globally. Because no comprehensive list of major dam initiatives (including storage capacities) is publicly available, we collected the number of dams that are currently planned, proposed, or under construction in upstream areas of riparian wetlands as a first indication of the threat of riparian wetland inundation. As a next step, new dam initiatives could be implemented in the WaterGAP3 model to quantitatively judge changes in flood volumes. This would account for operation, location in the upstream area, and storage volume of future dams, and thus improve analysis of future ecological and human water stress. Including the information on new dam initiatives in the model would be a huge step towards a comprehensive scenario assessment to quantitatively evaluate the combined effects of dam operation, water use, and climate change on river flow regimes. Additionally, riparian wetland inundation is also influenced by land-use changes (e.g., deforestation, land drainage, or sealing of large urban areas) and river construction (e.g., embankment, re-aligning, widening, or deepening). These influences interact with water resource management and climate change, but did not fall within the scope of this paper.

5 Conclusions and outlook

Freshwater demands of an exponentially growing world population, hydropower development as a new source of renewable energy, and projected climate change pose important challenges to the maintenance of riparian wetlands worldwide. Since riparian wetlands provide valuable ecosystem services and are disappearing at an alarming rate, assessing the alteration of ecologically important flood pulses addresses crucial research questions related to environmental, water, and flood management. Therefore, this study aimed at establishing a global screening tool to systematically identify hotspots and patterns of hydrological change and to flag riparian wetlands vulnerable to inundation regime modifications. The information provided by this tool can be useful for directing further hydro-ecological research that takes into account local information and expertise of site-specific ecological, social, and economic conditions.

A multitude of applications are possible with our proposed screening tool. The bankfull flow approach applied at grid cell level enables the assessment of all larger riparian wetlands worldwide and can be used to conduct a comprehensive global riparian wetland assessment. Considering the change in the extent of flooding, the quantification of specific ecosystem services from intact riparian wetlands could be performed. Examples comprise production of important resources such as wood, reed, hay, and fish, water purification by removing nutrients and toxins, as well as flood control and risk reduction for people, and how this is likely to change in the future under climate change and further dam construction.

The WaterGAP3 integrated global modeling framework allows scenario assessment considering different drivers of global change in renewable freshwater resources by allocating water resources to different water use sectors and evaluating the respective consequences under different management targets.

Overall, the screening tool based on quantitative and qualitative indicators could support policy makers at international level (e.g., at forums like UNEP, OECD, the European Union, the Convention on Wetlands of International Importance, and the Convention on Biological Diversity) in implementing global conservation efforts, targeting wetland conservation funds, planning of water infrastructure location and design, and balancing water allocations to humans and nature.

Competing interests. The authors declare that they have no conflict of interest.

Acknowledgements. Research funding for this study was partially provided by the Global Environment Facility Transboundary Water Assessment Programme (GEF TWAP, http://twap-rivers.org/) and is gratefully acknowledged. Data on riparian wetland location and extent were downloaded from the website of the Ramsar Convention Secretariat (http://www.ramsar.org/).

Edited by: N. Basu

References

Alcamo, J., Döll, P., Henrichs, T., Kaspar, F., Lehner, B., Rösch, T., and Siebert, S.: Global estimates of water withdrawals and availability under current and future "business-as-usual" conditions, Hydrol. Sci. J., 48, 339–349, 2003.

Allan, J. D., Palmer, M. A., and Poff, N. L.: Climate change and freshwater ecosystems, in: Climate Change and Biodiversity, Yale University Press, New Haven CT, 272–290, 2005.

Arthington, A. H., Bunn, S. E., Poff, N. L., and Naiman, R. J.: The challenge of providing environmental flow rules to sustain river ecosystems, Ecol. Appl., 16, 1311–1318, 2006.

aus der Beek, T., Flörke, M., Lapola, D. M., Schaldach, R., Voß, F., and Teichert, E.: Modelling historical and current irrigation water demand on the continental scale: Europe, Adv. Geosci., 27, 79–85, https://doi.org/10.5194/adgeo-27-79-2010, 2010.

Barbier, E. B. and Thompson, J. R.: The value of water: Floodplain versus large-scale irrigation benefits in northern Nigeria, AMBIO, 27, 434–440, 1998.

Bayley, P. B.: The Flood Pulse Advantage and the Restoration of River-Floodplain Systems, Regulated Rivers: Research & Management, 6, 75–86, 1991.

Bayley, P. B.: Understanding Large River: Floodplain Ecosystems, BioScience, 45, 153–158, 1995.

Beck, H. E., van Dijk, A. I. J. M., de Roo, A., Dutra, E., Fink, G., Orth, R., and Schellekens, J.: Global evaluation of runoff from ten state-of-the-art hydrological models, Hydrol. Earth Syst. Sci. Discuss., https://doi.org/10.5194/hess-2016-124, in review, 2016.

Bellman, R. E.: Dynamic Programming, Princeton, NJ, Princeton University Press, 340 pp., 1957.

Benítez Sanz, C., and Schmidt, G.: Analysis of the implementation of Environmental Flows in the wider context of the River Basin Management Plans, Report, INTECSA-INARSA, available at: http://ec.europa.eu/environment/archives/water/implrep2007/pdf/Waterabstractionanduse-Eflows.pdf (last access: 22 June 2015), 2012.

Chao, B. F., Wu, Y. H., and Li, Y. S.: Impact of artificial reservoir water impoundment on global sea level, Science, 320, 212–214, https://doi.org/10.1126/science.1154580, 2008.

Collier, M., Webb, R. H., and Schmidt, J. C.: Dams and Rivers – A Primer on the Downstream Effects of Dams, US Geological Survey, Circular 1126, 108 pp., 1996.

Costanza, R., d'Arge, R., de Groot, R., Faber, S., Grasso, M., Hannon, B., Limburg, K., Naeem, S., O'Neill, R. V., Paruelo, J., Raskin, R. G., Sutton, P., and van der Belt, M.: The value of the world's ecosystems and natural capital, Nature, 387, 253–260, 1997.

Cushing, C. E. and Allan, J. D.: Streams: Their Ecology and Life, Academic Press, San Diego, CA, 89 pp., 2001.

Dankers, R., Arnell, N. W., Clark, D. B., Falloon, P. D., Fekete, B. M., Gosling, S. N., Heinke, J., Kim, H., Masaki, Y., Satoh, Y., Stacke, T., Wada, Y., and Wisser, D.: First look at changes in flood hazard in the Inter-Sectoral Impact Model Intercomparison Project ensemble, P. Natl. Acad. Sci. USA, 111, 3257–3261, https://doi.org/10.1073/pnas.1302078110, 2013.

De Stefano, L., Duncan, J., Dinar, S., Stahl, K., Strzepek, K. M., and Wolf, A. T.: Climate change and the institutional resilience of international river basins, J. Peace Res., 49, 193–209, 2012.

Dixon, M. J. R., Loh, J., Davidson, N. C., Beltrame, C., Freeman, R., and Walpole, M.: Tracking global change in ecosystem area: The Wetland Extent Trends index, Biol. Cons., 193, 27–35, https://doi.org/10.1016/j.biocon.2015.10.023, 2016.

Döll, P. and Zhang, J.: Impact of climate change on freshwater ecosystems: a global-scale analysis of ecologically relevant river flow alterations, Hydrol. Earth Syst. Sci., 14, 783–799, https://doi.org/10.5194/hess-14-783-2010, 2010.

Döll, P., Fiedler, K., and Zhang, J.: Global-scale analysis of river flow alterations due to water withdrawals and reservoirs, Hydrol. Earth Syst. Sci., 13, 2413–2432, https://doi.org/10.5194/hess-13-2413-2009, 2009.

Doocy, S., Daniels, A., Murray, S., and Kirsch, T. D.: The Human Impact of Floods: a Historical Review of Events 1980-2009 and Systematic Literature Review, PLOS Currents Disasters, 1, https://doi.org/10.1371/currents.dis.f4deb457904936b07c09daa, 2013.

Dynesius, M. and Nilsson, C.: Fragmentation and flow regulation of river systems in the Northern third of the world, Science, 266, 753–62, 1994.

EEA (European Environment Agency): Corine Land Cover 2000 – Mapping a decade of change Document Actions, Tech. Rep. Brochure No 4/2004, 2004.

Eisner, S.: Comprehensive evaluation of the WaterGAP3 model across climatic, physiographic, and anthropogenic gradients, dissertation, KOBRA Dokumentenserver, University of Kassel, Germany, available at: http://nbn-resolving.de/urn:nbn:de:hebis:34-2016031450014 (last access: 6 June 2017), 2016.

Eisner, S., Flörke, M., Chamorro, A., Daggupati, P., Donnelly, C., Huang, J., Hundecha, Y., Koch, H., Kalugin, A., Krylenko, I., Mishra, V., Piniewski, M., Samaniego, L., Seidou, O., Wallner, M., and Krysanova, V.: An ensemble analysis of climate change impacts on streamflow seasonality across 11 large river basins, Climatic Change, 1–17, https://doi.org/10.1007/s10584-016-1844-5, 2017.

FAO (Food and Agriculture Organization of the United Nations): FAO Water Lex Legal Database, http://www.waterlex.org/waterlex-legal-database/, last access: 22 June 2015.

FAO (Food and Agriculture Organization of the United Nations): World Development Indicators, AQUASTAT data, available at: http://data.worldbank.org/indicator/ER.H2O.FWTL.K3 (last access: 20 January 2017), 2016.

Flörke, M., Bärlund, I., and Kynast, E.: Will climate change affect the electricity production sector? A European study, Water Climate Change, 3, 44–54, https://doi.org/10.2166/wcc.2012.066, 2012.

Flörke, M., Kynast, E., Bärlund, I., Eisner, S., Wimmer, F., and Alcamo, J.: Domestic and industrial water uses of the past 60 years as a mirror of socio-economic development: A global simulation study, Glob. Environ. Change, 23, 144–156, https://doi.org/10.1016/j.gloenvcha.2012.10.018, 2013.

Georgiyevsky, V. Y., Zhuravin, S. A., and Ezhov, A. V.: Assessment of trends in hydrometeorological situation on the Great Russian Plain under the effect of climate variations, in: Proceedings of American Geophysical Union, 15th Annual Hydrology Days, 47–58, 1995.

Georgiyevsky, V. Y., Yezhov, A. V., Shalygin, A. L., Shiklomanov, A. I., and Shiklomanov, I. A.: Evaluation of possible climate change impact on hydrological regime and water resources of the former USSR rivers, Russ. Meteorol. Hydrol., 11, 89–99, 1996.

Georgiyevsky, V. Y., Yezhov, A. V., and Shalygin, A. L.: An assessment of changing river runoff due to man's impact and global climate warming, in: River Runoff Calculations, Report at the International Symposium, UNESCO, 75–81, 1997.

GRDC: Long Term Mean Monthly Discharges and Annual Characteristics of Selected GRDC Stations, the Global Runoff Data Centre, Koblenz, Germany, 2004.

Gregory, S. V., Swanson, F. J., McKee, W. A., and Cummins, K. W.: An ecosystem perspective of riparian zones, BioScience, 41, 540–551, 1991.

Grill, G., Lehner, B., Lumsdon, A. E., MacDonald, G. K., Zarfl, C., and Liermann, C. R.: An index-based framework for assessing patterns and trends in river fragmentation and flow regulation by global dams at multiple scales, Environ. Res. Lett., 10, https://doi.org/10.1088/1748-9326/10/1/015001, 2015.

GWP (Global Water Partnership): Integrated water resource management in Central Asia: The challenges of managing large transboundary rivers, Technical Focus Paper, ISBN: 91-85321-99-0, 2014.

Haddeland, I., Heinke, J., Biemans, H., Eisner, S., Flörke, M., Hanasaki, N., Konzmann, M., Ludwig, F., Masaki, Y., Stacke, T., Tessler, Z. D., Wada, Y., and Wisser, D.: Global water resources affected by human interventions and climate change, Supporting Information, P. Natl. Acad. Sci. USA, 111, 3251–3256, 2014.

Hanasaki, N., Kanae, S., and Oki, T.: A reservoir operation scheme for global river routing models, J. Hydrol., 327, 22–41, 2006.

Hempel, S., Frieler, K., Warszawski, L., Schewe, J., and Piontek, F.: A trend-preserving bias correction – the ISI-MIP approach, Earth Syst. Dynam. Discuss., 4, 49–92, https://doi.org/10.5194/esdd-4-49-2013, 2013.

Hoekstra, A. Y., Chapagain, A. K., Aldaya, M. M., and Mekonnen, M. M.: The Water Footprint Assessment Manual: Setting the Global Standard, Earthscan, London, Washington, DC, ISBN 978-1-84971-279-8, 2011.

Hoekstra, A. Y., Mekonnen, M. M., Chapagain, A. K., Mathews, R. E., and Richter, B. D.: Global water scarcity: the monthly blue water footprint com- pared to blue water availability for the world's major river basins, P. Natl. Acad. Sci. USA, 109, 3232–3237, https://doi.org/10.1073/pnas.1109936109, 2012.

Hughes, F. M. R.: The ecology of African floodplain forests in semi-arid and arid zones: a review, J. Biogeogr., 15, 127–140, 1988.

ICOLD (International Commission of Large Dams): World Register of Dams, Int. Comm, Large Dams, Paris, France, 2007.

IPCC: Climate Change 2014: Synthesis Report, Contribution of Working Groups I, II and III to the Fifth Assessment Report of the Intergovernmental Panel on Climate Change edited by: Core Writing Team, Pachauri, R. K., and Meyer, L. A., IPCC, Geneva, Switzerland, 151 pp., 2014.

Jolly, I. D.: The effects of river management on the hydrology and hydroecology of arid and semi-arid floodplains, in: Floodplain Processes, edited by: Anderson, M. G., Walling, D. E., and Bates, P. D., Chichester, UK, Wiley, 577–609, 1996.

Jonkman, S. N.: Global Perspectives on Loss of Human Life Caused by Floods, Nat. Hazard., 34, 151–175, 2005.

Junk, W. J. and Wantzen, K. M.: The Flood Pulse Concept: New Aspects, Approaches and Applications – An Update, in Welcomme, R., Petr, T., (eds.), Proceedings of the 2nd Large River Symposium (LARS), Pnom Penh, Cambodia, Bangkok, RAP Publication, 117–140, 2004.

Junk, W. J., Bayley, P. B., and Sparks, R. E.: The flood pulse concept in river-floodplain systems, Can. Spec. Publ. Fish. Aquat. Sci., 110–127, 1989.

Kingsford, R. T.: Ecological impacts of dams, water diversions and river management on floodplain wetlands in Australia, Aust. Ecol., 25, 109–127, 2000.

Khublaryan, M. G.: Water Resources for Sustainable Development, with particular Reference to Russia, in: Encyclopedia of Life Support Systems (EOLSS), Vol. 1, Area Studies-Russia (Regional Sustainable Development Review, ISBN-978-1-84826-074-0, 2000.

Laize, C. L. R., Acreman, M. C., Schneider, C., Dunbar, M. J., Houghton-Carr, H., Flörke, M., and Hannah, D. M.: Projected flow alteration and ecological risk for pan-European rivers, River Res. Appl., 30, 299–314, https://doi.org/doi.org/10.1002/rra.2645, 2014.

Lehner, B., Verdin, K., and Jarvis., A.: New Global hydrography derived from spaceborne elevation data, Eos. Trans. 89, 93–94, 2008.

Lehner, B., Liermann, C. R., Revenga, C., Vörösmarty, C., Fekete, B., Crouzet, P., Döll, P., Endejan, M., Frenken, K., Magome, J., Nilsson, C., Robertson, J. C., Rödel, R., Sindorf, N., and Wisser, D.: High-resolution mapping of the world's reservoirs and dams for sustainable river-flow management, Front. Ecol. Environ., 9, 494–502, https://doi.org/10.1890/100125, 2011.

Le Quesne, T., Kendy, E., and Weston, D.: The Implementation Challenge: Taking stock of government policies to protect and restore environmental flows, WWF and Nature Conservancy, http://awsassets.panda.org/downloads/the_implementation_challenge.pdf (last access: 22 June 2015), 2010.

Lloyd, N., Quinn, G., Thoms, M., Arthington, A., Gawne, B., Humphries, P., and Walker, K.: Does flow modification cause geomorphological and ecological response in rivers?, a literature review from an Australian perspective, Technical report 1/2004, CRCFE, Canberra, 2004.

Magilligan, F. J. and Nislow, K. H.: Changes in hydrologic regime by dams, Geomorphology, 71, 61–78, 2005.

Maheshwari, B. L., Walker, K. F., and McMahon, T. A.: Effects of flow regulation on the flow regime of the River Murray, Australia, Regulated Rivers: Research & Management, 10, 15–38, 1995.

McCully, P.: Silenced rivers: the ecology and politics of large dams, Zed Books, London, UK, 1996.

McDonald, R., Weber, K., Padowski, J., Flörke, M., Schneider, C., Green, P., Gleeson, T., Eckman, S., Lehner, B., Balk, D., Boucher, T., Grill, G., and Montgomery, M.: Water on an urban planet: urbanization and the reach of urban water infrastructure, Glob. Enviro. Change, 27, 96–105, 2014.

Middelkoop, H., Alabyan, A. M., Babich, D. B., and Ivanov, V. V.: Post-dam Channel and Floodplain Adjustment along the Lower Volga River, Russia, Geomorphic Approaches to Integrated Floodplain Management of Lowland Fluvial Systems in North America & Europe, 245–264, https://doi.org/10.1007/978-1-4939-2380-9_10, 2015.

Milly, P. C. D., Betancourt, J., Falkenmark, M., Hirsch, R. M., Kundzewicz, Z. W., Lettenmaier, D. P., and Stouffer, R. J.: Stationarity Is Dead: Whither Water Management?, Science, 319, 573–574, https://doi.org/10.1126/science.1151915, 2008.

Moore, M.: Perceptions and interpretation of environmental flows and implications for future water resources management: a survey study, Thesis (MSc), Department of Water and Environment Studies, Linköping University, Sweden, 2004.

Naiman R. J., Décamps H., and Pollock, M.: The role of riparian corridors in maintaining regional biodiversity, Ecol. Appl., 3, 209–212, 1993.

Naiman, R. J., Bunn, S. E., Nilsson, C., Petts, G. E., Pinay, G., and Thompson, L. C.: Legitimizing fluvial ecosystems as users of water: an overview, Environ. Manage., 30, 455–467, https://doi.org/10.1007/s00267-002-2734-3, 2002.

Nilsson, C. and Berggren, K.: Alterations of riparian ecosystems caused by river regulation, Bioscience, 50, 783–792, 2000.

Nilsson, C., Reidy, C. A., Dynesius, M., and Revenga, C.: Fragmentation and flow regulation of the world's large river systems, Science 308, 405–408, 2005.

Nislow, K. H., Magilligan, F. J., Fassnacht, H., Bechtel, D., and Ruesink, A.: Effects of hydrologic alteration on flood regime of natural floodplain communities in the upper Connecticut River, J. Am. Water Resour. Assoc., 38, 1533–1548, 2002.

Nohara, D., Kitoh, A., Hosaka, M., and Oki, T.: Impact of Climate Change on River Discharge Projected by Multimodel Ensemble, J. Hydrometeorol., 7, 1076–1089, 2006.

Norris, R. H. and Thoms, M. C.: What is river health?, Freshwater Biol., 41, 197–209, 1999.

OECD (Organisation for Economic Co-operation and Development): Water Resources Allocation: Sharing Risks and Opportunities, OECD studies on Water, OECD Publishing, Paris, https://doi.org/10.1787/9789264229631-en, 2015.

Okruszko, T., Duel, H., Acreman, M., Grygoruk, M., Flörke, M., and Schneider, C.: Broad-scale ecosystem services of European wetlands – overview of the current situation and future perspectives under different climate and water management scenarios, Hydrol. Sci. J., 56, 1501–1517, 2011.

Pahl-Wostl, C., Arthington, A., Bogardi, J., Bunn, S. E., Hoff, H., Lebel, L., Nikitina, E., Palmer, M., Poff, L. N., Richards, K., Schlüter, M., Schulze, R., St-Hilaire, A., Tharme, R., Tockner, K., and Tsegai, D.: Environmental flows and water governance: managing sustainable water uses, Curr. Opin. Environ. Sust., 5, 341–351, https://doi.org/10.1016/j.cosust.2013.06.009, 2013.

Palau, A. and Alcazar, J.: The basic flow method for incorporating flow variability in environmental flows, River Res. Appl., 28, 93–102, 2012.

Pastor, A. V., Ludwig, F., Biemans, H., Hoff, H., and Kabat, P.: Accounting for environmental flow requirements in global water assessments, Hydrol. Earth Syst. Sci., 18, 5041–5059, https://doi.org/10.5194/hess-18-5041-2014, 2014.

Petersen-Perlman, J. D.: Mechanisms of cooperation for states' construction of large-scale water infrastructure in transboundary river basins, PhD Dissertation, Oregon State University, USA, 2014.

Petts, G. E.: Impounded Rivers: Perspectives for Ecological Management, John Wiley and Sons, Chichester, England, 326 pp., 1984.

Poff, N. L. and Hart, D. D.: How Dams Vary and Why It Matters for the Emerging Science of Dam Removal, BioScience, 52, 659–738, 2002.

Poff, N. L. and Matthews, J. H.: Environmental flows in the Anthropocence: past progress and future prospects, Current Opinion in Environmental Sustainability, 5, 667–675, 2013.

Poff, N. L. and Zimmerman, J. K. H.: Ecological responses to altered flow regimes: a literature review to inform the science and management of environmental flows, Freshwater Biol., 55, 194–205, 2010.

Poff, N. L., Allan, J. D., Bain, M. B., Karr, J. R., Prestegaard, K. L., Richter, B. D., Sparks, R. E., and Stromberg, J. C.: The natural flow regime, Bioscience 47, 769–784, 1997.

Poff, N. L., Olden, J. D., Merritt, D., M., and Pepin, D. M.: Homogenization of regional river dynamics by dams and global biodiversity implications, P. Natl. Acad. Sci. USA, 104, 5732–5737, 2007.

Poff, N. L., Richter, B. D., Arthington, A. H., Bunn, S. E., Naiman, R. J., Kendy, E., Acreman, M. C., Apse, C., Bledsoe, B. P., Freeman, M. C., Henriksen, J., Jacobson, R. B., Kennen, J. G., Merritt, D. M., O'Keeffe, J. H., Olden, J. D., Rogers, K., Tharme, R. E., and Warner, A.: The ecological limits of hydrologic alteration (ELOHA): a new framework for developing regional environmental flow standards, Freshwater Biology, 55, 147–170, https://doi.org/10.1111/j.1365-2427.2009.02204.x, 2009.

Postel, S. L. and Richter, B. D.: Rivers For Life: Managing Water for People and Nature, Island Press, 2003.

Riahi, K., Rao, S., Krey, V., Cho, C., Chirkov, V., Fischer, G., Kindermann, G., Nakicenovic, N., and Rafaj, P.: RCP8.5 – A scenario of comparatively high greenhouse gas emissions, Climatic Change, 109, 33–57, https://doi.org/10.1007/s10584-011-0149-y, 2011.

Richter, B. D.: Re-thinking environmental flows: from allocations and reserves to sustainability boundaries, River Res. Appl., 26, 1052–1063, https://doi.org/10.1002/rra.1320, 2009.

Richter, B. D. and Thomas, G. A.: Restoring environmental flows by modifying dam operations, Ecol. Soc., 12, 12, 2007.

Richter, B. D., Davis, M. M., Apse, C., and Konrad, C.: A presumptive standard for environmental flow protection, River Res. Appl., 28, 1312–1321, https://doi.org/10.1002/rra.1511, 2012.

Rosenberg, D. M., McCully, P., and Pringle, C. M.: Global-Scale Environmental Effects of Hydrological Alterations: Introduction, BioScience, 50, 746–751, 2000.

RSIS: Ramsar Sites Information Service, available at: https://rsis.ramsar.org/, last access: 27 January 2015.

Savarenskiy, A. D.: Metod rascheta regulirovaniya stoka (A method for streamflow control computation), Gidrotekh. Stroit., 2, 24–28, 1940.

Schmeier, S.: International RBO Database, Transboundary Freshwater Dispute Database (TFDD), Oregon State University, Corvallis, OR, available at: http://www.transboundarywaters.orst.edu/research/RBO/RBO_Database.html, last access: 27 July 2014.

Schneider, C.: River ecosystems at risk, Modelling and assessing the impacts of dam operation, water use and climate change on global and continental river flow regimes, dissertation, KOBRA Dokumentenserver, University of Kassel, Germany, 1–163, available at: http://nbn-resolving.de/urn:nbn:de:hebis:34-2015121849487 (last access: 6 June 2017), 2015.

Schneider, C., Flörke, M., Eisner, S., and Voss, F.: Large scale modelling of bankfull flow: an example for Europe, J. Hydrol., 408, 235–245, 2011a.

Schneider, C., Flörke, M., Geerling, G., Duel, H., Grygoruk, M., and Okruszko, T.: The future of European floodplain wetlands under a changing climate, J. Water Clim., 2, 106–122, 2011b.

Schneider, C., Laizé, C. L. R., Acreman, M. C., and Flörke, M.: How will climate change modify river flow regimes in Europe?, Hydrol. Earth Syst. Sci., 17, 325–339, https://doi.org/10.5194/hess-17-325-2013, 2013.

Siebert, S., Henrich, V., Frenken, K., and Burke, J.: Global Map of Irrigation Areas version 5, Rheinische Friedrich-Wilhelms-University, Bonn, Germany/Food and Agriculture Organization of the United Nations, Rome, Italy, 2013.

Smakhtin, V., Revenga, C., and Döll, P.: Taking into Account Environmental Water Requirements in Global-scale Water Resources Assessments, Comprehensive Assessment of Water Management in Agriculture Research Report 2 (IWMI, Colombo), 2004.

Smakhtin, V. U. and Eriyagama, N.: Developing a software package for global desktop assessment of environmental flows, Environ. Modell. Softw., 23, 1396–1406, 2008.

Sparks, R. E.: Need for Ecosystem Management of Large Rivers and Their Floodplains, Bioscience, 45, 168–182, 1995.

Swiss Re: World insurance in 2013: steering towards recovery, http://media.swissre.com/documents/sigma3_2014_en.pdf (last access: 17 May 2015), 2014.

Taylor P. J., Walker G. R., Hodgson G., Hatton T. J., and Correll, R. L.: Testing of a GIS Model of Eucalyptus largiflorens health on a semiarid, saline floodplain, Environ. Manage., 20, 553–564, 1996.

Tharme, R. E. and King, J. M.: Development of the Building Block Methodology for instream flow assessments, and supporting research on the effects of different magnitude flows on riverine ecosystems, Water Research Commission Report No. 576/1/98, 452 pp., 1998.

Tockner, K. and Stanford, J. A.: Riverine flood plains: present state and future trends, Environ. Conserv., 29, 308–30, 2002.

Tockner, K., Malard, F., and Ward, J. V.: An extention of the flood pulse concept, Hydrol. Process., 14, 2861–2883, 2000.

UDI (Utility Data Institute): World Electric Power Plants Database, Platts Energy InfoStore, available at: http://www.platts.com (last access: 17 April 2012), 2004.

Uluocha, N. O. and Okeke, I. C.: Implications of wetlands degradation for water resources management: Lessons from Nigeria, GeoJournal, 61, 151–154, 2004.

UNEP: Vital Water Graphics – An Overview of the State of the World's Fresh and Marine Waters, 2 nd Edn., UNEP, Nairobi, Kenya. ISBN: 92-807-2236-0, available at: http://www.grida.no/publications/vg/water2/page/3262.aspx (last access: 1 February 2017), 2008.

USGS: Global Land Cover Characterization (GLCC), available at: http://edc2.usgs.gov/glcc/glcc.php (last access: 27 June 2012), 2008.

van Beek, L. P. H., Wada, Y., and Bierkens, M. F. P.: Global monthly water stress: 1. Water balance and water availability, Water. Resour. Res., 47, W07517, https://doi.org/10.1029/2010wr009791, 2011.

Verzano, K. and Menzel, L.: Snow conditions in mountains and climate change – a global view, in: Hydrology in Mountain Regions: Observations, Processes and Dynamics, Proceedings of Symposium HS1003 at IUGG2007, Perugia, July 2007, 147–154, 2009.

Verzano, K., Bärlund, I., Flörke, M., Lehner, B., Kynast, E., and Voß, F.: Modeling variable river flow velocity on continental scale: Current situation and climate change impacts in Europe, J. Hydrol., 424–425, 238–251, 2012.

Vörösmarty, C. J., Sharma, K. P., Fekete, B. M., Copeland, A. H., Holden, J., Marble, J., and Lough, J. A.: The storage and aging of continental runoff in large reservoir systems of the world, Ambio, 26, 210–219, 1997.

Vörösmarty, C. J., McIntyre, P. B., Gessner, M. O., Dudgeon, D., Prusevich, A., Green, P., Glidden, S., Bunn, S. E., Sullivan, C. A., Reidy Liermann, C., and Davies, P. M.: Global threats to human water security and river biodiversity, Nature 467, 555–561, https://doi.org/10.1038/nature09440, 2010.

Watts, R. J., Richter, B. D., Opperman, J. J., and Bowmer, K. H.: Dam reoperation in an era of climate change, Mar. Freshwater Res., 62, 321–327, 2011.

WCD (World Commission on Dams): Dams and development: a new framework for decision-making, Earthscan, London, UK, 2000.

Weedon, G. P., Balsamo, G., Bellouin, N., Gomes, S., Best, M. J., and Viterbo, P.: The WFDEI meteorological forcing data set: WATCH Forcing Data methodology applied to ERA-Interim reanalysis data, Water Resour. Res., 50, 7505–7514, https://doi.org/10.1002/2014WR015638, 2014.

Welcomme, R.: Fisheries ecology of floodplain rivers, London: Longman, 1979.

Wilding, T. K. and Poff, N. L.: Flow-ecology relationships for the watershed flow evaluation tool, in: Watershed Flow Evaluation Tool Pilot Study for Roaring Fork and Fountain Creek Watersheds and Site-specific Quantification Pilot Study for Roaring Fork Watershed, edited by: Bledsoe, B. D., Miller, W. J., Poff, N. L., Sanderson, J. S., and Wilding, T. K., Colorado Water Conservation Board: Denver, Colorado, 2008.

WWF (World Wildlife Fund): Rivers at risk: dams and the future of freshwater ecosystems, available online at: http://assets.panda.org/downloads/riversatriskfullreport.pdf (last access: 11 May 2015), 2004.

WWF (World Wildlife Fund): Living Planet Report 2014: species and spaces, people and places, edited by: McLellan, R., Iyengar, L., Jeffries, B., and Oerlemans, N., WWF, Gland, Switzerland, 2014.

Zarfl, C., Lumsdon, A. E., Berlekamp, J., Tydecks, L., and Tockner, K.: A global boom in hydropower dam construction, Aquat. Sci., 77, 161–170, https://doi.org/10.1007/s00027-014-0377-0, 2014.

3

The effect of assimilating satellite-derived soil moisture data in SiBCASA on simulated carbon fluxes in Boreal Eurasia

M. K. van der Molen[1], R. A. M. de Jeu[2,3], W. Wagner[4], I. R. van der Velde[1], P. Kolari[5], J. Kurbatova[6], A. Varlagin[6], T. C. Maximov[7], A. V. Kononov[7], T. Ohta[8], A. Kotani[8], M. C. Krol[1], and W. Peters[1,9]

[1]Meteorology and Air Quality group, Wageningen University, P.O. Box 47, 6700 AA Wageningen, the Netherlands
[2]VU University Amsterdam, Department of Earth Sciences, De Boelelaan 1085, 1081 HV Amsterdam, the Netherlands
[3]Transmissivity B. V., Space Technology Centre, Huygensstraat 34, Noordwijk, the Netherlands
[4]Vienna University of Technology, Department of Geodesy and Geoinformation, Gusshausstrasse 27–29, 1040 Vienna, Austria
[5]Department of Physics, University of Helsinki, P.O. Box 48, Erik Palménin aukio 1, 00014 Helsinke, Finland
[6]A. N. Severtsov Institute of Ecology and Evolution of the Russian Academy of Sciences, Leninsky Prospekt 33, 119071 Moscow, Russia
[7]Institute for Biological Problems of the Cryolithozone, 677980, 41 Lenin Ave., Yakutsk, Republic of Sakha (Yakutia), Russia
[8]Graduate School of Bioagricultural Sciences, Nagoya University, Furo-cho, Chikusa-ku, Nagoya, 464-8601, Japan
[9]Centre for Isotope Research, Energy and Sustainability Research Institute Groningen, Groningen University, Groningen, the Netherlands

Correspondence to: M. K. van der Molen (michiel.vandermolen@wur.nl)

Abstract. Boreal Eurasia is a region where the interaction between droughts and the carbon cycle may have significant impacts on the global carbon cycle. Yet the region is extremely data sparse with respect to meteorology, soil moisture, and carbon fluxes as compared to e.g. Europe. To better constrain our vegetation model SiBCASA, we increase data usage by assimilating two streams of satellite-derived soil moisture. We study whether the assimilation improved SiBCASA's soil moisture and its effect on the simulated carbon fluxes. By comparing to unique in situ soil moisture observations, we show that the passive microwave soil moisture product did not improve the soil moisture simulated by SiBCASA, but the active data seem promising in some aspects. The match between SiBCASA and ASCAT soil moisture is best in the summer months over low vegetation. Nevertheless, ASCAT failed to detect the major droughts occurring between 2007 and 2013. The performance of ASCAT soil moisture seems to be particularly sensitive to ponding, rather than to biomass. The effect on the simulated carbon fluxes is large, 5–10 % on annual GPP and TER, tens of percent on local NEE, and 2 % on area-integrated NEE, which is the same order of magnitude as the inter-annual variations. Consequently, this study shows that assimilation of satellite-derived soil moisture has potentially large impacts, while at the same time further research is needed to understand under which conditions the satellite-derived soil moisture improves the simulated soil moisture.

1 Introduction

The interest of this publication is to explore the potential of assimilating satellite-derived soil moisture in the vegetation model SiBCASA over Boreal Eurasia with particular focus on the impact on simulated carbon fluxes. In remote regions such as Boreal Eurasia meteorological driver data for vegetation models (temperature, precipitation, etc.) are poorly constrained by surface observations, leaving room for improvement in the soil moisture content simulated in vegetation models. Boreal Eurasia is also a region with large carbon stocks in biomass and vegetation (Schepaschenko et al., 2013; McGuire et al., 2009; Tarnocai et al., 2009), which are subject to the fastest climatic change rates on Earth (Goetz et al., 2007), making it a relevant region in the context of ecosystem carbon sequestration. Furthermore, large parts of the region are located in continuous and discontinuous permafrost soils. In the spring, meltwater from the accumulated winter precipitation cannot percolate into the still-frozen soil – it runs off in hilly terrain, and forms ponds on the soil in flat terrain. This causes a bi-modal spatial distribution in soil moisture, with dry hills and wet plains. This process is probably hard to catch by land surface models and satellite observations alike. Satellite-derived soil moisture is observed to have large variability in the northern regions, both within and between different approaches (Mladenova et al., 2014). The derivation of satellite soil moisture is difficult in snow-, ice-, and surface-water-rich areas (Högström et al., 2014; Naeimi et al., 2012a). These aspects make a working soil moisture data assimilation system very relevant for vegetation modelling in the region, and challenging. The few tower sites running in the region now have fairly long measurement records, so that they can be used for validation of inter-annual variation (e.g. the 2010 drought). An effort specifically targeted at Boreal Eurasia and carbon fluxes has not been tried before.

Earlier efforts to assimilate satellite-derived soil moisture in vegetation models were often focused on the improvement of soil moisture itself and/or validated with in situ observations over short vegetation in temperate and Mediterranean climate zones (Reichle and Koster, 2005; Reichle et al., 2007; Draper et al., 2009a, b; Miralles et al., 2011b). Other studies focus on the effect on crop yield and carbon fluxes in Europe (Verstraeten et al., 2010; de Wit and van Diepen, 2007; Han et al., 2014). The Global Land Data Assimilation Systems (GLDAS, http://ldas.gsfc.nasa.gov/GLDAS/, see Chen et al., 2013) is worth mentioning here too. Our study is thus the first to assimilate satellite soil moisture in Boreal Eurasia and to evaluate the impact on simulated soil moisture and carbon fluxes with in situ data.

Soil moisture affects vegetation carbon fluxes through photosynthesis and respiration rates. Photosynthesis rates depend on the stomatal conductance, which the plants regulate according to the water potential in the leaf and the atmospheric vapour pressure deficit. The water potential in the leaf is a function of water supply by the roots and the water use by transpiration (Katul et al., 2010). Heterotrophic respiration depends on the soil moisture content, which is the substrate in which microbes and bacteria consume organic matter and release CO_2. The dependence of photosynthesis, transpiration, and respiration fluxes on soil moisture is implemented in virtually all contemporary vegetation models via similar types of drought sensitivity functions (Verhoef and Egea, 2014; Sellers et al., 1996; Vetter et al., 2008).

Based on simulations with global climate models, it is expected that global warming is associated with more extreme precipitation regimes, resulting in more frequent and more intense flooding and drought events (Dai, 2011). It is probable that this trend is already becoming visible in the large number of recent droughts, e.g. in Amazonia (2005 and 2010), the US (2008–2012, 2014), European Russia and Siberia (2010, 2013), West Europe (2003), East Africa (2011–2012), Australia (2003–2007), China (2010–2011), as well as floodings in Austria and Germany (2013), Southwest China (2013), India and Pakistan (2014), and the UK (2014). These extremes may have a large effect on the carbon balance of the affected regions, sometimes undoing 10 years of "normal" carbon uptake by natural vegetation and causing crop yield reduction or crop failure (van der Molen et al., 2011; Peters et al., 2010; Reichstein et al., 2007; Ciais et al., 2005).

Soil moisture in land surface models is the balance of precipitation and evaporation, transpiration, runoff, and lateral outflow. Land surface models are often primarily calibrated to simulate water, heat, and carbon exchange with the atmosphere correctly (Williams et al., 2009; Morales et al., 2005), and the water stress function is often one of the functions that is used for calibration. As a result, the absolute value of soil moisture in land surface models, and its variation, has often been subsidiary to simulating exchange fluxes correctly. In the perspective of the expected increasing occurrence of extremes in soil moisture, it is questionable whether this procedure results in satisfactory representation of droughts in land surface models, and their effects on the carbon balances and the disturbance of pools.

Two independent databases of remotely sensed soil moisture have been published recently: the passive microwave soil moisture data set, based on the land parameter retrieval model (LPRM) (Owe et al., 2008), and the active microwave data set, METOP ASCAT 25 (Wagner et al., 1999; Naeimi et al., 2009, 2012b). See Sect. 2.2 for details.

These data sets provide globally consistent, satellite-observed soil moisture data. As such, they provide ideal soil moisture information to analyse the development of droughts on a regional scale. In this paper, we describe the implementation of a data assimilation system into the land surface model SiBCASA (Schaefer et al., 2008). This scheme adjusts the simulated soil moisture towards the satellite-observed soil moisture, accounting for the errors in observation and model. We subsequently evaluate the performance of this soil moisture assimilation system by comparing the simulated soil moisture, and the resulting change in carbon fluxes against observations collected at four sites across Boreal Eurasia (defined here as $3° <$ latitude $< 180°$ E and longitude $> 50°$ N) between 2007 and 2013. Although the in situ observations have limited representability for the $1 \times 1°$ latitude/longitude satellite observations, we focus here on longer-term variability (e.g. droughts) which develop over larger areas. We also apply a normalization procedure (CDF matching technique, Sect. 2.3) which removes the impact of soil characteristics on soil moisture distributions. Therefore, the most important reason for mismatch is probably the difference between grid-size average and local precipitation.

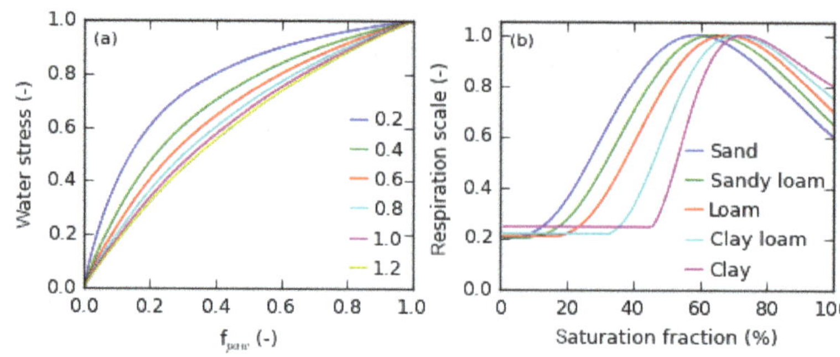

Figure 1. (a) The water stress function in SiBCASA as a function of plant available water fraction. The scaling function is shown for various shape parameters (see legend). The default value for the shape parameter is 0.2 for all biome types. **(b)** The heterotrophic respiration scaling function in SiBCASA as a function of soil moisture saturation fraction and soil type.

The objective of this paper is therefore to evaluate the use of satellite observed soil moisture for data assimilation in vegetation models in a data-poor region like Boreal Eurasia and to evaluate the impact on the simulated carbon fluxes.

The increasing availability of remote sensing products, e.g. the Soil Moisture and Ocean Salinity mission (SMOS) (Kerr et al., 2001), the Global Change Observation Mission – Water (GCOM-W) (Imaoka et al., 2010; Oki et al., 2010), the Global Land Evaporation: the Amsterdam Methodology (GLEAM) (Miralles, 2011; Miralles et al., 2011b), and the Soil Moisture Active Passive mission (SMAP) (Entekhabi et al., 2010), make this evaluation timely and relevant. If the assimilation of soil moisture proves worthwhile, it may also be applied in other land surface models, e.g. in mesoscale atmosphere models or in agro-meteorological models.

2 Methods

2.1 SiBCASA vegetation model

SiBCASA is an interactive vegetation–atmosphere model simulating how the growth of vegetation depends on the exchange of water, energy, and carbon between vegetation, atmosphere, and soil (Schaefer et al., 2008). SiBCASA consists of two coupled components, the SiB component simulates the exchange of heat and water and the uptake of carbon dioxide as a function of temperature, radiation density, humidity, and wind speed, as well as root-zone soil moisture conditions with several time steps per hour (Sellers et al., 1996; Jafarov and Schaefer, 2015). It uses the Farquhar photosynthesis parametrization (Farquhar et al., 1980) in combination with a Ball–Berry type stomatal conductance formulation (Collatz et al., 1991).

The CASA component simulates how the carbon taken up by photosynthesis is allocated to different parts of the vegetation, with specified residence times (Potter et al., 1993). The seasonal development of leaf area index is a function of the amount of carbon allocated to leaves, but the amount of absorbed photosynthetically active radiation (fPAR) is constrained by remote sensing. A more detailed summary of the general features of SiBCASA can be found in van der Velde et al. (2014).

SiBCASA is intended for use as a lower boundary condition for large-scale atmospheric transport models, such as general circulation models and data assimilation systems such as CarbonTracker (Peters et al., 2007). Therefore, it is specifically required to correctly simulate the regional effects of climate variations. In this paper we focus on the role of soil moisture in SiBCASA, with the aim to better describe the effects of climate extremes on the terrestrial carbon dioxide balance. Therefore, we will briefly describe SiBCASA's method of simulating soil water uptake and the relation with stomatal conductance.

SiBCASA is configured with 25 soil layers with a depth up to 15 m and thicknesses ranging from 2 cm near the surface to 3 m at depth. Roots extract water from the part of the soil they penetrate. The plant-available water fraction (f_{paw}) is computed as a function of root depth, porosity, wilting point, field capacity, and soil moisture content and varies between wilting point ($f_{paw} = 0$) and field capacity ($f_{paw} = 1$). The plant-available water fraction directly influences the photosynthesis capacity by means of the water stress function (Fig. 1), which equals one at field capacity and zero at wilting point. The shape parameter of the soil moisture is 0.2 by default (Fig. 1), implying an aggressive water use strategy, i.e. limited water stress in wet to medium dry soils and accelerating water stress with further drying. At low plant-available water fractions, the photosynthesis capacity is thus reduced by multiplication with the water stress function. This in turn reduces the stomatal conductance.

Soil moisture also affects the turnover times of organic matter in the soil. The turnover times are shortest at an optimal soil moisture saturation fraction (which varies around 60 % of the pore space) and from there increases towards drier and wetter soils (Fig. 1) (Raich et al., 1991). The respiration rates are a function of the carbon pools and turnover times, which are temperature and soil moisture dependent.

The net effect of soil moisture on Net Ecosystem Exchange (NEE) is the sum of the effect on photosynthesis (GPP, Gross Primary Production) and on respiration rates (TER, Total Ecosystem Respiration).

2.2 Satellite-derived soil moisture data

The satellite-observed soil moisture data used in this study come from two independent sources (Liu et al., 2011, 2012): the first is from passive microwave observations, the second from active radar observations. Passive microwave radiation sensors have been on board of various satellite platforms, e.g. Scanning Multichannel Microwave Radiometer (SMMR, 1978–1987), Special Sensor Microwave Imager (SSM/I, 1987–present), Tropical Rainfall Measuring Mission Microwave Imager (TMI, 1997–2015) and the Advanced Microwave Scanning Radiometer – Earth Observing System (AMSR-E, 2002–2011). The retrieval algorithm, the Land Parameter Retrieval Method (LPRM) (De Jeu et al., 2009; De Jeu and Owe, 2003; Owe et al., 2008), is based on a simple radiative transfer equation and used dual-polarized brightness temperature observations in an optimization routine to solve for soil moisture. The retrieved soil moisture is representative for the top few centimetres of the soil. LPRM accounts for the vegetation opacity in the microwave domain. Dense canopies attenuate the microwave signal from the underlying soil surface which results in a lower soil moisture retrieval accuracy. Therefore, the soil moisture retrieval is most reliable for bare and sparsely vegetated areas (de Jeu et al., 2008). As this study is focused on the period 2007–2013, we use only the AMSR-E data with version v05 of the LPRM with a 50 km spatial resolution, and a 2–3 day revisit time (Owe et al., 2008).

Complementarily, active radar soil moisture retrievals from the Advanced Scatterometer (ASCAT, 2006–present) in combination with the change detection algorithm is representative for the soil moisture in the top few centimetres (Bartalis et al., 2007; Wagner et al., 1999; Naeimi et al., 2009, 2012b). ASCAT soil moisture retrievals are reliable for sparse and moderately vegetated areas, and less for bare soils (Liu et al., 2012). We use METOP ASCAT25, version WARP5.5, release 2.1, with a 0.25° spatial resolution, and a 1-day temporal resolution) (Wagner et al., 1999; Naeimi et al., 2009, 2012b). The period of record of the ASCAT data constrains the study period to 2007–2013.

2.3 Assimilation method

The objective of this paper is to attempt an improvement of soil moisture dynamics in SiBCASA by assimilating the passive microwave and/or ASCAT satellite-derived data described above. We use the same assimilation method as used in GLEAM (Miralles et al., 2011a; Miralles, 2011):

$$dw = K_t\left(w_t^{\mathrm{obs}} - w_t^{\mathrm{sim}}\right) \tag{1}$$

with w_t^x the soil moisture content of the top soil layer, either satellite-observed ($x = \mathrm{obs}$) or simulated ($x = \mathrm{sim}$), dw the change in w after assimilation. The index t indicates the time in steps of days. K_t, the Kalman gain, describes how much of the difference ($w_t^{\mathrm{obs}} - w_t^{\mathrm{sim}}$) is applied to dw to update w^{sim}, and depends on the error in the model soil moisture σ^{sim} and the error in the satellite observed soil moisture σ^{obs}:

$$K_t = \frac{\sigma_t^{\mathrm{sim}}}{\sigma_t^{\mathrm{sim}} + \sigma_t^{\mathrm{obs}}}. \tag{2}$$

The error in model soil moisture depends on $d\sigma^{\mathrm{mod}}$, the uncertainty associated with model integration over a time step of one day:

$$\sigma_t^{\mathrm{mod}} = \sigma_{t-1}^{\mathrm{mod},+} + d\sigma^{\mathrm{mod}}. \tag{3}$$

We use a constant $d\sigma^{\mathrm{mod}} = 0.01\,\mathrm{m^3\,m^{-3}\,day^{-1}}$. The model soil moisture is updated according to

$$w = w + dw. \tag{4}$$

When observations were available, the model error σ_t^{mod} is reduced to σ_t^+ after the assimilation step:

$$\sigma_t^{\mathrm{mod},+} = \sigma_t^{\mathrm{mod}}\left(1 - K_t\right). \tag{5}$$

The error or noise in the satellite-observed soil moisture, σ^{obs}, depends on the vegetation optical depth, land surface heterogeneity of the pixel, and snow or ice in or on top of the soil, and is output by the retrieval algorithm. The noise is typically of the order of $0.1\,\mathrm{m^3\,m^{-3}}$ (standard deviation σ, e.g. Figs. 5–8).

Since the satellite-observed soil moisture data essentially carry information only about the temporal variations in soil moisture, and not about the absolute value of mean and the amplitude of the variations, we normalize the satellite data (see below) before assimilating the satellite data in SiBCASA. The entire assimilation procedure consists of the following steps:

- Step 1: run SiBCASA without data assimilation to equilibrium in 2007 and then run until 2013, storing daily model results.
- Step 2: take the spatial average of satellite-derived soil moisture within the 1° × 1° SiBCASA grid boxes. Normalize the satellite-observed soil moisture's mean, standard deviation, and higher moments towards the SiBCASA's soil moisture using the CDF matching technique (Liu et al., 2009, 2012; Reichle and Koster, 2004). We matched the distribution function at every 10th percentile between 10 and 90. Because the retrieval algorithms do not work under (partially) frozen and snow-covered conditions, we discarded periods with frozen soil in SiBCASA from building the CDF transformation coefficients.

Table 1. Characteristics of the flux tower sites used in this study.

Site	Hyytiälä	Tver wet forest	Yakutsk Larix	Elgeeii
Latitude	61°50′51″ N	56°26′52″ N	62°15′18″ N	60°01′01″ N
Longitude	24°17′37″ E	32°54′07″ E	129°37′08″ E	133°49′53″ E
Ecosystem description	taiga Pinus sylvestris	taiga Picea, on peat	taiga Larix cajanderii	taiga Larix cajanderii
Elevation (m a.s.l.)	181	263	220	202
Age (in 2015)	53	192	185	155
Years used	2007–2013	2007–2013	2007–2013	2010–2013
Maximum snow depth (cm)	50	50–100	40	50
Maximum LAI ($m^2\ m^{-2}$)	2.9	3	2.1	1.4
Annual precipitation (mm)	700	711	230	290
Depth of soil moisture sensors used (cm)	2.5	10	10	10
References	Rannik et al. (2004), Ilvesniemi et al. (2010)	Kurbatova et al. (2008), Milyukova et al. (2002)	Dolman et al. (2004), Ohta et al. (2008)	Kotani et al. (2014)

- Step 3: run SiBCASA from equilibrium in 2007 until 2013 with assimilation of the satellite-derived soil moisture.

- Step 4: evaluate the simulated soil moisture, water, and carbon fluxes with in situ flux tower data as described in Sect. 2.4.

2.4 In situ flux tower data

For evaluation of the model results, particularly the carbon fluxes, we use 25 site-years of data from four flux tower sites in Boreal Eurasia available in the period 2007–2013 (Table 1). The sites vary in vegetation type, continentality, and permafrost. Flux data were taken in Siberia at more locations, although predominantly in the period 1997–2005, when the ASCAT satellite was not yet launched (Dolman et al., 2012). The sites are in Hyytiälä, Finland (Ilvesniemi et al., 2010; Kolari et al., 2009; Mammarella et al., 2009; Rannik et al., 2004), Tver, European Russia (Kurbatova et al., 2008; Milyukova et al., 2002), Yakutsk, East Siberia (Dolman et al., 2004; Ohta et al., 2008), and Elgeeii, East Siberia (Kotani et al., 2014). The eddy covariance data have been processed according to the harmonized LaThuille FLUXNET database (Baldocchi et al., 2001; Reichstein et al., 2005; Papale et al., 2006; Moffat et al., 2007). All sites are covered with needle leaf forests, evergreen in the western sites, and deciduous in the eastern sites. The stations Yakutsk and Elgeeii are on permafrost and have maximum snow depths of the order of 40 and 50 cm, respectively.

3 Results

We will first evaluate the spatial coherence between the simulated and satellite-observed soil moisture, then the temporal coherence. All satellite and in situ data are CDF-matched to the SiBCASA soil moisture. Next, we will compare model and satellite soil moisture data with in situ observations. Finally, we will evaluate the impact of satellite soil moisture assimilation on the simulated carbon fluxes.

3.1 Reliability of passive microwave and ASCAT soil moisture in Boreal Eurasia

The spatial coherence between SiBCASA and satellite-observed soil moisture is studied by comparing maps of

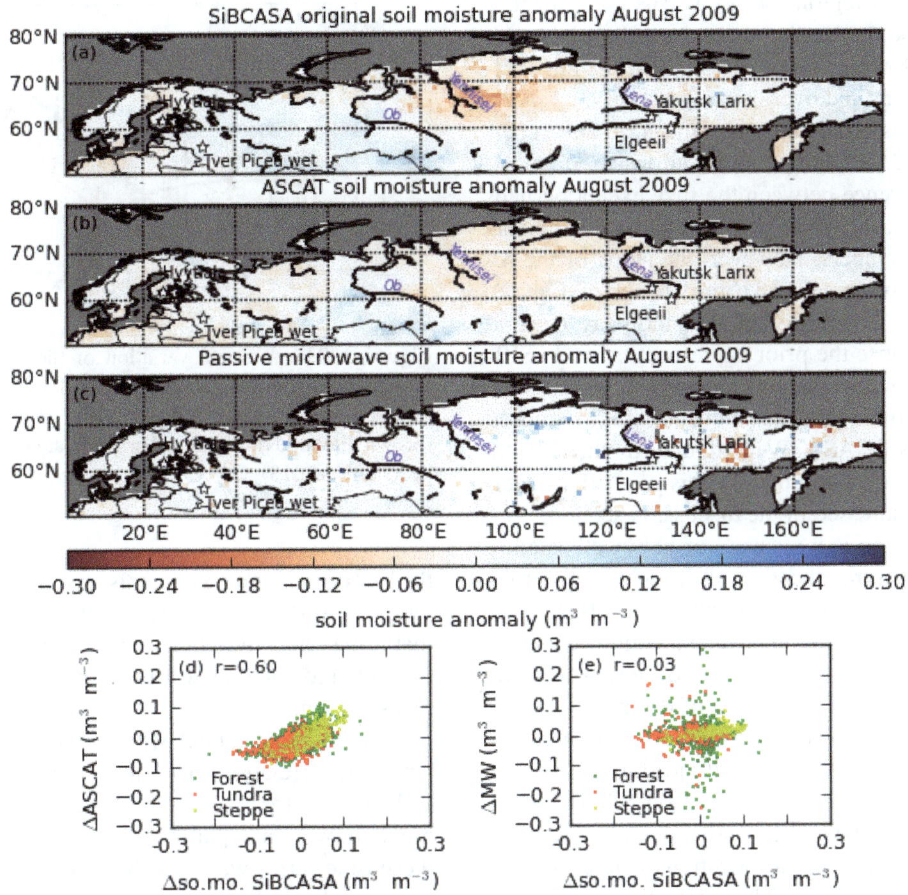

Figure 2. Soil moisture anomalies in August 2009 with respect to the average soil moisture in the months August 2007–2011. **(a)–(c)** The anomaly in SiBCASA, ASCAT and passive microwave soil moisture. **(d)** The spatial correlation between the anomalies in ASCAT and SiBCASA soil moisture; each point represents one grid point in the maps **(a)** and **(b)**. **(e)** The spatial correlation between passive microwave and SiBCASA soil moisture. The location of the four field sites is shown with open asterisks.

monthly soil moisture anomalies and by quantifying the spatial correlation between the anomalies. We compute the anomalies for each month with respect to the average soil moisture in that month in the years 2007 to 2011. We use the reference period until 2011 (and not 2013), because the AMSR-E satellite became dysfunctional in November 2011, and we have no passive microwave soil moisture after that date. Figure 2 shows an example of the spatial coherence in August 2009. This is the month with the largest spatial correlation between SiBCASA and ASCAT soil moisture in the period that AMSR-E data are available ($r = 0.60$). A dry anomaly in North-central Siberia is apparent in both SiBCASA and ASCAT and a wet anomaly in South and West Siberia, showing that there is coherence in the spatial structure of both data sources. But there are also striking differences. The drought is more intensive and confined to a smaller area in SiBCASA as compared to ASCAT. In addition, east from the Lena River, SiBCASA tends to have a wet anomaly, where ASCAT has a light dry anomaly. Nevertheless, the spatial correlation coefficient is good with $r = 0.60$ (second panel in Fig. 2). The correlation appears to be better for wet anomalies than for dry ones. If we compare the soil moisture from the passive microwave data to SiBCASA and ASCAT, no coherent pattern emerges, neither in August 2009 nor in other months. This is reflected in the low spatial correlation coefficient ($r = 0.03$).

These findings are also quite typical for August months in other years. Only in August 2013 was the spatial correlation coefficient between SiBCASA and ASCAT soil moisture slightly larger, $r = 0.62$ (figure not shown), but the corresponding AMSR-E data were no longer available. For other months, the spatial patterns are usually less pronounced, and the correlation coefficients smaller. Figure 3 shows the seasonal evolution of the spatial correlation coefficients, also for the land cover types tundra, forest, and steppe (i.e. grasslands and croplands) separately. The correlation coefficients are generally small outside the summer months. Steppe regions have larger correlation coefficients, and tundra regions smaller ones. The overall correlation coefficient is strongly dominated by the forests, because forests cover by far the largest part of the study region (66 %), versus 24 % for tundra and 9 % for steppe. In the discussion we will provide potential explanations for the variation of the correlation coefficients over the seasons and over land cover types.

Considering the seasonal evolution of the correlation coefficients between SiBCASA and passive microwave soil moisture, there is no coherence between the two, except perhaps for the steppe regions, for which the correlation coefficients reach to 0.50 in Septembers. However, the slope of the regression curve is only about 1 : 3 (third panel of Fig. 2), whereas it is near 1 : 1 for ASCAT soil moisture (second panel of Fig. 2). Because the prior agreement between SiBCASA and passive microwave soil moisture is too low for Boreal Eurasia, we do not consider it meaningful to proceed with assimilation in SiBCASA. We will focus on the ASCAT soil moisture alone in the remainder of this paper. This decision will be further addressed in the discussion section.

The spatial correlation is a measure of how well satellite and SiBCASA agree on the location of drought and wet *regions*. For assimilation purposes it is also interesting to investigate the temporal correlation at each location. The temporal correlation coefficient is a measure of how well satellite and SiBCASA agree on the timing of dry and wet *periods*. Figure 4 shows the temporal correlation coefficient between SiBCASA and ASCAT daily soil moisture for all August months between 2007 and 2013 (31 days × 7 days). We discarded all grid points where the associated observational error exceeds 0.25 $m^3\ m^{-3}$ and all locations where time series had less than 50 % coverage. The correlation coefficients are quite large, up to 0.80 in the West Siberian Plains southwest of the Ob River, with a transition zone via the Yenissei River to the West Siberian tundra region. The Yenissei River marks the western border of the Central Siberian Plateau, where the correlations are much smaller ($0 < r < 0.2$). In eastern Siberia, east of the Lena River, the correlations are variable, but generally small and sometimes even negative ($-0.3 < r < 0.3$). This pattern is somewhat representative for July, August, and September (see Figs. S1 and S2 in the Supplement, except that in September the area northeast of 60° N, 90° E is masked out for the lack of good-quality satellite data. Since we do not calculate correlations when SiBCASA simulates frozen soil or when good-quality data are lacking, an apparent "winter mask" advances from the northeast in September to cover all of the region by December. In April this winter mask regresses into Siberia and disappears in June. As expected, at the front of the winter mask, which is generally 5° wide, the correlations are low and patchy. We will consider potential underlying reasons for these patterns in the discussion.

The temporal correlations shown in Fig. 4 were computed with soil moisture on a daily basis. When computed on a monthly average basis, the correlation coefficients generally improve considerably, but the variability in soil moisture is smaller accordingly. This shows that day-to-day noise, particularly in the satellite soil moisture data, is responsible for

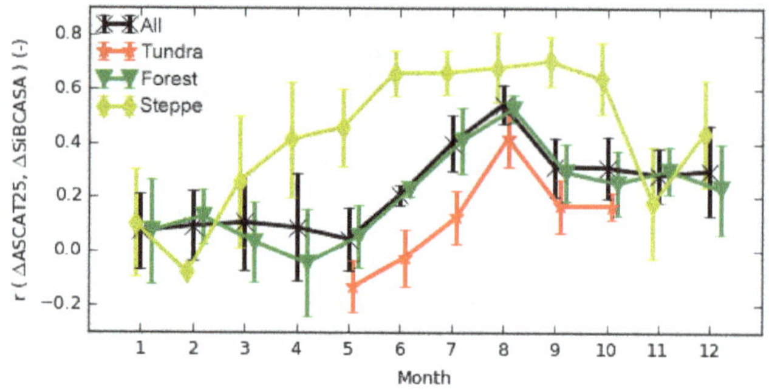

Figure 3. The seasonal variation of the spatial correlation coefficient of SiBCASA and ASCAT soil moisture for all grid cells (black x), tundra cells (red *), forest (green triangles), and steppe (brown diamonds). The errorbars indicate the variation between the years 2007–2013.

the low correlation coefficients, and that the remaining correlation is dominated by inter-annual variations. This suggests that it may be worthwhile to investigate assimilation of low-pass filtered satellite soil moisture instead of instantaneous measurements.

To better understand what these large-scale spatial and temporal correlation coefficients imply for the use of satellite soil moisture for assimilation in SiBCASA, time series of soil moisture were compared for four stations across Eurasia. In these time series we show the original SiBCASA soil moisture, the ASCAT soil moisture (CDF-matched), and in situ soil moisture. In addition, we show the SiBCASA soil moisture after assimilation of ASCAT soil moisture according to Sect. 2.3. The first station is Hyytiälä in Finland. The time series are shown in Fig. 5. Simulated and in situ observed soil moisture generally change slowly in time, because of the soil moisture retention in the soil. Since satellite observations lack this memory effect, the satellite observations are noisier. Over the years ASCAT and in situ soil moisture are larger than SiBCASA soil moisture in the spring period (May to early June). This causes the assimilation procedure to increase soil moisture in SiBCASA (red line is higher than the blue line). This increase of soil moisture in the spring period improves the match with in situ observed soil moisture to the degree specified by the uncertainties (Eq. 2).

We loosely define a drought as a period when in situ observed soil moisture is more below the average soil moisture in that period (see bottom panels of Figs. 5–8) than the day-to-day variation during three or more weeks. Subsequently, two distinct drought periods in the time series occurred in July/August 2010 and in July/August 2013. In 2010 the original SiBCASA simulation also "saw" the drought, but ASCAT did not. The assimilation therefore decreased the match with in situ soil moisture. In 2013 the in situ observed drought in Hyytiälä was picked up by neither SiBCASA nor ASCAT.

The second site we analyse is "Tver wet forest", for which the time series are shown in Fig. 6. In Tver, a

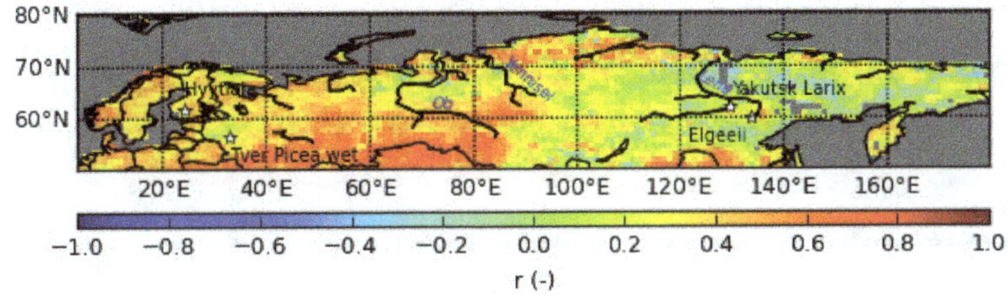

Figure 4. The temporal correlation coefficient of SiBCASA and ASCAT soil moisture for all August months in the period 2007–2013 (7 years × 31 days).

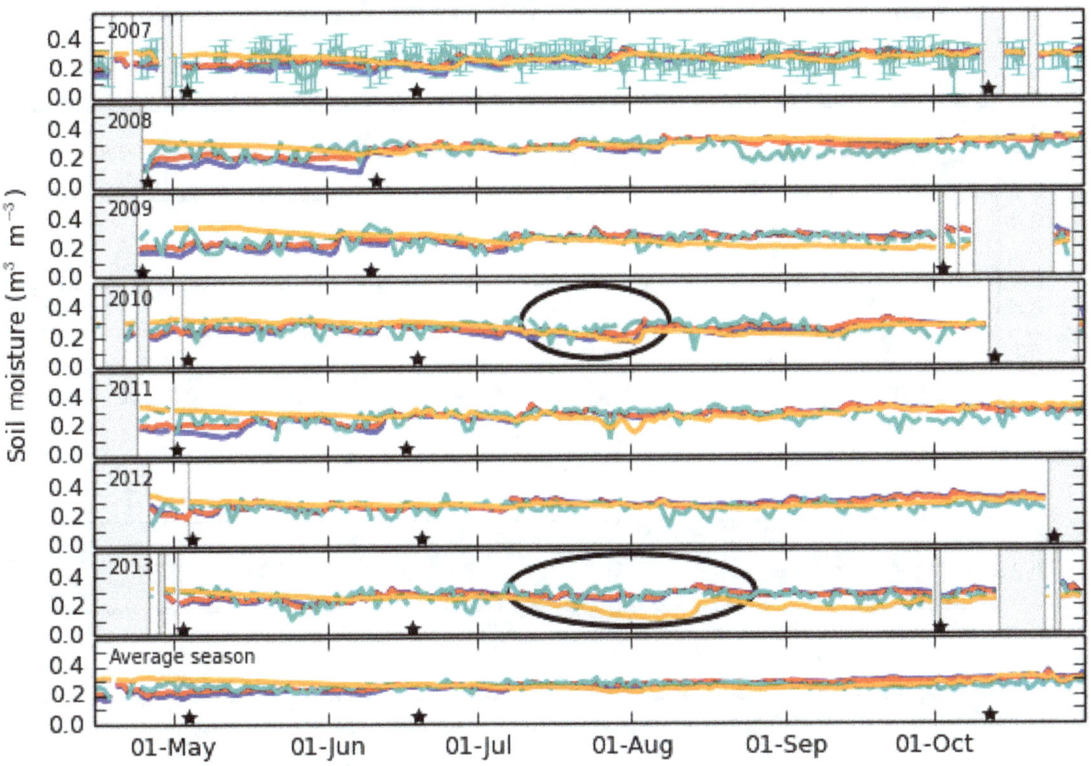

Figure 5. Time series of soil moisture in SiBCASA original (without assimilation) (blue), and with assimilation (red), in situ soil moisture (orange), and ASCAT soil moisture (marine) in Hyytiälä, Finland. Each panel shows 1 year of soil moisture. Grey shades indicate periods when the top soil is frozen. The three asterisks indicate the date when the top soil is last frozen in the spring, 46 days after that, and the date when the top soil is first frozen again in the fall. Error bars in the top panel indicate the uncertainty in ASCAT soil moisture, which is for clarity only shown for 1 year. The bottom panel shows the average seasonal cycle of the each soil moisture type. In situ and ASCAT soil moisture are CDF-matched to SiBCASA soil moisture, which explains why they have the same mean and standard deviation.

similar springtime behaviour emerges. ASCAT is generally larger than SiBCASA soil moisture in April to early May, and the assimilation improves the match with in situ observed soil moisture. Field workers confirm that the water table is generally high or even above the soil surface after snowmelt (April) and decreases quickly in May. In the summers, soil moisture is generally quite constant, except during the 2010 drought, which caused extensive fires in European Russia. The drought was picked up by in situ observations and SiBCASA, but not by ASCAT. As a consequence, the assimilation decreased the match of SiBCASA with in situ soil moisture.

The third site is Yakutsk Larix. At this site, ASCAT soil moisture is noisier than for the other sites (Fig. 7). There is a tendency that in situ soil moisture is high in spring, due to melted snow, and decreasing during the season. This trend is reproduced by SiBCASA in 2007 and 2008 and perhaps in 2009, but not in 2011 to 2013. The ASCAT signal tends to be smaller than average in the spring, and increases somewhat in the summer. There are four intense “droughts” in the in situ time series – in 2008, 2011, 2012, and 2013. Ohta et

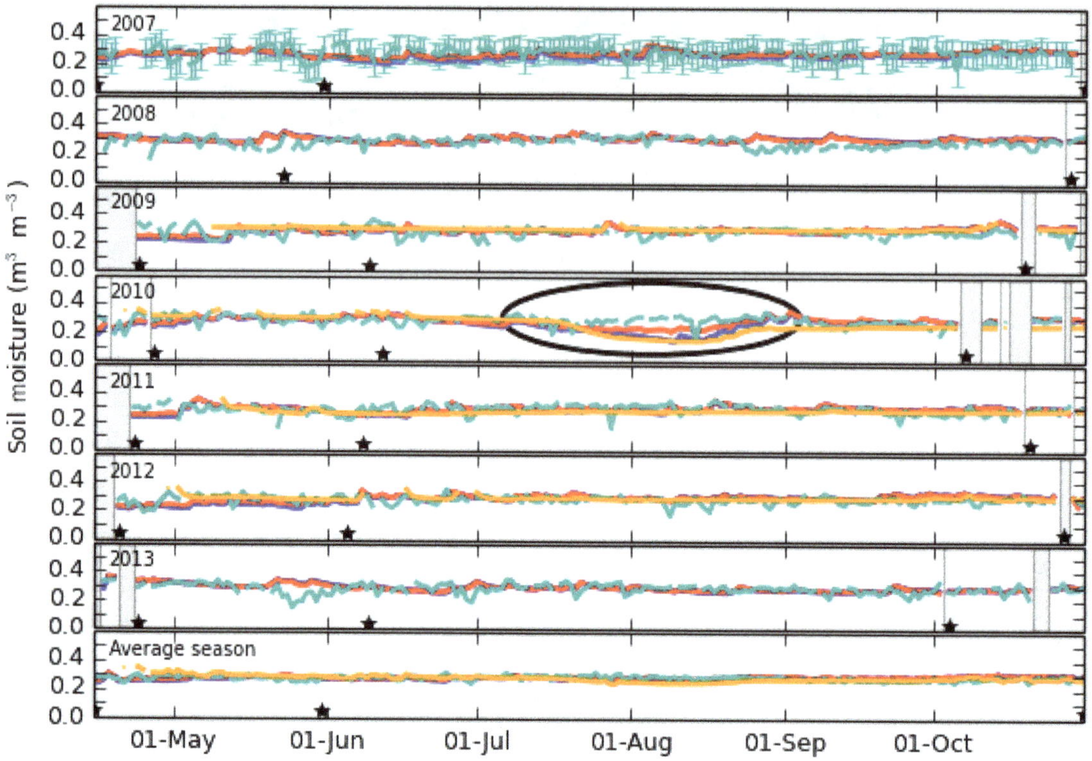

Figure 6. As Fig. 5, but for the Tver wet forest site.

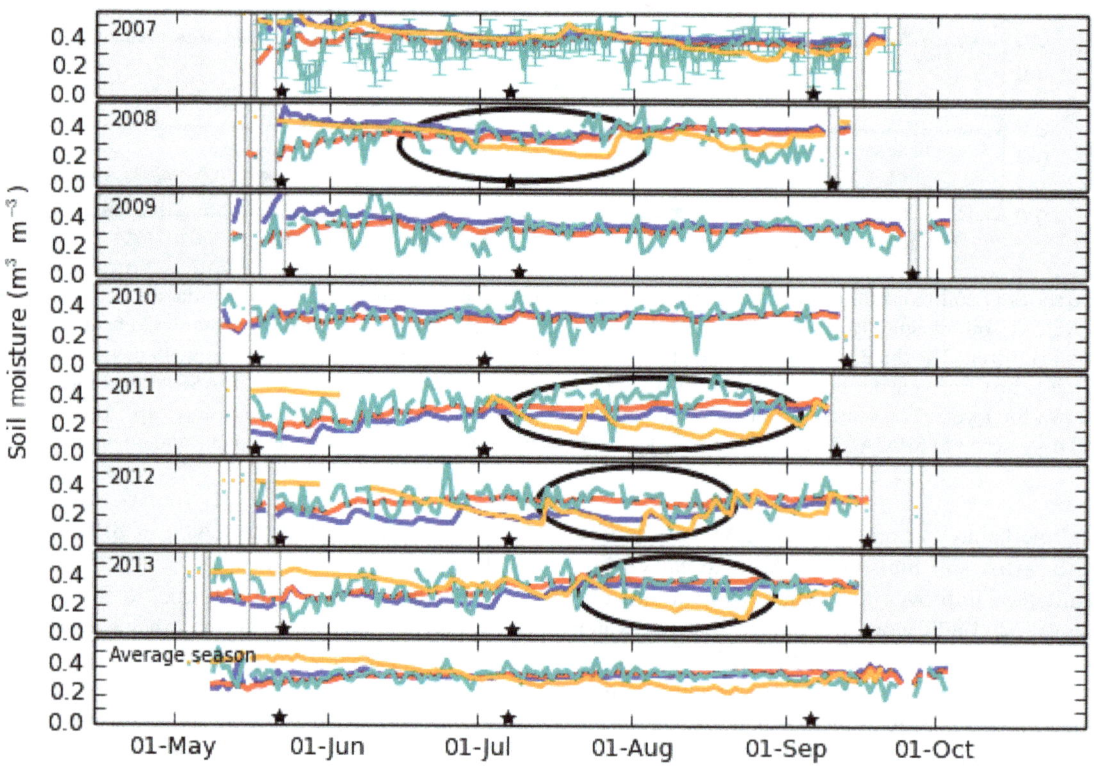

Figure 7. As Fig. 5, but for the Yakutsk Larix site.

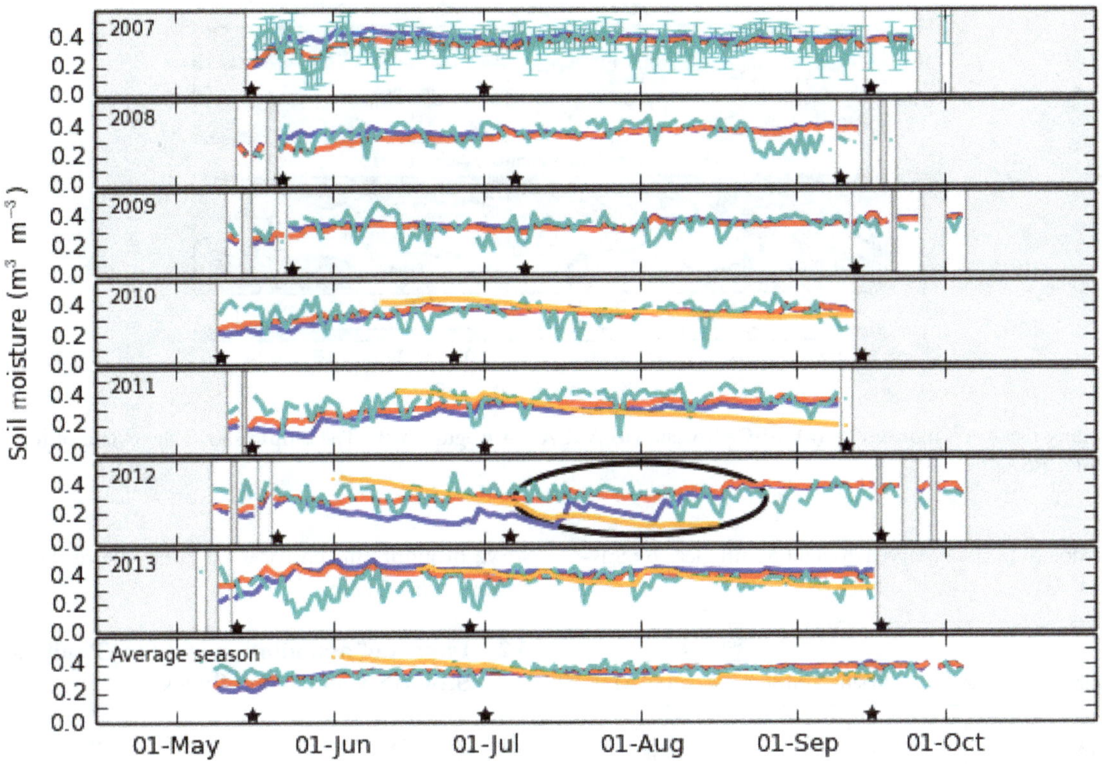

Figure 8. As Fig. 5, but for the Elgeeii site.

al. (2014) show that drought conditions at the site occurred between 2001 and 2004, and that the site was actually water logged from 2005 to 2009, returning to normal conditions afterwards. The water logging had severe impact on the ecosystem, with reduced photosynthesis rates and tree browning and mortality. This water logging may be a larger-scale process in eastern Siberia (Muskett and Romanovsky, 2009; Vey et al., 2013). Therefore the term "drought" is relative to the studied period. SiBCASA sees the in situ observed droughts in 2008 and 2012 to some extent, but not the 2011 and 2013 ones. This inter-annual variation in soil moisture data is reflected in the minimum summertime in situ observed soil moisture in Fig. 7, which is much lower in 2011–2013 than in 2007–2008. ASCAT does not detect this inter-annual variation. ASCAT also does not observe the droughts, whereas the ASCAT soil moisture is generally even larger than in SiBCASA, causing the assimilation to decrease the match with in situ soil moisture.

The fourth site is Elgeeii, for which the time series are shown in Fig. 8. SiBCASA soil moisture is again low in the springtime (May), and ASCAT soil moisture is larger. The assimilation increases the soil moisture in SiBCASA, and this seems to improve the match with in situ observed soil moisture, although the early springtime in situ observations were unreliable. In 2012 a drought occurred in July and August. SiBCASA sees the drought too, although with a much earlier development. ASCAT does not see the drought, and as a consequence, the assimilation moves the soil moisture in SiBCASA away from the in situ observations. It is interesting to see that on 5 August 2012 the soil moisture in SiBCASA increases due to a precipitation event, but this is not seen in the in situ observations. However, an increase is seen at that time in the in situ observations in Yakutsk, some 340 km to the northeast, possibly suggesting a displacement of the precipitation event in the SiBCASA driver data from ECMWF ERA-interim.

Of the 11 droughts observed in the in situ time series at the four sites, SiBCASA reproduces four, and ASCAT none (Table 2). The poor skills of ASCAT to reproduce local drought conditions are an apparent contradiction given the good skills in positioning the major drought regions shown in Figs. 2 and 3. Particularly at the locations of our in situ observation sites ASCAT does not perform well in reproducing the temporal variability (see Fig. 4), which is confirmed at the site level in Figs. 5–8. To explain this better, we look at monthly mean soil moisture maps for the five drought occurrences which SiBCASA observes, but ASCAT does not.

In August 2010 there was a large drought and heat wave in European Russia and western Siberia, which was caused by a strong blocking situation. The drought was accompanied with many wildfires (Miralles et al., 2014; Krol et al., 2013). The drought was also apparent in the in situ measurements performed in Hyytiälä and Tver. Figure 9 shows that SiBCASA simulates a drought extending from Scandi-

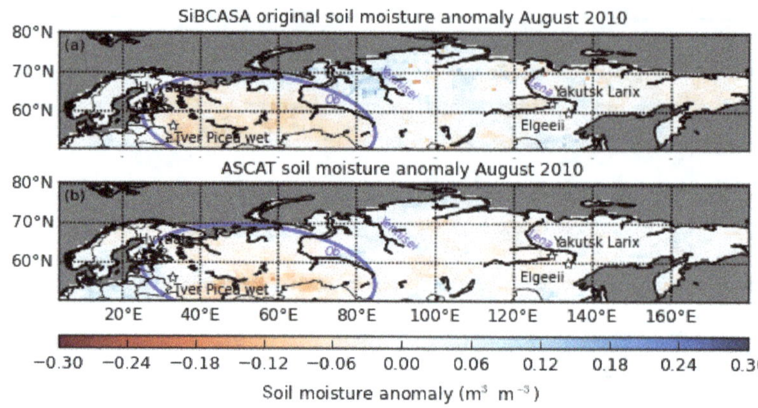

Figure 9. Monthly mean soil moisture in **(a)** SiBCASA and **(b)** ASCAT in August 2010. The ellipse shows the extent of the 2010 drought according to SiBCASA.

Table 2. Detection of in situ observed drought occurrences by SiBCASA and ASCAT.

			Drought seen by		
Year	Month	Site	In situ	SiBCASA	ASCAT
2008	July	Yakutsk	yes	–	no
2010	July/August	Hyytiälä	yes	yes	no
2010	August	Tver	yes	yes	no
2011	July/August	Yakutsk	yes	no	no
2012	July/August	Yakutsk	yes	yes	no
2012	July/August	Elgeeii	yes	yes	no
2013	July/August	Hyytiälä	yes	no	no
2013	July/August	Yakutsk	yes	no	no

navia to Novosibirsk (55° N, 80° E), with the Hyytiälä and Tver sites on the western rim of the drought region. ASCAT locates a drought in roughly the same region, although less intense and with a smaller geographical extent. The ASCAT wet anomaly over Europe expands further into Russia and Scandinavia. As a result, the sites Hyytiälä and Tver are just outside of the drought region as observed by ASCAT and this is most likely attributable to the ASCAT soil moisture retrieval skills. Figure 4 shows that the ASCAT performance is low around those sites. In Sect. 5 (and Fig. 12) we will discuss the performance at the sites in more detail.

In July 2012 SiBCASA simulates an intense drought that was located around the city Yakutsk, extending eastward to the region between the Lena and Aldan rivers (Fig. S3). The Elgeeii site was located just on the eastern border of the simulated drought region. ASCAT does not observe this drought region, not in July, nor in earlier or later months. Where Elgeeii was on the perimeter of the 2012 drought region, as were Hyytiälä and Tver in 2010, Yakutsk was in the centre of the drought region, which ASCAT does not observe at all. Therefore a site's location on the rim of a drought does not explain why ASCAT does not observe the drought. Rather it appears that ASCAT has limited capability to observe droughts in the forested zone where the in situ observations were made.

3.2 Impact of assimilation of ASCAT soil moisture in SiBCASA on carbon fluxes

The changes in soil moisture in springtime and during drought, induced by the assimilation of ASCAT observed soil moisture in SiBCASA (Sect. 3.1, Figs. 5–8), may have substantial effects on the representation of the carbon fluxes, which we will look at next. We will show separately how GPP, TER, and NEE depend on the change in soil moisture and the season.

An interesting case is presented at the Yakutsk Larix site (Fig. 10). At this location, the change in soil moisture due to assimilation of ASCAT soil moisture was large relative to the other sites (Fig. 7). However, although the absolute value of the change in soil moisture was more or less constant throughout the years, the change in GPP shows a distinct seasonal cycle, with large changes in spring, small changes in summer, and hardly any change in fall. This is because of two reasons: (i) the sensitivity of GPP to soil moisture is simulated as a function of the plant-available water fraction (Sect. 2.1, Fig. 1a). In Yakutsk in spring, the permafrost soil has only thawed for a couple of centimetres, resulting in a very small plant available water fraction and very strong soil moisture sensitivity (Fig. 1a: the soil moisture sensitivity curve is steepest on the low plant-available water fraction side). This results in a strong GPP effect of assimilating ASCAT soil moisture in SiBCASA. Note that soil thawing does not automatically mean that more soil moisture becomes available for root uptake. The soils in Yakutsk often freeze after a relatively dry summer, so that the frozen soil may be quite dry. In the spring, the snow meltwater cannot penetrate the soil, which is still frozen, and may run off; (ii) the Yakutian spring is almost simultaneous with the solar maximum on 21 June, so that the potential GPP is large. In the course of the growing season, the permafrost active

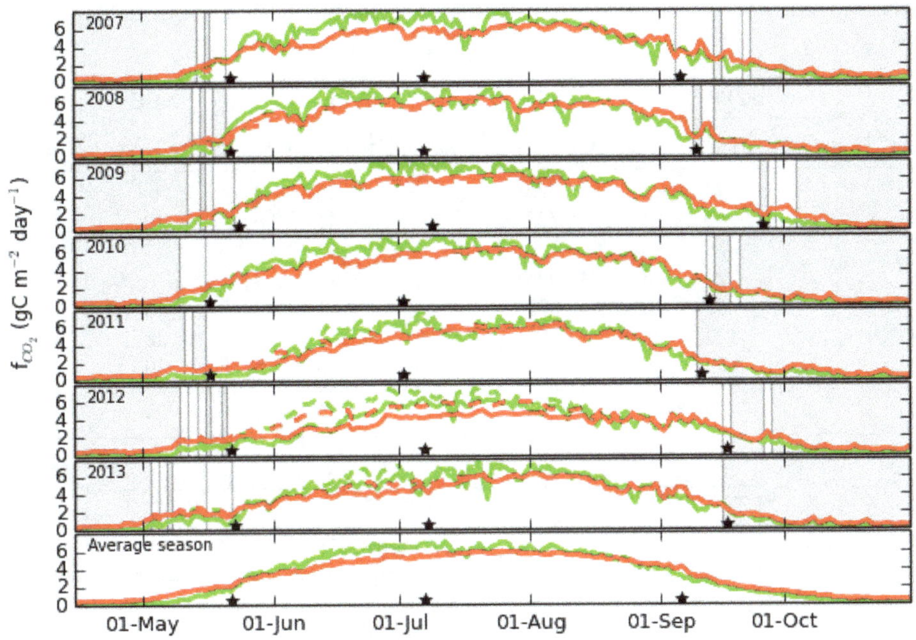

Figure 10. GPP (green) and TER (red) simulated with SiBCASA without (solid lines) and with assimilation (dashed lines) of ASCAT soil moisture for Yakutsk Larix.

layer thaws deeper, resulting in a larger plant-available water fraction, reducing the drought sensitivity. This explains the smaller change in GPP in the summer. In the fall, GPP is limited more by the lack of available sunlight than by water stress, explaining the absence of change in GPP with assimilation of satellite soil moisture data.

In a similar way, the change in TER (Fig. 10) does not only depend on the change in soil moisture with satellite soil moisture assimilation, but also in the absolute value of soil moisture (Fig. 1b) and temperature limitation on TER. In June, when the soil is still cold, the changes in TER are small. In July and August the changes in TER are larger than in GPP, because the soil is warm and TER is a function of absolute soil moisture change. In this example, the changes in GPP and TER have the same direction. Figure 1 shows that this is always the case when the soil moisture saturation fraction is below its minimum value of $\sim 60\,\%$. Consequently, the changes in GPP and TER compensate each other partly in the NEE.

Accumulated over a year (Table 3), the changes in GPP, TER, and NEE are of the order of tens of $\mathrm{gC\,m^{-2}\,yr^{-1}}$, amounting to a few percent of GPP and TER. For NEE however, the changes can amount to tens of percent and a 7-year mean of $-34\,\%$. We note that the changes in GPP and TER are larger in Yakutsk than in Hyytiälä, Tver, and Elgeeii. This is because the plant-available water fraction is smaller in Yakutsk than for the other sites, creating a strong drought sensitivity, and because the change in soil moisture is larger. While the relative changes in GPP and TER for these sites is generally small, and they partly compensate, the 7-year mean changes in NEE are $+52\,\%$ at Hyytiälä, $-105\,\%$ at Tver, and $-38\,\%$ at Elgeeii.

The effects of ASCAT soil moisture assimilation in SiBCASA are also significant when integrated over the entire study domain ($27.8 \times 10^6\,\mathrm{km^2}$) and the year (Fig. 11). The mean simulated NEE is $-1.91\,\mathrm{PgC\,yr^{-1}}$ with an interannual variation of $0.12\,\mathrm{PgC\,yr^{-1}}$ (RMSD). Assimilation of ASCAT soil moisture in SiBCASA causes a change of $0.045\,\mathrm{PgC\,yr^{-1}}$ (RMSD). This is 41 % of the normal interannual variation of $0.11\,\mathrm{PgC\,yr^{-1}}$ (RMSD), and 2.4 % of the mean NEE. The effect of satellite soil moisture assimilation is negligible until May; it then grows in the months June and July. After August, the net effect does not change much. This is in line with the observation that the effect of assimilation on soil moisture and carbon fluxes is largest in springtime (Sects. 3.1 and 3.2, Figs. 5–8 and 10). The effect of assimilation was largest in 2010 with an extra anomaly in NEE of $+0.08\,\mathrm{PgC\,yr^{-1}}$ (less uptake). This anomaly grew between May and September. In this year, a widespread drought occurred in European Russia and West Siberia, which ASCAT captures quite well. The assimilation effect could have been even larger if ASCAT had not wrongfully detected a wet anomaly over far eastern Siberia, where SiBCASA simulates a second drought region (Fig. 9).

The second largest effect of soil moisture assimilation occurred in 2012, with an extra anomaly of $-0.07\,\mathrm{PgC\,yr^{-1}}$ (more uptake). This anomaly grew mostly in June and July, when ASCAT soil moisture was much higher in June over large parts of Siberia, and the July drought in Central Siberia was confined to a smaller region in ASCAT data.

Table 3. GPP, TER, and NEE in SiBCASA without ASCAT soil moisture assimilation, and the changes dGPP, dTER, dNEE with assimilation for the site Yakutsk Larix.

	GPP	dGPP		TER	dTER		NEE	dNEE	
	$gC\,m^{-2}\,yr^{-1}$	$gC\,m^{-2}\,yr^{-1}$	%	$gC\,m^{-2}\,yr^{-1}$	$gC\,m^{-2}\,yr^{-1}$	%	$gC\,m^{-2}\,yr^{-1}$	$gC\,m^{-2}\,yr^{-1}$	%
2007	719	−24	−3	725	−25	−3	6	−1	−22
2008	697	−45	−6	727	−45	−6	30	0	0
2009	710	−39	−6	733	−44	−6	23	−4	−19
2010	688	−20	−3	695	−28	−4	7	−9	−123
2011	556	53	10	601	41	7	45	−12	−27
2012	524	121	23	553	110	20	28	−11	−38
2013	612	54	9	642	33	5	30	−21	−68
Mean	644	14	2	668	6	1	24	−8	−34

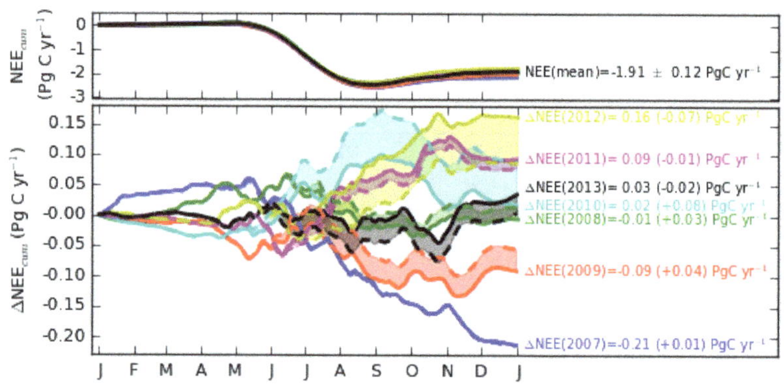

Figure 11. Top panel: cumulative NEE in Boreal Eurasia for the years 2007 to 2013 according to SiBCASA without assimilation of satellite-observed soil moisture. The text describes the 7-year mean NEE and the inter-annual variation (as standard deviation). Bottom panel: cumulative NEE anomaly relative to the 7-year mean. Solid lines represent the SiBCASA NEE anomaly (ΔNEE) without assimilation of ASCAT soil moisture, dashed lines represent the NEE anomaly with assimilation of ASCAT soil moisture. The text behind the lines describes the NEE anomaly relative to the inter-annual mean NEE and in parentheses the change caused by assimilation of ASCAT soil moisture.

4 Discussion

4.1 Soil moisture

The spatial and temporal correlation coefficients between SiBCASA and satellite-observed soil moisture shown in Sect. 3 suggest that ASCAT and passive microwave satellite signals have a certain skill in observing land surface soil moisture. The absence of perfect correlations implies that assimilating the satellite-observed soil moisture in SiBCASA will have an effect. The question is whether that effect is an improvement.

The performance of passive microwave data was low over the entire study region and in all months (Fig. 2, Sect. 3.1). Only in steppe regions were the temporal correlations large ($r = 0.8$). The spatial correlation is smaller than that ($r \sim 0.5$) and with a smaller sensitivity (a slope of $\sim 1:3$, Fig. 2), probably because of the absence of significant spatial patterns in the small extent of the steppe zone. The poor performance of the microwave soil moisture in Boreal Eurasia is not entirely surprising: the passive microwave radiation emitted by the soil moisture is known to be disturbed by vegetation, surface water, snow, and ice (de Jeu et al., 2008; Mladenova et al., 2014; Champagne et al., 2010), which are abundant in Boreal Eurasia. The microwave soil moisture product has been validated extensively (Miralles, 2011; Miralles et al., 2011b; de Jeu et al., 2008; Liu et al., 2011, 2012; Owe et al., 2008; Griesfeller et al., 2016; Champagne et al., 2010). However, the vast majority of the validation sites were located on grasslands and croplands, and in temperate and (semi)arid climate zones. Therefore, the poor performance of microwave soil moisture in Boreal Eurasia, except perhaps the steppe zone, is probably related to the canopy, which is too dense, as well as to the presence of snow, ice, and surface water. Our results are therefore specific to our region, and cannot be simply extrapolated to other climate zones and land covers.

The spatial and temporal correlation coefficients vary with the months and with land cover. The spatial correlation between SiBCASA and ASCAT soil moisture is largest in August and quickly decreases towards the spring and fall. What processes may cause this? Ecologically there are large differences between the seasons in Siberia. Large parts of Siberia are snow covered and particularly the region north of Mongolia and east of the Yenissei River is subject to continuous permafrost. This hampers a correct retrieval of soil moisture from satellite-observed signals (Naeimi et al., 2012a; Högström et al., 2014), while correctly simulating soil moisture under snow conditions is also difficult in vegetation models. However, even in the northern tundra regions most snow and ice have disappeared by June. Considering that the grid cells with frozen top soil in SiBCASA and snow/ice detection in ASCAT have been excluded from the statistical analysis, the lower correlations in June, July, and September (Fig. 3) are probably not only caused by the presence of snow and ice on the land surface.

Other important changes from May to July are the expansion of leafs, the drying out of the topsoil after snowmelt on frozen ground, and the deeper thawing of the permafrost active layer. The increase of the leaf area index (LAI) does not seem beneficial for better satellite soil moisture retrievals, as is also suggest by the smaller correlation coefficients for forests than for steppe zone (Fig. 3). The decrease in the ponded area fraction after snowmelt on frozen ground is a potential explanation for the improving correlation coefficients (Högström et al., 2014), since they occur particularly in the forest and the tundra zones, which contain the wettest parts of the region, and not for the steppe zone, which is drier and outside the permafrost zone.

With the same arguments the increasing depth of the permafrost ice front may also be a potential explanation of the improving spatial correlation coefficients towards August. Indeed, ice and frozen soil at some depth may disturb the satellite signal (Way et al., 1997; Wegmüller, 1990). Maximum active layer thicknesses of a mere 10–20 cm are not uncommon in the northern tundra, although the penetration depth of microwave radiation in the soil is of the order of one to a few centimetres.

It is interesting that the spatial correlation coefficients for steppe zones are larger and for tundra zones smaller than average. Both steppe and tundra vegetation are characterized by short vegetation, but tundra regions are generally much wetter than steppe regions and with continuous permafrost. This implies that the presence of short vegetation alone is not the only prerequisite to obtain a good match between SiBCASA and ASCAT soil moisture.

On the site level, Figs. 5–8 show that ASCAT soil moisture has much more day-to-day variability than SiBCASA soil moisture. While SiBCASA soil moisture has a significant, physically meaningful auto-correlation with lag times up to 10–17 days ($r > 0.3$), ASCAT observations and associated errors are independent in time, which indicates that the signal is compromised by measurement noise. On top of this, ASCAT was not able to detect the eight large drought occurrences observed in situ soil moisture time series, nor the pronounced inter-annual variation associated with recovery after water logging in Yakutsk. This is reflected in small site-level temporal correlation coefficients between in situ soil moisture and ASCAT soil moisture ($r < 0.06$ at all sites), while the June–September correlation between in situ soil moisture and SiBCASA soil moisture is much larger (0.49 at Hyytiälä, 0.63 at Tver, 0.74 at Yakutsk and 0.76 at Elgeeii). The applicability of in situ soil moisture observations for this purpose is supported by Robock et al. (2000) and Mittelbach and Seneviratne (2012). This suggests that SiBCASA soil moisture is more reliable than ASCAT soil moisture at these sites. This is not entirely surprising, because Fig. 4 shows that the in situ observations were made at locations outside the area of high temporal correlations between SiBCASA and ASCAT. However, it suggests that the low correlations outside the steppe zone are more likely to be due to poor performance of ASCAT soil moisture than to SiBCASA soil moisture. It is unfortunate that there were no in situ observations during the ASCAT period of operation to evaluate ASCAT observations in the core drought regions. Now the added-value of ASCAT observations remains limited, because of the remaining questions about their accuracy.

On the positive side, in Figs. 5–8 we found a quite consistent pattern in springtime, which indicates that ASCAT soil moisture was larger than SiBCASA soil moisture, and assimilation seemed to improve the match with in situ observed soil moisture. Is this a realistic pattern? Experimentalists confirm that ponding after snowmelt occurs on the sites. However, it is known that ASCAT soil moisture is unreliable when the footprint of the observation is (partially) covered with snow, ice, or surface water, which is likely to happen in springtime. At the same time, SiBCASA soil moisture in spring depends on the amount of snow accumulated in the winter, the time of snowmelt, the fate of the meltwater on frozen ground (runoff or ponding). Since it is hard to simulate these processes correctly, also considering the coarse resolution of SiBCASA relative to dependency of these processes on topography, springtime soil moisture in SiBCASA may also be questioned. Nevertheless, this springtime underestimation pattern is also observed at other steppe and forest grid cells where the temporal correlations are large. Thus there are indications that the spring wetting with assimilation of ASCAT data in SiBCASA improves the soil moisture. Field workers (see author contributions) confirm the springtime water logging and ponding at the four sites. High water tables during springtime are followed by drying out of the soil, depending on the weather conditions. The low soil moisture in SiBCASA could be caused by overestimation of the evapo-transpiration rates in the spring.

In an attempt to explain the variation in temporal correlation coefficients over the region, Fig. 12 shows the temporal correlation coefficient of SiBCASA and ASCAT soil moisture in August 2013 as a function of several variables. Each dot in the figures represents a grid point. With increasing LAI the correlation coefficient r indeed decreases, which is physically logical, because water in leafs disturbs the soil moisture signal. Similarly, the aboveground carbon

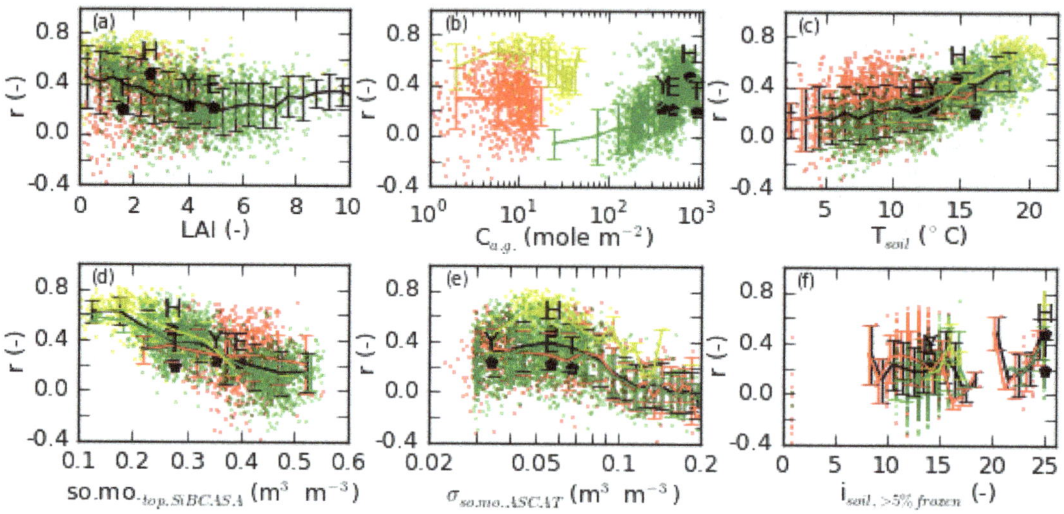

Figure 12. Variables possibly explaining the temporal correlation coefficient of SiBCASA and ASCAT soil moisture: **(a)** leaf area index; **(b)** aboveground carbon; **(c)** soil temperature; **(d)** top soil moisture in SiBCASA; **(e)** the uncertainty in ASCAT soil moisture; **(f)** the first soil layer with frozen fraction larger than 5 %. Red colours represent tundra pixels, green ones forest pixels, and yellow dots steppe pixels. The r represents the correlation coefficient between SiBCASA and ASCAT in all August days in the period of record (31 days × 7 years). The four black marks indicate the characteristics of the four sites Hyytiälä (H), Tver (T), Yakutsk (Y) and Elgeeii (E).

in biomass has a negative relationship with r for steppe, but not for forests and tundra zones. For forests, the relationship is, counterintuitively, positive. This may be explained by a cross-correlation between carbon in biomass and temperature: the forest biomass decreases towards the northern treeline, where temperatures are lower. Apparently, aboveground biomass itself does not necessarily disturb the satellite signal. Soil temperature has a positive relation with r, and there is no indication that the relationship saturates at higher temperatures. This is a somewhat puzzling observation. We would have expected low correlation coefficients at low temperatures, due to the presence of snow and ice, but at temperatures higher than 10 °C the ice would have disappeared, and we would not have expected an increase in r with temperature. Possibly, higher temperatures are indicative of a longer period into the local growing season, when soil ponding has diminished after snowmelt and the performance of SiBCASA is consequently better. This is confirmed by the negative relation between top soil moisture in SiBCASA with r. At large soil moisture contents, the chance of (partial) ponding is larger, with subsequent disturbances of the satellite signal (see Naeimi et al., 2012b; Högström et al., 2014; but also Griesfeller et al., 2016). The correlation coefficients between SiBCASA and ASCAT are best when the error estimate of the retrieved ASCAT soil moisture is smaller than 10^{-1} $m^3 m^{-3}$. Finally, the top most soil layer which contains ice is a poor predictor of r. Where the first layer is frozen, the r are indeed near 0, but all other grid points have ice only much deeper than the eighth soil layer, and there is no relation with r. This essentially means that permafrost does not disturb the satellite signal in August in Siberia. The characteristics of the four field sites are indicated by black marks in Fig. 12. This shows that the performance at the Yakutsk and Elgeeii sites may be expected to be low, because of the large LAI, low temperatures, and relatively large soil moisture. At the Tver and Hyytiälä sites, the expected performance is better, although the Tver site performs below average. We can only guess what might explain this difference. The region around the Tver site is quite heterogeneous, with a mixture of Spruce and deciduous forests and peat bogs, rivers and Lake Seliger. Perhaps the LAI is in reality larger than SiBCASA predicts, and the satellite retrieval is hampered by surface water.

In conclusion, (partial) ponding of the soil appears to be a good potential explanation of why the poor performance of ASCAT soil moisture improves into the summer months in Boreal Eurasia. The presence of dense leafs rather than aboveground biomass disturbs the satellite signal.

4.2 Carbon effects

It has been shown in Sect. 3.2 that assimilation of ASCAT soil moisture in SiBCASA has an effect of 5 to 10 % on GPP and TER, and of a few tens of percent on NEE, at the site of Yakutsk, over the entire year. This represents the higher end of the range, since the effect of assimilation on soil moisture and carbon fluxes was relatively large in Yakutsk. The reason why Yakutsk is so sensitive is because the plant-available water fraction is small there, so that the drought sensitivity is large (Fig. 1a). Integrated over the entire region, assimilation causes changes in the order of half the inter-annual variability, or 2 % of the mean annual NEE. We consider this quite large, given the fact that we only applied the assimilation to the top soil moisture. However, the temporal correlation coefficients were quite low in large parts of the region (Fig. 4). This implies that simulated and observed soil moisture are quite different. Assimilation will thus have a large effect when the observational errors are small. A comparison between observed and simulated NEE is made and discussed in the Supplement (Fig. S4).

The effect of changing soil moisture on GPP is largest in SiBCASA when the plant-available water fraction is smaller than 0.3. The area where this occurs is confined to the steppe zone in Southwest Siberia in South European Russia, where it is dry and in the northeast Siberian forest zone, where water availability is limited by permafrost. If the drought stress function in Fig. 1a were to be defined more linearly, the effects of soil moisture would be spread more evenly over the study domain. Note that it may not be realistic to prescribe identical water stress formulations for all biome types, as SiBCASA does. Furthermore, Fig. 10 shows that the drought sensitivity in Fig. 1 only represents the potential drought sensitivity. The actual sensitivity of GPP to change in soil moisture also depends on the temperature, radiation, and vapour pressure deficit (Fig. 10). This applies to TER in a similar way too. As a result, changes in NEE are not linearly dependent on the change in soil moisture due to assimilation of satellite observed soil moisture. Consequently, local effects may be much larger than 2 % of the mean annual NEE. Furthermore, Ohta et al. (2014) show that, in reality, water logging at high plant-available water fractions may also reduce photosynthesis rates and affect the water use efficiency.

5 Conclusions

The spatial and temporal correlation between SiBCASA and ASCAT soil moisture are considerable in the summer period and the steppe zone. However, ASCAT-derived soil moisture fails to detect the eight major droughts observed in situ at four sites during 7 years, while SiBCASA reproduces half of those droughts. At site level, temporal correlations between SiBCASA and in situ observed soil moisture are larger than between SiBCASA and ASCAT soil moisture. These facts suggest that SiBCASA soil moisture is more reliable than ASCAT soil moisture at those four locations and that assimilation of ASCAT soil moisture does not improve SiBCASA soil moisture.

The temporal correlation between SiBCASA and ASCAT soil moisture is best in the steppe zone, and in a selection of forest locations where LAI is low, soil temperature is high, and soil moisture is low (Fig. 12). Unfortunately, we do not have ground observations to prove whether assimilation in such conditions would lead to improved soil moisture in SiBCASA.

There is evidence that assimilation of ASCAT soil moisture improves the match of SiBCASA soil moisture with in situ observations in springtime (Figs. 5–8).

However, these results should be taken carefully, because ice and ponding occur often in the spring.

Irrespective of the question of whether assimilation improves soil moisture in SiBCASA, assimilation of ASCAT soil moisture causes considerable changes in GPP, TER, and NEE. At individual locations these changes may reach up to 5 to 10 % of annual GPP and TER, and tens of percent of annual NEE, and integrated over the entire region, the changes cause changes of the order of half the interannual variability in NEE or 2 % of annual NEE.

Ultimately, this study shows that assimilation of satellite-observed soil moisture in vegetation models potentially has large impacts on the simulated carbon fluxes, but that further research is needed to clarify when, where, and in which conditions assimilation leads to more reliable soil moisture simulations. In the near future, important improvements in the quality and spatial resolution of soil moisture are expected to be realized with the SMAP L-band instrument and Sentinel-1. Additionally, the benefit of more advanced assimilation techniques, e.g. by assimilating low-pass filtered satellite signals, may be investigated.

Author contributions. M. K. van der Molen implemented the assimilation scheme and performed the model runs and analysis together with I. R. van der Velde. R. A. M. de Jeu and W. Wagner contributed by discussing the reliability of the satellite-observed soil moisture data. W. Peters and M. C. Krol contributed by discussing the research plan and results. The in situ soil moisture observations were collected and provided by P. Kolari (Hyytiälä), J. Kurbatova, A. Varlagin (Tver), T. C. Maximov, A. V. Kononov, T. Ohta, A. Kotani (Yakutsk and Elgeeii), who also contributed by discussing the reliability of the in situ observations.

Acknowledgements. The research on which this publication is based has been partly funded by the Dutch Organization for Scientific Research (NWO) under grant number 864.08.012 (VIDI WP "A multiple constraint data assimilation system for the carbon cycle"). J. Kurbatova and A. Varlagin, and the in situ observations in Tver were funded by the Russian Science Foundation, Grant 14-27-00065.

Edited by: P. Saco

References

Baldocchi, D., Falge, E., Gu, L. H., Olson, R., Hollinger, D., Running, S., Anthoni, P., Bernhofer, C., Davis, K., Evans, R., Fuentes, J., Goldstein, A., Katul, G., Law, B., Lee, X. H., Malhi, Y., Meyers, T., Munger, W., Oechel, W., U, K. T. P., Pilegaard, K., Schmid, H. P., Valentini, R., Verma, S., Vesala, T., Wilson, K., and Wofsy, S.: FLUXNET: A new tool to study the temporal and spatial variability of ecosystem-scale carbon dioxide, water vapor, and energy flux densities, B. Am. Meteorol. Soc., 82, 2415–2434, 2001.

Bartalis, Z., Wagner, W., Naeimi, V., Hasenauer, S., Scipal, K., Bonekamp, H., Figa, J., and Anderson, C.: Initial soil moisture retrievals from the METOP-A Advanced Scatterometer (ASCAT), Geophys. Res. Lett., 34, L20401, doi:10.1029/2007gl031088, 2007.

Champagne, C., Berg, A., Belanger, J., McNairn, H., and De Jeu, R.: Evaluation of soil moisture derived from passive microwave remote sensing over agricultural sites in Canada using ground-based soil moisture monitoring networks, Int. J. Remote Sens., 31, 3669–3690, doi:10.1080/01431161.2010.483485, 2010.

Chen, Y., Yang, K., Qin, J., Zhao, L., Tang, W., and Han, M.: Evaluation of AMSR-E retrievals and GLDAS simulations against observations of a soil moisture network on the central Tibetan Plateau, J. Geophys. Res.-Atmos., 118, 4466–4475, doi:10.1002/jgrd.50301, 2013.

Ciais, P., Reichstein, M., Viovy, N., Granier, A., Ogee, J., Allard, V., Aubinet, M., Buchmann, N., Bernhofer, C., Carrara, A., Chevallier, F., De Noblet, N., Friend, A. D., Friedlingstein, P., Grunwald, T., Heinesch, B., Keronen, P., Knohl, A., Krinner, G., Loustau, D., Manca, G., Matteucci, G., Miglietta, F., Ourcival, J. M., Papale, D., Pilegaard, K., Rambal, S., Seufert, G., Soussana, J. F., Sanz, M. J., Schulze, E. D., Vesala, T., and Valentini, R.: Europe-wide reduction in primary productivity caused by the heat and drought in 2003, Nature, 437, 529–533, doi:10.1038/nature03972, 2005.

Collatz, G. J., Ball, J. T., Grivet, C., and Berry, J. A.: Physiological and environmental regulation of stomatal conductance, photosynthesis and transpiration: a model that includes a laminar boundary layer, Agr. Forest Meteorol., 54, 107–136, 1991.

Dai, A.: Drought under global warming: A review, Wiley Interdisciplinary Reviews: Climate Change, 2, 45–65, doi:10.1002/wcc.81, 2011.

De Jeu, R. A. M. and Owe, M.: Further validation of a new methodology for surface moisture and vegetation optical depth retrieval, Int. J. Remote Sens., 24, 4559–4578, doi:10.1080/0143116031000095934, 2003.

de Jeu, R. A. M., Wagner, W., Holmes, T., Dolman, A., van de Giesen, N., and Friesen, J.: Global Soil Moisture Patterns Observed by Space Borne Microwave Radiometers and Scatterometers, Surv. Geophys., 29, 399–420, doi:10.1007/s10712-008-9044-0, 2008.

De Jeu, R. A. M., Holmes, T. R. H., Panciera, R., and Walker, J. P.: Parameterization of the land parameter retrieval model for L-band observations using the NAFE'05 data set, IEEE Geosci. Remote Sens. Lett., 6, 630–634, doi:10.1109/lgrs.2009.2019607, 2009.

de Wit, A. J. W. and van Diepen, C. A.: Crop model data assimilation with the Ensemble Kalman filter for improving regional crop yield forecasts, Agr. Forest Meteorol., 146, 38–56, doi:10.1016/j.agrformet.2007.05.004, 2007.

Dolman, A. J., Maximov, T. C., Moors, E. J., Maximov, A. P., Elbers, J. A., Kononov, A. V., Waterloo, M. J., and van der Molen, M. K.: Net ecosystem exchange of carbon dioxide and water of far eastern Siberian Larch (*Larix cajanderii*) on permafrost, Biogeosciences, 1, 133–146, doi:10.5194/bg-1-133-2004, 2004.

Dolman, A. J., Shvidenko, A., Schepaschenko, D., Ciais, P., Tchebakova, N., Chen, T., van der Molen, M. K., Belelli Marchesini, L., Maximov, T. C., Maksyutov, S., and Schulze, E.-D.: An estimate of the terrestrial carbon budget of Russia using inventory-based, eddy covariance and inversion methods, Biogeosciences, 9, 5323–5340, doi:10.5194/bg-9-5323-2012, 2012.

Draper, C. S., Mahfouf, J. F., and Walker, J. P.: An EKF assimilation of AMSR-E soil moisture into the ISBA land surface scheme, J. Geophys. Res.-Atmos., 114, D20104, doi:10.1029/2008jd011650, 2009a.

Draper, C. S., Walker, J. P., Steinle, P. J., de Jeu, R. A. M., and Holmes, T. R. H.: An evaluation of AMSR-E derived soil moisture over Australia, Remote Sens. Environ., 113, 703–710, doi:10.1016/j.rse.2008.11.011, 2009b.

Entekhabi, D., Njoku, E. G., O'Neill, P. E., Kellogg, K. H., Crow, W. T., Edelstein, W. N., Entin, J. K., Goodman, S. D., Jackson, T. J., Johnson, J., Kimball, J., Piepmeier, J. R., Koster, R. D., Martin, N., McDonald, K. C., Moghaddam, M., Moran, S., Reichle, R., Shi, J. C., Spencer, M. W., Thurman, S. W., Tsang, L., and Van Zyl, J.: The soil moisture active passive (SMAP) mission, Proc. IEEE, 98, 704–716, doi:10.1109/jproc.2010.2043918, 2010.

Farquhar, G. D. von Caemmerer, S., and Berry, J. A.: A biochemical model of photosynthetic CO_2 assimilation in leaves of C3 species, Planta, 149, 78–90, doi:10.1007/bf00386231, 1980.

Goetz, S. J., MacK, M. C., Gurney, K. R., Randerson, J. T., and Houghton, R. A.: Ecosystem responses to recent climate change and fire disturbance at northern high latitudes: Observations and model results contrasting northern Eurasia and North America, Environ. Res. Lett., 2, 045031, doi:10.1088/1748-9326/2/4/045031, 2007.

Griesfeller, A., Lahoz, W. A., d. Jeu, R. A. M., Dorigo, W., Haugen, L. E., Svendby, T. M., and Wagner, W.: Evaluation of satellite soil moisture products over Norway using ground-based observations, Int. J. Appl. Earth Obs. Geoinf., 45, 155–164, doi:10.1016/j.jag.2015.04.016, 2016.

Han, E., Crow, W. T., Holmes, T., and Bolten, J.: Benchmarking a soil moisture data assimilation system for agricultural drought monitoring, J. Hydrometeorol., 15, 1117–1134, doi:10.1175/jhm-d-13-0125.1, 2014.

Högström, E., Trofaier, A., Gouttevin, I., and Bartsch, A.: Assessing Seasonal Backscatter Variations with Respect to Uncertainties in Soil Moisture Retrieval in Siberian Tundra Regions, Remote Sensing, 6, 8718–8738, 2014.

Ilvesniemi, H., Pumpanen, J., Duursma, R., Hari, P., Keronen, P., Kolari, P., Kulmala, M., Mammarella, I., Nikinmaa, E., Rannik, Ü., Pohja, T., Siivola, E., and Vesala, T.: Water balance of a boreal scots pine forest, Boreal Environ. Res., 15, 375–396, 2010.

Imaoka, K., Kachi, M., Fujii, H., Murakami, H., Hori, M., Ono, A., Igarashi, T., Nakagawa, K., Oki, T., Honda, Y., and Shimoda, H.: Global change observation mission (GCOM) for monitoring carbon, water cycles, and climate change, Proc, IEEE, 98, 717–734, doi:10.1109/jproc.2009.2036869, 2010.

Jafarov, E. and Schaefer, K.: The importance of a surface organic layer in simulating permafrost thermal and carbon dynamics, The Cryosphere Discuss., 9, 3137–3163, doi:10.5194/tcd-9-3137-2015, 2015.

Katul, G., Manzoni, S., Palmroth, S., and Oren, R.: A stomatal optimization theory to describe the effects of atmospheric CO_2 on leaf photosynthesis and transpiration, Ann. Bot., 105, 431–442, doi:10.1093/aob/mcp292, 2010.

Kerr, Y. H., Waldteufel, P., Wigneron, J. P., Martinuzzi, J. M., Font, J., and Berger, M.: Soil moisture retrieval from space: The Soil Moisture and Ocean Salinity (SMOS) mission, IEEE T. Geosci. Remote Sens., 39, 1729–1735, doi:10.1109/36.942551, 2001.

Kolari, P., Kulmala, L., Pumpanen, J., Launiainen, S., Llvesniemi, H., Hari, P., and Nikinmaa, E.: CO_2 exchange and component CO_2 fluxes of a boreal Scots pine forest, Boreal Environ. Res., 14, 761–783, 2009.

Kotani, A., Kononov, A. V., Ohta, T., and Maximov, T. C.: Temporal variations in the linkage between the net ecosystem exchange of water vapour and CO_2 over boreal forests in eastern Siberia, Ecohydrology, 7, 209–225, 2014.

Krol, M. C., Peters, W., Hooghiemstra, P., George, M., Clerbaux, C., Hurtmans, D., McInerney, D., Sedano, F., Bergamaschi, P., El Hajj, M., Kaiser, J. W., Fisher, D., Yershov, V., and Muller, J. P.: How much CO was emitted by the 2010 fires around Moscow?, Atmos. Chem. Phys., 13, 4737–4747, doi:10.5194/acp-13-4737-2013, 2013.

Kurbatova, J., Li, C., Varlagin, A., Xiao, X., and Vygodskaya, N.: Modeling carbon dynamics in two adjacent spruce forests with different soil conditions in Russia, Biogeosciences, 5, 969–980, doi:10.5194/bg-5-969-2008, 2008.

Liu, Y. Y., van Dijk, A., de Jeu, R. A. M., and Holmes, T. R. H.: An analysis of spatiotemporal variations of soil and vegetation moisture from a 29-year satellite-derived data set over mainland Australia, Water Resour. Res., 45, W07405, doi:10.1029/2008wr007187, 2009.

Liu, Y. Y., Parinussa, R. M., Dorigo, W. A., De Jeu, R. A. M., Wagner, W., M. Van Dijk, A. I. J., McCabe, M. F., and Evans, J. P.: Developing an improved soil moisture dataset by blending passive and active microwave satellite-based retrievals, Hydrol. Earth Syst. Sci., 15, 425–436, doi:10.5194/hess-15-425-2011, 2011.

Liu, Y. Y., Dorigo, W. A., Parinussa, R. M., De Jeu, R. A. M., Wagner, W., McCabe, M. F., Evans, J. P., and Van Dijk, A. I. J. M.: Trend-preserving blending of passive and active microwave soil moisture retrievals, Remote Sens. Environ., 123, 280–297, doi:10.1016/j.rse.2012.03.014, 2012.

Mammarella, I., Launiainen, S., Gronholm, T., Keronen, P., Pumpanen, J., Rannik, Ü., and Vesala, T.: Relative humidity effect on the high-frequency attenuation of water vapor flux measured by a closed-path eddy covariance system, J. Atmos. Ocean. Tech., 26, 1856–1866, doi:10.1175/2009jtecha1179.1, 2009.

McGuire, A. D., Anderson, L. G., Christensen, T. R., Scott, D., Laodong, G., Hayes, D. J., Martin, H., Lorenson, T. D., Macdonald, R. W., and Nigel, R.: Sensitivity of the carbon cycle in the Arctic to climate change, Ecol. Monogr., 79, 523–555, doi:10.1890/08-2025.1, 2009.

Milyukova, I. M., Kolle, O., Varlagin, A. V., Vygodskaya, N. N., Schulze, E. D., and Lloyd, J.: Carbon balance of a southern taiga spruce stand in European Russia, Tellus B, 54, 429–442, doi:10.1034/j.1600-0889.2002.01387.x, 2002.

Miralles, D. G.: Evaporation in the global water cycle: Analysing land evaporation using satellite observations PhD, Department of Hydrology and Geo-Environmental Sciences, VU University Amsterdam, Amsterdam, 126 pp., 2011.

Miralles, D. G., De Jeu, R. A. M., Gash, J. H., Holmes, T. R. H., and Dolman, A. J.: Magnitude and variability of land evaporation and its components at the global scale, Hydrol. Earth Syst. Sci., 15, 967–981, doi:10.5194/hess-15-967-2011, 2011a.

Miralles, D. G., Holmes, T. R. H., De Jeu, R. A. M., Gash, J. H., Meesters, A. G. C. A., and Dolman, A. J.: Global land-surface evaporation estimated from satellite-based observations, Hydrol. Earth Syst. Sci., 15, 453–469, doi:10.5194/hess-15-453-2011, 2011b.

Miralles, D. G., Teuling, A. J., Van Heerwaarden, C. C., and De Arellano, J. V. G.: Mega-heatwave temperatures due to combined soil desiccation and atmospheric heat accumulation, Nat. Geosci., 7, 345–349, doi:10.1038/ngeo2141, 2014.

Mittelbach, H. and Seneviratne, S. I.: A new perspective on the spatio-temporal variability of soil moisture: temporal dynamics versus time-invariant contributions, Hydrol. Earth Syst. Sci., 16, 2169–2179, doi:10.5194/hess-16-2169-2012, 2012.

Mladenova, I. E., Jackson, T. J., Njoku, E., Bindlish, R., Chan, S., Cosh, M. H., Holmes, T. R. H., de Jeu, R. A. M., Jones, L., Kimball, J., Paloscia, S., and Santi, E.: Remote monitoring of soil moisture using passive microwave-based techniques – Theoretical basis and overview of selected algorithms for AMSR-E, Remote Sens. Environ., 144, 197–213, doi:10.1016/j.rse.2014.01.013, 2014.

Moffat, A. M., Papale, D., Reichstein, M., Hollinger, D. Y., Richardson, A. D., Barr, A. G., Beckstein, C., Braswell, B. H., Churkina, G., Desai, A. R., Falge, E., Gove, J. H., Heimann, M., Hui, D., Jarvis, A. J., Kattge, J., Noormets, A., and Stauch, V. J.: Comprehensive comparison of gap-filling techniques for eddy covariance net carbon fluxes, Agr. Forest Meteorol., 147, 209–232, doi:10.1016/j.agrformet.2007.08.011, 2007.

Morales, P., Sykes, M. T., Prentice, I. C., Smith, P., Smith, B., Bugmann, H., Zierl, B., Friedlingstein, P., Viovy, N., Sabaté, S., Sánchez, A., Pla, E., Gracia, C. A., Sitch, S., Arneth, A., and Ogee, J.: Comparing and evaluating process-based ecosystem model predictions of carbon and water fluxes in major European forest biomes, Global Change Biol., 11, 2211–2233, doi:10.1111/j.1365-2486.2005.01036.x, 2005.

Muskett, R. R. and Romanovsky, V. E.: Groundwater storage changes in arctic permafrost watersheds from GRACE and insitu measurements, Environ. Res. Lett., 4, 045009, doi:10.1088/1748-9326/4/4/045009, 2009.

Naeimi, V., Scipal, K., Bartalis, Z., Hasenauer, S., and Wagner, W.: An Improved Soil Moisture Retrieval Algorithm for ERS and METOP Scatterometer Observations, IEEE T. Geosci. Remote, 47, 1999–2013, doi:10.1109/tgrs.2008.2011617, 2009.

Naeimi, V., Paulik, C., Bartsch, A., Wagner, W., Kidd, R., Park, S. E., Elger, K., and Boike, J.: ASCAT surface state flag (SSF): Extracting information on surface freeze/thaw conditions from backscatter data using an empirical threshold-analysis algorithm, IEEE T. Geosci. Remote, 50, 2566–2582, doi:10.1109/tgrs.2011.2177667, 2012a.

Naeimi, V., Paulik, C., Bartsch, A., Wagner, W., Kidd, R., Sang-Eun, P., Elger, K., and Boike, J.: ASCAT Surface State Flag (SSF): Extracting Information on Surface Freeze/Thaw Conditions From Backscatter Data Using an Empirical Threshold-Analysis Algorithm, IEEE T. Geosci. Remote, 50, 2566–2582, doi:10.1109/tgrs.2011.2177667, 2012b.

Ohta, T., Maximov, T. C., Dolman, A. J., Nakai, T., Molen, M. K. v. d., Kononov, A. V., Maximov, T., Hiyama, T., Iijima, Y., Moors, E. J., and Tanaka, H.: Interannual variation of water balance and summer evapotranspiration in an eastern Siberian larch forest over a 7-year period (1998–2006), Agr. Forest Meteorol., 148, 1941–1953, 2008.

Ohta, T., Kotani, A., Iijima, Y., Maximov, T. C., Ito, S., Hanamura, M., Kononov, A. V., and Maximov, A. P.: Effects of waterlogging on water and carbon dioxide fluxes and environmental variables in a Siberian larch forest, 1998–2011, Agr. Forest Meteorol., 188, 64–75, doi:10.1016/j.agrformet.2013.12.012, 2014.

Oki, T., Imaoka, K., and Kachi, M.: AMSR instruments on GCOM-W1/2: Concepts and applications, International Geoscience and Remote Sensing Symposium (IGARSS), Honolulu, Hawaii, USA, 1363–1366, 2010.

Owe, M., de Jeu, R., and Holmes, T.: Multisensor historical climatology of satellite-derived global land surface moisture, J. Geophys. Res.-Earth, 113, F01002, doi:10.1029/2007JF000769, 2008.

Papale, D., Reichstein, M., Aubinet, M., Canfora, E., Bernhofer, C., Kutsch, W., Longdoz, B., Rambal, S., Valentini, R., Vesala, T., and Yakir, D.: Towards a standardized processing of Net Ecosystem Exchange measured with eddy covariance technique: algorithms and uncertainty estimation, Biogeosciences, 3, 571–583, doi:10.5194/bg-3-571-2006, 2006.

Peters, W., Jacobson, A. R., Sweeney, C., Andrews, A. E., Conway, T. J., Masarie, K., Miller, J. B., Bruhwiler, L. M. P., Petron, G., Hirsch, A. I., Worthy, D. E. J., van der Werf, G. R., Randerson, J. T., Wennberg, P. O., Krol, M. C., and Tans, P. P.: An atmospheric perspective on North American carbon dioxide exchange: CarbonTracker, P. Natl. Acad. Sci. USA, 104, 18925–18930, doi:10.1073/pnas.0708986104, 2007.

Peters, W., Krol, M. C., van der Werf, G. R., Houweling, S., Jones, C. D., Hughes, J., Schaefer, K., Masarie, K. A., Jacobson, A. R., Miller, J. B., Cho, C. H., Ramonet, M., Schmidt, M., Ciattaglia, L., Apadula, F., Helta, D., Meinhardt, F., di Sarra, A. G., Piacentino, S., Sferlazzo, D., Aalto, T., Hatakka, J., Strom, J., Haszpra, L., Meijer, H. A. J., van de Laan, S., Neubert, R. E. M., Jordan, A., Rodo, X., Morgui, J. A., Vermeulen, A. T., Popa, E., Rozanski, K., Zimnoch, M., Manning, A. C., Leuenberger, M., Uglietti, C., Dolman, A. J., Ciais, P., Heimann, M., and Tans, P. P.: Seven years of recent European net terrestrial carbon dioxide exchange constrained by atmospheric observations, Global Change Biol., 16, 1317–1337, doi:10.1111/j.1365-2486.2009.02078.x, 2010.

Potter, C. S., Randerson, J. T., Field, C. B., Matson, P. A., Vitousek, P. M., Mooney, H. A., and Klooster, S. A.: Terrestrial ecosystem production: A process model based on global satellite and surface data, Global Biogeochem. Cy., 7, 811–841, doi:10.1029/93gb02725, 1993.

Raich, J. W., Rastetter, E. B., Melillo, J. M., Kicklighter, D. W., Steudler, P. A., Peterson, B. J., Grace, A. L., Iii, B. M., and Vörösmarty, C. J.: Potential Net Primary Productivity in South America: Application of a Global Model, Ecol. Appl., 1, 399–429, doi:10.2307/1941899, 1991.

Rannik, U., Keronen, P., Hari, P., and Vesala, T.: Estimation of forest-atmosphere CO_2 exchange by eddy covariance and profile techniques, Agr. Forest Meteorol., 126, 141–155, doi:10.1016/j.agrformet.2004.06.010, 2004.

Reichle, R. H. and Koster, R. D.: Bias reduction in short records of satellite soil moisture, Geophys. Res. Lett., 31, L19501, doi:10.1029/2004gl020938, 2004.

Reichle, R. H., and Koster, R. D.: Global assimilation of satellite surface soil moisture retrievals into the NASA catchment land surface model, Geophys. Res. Lett., 32, 1–4, doi:10.1029/2004gl021700, 2005.

Reichle, R. H., Koster, R. D., Liu, P., Mahanama, S. P. P., Njoku, E. G., and Owe, M.: Comparison and assimilation of global soil moisture retrievals from the Advanced Microwave Scanning Radiometer for the Earth Observing System (AMSR-E) and the Scanning Multichannel Microwave Radiometer (SMMR), J. Geophys. Res.-Atmos., 112, D09108, doi:10.1029/2006jd008033, 2007.

Reichstein, M., Falge, E., Baldocchi, D., Papale, D., Aubinet, M., Berbigier, P., Bernhofer, C., Buchmann, N., Gilmanov, T., Granier, A., Grünwald, T., Havránková, K., Ilvesniemi, H., Janous, D., Knohl, A., Laurila, T., Lohila, A., Loustau, D., Matteucci, G., Meyers, T., Miglietta, F., Ourcival, J. M., Pumpanen, J., Rambal, S., Rotenberg, E., Sanz, M., Tenhunen, J., Seufert, G., Vaccari, F., Vesala, T., Yakir, D., and Valentini, R.: On the separation of net ecosystem exchange into assimilation and ecosystem respiration: Review and improved algorithm, Global Change Biol., 11, 1424–1439, doi:10.1111/j.1365-2486.2005.001002.x, 2005.

Reichstein, M., Ciais, P., Papale, D., Valentini, R., Running, S., Viovy, N., Cramer, W., Granier, A., Ogee, J., Allard, V., Aubinet, M., Bernhofer, C., Buchmann, N., Carrara, A., Grunwald, T., Heimann, M., Heinesch, B., Knohl, A., Kutsch, W., Loustau, D., Manca, G., Matteucci, G., Miglietta, F., Ourcival, J. M., Pilegaard, K., Pumpanen, J., Rambal, S., Schaphoff, S., Seufert, G., Soussana, J. F., Sanz, M. J., Vesala, T., and Zhao, M.: Reduction of ecosystem productivity and respiration during the European summer 2003 climate anomaly: a joint flux tower, remote sensing and modelling analysis, Global Change Biol., 13, 634–651, doi:10.1111/j.1365-2486.2006.01224.x, 2007.

Robock, A., Vinnikov, K. Y., Srinivasan, G., Entin, J. K., Hollinger, S. E., Speranskaya, N. A., Liu, S., and Namkhai, A.: The Global Soil Moisture Data Bank, B. Am. Meteorol. Soc., 81, 1281–1299, 2000.

Schaefer, K., Collatz, G. J., Tans, P., Denning, A. S., Baker, I., Berry, J., Prihodko, L., Suits, N., and Philpott, A.: Combined simple biosphere/carnegie-ames-stanford approach terrestrial carbon cycle model, J. Geophys. Res.-Biogeo., 113, G03034, doi:10.1029/2007jg000603, 2008.

Schepaschenko, D. G., Mukhortova, L. V., Shvidenko, A. Z., and Vedrova, E. F.: The pool of organic carbon in the soils of Russia, Eurasian Soil Sci., 46, 107–116, doi:10.1134/s1064229313020129, 2013.

Sellers, P. J., Randall, D. A., Collatz, G. J., Berry, J. A., Field, C. B., Dazlich, D. A., Zhang, C., Collelo, G. D., and Bounoua, L.: A revised land surface parameterization (SiB2) for atmospheric GCMs. Part I: Model formulation, J. Climate, 9, 676–705, 1996.

Tarnocai, C., Canadell, J. G., Schuur, E. A. G., Kuhry, P., Mazhitova, G., and Zimov, S. C. G.: Soil organic carbon pools in the northern circumpolar permafrost region, Global Biogeochem. Cy., 23, GB2023, doi:10.1029/2008gb003327, 2009.

van der Molen, M. K., Dolman, A. J., Ciais, P., Eglin, T., Gobron, N., Law, B. E., Meir, P., Peters, W., Phillips, O. L., Reichstein, M., Chen, T., Dekker, S. C., Doubková, M., Friedl, M. A., Jung, M., van den Hurk, B. J. J. M., de Jeu, R. A. M., Kruijt, B., Ohta, T., Rebel, K. T., Plummer, S., Seneviratne, S. I., Sitch, S., Teuling, A. J., van der Werf, G. R., and Wang, G.: Drought and ecosystem carbon cycling, Agr. Forest Meteorol., 151, 765–773, 2011.

van der Velde, I. R., Miller, J. B., Schaefer, K., van der Werf, G. R., Krol, M. C., and Peters, W.: Terrestrial cycling of $^{13}CO_2$ by photosynthesis, respiration, and biomass burning in SiBCASA, Biogeosciences, 11, 6553–6571, doi:10.5194/bg-11-6553-2014, 2014.

Verhoef, A. and Egea, G.: Modeling plant transpiration under limited soil water: Comparison of different plant and soil hydraulic parameterizations and preliminary implications for their use in land surface models, Agr. Forest Meteorol., 191, 22–32, doi:10.1016/j.agrformet.2014.02.009, 2014.

Verstraeten, W. W., Veroustraete, F., Wagner, W., van Roey, T., Heyns, W., Verbeiren, S., and Feyen, J.: Remotely sensed soil moisture integration in an ecosystem carbon flux model. The spatial implication, Climatic Change, 103, 117–136, doi:10.1007/s10584-010-9920-8, 2010.

Vetter, M., Churkina, G., Jung, M., Reichstein, M., Zaehle, S., Bondeau, A., Chen, Y., Ciais, P., Feser, F., Freibauer, A., Geyer, R., Jones, C., Papale, D., Tenhunen, J., Tomelleri, E., Trusilova, K., Viovy, N., and Heimann, M.: Analyzing the causes and spatial pattern of the European 2003 carbon flux anomaly using seven models, Biogeosciences, 5, 561–583, doi:10.5194/bg-5-561-2008, 2008.

Vey, S., Steffen, H., Müller, J., and Boike, J.: Inter-annual water mass variations from GRACE in central Siberia, J. Geodesy, 87, 287–299, doi:10.1007/s00190-012-0597-9, 2013.

Wagner, W., Lemoine, G., and Rott, H.: A method for estimating soil moisture from ERS Scatterometer and soil data, Remote Sens. Environ., 70, 191–207, doi:10.1016/s0034-4257(99)00036-x, 1999.

Way, J., Zimmermann, R., Rignot, E., McDonald, K., and Oren, R.: Winter and spring thaw as observed with imaging radar at BOREAS, J. Geophys. Res.-Atmos., 102, 29673–29684, doi:10.1029/96jd03878, 1997.

Wegmüller, U.: The effect of freezing and thawing on the microwave signatures of bare soil, Remote Sens. Environ., 33, 123–135, doi:10.1016/0034-4257(90)90038-N, 1990.

Williams, M., Richardson, A. D., Reichstein, M., Stoy, P. C., Peylin, P., Verbeeck, H., Carvalhais, N., Jung, M., Hollinger, D. Y., Kattge, J., Leuning, R., Luo, Y., Tomelleri, E., Trudinger, C. M., and Wang, Y. -P.: Improving land surface models with FLUXNET data, Biogeosciences, 6, 1341–1359, doi:10.5194/bg-6-1341-2009, 2009.

4

Variations of deep soil moisture under different vegetation types and influencing factors in a watershed of the Loess Plateau, China

Xuening Fang[1], Wenwu Zhao[1,2], Lixin Wang[2], Qiang Feng[1,3], Jingyi Ding[1], Yuanxin Liu[1], and Xiao Zhang[1]

[1]State Key Laboratory of Earth Surface Processes and Resource Ecology, College of Resource and Technology, Beijing Normal University, Beijing 100875, P. R. China
[2]Department of Earth Sciences, Indiana University–Purdue University Indianapolis (IUPUI), Indianapolis, Indiana 46202, USA
[3]College of Forestry, Shanxi Agricultural University, Taigu, Shanxi 030801, P. R. China

Correspondence to: Wenwu Zhao (zhaoww@bnu.edu.cn) and Lixin Wang (lxwang@iupui.edu)

Abstract. Soil moisture in deep soil layers is a relatively stable water resource for vegetation growth in the semi-arid Loess Plateau of China. Characterizing the variations in deep soil moisture and its influencing factors at a moderate watershed scale is important to ensure the sustainability of vegetation restoration efforts. In this study, we focus on analyzing the variations and factors that influence the deep soil moisture (DSM) in 80–500 cm soil layers based on a soil moisture survey of the Ansai watershed in Yan'an in Shanxi Province. Our results can be divided into four main findings. (1) At the watershed scale, higher variations in the DSM occurred at 120–140 and 480–500 cm in the vertical direction. At the comparable depths, the variation in the DSM under native vegetation was much lower than that in human-managed vegetation and introduced vegetation. (2) The DSM in native vegetation and human-managed vegetation was significantly higher than that in introduced vegetation, and different degrees of soil desiccation occurred under all the introduced vegetation types. *Caragana korshinskii* and black locust caused the most serious desiccation. (3) Taking the DSM conditions of native vegetation as a reference, the DSM in this watershed could be divided into three layers: (i) a rainfall transpiration layer (80–220 cm); (ii) a transition layer (220–400 cm); and (iii) a stable layer (400–500 cm). (4) The factors influencing DSM at the watershed scale varied with vegetation types. The main local controls of the DSM variations were the soil particle composition and mean annual rainfall; human agricultural management measures can alter the soil bulk density, which contributes to higher DSM in farmland and apple orchards. The plant growth conditions, planting density, and litter water holding capacity of introduced vegetation showed significant relationships with the DSM. The results of this study are of practical significance for vegetation restoration strategies, especially for the choice of vegetation types, planting zones, and proper human management measures.

1 Introduction

Soil moisture is an indispensable component of terrestrial systems and plays a critical role in surface hydrological processes, especially runoff generation, soil evaporation and plant transpiration (Cheema et al., 2011; Legates et al., 2011; Wang et al., 2012a; Zhao et al., 2013). Soil moisture in different soil layers is usually related to different hydrological processes and ecological functions (Yang et al., 2012a). Surface or shallow layer soil moisture is usually greatly influenced by rainfall infiltration or evapotranspiration and is a regular water source for vegetation growth, while the moisture in deep soil layers functions as a soil reservoir. During the years with heavy rainfall, deep soil moisture (DSM) can be replenished by rainfall infiltration; during drought years, DSM can provide necessary water for plant growth and is thus important for plant growth during dry seasons (Yang et al., 2012c; Jia and Shao, 2014). This relationship is particularly true in semi-arid areas, such as the Loess Plateau of China, where water resources are incredibly scarce. In such regions, DSM even becomes the main constraining factor of plant productivity and ecosystem sustainability (Wang et al., 2010c, 2011a). Traditionally the boundary of deep soil moisture is not clearly defined. In this study, we define the deep soil layer as the layer whose soil moisture is not sensitive to daily evapotranspiration and regular rainfall events.

The Loess Plateau of China is located in a semi-arid area. The average annual rainfall in this region ranges from 150 to 800 mm, which is far lower than the average annual pan evaporation (1400–2000 mm; Wang et al., 2010b). Low precipitation and high evaporation results in lower soil moisture content in this region. The shallow soil moisture is not sufficient to meet the growth needs of introduced vegetation (Yang et al., 2014b). Moreover, the thickness of loess soil in this area ranges from 30 to 80 m; at these depths, groundwater is not available for plants (Wang et al., 2013). Therefore, DSM that is stored in unsaturated soil becomes an important water resource for plant growth (Yang et al., 2012c). However, the vegetation that has been introduced by the national Grain for Green project tends to have strong water consumption. Large-scale afforestation has resulted in the excessive consumption of DSM, and a large range of soil desiccation has been reported (Wang et al., 2008b, 2010b, c, 2011b). Soil desiccation greatly reduces the capability of a "soil reservoir" to supply water to deep soil layers for plant growth in the Loess Plateau (Chen et al., 2008a). Introduced vegetation in desiccated land is easily degraded with low productivity, and "dwarf aged trees" with heights of 3–5 m have widely appeared. Therefore, the sustainability of the restored ecosystem is being challenged. Moreover, traditional soil moisture studies, which have mainly focused on shallow soil layers (Gómez-Plaza et al., 2000; Bi et al., 2009; Baroni et al., 2013), cannot clearly reveal the sustainability needs for vegetation restoration.

In recent years, several studies have been conducted on the variations in DSM and factors influencing DSM in the Loess Plateau (Liu et al., 2010; Wang et al., 2012b, 2013; Yang et al., 2014a; Jia and Shao, 2014; Sun et al., 2014). Deep soil moisture is an indispensable water source for vegetation growth in the semi-arid Loess Plateau; understanding the variations and influencing factors of DSM is important for "timely, suitable, and moderate" vegetation restoration and can also help in developing proper measures to control soil desiccation. In fact, DSM is a result of long-term biophysical processes that are controlled by multiple factors (Vereecken et al., 2007). Several factors may impact DSM variations, such as vegetation traits, soil properties, topographical factors, climate factors, and human landscape management measures (Qiu et al., 2001; Lu et al., 2007, 2008; Vivoni et al., 2008; Zhu and Lin, 2011; Montenegro and Ragab, 2012). The dominant factors that affect DSM variations depend on the research scale (Entin et al., 2000). For instance, DSM variations are mainly dominated by the type of vegetation at the slope scale (0.1–1 km^2; Jia et al., 2013). Additionally, vegetation and topography are key factors that contribute to DSM variations at the small catchment scale (1–100 km^2; Yang et al., 2012a). Meanwhile, Wang et al. (2012b) reported that DSM variations at the regional scale (i.e., the Loess Plateau, covering 640 000 km^2) are mainly determined by plant types and climatic conditions. Vegetation factors play an important role in spatial variations of DSM at all scales (Western et al., 2004).

While all spatial scales, from slopes and small catchments to regions, are relevant to understanding DSM variations, some scales are more operational and meaningful than others. For example, slope- and small catchment-based studies tend to be too small in spatial extent to incorporate all environmental factors and human-managed measures because soil traits, climate characteristics, and human-managed measures are usually homogeneous at one slope or small catchment (Gómez-Plaza et al., 2000; Bi et al., 2009; Zhu et al., 2014a, b). At the same time, it is often impossible to assess the mechanistic details of DSM variations to guide local policies at the regional scale since high variations in rainfall and temperature can cover the influencing effects of other factors (Wang et al., 2010a, d, 2012c). A moderate scale that covers an area of approximately 100–1000 km^2 over a watershed or a geopolitically defined area represents a pivotal scale domain for research on DSM variation mechanisms. In particular, people and nature interact most acutely at this scale (Zhao and Fang, 2014; Fang et al., 2015), which is a more operational scale for sustainable vegetation restoration policy making. To date, however, minimal research on DSM variations has centered on such a moderate scale, and the variation mechanisms of DSM at this scale are still unclear.

In this study, we aim to reveal the variations in DSM and its influencing factors at a moderate watershed scale. This study includes shallow root vegetation and deep root vegetation, covering eight specific vegetation types. We first identify the deep soil layer in the Ansai watershed, whose soil moisture is not sensitive to regular rainfall and daily evapotranspiration. Then, we explore the overall variation in DSM in this area, compare the DSM of various vegetation types with different root systems, and identify variations in their profiles. Finally, the influence of various environmental factors on the DSM under different vegetation types is discussed. The objectives of this study are to (1) quantify the variation characteristics of DSM, (2) explore the mechanisms for controlling DSM variability among different vegetation types at the watershed scale, and (3) develop recommendations for land use management and the sustainability of vegetation recovery for the Loess Plateau.

2 Materials and methods

2.1 Study area

The Yanhe watershed lies in the middle of the Loess Plateau in northern Shaanxi Province. The Ansai watershed

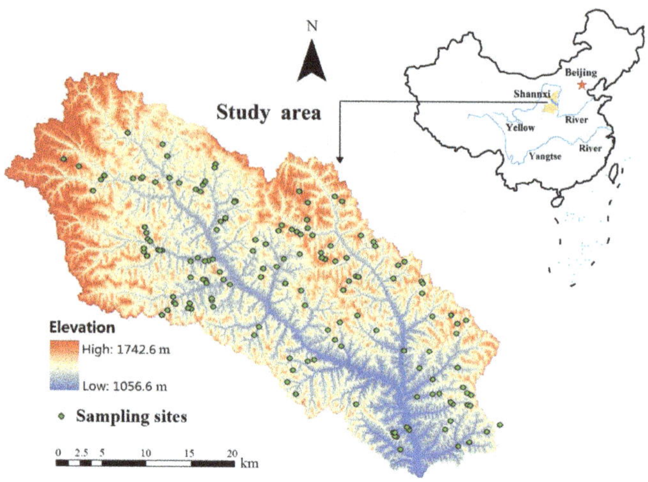

Figure 1. Location of the study area and sampling sites.

(108°47′–109°25′ E, 36°52′–37°19′ N; Fig. 1) is located in the upstream section of the Yanhe river and covers an area of approximately 1334 km^2. This watershed has a highly fragmented terrain, and the elevation here ranges from 1057 to 1743 m above sea level. This typical semi-arid loess hilly region has a mean annual temperature of 8.8 °C, and the average annual precipitation ranges from 375 to 546 mm across the watershed (Fig. S1 in the Supplement). Most rainfall occurs in the form of thunderstorms during the summer months from June to September. The soil types in this study area include loess soil with low fertility and vulnerability to soil erosion (Zhao et al., 2012). The soil texture is different across the watershed, with the sand content ranging from 24 to 57 %, the silt content ranging from 40 to 65 %, and the clay content ranging from 6 to 10 % (Fig. S2).

The predominant land use types in the Ansai watershed are rain-fed farmland, orchard land, sparse native grassland, pasture grassland, shrubland, and forest (Feng et al., 2013). The native vegetation in the study area consists of sparse grasses with shallow roots that are dominated by species such as bunge needlegrass, common leymus, and Altai heteropappus (*Heteropappus altaicus* (Willd.) Novopokr). Non-native species, such as alfalfa, black locust, David peach, sea buckthorn, and *Caragana korshinskii*, were predominantly used in the study area under the national Grain for Green project. The cultivated crops are predominantly maize, millet and broom corn millet. Because this region is located in a semi-arid climatic zone, water resources represent the major constraint of vegetation growth and agricultural crop production.

2.2 Sampling locations and description

Two vegetation groups with different root systems were selected: (1) shallow root vegetation, including native grasses (NGs) and farmland (FL) with human agricultural measures; and (2) deep root vegetation, including pasture grasses (PGs), sea buckthorn (SB), *Caragana korshinskii* (CK), David peach (DP), black locust (BL) and apple orchard (AO) with human agricultural measures. Descriptions of the root distribution of the selected vegetation types are provided in Table 1.

To fully explore the influencing factors of deep soil moisture, we examined topographic factors, surface soil properties, vegetation traits, and climate factors. These factors further included 23 independent variables: average annual rainfall (AAR), altitude (Al), slope position (SP), slope aspect (SA), slope gradient (SG), clay (Cl), silt (Sl), sand (Sa), organic matter (OM), capillary porosity (CP), soil bulk density (SBD), vegetation coverage (VC), grass biomass (GB), grass height (GH), planting density (PD), plant height (PH), diameter at breast height (DBH), crown width (CW), basal diameter (BD), litter maximum water holding (LMWH), litter biomass (LB), and clear bole height (CBH). The distance between each vegetation sampling site was at least 2 km. The sampling locations are shown in Fig. 1. The main characteristics and sampling numbers for each vegetation type are shown in Table 2.

2.3 Data collection and analysis

Soil samples were collected from depths of 0–500 cm using a soil drill (5 cm in diameter) with 20 cm increments within 28 days (from 10 July to 6 August 2014). The soil samples were sealed and taken to the laboratory, and the gravimetric soil moisture content was determined via oven drying at 105 °C to a constant weight. Three sampling profiles were randomly chosen to obtain the average soil moisture content for each sampling site. Native grasses and *Caragana korshinskii* were selected as representatives of shallow root vegetation and deep root vegetation, respectively. Soil moisture dynamic data (0–200 cm) of these vegetation types were monitored by EM50 (109°19′23 E, 36°51′26 N) from the same time period in 2015. The average annual rainfall (2006–2013) was provided by 29 rain gauges in or around the Ansai watershed, and the inverse distance weighted (IDW) interpolation method was performed in ArcGIS10.0 to obtain the average annual rainfall at each sampling site (Fig. S1).

The longitude, latitude and altitude were collected for each experimental site by using a Garmin GPS (version eTrex 30). The slope gradients and slope aspects were determined by using the compass method in field investigations; the slope gradients were transformed into tan (slope), and the slope aspects (clockwise from north) were transformed into cos (aspect). At each sampling site, six undisturbed soil cores were collected from the soil surface in metal cylinders (diameter of 5 cm, length of 5 cm) to measure the bulk density and capillary porosity (Wang et al., 2008a). The bulk density was determined from the volume–mass relationship for each core sample, and the capillary porosity was measured by the "cylinder soak method". Soil samples were also collected at each sampling site. The soil particle size distri-

Table 1. Descriptions of the root distributions of the selected vegetation species.

Vegetation type	Root distribution traits	Source
NG	The roots of native grasses are usually distributed at depths of 0–50 cm.	Han et al. (2009)
PG	The fibrous roots of pasture grasses are mainly distributed at depths of 0–50 cm, while taproot systems can extend to 3 m.	Wei et al. (2006); Wang et al. (2010d)
FL	The roots of farmland are mainly distributed at depths of 0–40 cm.	Feng et al. (2007)
AO	Ninety percent of apple tree roots are distributed at depths of 0–120 cm, while deep roots can reach 160 cm.	Lei (2013)
CK	The fibrous roots of *Caragana korshinskii* are mainly distributed at depths of 0–100 cm, while taproot systems can extend to 6.4 m.	Wang et al. (2010d)
SB	The fibrous roots of sea buckthorn are mainly distributed at soil depths of 0–160 cm, while deep roots can reach 200–300 cm.	Chong and Liang (1990)
DP	Ninety percent of David peach roots are distributed within depths of 100 cm, while deep roots can reach 150 cm.	Shi et al. (1989)
BL	Coarse roots are mainly distributed within depths of 260 cm, while fine roots can reach 350 cm.	Zhang and Xu (2011)

Table 2. Main characteristics and sampling numbers for different vegetation types.

Vegetation conditions	Shallow root vegetation			Deep root vegetation				
	NG*	FL	AO	PG	CK	SB	DP	BL
Sampling number	25	22	10	11	18	15	12	38
Altitude (m)	1392.6	1380.1	1370.1	1401.0	1350.6	1435.7	1377.6	1326.5
Slope aspect (°)	170.7	200.6	173.5	195.4	161.8	195.8	128.1	156.4
Slope gradient (°)	16.7	6.3	19.9	13.1	17.6	16.4	24.2	27.2
Sand (%)	44.9	39.4	38.2	55.3	46.4	46.2	52.7	40.0
Silt (%)	47.1	52.6	53.6	38.2	46.6	46.9	47.3	51.8
Clay (%)	8.1	7.9	8.2	6.5	7.0	7.0	7.4	8.3
Organic matter ($g\,kg^{-1}$)	7.0	5.3	5.8	6.3	13.3	8.9	6.0	8.1
Soil bulk density ($g\,cm^{-3}$)	1.3	1.3	1.3	1.3	1.3	1.2	1.3	1.2
Capillary porosity (%)	48	46	48	47	49	48	49	49
Mean canopy coverage (%)	57.4	53.3	39.7	67.8	45.6	66.1	33.8	59.6
Mean canopy height (m)	0.6	1.8	3.6	0.7	1.7	1.9	3.0	11.8
Mean tree DBH (cm)	–	–	6.3	–	–	–	5.0	10.4
Mean crown (cm)	–	–	398.4	–	199.7	184.9	293.4	455.3
Basal diameter (cm)	–	–	10.2	–	1.3	3.8	8.1	12.9
Planting density (m^2)	–	–	30.5	–	129.7	262.4	36.2	58.7

* NG, FL, AO, PG, CK, SB, DP and BL refer to native grasses, farmland, apple orchard, pasture grasses, *Caragana korshinskii*, sea buckthorn, David peach and black locust, respectively.

butions were measured by using a laser scattering particle size distribution analyzer (BT-9300H, Dandong, China). The proportions of clay (< 0.002 mm), silt (0.002–0.02 mm), and sand (> 0.02 mm) content were then calculated. The soil organic matter content was determined by using the dichromate oxidation method (Hu et al., 2010). A vegetation investigation was also conducted at each sampling site. The stand density (plants ha^{-1}), tree height (m), diameter at breast height (DHB, cm), basal diameter (cm), under branch height (m), canopy width in a 20 m × 20 m quadrat, and total canopy or coverage of each quadrat were recorded at the forest sites. The stand density (plants ha^{-1}), plant height (m), basal diameter (cm), and canopy width in a 10 m × 10 m quadrat were measured at the shrub sites. The species composition, total herbaceous coverage, grass height (m), litter and grass biomass were measured in each herbaceous quadrat. The canopy cover was measured by visual estimation, and the lit-

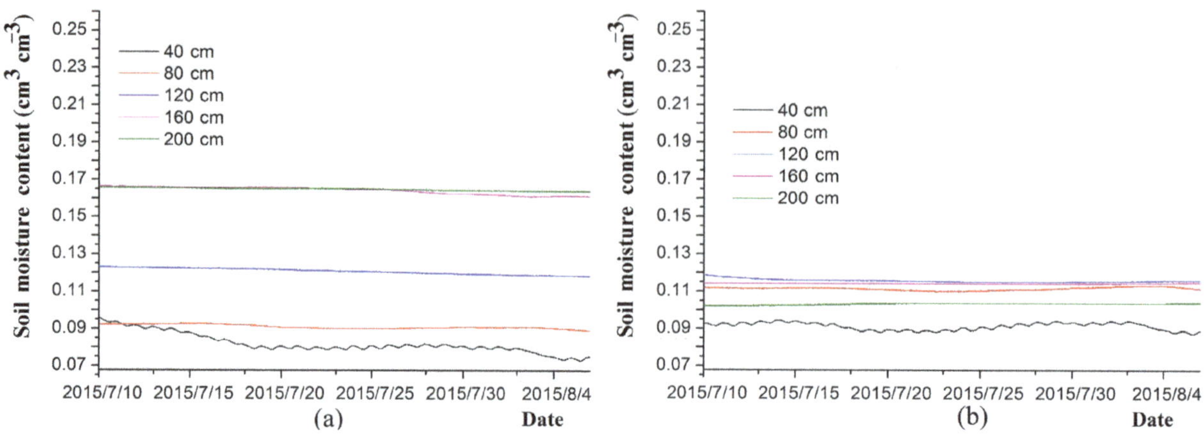

Figure 2. Soil moisture (0–200 cm) dynamics during the sampling period: **(a)** native grasses with a shallow root system and **(b)** *Caragana korshinskii* with a deep root system.

ter maximum water holdup was measured by using the immersion method.

2.4 Statistical analyses

The DSM from each layer was pooled together for the 151 sampling locations to conduct a descriptive analysis. Basic population statistics, such as minimum values (Min), maximum values (Max), mean values (Mean), standard deviations (SD), and coefficients of variation (CV), were reported for both the overall soil moisture datasets and for each vegetation type. The SD and CV values were employed to reflect the degree of variability of the DSM in different layers and for different vegetation types (Ruan and Li, 2002). One-way ANOVA was used to assess the contribution of different vegetation cover types to the overall variation in the DSM. Multiple comparisons were made by using the least significant difference (LSD) method. First, Spearman correlation analysis was used to examine the relationships between the DSM and environmental variables to determine the factors contributing to the DSM dynamics. Then, principle component analysis was performed to reduce the linear correlation that may exist among selected environmental variables and to further identify a minimum data set (MDS) of environmental variables for each vegetation type. All the statistical analyses were performed by using SPSS (Version 20.0).

3 Results

3.1 Deep soil moisture identification

The soil moisture dynamics at 0–200 cm during the sampling period are reported in Fig. 2. The soil moisture under both deep and shallow root vegetation types fluctuated daily at a depth of 40 cm, while the soil moisture at 80–200 cm remained constant with time. Thus, the evapotranspiration during the sampling period influences the soil moisture no deeper than 80 cm under both shallow and deep root vegetation types. According to both Figs. 2 and S3, the soil moisture at 40 cm did not change obviously with rainfall events, which indicates that rainfall during the monitoring period influenced the soil moisture no deeper than 40 cm.

According to Fig. S3, the mean air temperature and rainfall of the monitoring period in 2015 were similar to those observed during the sampling time period in 2014; thus, rainfall and evapotranspiration during the sampling time period influenced the soil moisture no deeper than 80 cm. In this study, we consider soil moisture at depths of 80–500 cm as deep soil moisture.

3.2 Summary statistics of deep soil moisture

The summary statistics of the DSM at various depths are listed in Table 3. In general, the mean soil moisture, SD, and CV highly depended on the depth. The profile distributions of the mean DSM, SD, and CV are listed in Table 3 and Fig. 3. The highest mean value (9.45 %) was observed at depths of 400–500 cm, the lowest (8.15 %) was observed at depths of 120–140 cm, and the mean DSM below 300 cm was almost constant. However, both the SD and CV showed varying trends with increasing depth (Fig. 3). The profile distributions of the SD and CV were consistent. The highest values of both parameters occurred at 100–120 and 480–500 cm (Table 3), which indicated that the DSM at these depth ranges had relatively higher variability. Meanwhile, the lowest values occurred at 260–300 cm, which indicated lower variability of DSM at these depth ranges. Most of the kurtosis (except for 80–120 cm) and skewness values were positive, and the highest values of both factors occurred at depths of 200–240 cm. The Kolmogorov–Smirnov test indicated that the soil moisture data sets were normally distributed. Thus, statistical analysis could be performed without data transformation.

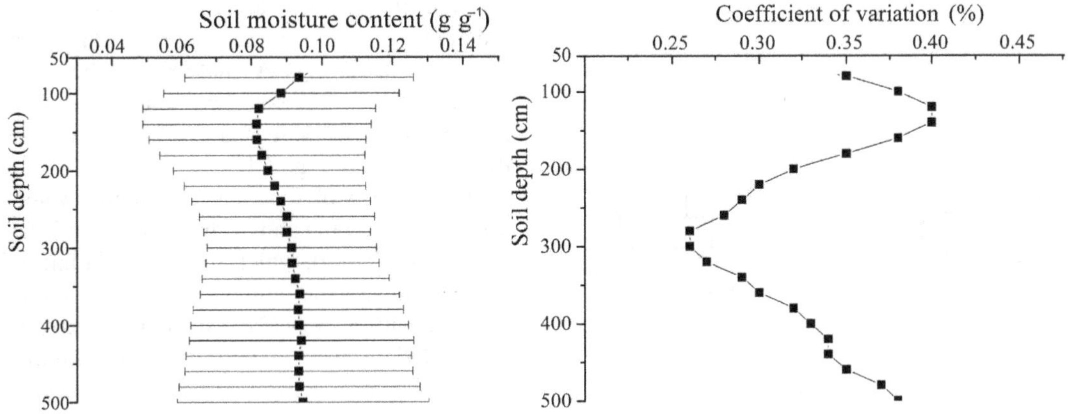

Figure 3. Profile distribution of the deep soil moisture contents and coefficients of variation. The error bars indicate the standard deviation.

Table 3. Summary statistics of the deep soil moisture at various depths in the Ansai watershed.

Depth (cm)	n[a]	Mean (%)	SD[b] (%)	Minimum (%)	Maximum (%)	CV[c]	K[d]	S	K–S
80–100	151	8.84	3.35	2.60	18.29	0.38	−0.45	0.28	N(0.95)
100–120	151	8.21	3.31	3.29	18.23	0.40	−0.27	0.57	N(1.36)
120–140	151	8.15	3.25	3.22	18.95	0.40	0.30	0.75	N(0.93)
140–160	151	8.16	3.09	3.37	18.56	0.38	0.40	0.80	N(0.99)
160–180	151	8.30	2.92	3.14	17.85	0.35	0.70	0.85	N(1.06)
180–200	151	8.47	2.70	3.22	17.89	0.32	1.48	1.01	N(1.13)
200–220	151	8.66	2.58	3.47	19.19	0.30	2.35	1.06	N(1.23)
220–240	151	8.83	2.54	3.59	19.72	0.29	2.99	1.05	N(1.02)
240–260	151	9.00	2.49	3.92	19.47	0.28	2.33	0.88	N(0.94)
260–280	151	9.00	2.37	4.08	18.46	0.26	1.94	0.74	N(1.11)
280–300	151	9.14	2.41	3.56	18.72	0.26	1.35	0.53	N(0.65)
300–320	151	9.15	2.46	3.26	18.08	0.27	1.45	0.54	N(0.73)
320–340	151	9.24	2.66	3.09	19.56	0.29	1.92	0.67	N(0.81)
340–360	151	9.36	2.83	2.98	19.38	0.30	1.31	0.59	N(0.91)
360–380	151	9.32	2.99	3.13	19.88	0.32	1.49	0.61	N(0.91)
380–400	151	9.35	3.09	2.81	20.85	0.33	1.99	0.60	N(1.00)
400–420	151	9.41	3.19	2.68	21.92	0.34	2.09	0.60	N(0.80)
420–440	151	9.33	3.21	2.70	20.97	0.34	1.43	0.55	N(0.57)
440–460	151	9.33	3.24	2.65	19.63	0.35	0.20	0.23	N(0.73)
460–480	151	9.35	3.43	2.67	19.88	0.37	−0.08	0.26	N(0.84)
480–500	151	9.45	3.58	2.43	19.98	0.38	−0.22	0.23	N(0.87)

[a] n refers to the number of sampling points. [b] SD refers to the standard deviation. [c] CV refers to the coefficient of variation. [d] K, S, and K–S refer to the kurtosis, skewness, and Kolmogorov–Smirnov test values, respectively. N refers to the normal distribution (significance level is 0.05, Kolmogorov–Smirnov value is in parentheses).

Moreover, different vegetation types greatly determined the variation in the DSM; the DSM statistics of various vegetation types under different vegetation types are reported in Fig. 4. The results showed that the depth-averaged DSM of native grasses and human-managed vegetation (farmland and apple orchard) were significantly higher than that of introduced deep root vegetation. In general, the mean DSM of different vegetation covers could be organized as follows: FL > NG > AO > DP > SB > PG > BL > CK. The highest DSM existed in farmland and the lowest in *Caragana korshinskii*. This result indicated that human agricultural management measures can significantly improve DSM conditions and that *Caragana korshinskii* showed the greatest water consumption among the selected introduced vegetation types.

3.3 Profile distribution of deep soil moisture by vegetation type

According to a previous study, the soil moisture profile characteristics in vegetation covering zones are usually complex

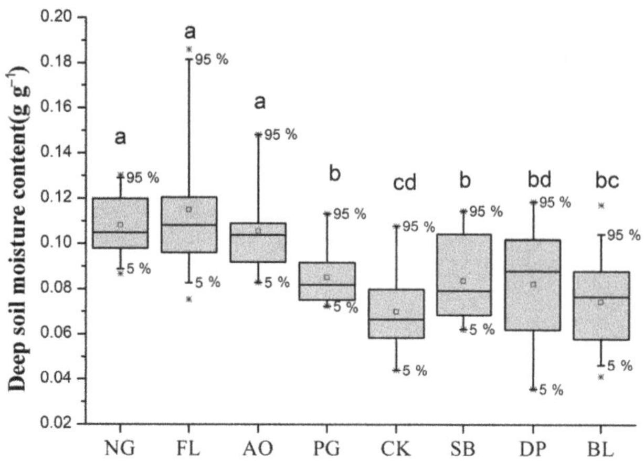

Figure 4. Deep soil moisture statistics for different vegetation types. Means with the same letter above the box are not significantly different at the 0.05 significance level (LSD test); NG, FL, AO, PG, CK, SB, DP and BL refer to native grasses, farmland, apple orchard, pasture grasses, *Caragana korshinskii*, sea buckthorn, David peach and black locust, respectively.

(Jia et al., 2013). Thus, soil moisture profiles for each vegetation type were chosen for analysis. As expected, the profile distribution characteristics of the DSM varied with the vegetation type (Fig. 5).

The DSM in native grassland zones is seldom affected by vegetation because of shallow root systems; thus, the DSM in native grasslands can be regarded as a reference (Yang et al., 2012c). The 80–500 cm soil moisture profile of native grasslands can be divided into three layers based on the inflection point of the DSM and the changes in the SD:

1. 80–220 cm; in this layer, both the DSM and SD decreased as the soil depth increased, which indicates that this layer may be a main rainfall infiltration layer. Furthermore, the level of rainfall infiltration decreased as the depth increased;

2. 220–400 cm; the DSM in this layer remained relatively constant as the soil depth increased, but its SD increased with soil depth, which indicates that this layer is unstable. We characterize this layer as a transition layer;

3. 400–500 cm; this section is a relatively stable layer whose SD is constant as the soil depth increases, despite the increasing DSM with soil depth. In this layer, the DSM is seldom influenced by rainfall infiltration. The profile distribution characteristics of farmland and apple orchards were similar to those of native grasslands, except for the 300–500 cm layer, possibly because management measures increased the ranges of the rainfall infiltration layer.

The DSM of all the introduced vegetation types reached a minimum at depths of 80–220 cm. The DSM of different introduced vegetation at 220–500 cm could be generally divided into three categories: (1) the DSM increased as the soil depth increased (e.g., PG and SB), (2) the DSM remained relatively stable as the soil depth increased (e.g., DP and BL), and (3) the DSM increased first before decreasing as the soil depth increased (e.g., CK).

We generally divided the DSM in this watershed into three layers based on the above analysis: (i) a rainfall transpiration layer (80–220 cm), which is a main rainfall infiltration layer and can be greatly influenced by vegetation transpiration; (ii) a transition layer (220–400 cm), which can be recharged by rainfall infiltration during heavy rainfall years and can supply deep root vegetation with DSM during drought years; and (iii) a stable layer (400–500 cm), which has a DSM that is seldom influenced by rainfall infiltration during regular years but can be influenced by extremely deep root vegetation such as CK and BL.

The profile variation of DSM under different vegetation types displayed different characteristics as well (Fig. S4). The vertical variation of native grassland was clearly less than that in human-managed farmland, orchard and introduced vegetation, and the variation was relatively stable as depth increased. In human-managed vegetation (farmland and orchard), the variation was relatively higher and had a complex profile distribution due to different management measures. However, the variation in introduced vegetation was, to some extent, consistent with the overall variation characteristics in this area (Fig. 3), which indicates that introduced vegetation plays an important role in the vertical variation of DSM in this area.

3.4 Comparison of deep soil moisture contents under different vegetation types

Generally, the DSM at comparable soil depths was lower in introduced deep root vegetation (pasture grassland, shrubland and forestland) compared to native grassland and human-managed vegetation (farmland and orchard). Farmland (11.07–11.79 %) had the highest DSM, followed by native grasses (10.52–11.19 %). The LSD test indicated that the DSM in native grasses and farmland was significantly higher than that in introduced vegetation ($P < 0.05$, Table 4) at almost every soil depth. The DSM varied from 6.81 to 10.4 % in pasture grassland, 6.85–9.75 % in sea buckthorn, 6.10–8.07 % in *Caragana korshinskii*, 7.19–7.66 % in black locust, and 7.71–8.51 % in David peach at depths of 80–500 cm. The LSD test indicated significant differences in the DSM at depths of 400–500 cm between different introduced vegetation types. For example, *Caragana korshinskii* was significantly different from pasture grassland, sea buckthorn, and David peach, while black locust was significantly different from pasture grassland and sea buckthorn ($P < 0.05$, Table 4).

As shown in Fig. 6, the DSM in farmland was higher than that in native grassland, and soil desiccation occurred in all the introduced vegetation. However, soil desiccation varied

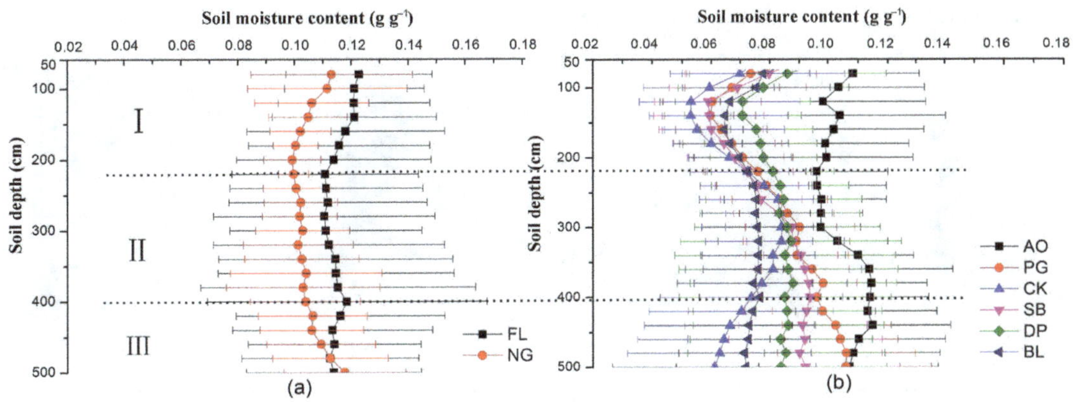

Figure 5. Profile distribution of DSM for different vegetation types: **(a)** shallow root system vegetation (NG – native grass; FL – farmland) and **(b)** deep root system vegetation (PG – pasture grasses; CK – *Caragana korshinskii*; SB – sea buckthorn; DP – David peach; BL – black locust; AO – apple orchard). The error bars indicate the standard deviation. I–III: DSM at different soil layer depth ranges (I: 80–220 cm, II: 220–400 cm, and III: 400–500 cm). The dashed lines are the boundaries of different soil layer depth ranges.

Table 4. Deep soil moisture of the 80–500 cm soil layers for different vegetation types.

Root	Vegetation	80–220 cm				220–400 cm				400–500 cm			
types	types	Min %	Max %	Mean %	SD %	Min %	Max %	Mean %	SD %	Min %	Max %	Mean %	SD %
Shallow root vegetation	NG*	8.4	13.8	10.9a	1.5	8.4	12.8	10.5ab	1.6	8.2	14.7	11.2ab	2.0
	FL	7.3	17.9	11.8a	2.8	7.8	18.6	11.1a	3.2	7.5	20.0	11.8a	3.6
Deep root vegetation	AO	6.9	15.4	10.1ab	2.7	7.7	14.1	10.5abc	1.7	7.4	15.3	11.4ab	2.3
	PG	4.8	8.9	6.8c	1.3	7.7	13.1	9.0bcd	1.6	8.5	14.3	10.4abc	1.9
	SB	4.9	11.3	6.9c	1.8	7.1	12.1	8.9cd	1.6	5.1	14.7	9.8bc	2.6
	CK	4.2	8.8	6.1c	1.4	4.9	11.6	8.1d	2.1	2.6	12.5	6.5e	2.9
	BL	3.8	12.8	7.2c	2.1	4.2	10.9	7.7d	1.8	4.0	13.3	7.5de	2.5
	DP	3.7	10.7	7.7bc	2.1	3.8	14.0	8.5d	3.2	3.2	13.1	8.5cd	3.2

* NG, FL, AO, PG, CK, SB, DP, and BL refer to native grasses, farmland, apple orchard, pasture grasses, *Caragana korshinskii*, sea buckthorn, David peach and black locust, respectively. Means with the same letter in the same column are not significantly different at the 0.05 significance level (LSD).

among the vegetation types. In general, the DSM in layer I (80–220 cm) was heavily consumed in almost all the introduced vegetation types. PG and SB consumed less DSM in layers II–III (220–500 cm) compared to the three other introduced vegetation types, while the DSM in layers II–III (220–500 cm) of DP and BL was consumed more consistently. Double-layer soil desiccation occurred in CK, which indicates that the DSM in layers I and III of CK was heavily consumed, while the DSM in layer II was less consumed. Despite the deep root system of the apple orchards, soil desiccation did not occur across the soil profile from 80 to 500 cm; even in the 320–450 cm layer, the DSM in the apple orchards was higher than that in native grasses.

3.5 Spearman correlation coefficients between deep soil moisture and selected environmental variables

Although the DSM data were normally distributed, significant correlations existed between the soil moisture contents at different soil depth ranges. Thus, non-parametric correlation tests (Spearman) were used to determine the strength of possible relationships between the DSM and selected variables. The correlation analysis results are presented in Tables S1, S2, and S3 in the Supplement. The correlation between the DSM and environmental variations changed with the soil depth and vegetation type. The DSM in native grassland showed significant correlations with the altitude (80–500 cm), slope gradient (220–500 cm), soil particle composition (80–500 cm), and average annual rainfall (220–400 cm). The DSM in farmland (80–220 cm) was only influenced by the bulk density. The DSM in areas of introduced vegetation, apart from the significant correlations with the topography, soil properties, and average annual rainfall, showed different correlations with vegetation growth traits. For instance, the DSM of BL showed significant negative correlations with the plant height (80–220 cm) and diameter at breast height (400–500 cm), and the DSM of SB showed a significant negative correlation with the plant density (80–500 cm). A significant correlation was found between the aspect and DSM in some introduced vegetation in PG (400–500 cm) and BL (80–400 cm). Moreover, positive correlations existed between the DSM and soil surface conditions; for instance, the DSM of DP showed significant correlations with the grass biomass (400–500 cm), the DSM of AO showed significant

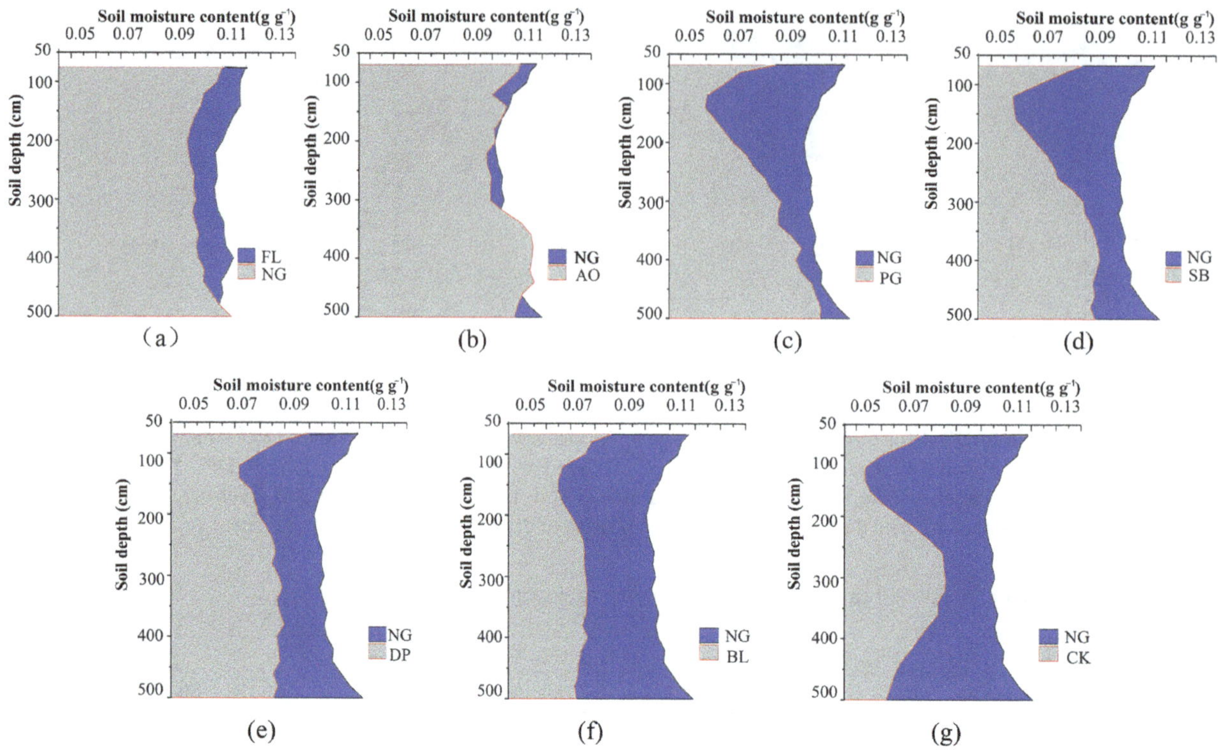

Figure 6. Comparison of deep soil moisture among human-managed vegetation, introduced vegetation and native grasslands: **(a)** farmland (FL) and native grasses (NG), **(b)** apple orchard (AO) and native grasses (NG), **(c)** pasture grasslands (PG) and native grasses (NG), **(d)** sea buckthorn (SB) and native grasses (NG), **(e)** David peach (DP) and native grasses (NG), **(f)** black locust (BL) and native grasses (NG), **(g)** *Caragana korshinskii* (CK) and native grasses (NG).

correlations with the litter biomass (400–500 cm), and the DSM of CK showed significant correlations with the litter maximum water holding (220–500 cm). Furthermore, both the soil bulk density (80–400 cm) and capillary porosity (80–220 cm) in the apple orchards showed significant correlations with the DSM.

3.6 Principal component analysis (PCA)

Only the environmental variables that showed significant correlations ($P < 0.05$) with the DSM were retained for Spearman correlation analysis, including 9 environmental variables for grassland and farmland (Group 1), 7 environmental variables for shrubland (Group 2), and 15 environmental variables for forestland and orchards (Group 3). Some of these variables were linearly correlated, so the dimensionality of these data sets could be reduced. Principal component analysis was performed following Hu et al. (2010) and Xu et al. (2008) to obtain an MDS of the environmental variables; the results are listed in Table 5. Only principal components (PCs) with eigenvalues $N > 1.0$ and only variables with highly weighted factor loading (i.e., those with absolute values for factor loading within 10 % of the highest value) were retained for the MDS (Mandal et al., 2008; Shi et al., 2014). For Group 1, the PCA identified four PCs that comprised 80 % of the variance, of which the first three PCs comprised most of this variance (68 %). For Group 2, four PCs comprised 84 % of the variance. For Group 3, five PCs comprised 75 % of the variance. In grassland and farmland, PC 1 included three variables that had highly weighted factor loadings, including clay, silt, and sand, which indicates that the soil particle composition was the most important factor that influenced variations in the DSM. For PC 2, PC 3, and PC 4, only one variable for each principal component had a high factor loading: the slope aspect, annual average rainfall, and soil bulk density, respectively. In shrubland, the highly weighted factor loadings of PC 1 were silt and sand, while the altitude and plant density were the highly weighted factor loadings for PC 2. For PC 3, only organic matter had a high factor loading. In forestland and orchard land, the diameter at breast height, basal diameter, and sand content comprised the highly weighted factor loadings of PC 1. The capillary porosity was the only variation that constituted the highly weighted factor loadings of PC 2. Four variations for PC 3 were highly weighted: clay, silt, soil bulk density, and litter maximum water holding. For PC 4 and PC 5, only one variable had a high factor loading: the slope aspect and slope gradient, respectively.

In total, 6 out of 9 environmental variables for grassland and farmland (Group 1), 5 out 7 for shrubs (Group 2), and

Table 5. Principle component analysis (PCA) of environmental attributes.

	Group 1: grassland and farmland				Group 2: shrubland			Group 3: orchard and forest				
Principal component	PC* 1	PC 2	PC 3	PC 4	PC 1	PC 2	PC 3	PC 1	PC 2	PC 3	PC 4	PC 5
Eigenvalue	3.6	1.5	1.1	1.1	2.2	1.9	1.0	4.5	2.6	1.8	1.2	1.1
% of variance	39.8	16.1	12.5	11.7	31.5	26.9	14.5	30.1	17.5	12.0	7.8	7.2
Cumulative %	39.8	55.8	68.3	80.0	31.5	58.9	72.9	30.1	47.6	59.5	67.3	74.5
Factor loading/eigenvector												
Annual average rainfall	0.21	−0.50	**0.71**	0.06	−0.39	−0.59	−0.39					
Altitude	−0.46	0.23	0.53	0.51	0.24	**0.83**	0.05					
Slope aspect	−0.16	**0.81**	0.24	−0.07				−0.06	−0.05	0.21	**0.64**	−0.10
Slope gradient	*0.64*	−0.43	−0.04	−0.23				0.55	0.22	0.07	0.01	**−0.70**
Clay	**0.86**	0.25	−0.13	0.15				*0.64*	−0.44	**−0.50**	−0.03	−0.04
Silt	**0.93**	0.27	0.05	0.09	**−0.94**	0.16	0.27	*0.68*	−0.44	**−0.45**	0.28	0.08
Sand	**−0.94**	−0.27	−0.02	−0.11	**0.95**	−0.16	−0.24	**−0.77**	0.39	0.37	−0.23	−0.17
Organic matter	−0.48	−0.03	−0.50	0.49	0.15	−0.53	**0.70**	0.54	−0.23	0.28	−0.16	−0.12
Soil bulk density	−0.41	0.31	0.07	**−0.67**				−0.25	*0.61*	**−0.49**	0.29	−0.05
Capillary porosity								0.38	**−0.74**	0.22	−0.37	−0.13
Litter biomass								0.54	0.43	−0.29	−0.04	0.20
Litter maximum water holding					0.42	−0.27	0.45	−0.24	−0.48	**0.46**	0.31	0.50
Grass biomass								0.27	0.49	−0.17	−0.37	0.39
Plant height								*0.69*	0.31	0.34	−0.22	0.24
Diameter at breast height								**0.80**	0.34	0.31	0.14	0.05
Crown width								0.39	0.20	0.43	0.22	−0.09
Basal diameter								**0.75**	0.37	0.26	0.18	0.07
Plant density					0.10	**0.77**	0.22					

* PC refers to principal components. Significant correlations ($P < 0.05$) are shown in italics, and very significant correlations ($P < 0.01$) are shown in bold. The factor loadings in bold are considered highly weighted when within 10 % of the variation of the absolute values of the highest factor loading in each PC.

10 out of 15 for forests and apple orchards (Group 3) were selected as MDS variables. Moreover, the MDS variables for each vegetation type were selected (Table 6). At the watershed scale, the main influencing factors of variations in the DSM under native grasslands were the soil particle composition (clay, silt, and clay content) and average annual rainfall. In farmland, the dominant influencing factors were the clay content and soil bulk density. The main influencing factors for introduced vegetation types were more complex; apart from the soil texture and physical characteristics, topographical factors and vegetation traits also strongly affected variations in the DSM. Moreover, the main influencing depth ranges of different environmental factors varied with vegetation types. For example, the soil particle size composition in native grasslands and apple orchard land mainly influenced the DSM at 80–220 cm, while the most significant influencing depths in pasture grassland were 220–400 cm. These results indicate that vegetation coverage or human management measures can alter the depths of the environmental factors that influence the DSM.

Table 6. Minimum data set of environmental variables.

Vegetation types	Influencing variables
Native grasses	Cl*, Sl, Sa, AAR
Farmland	Cl, SBD
Apple orchard	SG, Cl, Sl, Sa, SBD, CP
Pasture grasses	SA, Sl
Sea buckthorn	Al, Sl, Sa, PD
Caragana korshinskii	Al, Sl, OM
Black locust	SA, SG, DBH
David peach	Cl, Sa, BD, LMWH

* Cl, SA, SG, Sl, Sa, OM, CP, SBD, DBH, BD, and LMWH refer to clay, slope aspect, slope gradient, silt, sand, organic matter, capillary porosity, soil bulk density, diameter at breast height, basal diameter, and litter maximum water holding, respectively.

4 Discussion

4.1 Variation characteristics of deep soil moisture at the watershed scale

The variations in the DSM at the watershed scale varied with the soil depth (Figs. 3 and 5). At a soil depth of 80–220 cm, the influence of soil evaporation was relatively weak; rainfall infiltration could be stored in soil without strong consumption from vegetation, and rainfall infiltration decreased as the soil depth increased. Meanwhile, the soil layer at 220–400 cm was less influenced by rainfall infiltration; thus, the DSM remained constant as the soil depth increased. Soil depths below 400 cm comprised a deep stable DSM storage layer (Fig. 5). However, the existence of deep root vegetation and human agricultural management measures altered the vertical DSM distribution rules, resulting in more complex variations (Fig. 3). The highest variations in the DSM at this watershed occurred at 100–120 and 480–500 cm. However, the DSM was lowest in the 120–140 cm layer and had high variation. This result is inconsistent with previous studies, which reported that high variations usually appear with higher DSM and decrease when the DSM decreases (Ibrahim and Huggins, 2011). This result likely occurred because the most serious soil desiccation occurred in this layer for all the introduced vegetation types (Fig. 6), increasing their difference with native grasses and human management vegetation types and eventually resulting in high variation. While the high variation at 400–500 cm may have been caused by the different water consuming capacities of different vegetation types, this depth range is rarely influenced by rainfall event infiltration and soil evaporation (Chen et al., 2008b; Wang et al., 2009).

The DSM variation traits also varied with the vegetation types. The DSM variations in native grassland were relatively low and stable. Usually, the roots of native grasses are distributed at depths of 0–50 cm (Han et al., 2009). Thus, the DSM below this depth is seldom influenced by vegetation transpiration, and local controls, such as topographic factors, soil factors, and climate conditions, may contribute to variations in the DSM. The DSM and its variation in farmland were higher than those in native grassland, which indicates that human agricultural measures can greatly increase DSM and its variation. The DSM for introduced vegetation was significantly lower than that in native grassland, which indicates that soil desiccation occurred for all the introduced vegetation. Moreover, different introduced vegetation showed different soil desiccation traits (Fig. 6). This result is different from that of a previous study (Yang et al., 2012c), which reported that no significant differences existed among different introduced vegetation. This difference was probably caused by differences in the annual precipitation; the mean annual precipitation of Yang's study area was 386 mm, which is far less than that in our study area (505 mm). This lower annual precipitation resulted in plants not receiving enough water, eventually leading to more homogeneous soil desiccation among the different introduced vegetation. Moreover, the main difference in soil desiccation under introduced vegetation occurred at 80–220 and 400–500 cm (Table 4, Fig. 6), which contributed to the higher variations in DSM in these two layers (Fig. 4).

4.2 Mechanisms of deep soil moisture variability

Variations in DSM are the combined result of topographic factors, soil factors, vegetation factors, and climate conditions. In this study, the vegetation type was an important factor that influenced variations in the DSM. The effect of vegetation on the DSM is shown in many aspects. First, different root systems determine different DSM consumption traits for various vegetation types (Fig. 6). Introduced vegetation with a deep root system consumes more and deeper DSM than farmland and native grasses (Table 4). Individual vegetation growth conditions and planting density can also influence variations in the DSM. The individual growth conditions in the deeper root systems of forests mainly explain the DSM consumption, while the planting density mainly accounts for the DSM consumption in the shallower root systems of shrubs (Tables S2 and S3). In addition to DSM consumption, canopy interception and surface coverage systems can positively affect the DSM (Martínez-Fernández and Ceballos, 2003; Starks et al., 2006). In this study, the litter biomass, water holding capacity, and forest grasses all showed different degrees of significant positive correlations with the DSM for different vegetation types (Tables S1, S2, and S3), probably because thick litter, humus layers and forest grasses can reduce surface runoff, which may help retain more rainfall for infiltration into deep soil layers. Additionally, these factors can reduce soil evaporation, which may decrease DSM consumption (Vivoni et al., 2008).

Climate factors that affect DSM are mainly determined by differences in rainfall infiltration and solar radiation (Savva et al., 2013). According to previous studies, DSM is relatively stable compared to the shallow layer, especially at depths below 200 cm, and soil moisture in deeper layers is seldom influenced by rainfall events (Wang et al., 2009). However, the DSM in this study (80–220 cm in *Caragana korshinskii* and 220–400 cm in native grasslands) showed significant positive correlations with the 6-year average annual rainfall, which indicates that DSM may be a long-term result of a water budget surplus.

Topography is another important factor that greatly affects the redistribution and consumption of DSM (Qiu et al., 2001; Zhu et al., 2014b). The slope position, altitude, and slope gradient mainly affect the lateral flow of soil moisture. Lower positions or latitudes usually have higher soil moisture contents (He et al., 2003; Zhu et al., 2014a), while the slope gradient usually shows a negative correlation with the soil moisture content; thus, a steep slope usually has a lower soil moisture content than a gentle slope (Kim et al., 2007). Different slope aspects are usually caused by changes in solar radiation (Yang et al., 2012b; Zhao et al., 2007; Wang et al., 2008a), resulting in different rates of soil moisture evaporation. In this study, the altitude had a negative correlation with the DSM; the slope gradient showed significant positive correlations with the DSM in grasslands (220–500 cm), while significant negative correlations were found in black locust (400–500 cm). These results indicate that introduced vegetation can alter the topographic factors' influence on DSM variations. This result was also verified by Yang et al. (2012b), who found that introduced vegetation can lead to homogeneity in the DSM. This relationship was true for the slope aspect, which only showed a positive correlation with the DSM in pasture grasses (400–500 cm) and black locust (80–400 cm).

Moreover, different soil traits determine different water transmission and conservancy characteristics, which may greatly influence the flow or storage of water in soil (Western et al., 2004). The DSM in the Loess Plateau is mainly determined by land surface rainfall infiltration and evapotranspiration. Surface soil properties usually have a greater influence on surface rainfall infiltration and evaporation compared to deep soil properties; thus, we mainly analyzed the influence of surface soil properties on DSM. The result indicated that the soil particle composition was an important factor influencing variations in DSM at the watershed scale. Both the clay and silt content showed significantly positive correlations with the DSM, and the sand content showed a negative correlation with the DSM for most vegetation types. However, the soil bulk density and capillary porosity only showed significant correlations with the DSM in farmland (80–220 cm) and apple orchards (80–400 cm). This result indicates that human agricultural management measures or other factors that result in lower soil bulk density and higher capillary porosity conditions can significantly improve the infiltration capacity and thus increase the DSM.

4.3 Implications for land use management and vegetation recovery

A balance between soil water availability and water utilization by plants is crucial to maintaining ecosystem health, particularly in the arid and semi-arid regions of the Loess Plateau. The implementation of the "Grain for Green" program has effectively controlled soil erosion (Chen et al., 2010; Wang et al., 2015). However, according to this study, soil desiccation occurred in almost all the introduced vegetation, while higher soil moisture contents were found in native grassland and farmland (Fig. 6). These phenomena indicate that the improper selection of vegetation type is a dominant reason for soil desiccation in this area. Thus, more attention should be paid to the selection of vegetation types based on the interactions between soil moisture and vegetation. Among these selected vegetation types, CK and BL caused the most serious soil desiccation (Figs. 4 and 6); thus, these two types are especially unsuitable for large-scale plants in the study area, while SB, PG, and DP can be properly planted under good soil moisture conditions with suitable planting density and human management measures.

Furthermore, proper planting locations should also be considered based on DSM conditions. Annual average rainfall spatial variations can significantly influence DSM conditions (Tables S1 and S2). Thus, the annual average rainfall is another important factor for determining planting locations. Vegetation enclosure and natural restoration may be good choices in lower rainfall zones, while shrubs and forests could be rationally arranged in higher rainfall zones. DSM was not evenly distributed even within the same rainfall regions: lower altitudes (such as a gully bottom or lower slope) usually had higher DSM (Tables S1 and S2), while the DSM of native grasslands along steeper slopes was higher than that along gentle slopes (Table S1). Thus, shrubs or trees with high water consumption capacities can be arranged at these locations. Native grass and low-moisture-consuming shrubs can be arranged at higher altitudes or upper slopes, where the DSM is lower.

The results of this study also indicate that human agricultural management measures can effectively improve DSM conditions. The DSM of farmland was the highest among the selected vegetation types (Fig. 4); even though introduced vegetation has deep root systems, no soil desiccation was found in apple orchards (Fig. 6). Most of the farmlands that we surveyed were level terraces and back-slope level benches with cultivation practices, while apple orchards were equipped with artificial rainwater gathering measures. All these agricultural measures can significantly increase rainwater infiltration, eventually resulting in higher DSM in these vegetation zones. Moreover, forest grasses, litter biomass, and litter maximum water holding showed significant correlations with the DSM in this study (Tables S2 and S3). Thus, increasing the land surface cover (such as crop straw coverage and mix-sowing shrubs and grass) can be another effective measure to improve DSM recharge. Likewise, vegetation control (when artificial forests and shrubs are mature, the density should be reduced according to deep soil water conditions) may be an effective measure to reduce soil desiccation because the plant density has a significantly negative correlation with the DSM.

5 Conclusions

At watershed scale, the variations in the DSM changed with the soil depth and vegetation types. Higher variations in DSM occurred at two depth ranges, namely, 120–140 and 480–500 cm, while variations in native grasses in the horizontal direction were far lower than those of farmland, apple orchards, and introduced vegetation at comparable depths. Soil desiccation occurred for almost all the vegetation types; among them, CK and BL were the most serious. Moreover, 80–220 cm soil depth had the most serious desiccation layer. The high DSM in farmland and apple orchards indicated that human management measures could greatly improve DSM. Although the vegetation type was a dominant factor, the variations in the DSM in this area were actually the combined result of the climate, vegetation, topography, soil, and human management measures. The DSM in native grassland was found to be significantly related to the topography, soil traits and annual average rainfall. The plant growth conditions, planting density, and litter water holding capacity showed significant relationships with the DSM for introduced vegetation. Human management measures greatly increased the influence of soil traits on the DSM in farmland and orchards, which increased rainfall infiltration and improved the DSM. Proper selection of the vegetation type, planting locations, and landscape management measures can be suggested based on the results of this study. The results of this study are of practical significance for vegetation restoration strategies and the sustainability of restored ecosystems.

6 Data availability

All the data used in this study are available on request.

Acknowledgements. This work was supported by the National Natural Science Foundation of China (no. 41390462) and the Program for Changjiang Scholars and Innovative Research Team in University (no. IRT_15R06). Lixin Wang acknowledges support from USDA grant 2014-51130-22492. Wenwu Zhao acknowledges the support from China Scholarship Council (no. 201406045031). We are also grateful to the Ansai Research Station of Soil and Water Conservation, Chinese Academy of Sciences, for their support and contributions to this fieldwork. We are thankful for reviews from the editor, N. Ursino, and five anonymous reviewers, which significantly improved the quality of this article.

Edited by: N. Ursino

References

Baroni, G., Ortuani, B., Facchi, A., and Gandolfi, C.: The role of vegetation and soil properties on the spatio-temporal variability of the surface soil moisture in a maize-cropped field, J. Hydrol., 489, 148–159, 2013.

Bi, H., Li, X., Liu, X., Guo, M., and Li, J.: A case study of spatial heterogeneity of soil moisture in the Loess Plateau, western China: A geostatistical approach, Int. J. Sediment Res., 24, 63–73, 2009.

Cheema, M. J. M., Bastiaanssen, W. G. M., and Rutten, M. M.: Validation of surface soil moisture from AMSR-E using auxiliary spatial data in the transboundary Indus basin, J. Hydrol., 405, 137–149, 2011.

Chen, H., Shao, M., and Li, Y.: Soil desiccation in the Loess Plateau of China, Geoderma, 143, 91–100, 2008a.

Chen, H., Shao, M., and Li, Y.: The characteristics of soil water cycle and water balance on steep grassland under natural and simulated rainfall conditions in the Loess Plateau of China, J. Hydrol., 360, 242–251, 2008b.

Chen, L., Wang, J., Wei, W., Fu, B., and Wu, D.: Effects of landscape restoration on soil water storage and water use in the Loess Plateau region, China, Forest Ecol. Manag., 259, 1291–1298, 2010.

Chong, X. and Liang, Y.: Root distribution characteristics and soil moisture dynamics of Hippophae rhamnoides in semiarid region of Loess Plateau, B. Soil Water Conserv., 98–103, 1990 (in Chinese).

Entin, J. K., Robock, A., Vinnikov, K. Y., Hollinger, S. E., Liu, S., and Namkhai, A.: Temporal and spatial scales of observed soil moisture variations in the extratropics, J. Geophys. Res.-Atmos., 105, 11865–11877, 2000.

Fang, X., Zhao, W., Fu, B., and Ding, J.: Landscape service capability, landscape service flow and landscape service demand: a new framework for landscape services and its use for landscape sustainability assessment, Prog. Phys. Geogr., 39, 817–836, 2015.

Feng, D., Zongsuo, L., Xuexuan, X., Lun, S., and Xingchang, Z.: Community biomass of abandoned farmland and its effects on soil nutrition in the Loess hilly region of Northern Shaanxi, China, Acta Ecol. Sin., 27, 1673–1683, 2007 (in Chinese).

Feng, Q., Zhao, W., Qiu, Y., Zhao, M., and Zhong, L.: Spatial heterogeneity of Soil moisture and the scale variability of its influencing factors: a Case Study in the Loess Plateau of China, Water, 5, 1226–1242, 2013.

Gómez-Plaza, A., Alvarez-Rogel, J., Albaladejo, J., and Castillo, V. M.: Spatial patterns and temporal stability of soil moisture across a range of scales in a semi-arid environment, Hydrol. Process., 14, 1261–1277, 2000.

Gómez-Plaza, A., Martınez-Mena, M., Albaladejo, J., and Castillo, V. M.: Factors regulating spatial distribution of soil water content in small semiarid catchments, J. Hydrol., 253, 211–226, 2001.

Han, F., Zheng, J., and Zhang, X.: Plant root system distribution and its effect on soil nutrient on slope land converted from farmland in the Loess Plateau, Transactions of the Chinese Society of Agricultural Engineering, 25, 50–55, 2009.

He, F., Huang, M., and Dang, T.: Distribution characteristic of dried soil layer in Wangdonggou watershed in gully region of the Loess Plateau, J. Nat. Resour., 18, 30–36, 2003.

Hu, W., Shao, M., Han, F., Reichardt, K., and Tan, J.: Watershed scale temporal stability of soil water content, Geoderma, 158, 181–198, 2010.

Ibrahim, H. M. and Huggins, D. R.: Spatio-temporal patterns of soil water storage under dryland agriculture at the watershed scale, J. Hydrol., 404, 186–197, 2011.

Jia, Y. and Shao, M.: Dynamics of deep soil moisture in response to vegetational restoration on the Loess Plateau of China, J. Hydrol., 519, 523–531, 2014.

Jia, Y., Shao, M., and Jia, X.: Spatial pattern of soil moisture and its temporal stability within profiles on a loessial slope in northwestern China, J. Hydrol., 495, 150–161, 2013.

Kim, S., Lee, H., Woo, N. C., and Kim, J.: Soil moisture monitoring on a steep hillside, Hydrol. Process., 21, 2910–2922, 2007.

Legates, D. R., Mahmood, R., Levia, D. F., DeLiberty, T. L., Quiring, S. M., Houser, C., and Nelson, F. E.: Soil moisture: A central and unifying theme in physical geography, Prog. Phys. Geogr., 35, 65–86, 2011.

Lei, S.: Root distribution traits of apple trees in platform terrain, Journal Vocational College, 97–99, 2013 (in Chinese).

Liu, W., Zhang, X., Dang, T., Ouyang, Z., Li, Z., Wang, J., Wang, R., and Gao, C.: Soil water dynamics and deep soil recharge in a record wet year in the southern Loess Plateau of China, Agric. Water Manag., 97, 1133–1138, 2010.

Lu, Y., Fu, B., Chen, L., Liu, G., and Wei, W.: Nutrient transport associated with water erosion: progress and prospect, Prog. Phys. Geogr., 31, 607–620, 2007.

Lu, Y., Chen, L., and Fu, B.: Land-cover effects on red soil rehabilitation in China: a meta-analysis, Prog. Phys. Geogr., 32, 491–502, 2008.

Mandal, U. K., Warrington, D. N., Bhardwaj, A. K., Bar-Tal, A., Kautsky, L., Minz, D., and Levy, G. J.: Evaluating impact of irrigation water quality on a calcareous clay soil using principal component analysis, Geoderma, 144, 189–197, 2008.

Martínez-Fernández, J. and Ceballos, A.: Temporal stability of soil moisture in a large-field experiment in Spain, Soil Sci. Soc. Am. J., 67, 1647–1656, 2003.

Montenegro, S. and Ragab, R.: Impact of possible climate and land use changes in the semi arid regions: a case study from North Eastern Brazil, J. Hydrol., 434–435, 55–68, 2012.

Qiu, Y., Fu, B., Wang, J., and Chen, L.: Soil moisture variation in relation to topography and land use in a hillslope catchment of the Loess Plateau, China, J. Hydrol., 240, 243–263, 2001.

Ruan, C. and Li, D.: Community characteristics of Hippophae rhamnoides forest and water and nutrient condition of the woodland in Loess hilly region, Ying Yong Sheng Tai Xue Bao, 13, 1061–1064, 2002 (In Chinese).

Savva, Y., Szlavecz, K., Carlson, D., Gupchup, J., Szalay, A., and Terzis, A.: Spatial patterns of soil moisture under forest and grass land cover in a suburban area, in Maryland, USA, Geoderma, 192, 202–210, 2013.

Shi, Y., Liu, Y., and Zhang, J.: Study on the root distribution of peach trees, Journal Fruit Science, 4, 232–235, 1989 (in Chinese).

Shi, Y., Wu, P., Zhao, X., Li, H., Wang, J., and Zhang, B.: Statistical analyses and controls of root-zone soil moisture in a large gully of the Loess Plateau, Environ. Earth Sci., 71, 4801–4809, 2014.

Starks, P. J., Heathman, G. C., Jackson, T. J., and Cosh, M. H.: Temporal stability of soil moisture profile, J. Hydrol., 324, 400–411, 2006.

Sun, F., Lü, Y., Fu, B., Ma, Z., and Yao, X.: Spatial explicit soil moisture analysis: pattern and its stability at small catchment scale in the loess hilly region of China, Hydrol. Process., 28, 4091–4109, 2014.

Vereecken, H., Kamai, T., Harter, T., Kasteel, R., Hopmans, J., and Vanderborght, J.: Explaining soil moisture variability as a function of mean soil moisture: a stochastic unsaturated flow perspective, Geophys. Res. Lett., 34, 315–324, 2007.

Vivoni, E. R., Rinehart, A. J., Méndez-Barroso, L. A., Aragón, C. A., Bisht, G., Cardenas, M. B., Engle, E., Forman, B. A., Frisbee, M. D., Gutiérrez-Jurado, H. A., Hong, S., Mahmood, T. H., Tai, K., and Wyckoff, R. L.: Vegetation controls on soil moisture distribution in the Valles Caldera, New Mexico, during the North American monsoon, Ecohydrology, 1, 225–238, 2008.

Wang, L., Wang, Q., Wei, S., Shao, M. A., and Li, Y.: Soil desiccation for Loess soils on natural and regrown areas, Forest Ecol. Manag., 255, 2467–2477, 2008a.

Wang, Y., Shao, M., and Zhang, X.: Soil moisture ecological environment of artificial vegetation on steep slope of loess region in North Shaanxi Province, China, Acta Ecol. Sin., 28, 885–894, 2008b (in Chinese).

Wang, Z., Liu, B., and Zhang, Y.: Soil moisture of different vegetation types on the Loess Plateau, J. Geogr. Sci., 19, 707–718, 2009.

Wang, Y., Shao, M., and Liu, Z.: Large-scale spatial variability of dried soil layers and related factors across the entire Loess Plateau of China, Geoderma, 159, 99–108, 2010a.

Wang, Y., Shao, M., and Liu, Z.: Large-scale spatial variability of dried soil layers and related factors across the entire Loess Plateau of China, Geoderma, 159, 99–108, 2010b.

Wang, Y., Shao, M., and Shao, H.: A preliminary investigation of the dynamic characteristics of dried soil layers on the Loess Plateau of China, J. Hydrol., 381, 9–17, 2010c.

Wang, Y., Shao, M., and Shao, H.: A preliminary investigation of the dynamic characteristics of dried soil layers on the Loess Plateau of China, J. Hydrol., 381, 9–17, 2010d.

Wang, X. C., Muhammad, T. N., Hao, M. D., and Li, J.: Sustainable recovery of soil desiccation in semi-humid region on the Loess Plateau, Agric. Water Manag., 98, 1262–1270, 2011a.

Wang, Y., Shao, M., Zhu, Y., and Liu, Z.: Impacts of land use and plant characteristics on dried soil layers in different climatic regions on the Loess Plateau of China, Agr. Forest Meteorol., 151, 437–448, 2011b.

Wang, L., D'Odorico, P., Evans, J. P., Eldridge, D. J., McCabe, M. F., Caylor, K. K., and King, E. G.: Dryland ecohydrology and climate change: critical issues and technical advances, Hydrol. Earth Syst. Sci., 16, 2585–2603, doi:10.5194/hess-16-2585-2012, 2012a.

Wang, Y., Shao, M., Liu, Z., and Warrington, D. N.: Regional spatial pattern of deep soil water content and its influencing factors, Hydrol. Sci. J., 57, 265–281, 2012b.

Wang, Y., Shao, M., Liu, Z., and Warrington, D. N.: Regional spatial pattern of deep soil water content and its influencing factors, Hydrol. Sci. J., 57, 265–281, 2012c.

Wang, Y., Shao, M., and Liu, Z.: Vertical distribution and influencing factors of soil water content within 21-m profile on the Chinese Loess Plateau, Geoderma, 193–194, 300–310, 2013.

Wang, S., Fu, B., Piao, S., Lü, Y., Ciais, P., Feng, X., and Wang, Y.: Reduced sediment transport in the Yellow River due to anthropogenic changes, Nat. Geosci., 9, 38–41, 2015.

Wei, S., Ren, S., Yang, P., and Yan, M.: Soil moisture dynamic and root growth of alfalfa in Weichang region, Chinese Agricultural Science Bulletin, 5, 448–451, 2006 (in Chinese).

Western, A. W., Zhou, S., Grayson, R. B., McMahon, T. A., Blöschl, G., and Wilson, D. J.: Spatial correlation of soil moisture in small catchments and its relationship to dominant spatial hydrological processes, J. Hydrol., 286, 113–134, 2004.

Xu, X., Ma, K., Fu, B., Song, C., and Liu, W.: Relationships between vegetation and soil and topography in a dry warm river valley, SW China, Catena, 75, 138–145, 2008.

Yang, L., Wei, W., Chen, L., Jia, F., and Mo, B.: Spatial variations of shallow and deep soil moisture in the semi-arid Loess Plateau, China, Hydrol. Earth Syst. Sci., 16, 3199–3217, doi:10.5194/hess-16-3199-2012, 2012a.

Yang, L., Wei, W., Chen, L., and Mo, B.: Response of deep soil moisture to land use and afforestation in the semi-arid Loess Plateau, China, J. Hydrol., 475, 111–122, 2012b.

Yang, L., Wei, W., Chen, L., and Mo, B.: Response of deep soil moisture to land use and afforestation in the semi-arid Loess Plateau, China, J. Hydrol., 475, 111–122, 2012c.

Yang, L., Chen, L., Wei, W., Yu, Y., and Zhang, H.: Comparison of deep soil moisture in two re-vegetation watersheds in semi-arid regions, J. Hydrol., 513, 314–321, 2014a.

Yang, L., Wei, W., Chen, L., Chen, W., and Wang, J.: Response of temporal variation of soil moisture to vegetation restoration in semi-arid Loess Plateau, China, Catena, 115, 123–133, 2014b.

Zhang, L. and Xu, X.: The distribution characteristics of Robinia pseudoacacia root in Yangou watershed in Yanan, Journal Northwest, Forestry College, 26, 9–14, 2011 (in Chinese).

Zhao, J., Du, J., and Chen, B.: Dried earth layers of artificial forestland in the Loess Plateau of Shaanxi Province, J. Geogr. Sci., 17, 114–126, 2007.

Zhao, W. and Fang, X.: Landscape sustainability and landscape sustainability science, Acta Ecol. Sin., 34, 2453–2459, 2014 (in Chinese).

Zhao, W., Fu, B., and Chen, L.: A comparison between soil loss evaluation index and the C-factor of RUSLE: a case study in the Loess Plateau of China, hydrology and Earth Systerm, Science, 16, 2739–2748, 2012.

Zhao, W., Fu, B., and Qiu, Y.: An upscaling method for cover-management factor and its application in the Loess Plateau of China, Int. J. Environ. Res. Public Health, 10, 4752–4766, 2013.

Zhu, Q. and Lin, H.: Influences of soil, terrain, and crop growth on soil moisture variation from transect to farm scales, Geoderma, 163, 45–54, 2011.

Zhu, H. D., Shi, Z. H., Fang, N. F., Wu, G. L., Guo, Z. L., and Zhang, Y.: Soil moisture response to environmental factors following precipitation events in a small catchment, Catena, 120, 73–80, 2014a.

Zhu, Q., Nie, X., Zhou, X., Liao, K., and Li, H.: Soil moisture response to rainfall at different topographic positions along a mixed land-use hillslope, Catena, 119, 61–70, 2014b.

5

Environmental controls on seasonal ecosystem evapotranspiration/potential evapotranspiration ratio as determined by the global eddy flux measurements

Chunwei Liu[1], Ge Sun[2], Steven G. McNulty[2], Asko Noormets[3], and Yuan Fang[3]

[1]Jiangsu Provincial Key Laboratory of Agricultural Meteorology, College of Applied Meteorology, Nanjing University of Information Science and Technology, Nanjing 210044, China
[2]Eastern Forest Environmental Threat Assessment Center, Southern Research Station, USDA Forest Service, Raleigh, NC 27606, USA
[3]Department of Forestry and Environmental Resources, North Carolina State University, Raleigh, NC 27695, USA

Correspondence to: Ge Sun (gesun@fs.fed.us)

Abstract. The evapotranspiration / potential evapotranspiration (AET / PET) ratio is traditionally termed as the crop coefficient (K_c) and has been generally used as ecosystem evaporative stress index. In the current hydrology literature, K_c has been widely used as a parameter to estimate crop water demand by water managers but has not been well examined for other types of ecosystems such as forests and other perennial vegetation. Understanding the seasonal dynamics of this variable for all ecosystems is important for projecting the ecohydrological responses to climate change and accurately quantifying water use at watershed to global scales. This study aimed at deriving monthly K_c for multiple vegetation cover types and understanding its environmental controls by analyzing the accumulated global eddy flux (FLUXNET) data. We examined monthly K_c data for seven vegetation covers, including open shrubland (OS), cropland (CRO), grassland (GRA), deciduous broad leaf forest (DBF), evergreen needle leaf forest (ENF), evergreen broad leaf forest (EBF), and mixed forest (MF), across 81 sites. We found that, except for evergreen forests (EBF and ENF), K_c values had large seasonal variation across all land covers. The spatial variability of K_c was well explained by latitude, suggesting site factors are a major control on K_c. Seasonally, K_c increased significantly with precipitation in the summer months, except in EBF. Moreover, leaf area index (LAI) significantly influenced monthly K_c in all land covers, except in EBF. During the peak growing season, forests had the highest K_c values, while croplands (CRO) had the lowest. We developed a series of multivariate linear monthly regression models for K_c by land cover type and season using LAI, site latitude, and monthly precipitation as independent variables. The K_c models are useful for understanding water stress in different ecosystems under climate change and variability as well as for estimating seasonal ET for large areas with mixed land covers.

1 Introduction

Evapotranspiration (ET) is one of the major hydrological processes that link energy, water, and carbon cycles in terrestrial ecosystems (Sun et al., 2010, 2011a, b; Fang et al., 2015). In contrast to potential ET (PET), which depends only on atmospheric water demand (Lu et al., 2005), actual evapotranspiration (AET) is arguably the most uncertain ecohydrologic variable for quantifying watershed water budgets (Baldocchi and Ryu, 2011; Fang et al., 2015; Hao et al., 2015a) and for understanding the ecological impacts of climate and land use change (Budyko, 1974; Hao et al., 2015b), and climate variability (Hao et al., 2014). In recent years, one of the most important research questions of ecohydrology focused on how ecosystem dynamics, precipitation, AET, and PET interact in different ecosystems at seasonal and long-term scales under a changing environment (Vose et al., 2011).

The ratio of AET to PET is traditionally termed as crop coefficient (K_c), and has been widely used to as a parameter to estimate crop water demand by water managers (Allen and Pereira, 2009; Irmak et al., 2013a). However, this parameter has not been well examined for other ecosystems (Zhou et al., 2010; Zhang et al., 2012). The ratio of AET to PET has also been used as an indicator of regional terrestrial water availability, wetness or drought index, and plant water stress (Anderson et al., 2012; Mu et al., 2012). When the annual AET / PET ratio is close to 1.0, the soil water meets ecosystem water use demand. The ratio of AET / PET or water stress level can be drastically different among different ecosystems in different environmental conditions, because AET is mainly controlled by climate (precipitation and PET) (Zhang et al., 2001; Jaramillo et al., 2013) and ecosystem species composition and structure (i.e., leaf area index, rooting depth) (Sun et al., 2011a; Hasper et al., 2016). The same seasonal PET values for a particular region are generally stable among different years (Lu et al., 2005; Rao et al., 2011), and deviation of AET / PET from the norm indicates variability in AET, which responds to precipitation and water availability when PET is stable (Rao et al., 2011). However, under a changing climate, the monthly AET / PET patterns can be rather complex since both AET and PET are affected by air temperature and precipitation (Sun et al., 2015a, b) and corresponding changes in ecosystem characteristics (e.g., plant species shift) (Donohue et al., 2007; Vose et al., 2011; Sun et al., 2014).

In the agricultural water management community, the crop coefficient method remains a popular one for approximating crop water use, despite recent advances in direct ET measurement methods (Allen et al., 1998; Baldocchi et al., 2001; Allen and Pereira, 2009; Fang et al., 2015). The K_c is termed a single-crop coefficient (Allen et al., 2006; Tabari et al., 2013) which is affected by growing periods, crop species, canopy conductance, and soil evaporation in the field scale (Shukla et al., 2014b; Ding et al., 2015). Moreover, K_c can be influenced by soil characteristics, vegetative soil cover, height, plant species distribution, and leaf area index in a larger spatial scale (Descheemaeker et al., 2011; Anda et al., 2014; Consoli and Vanella, 2014). Although the Food and Agriculture Organization of the United Nations (FAO) provides various guidelines for several crops (Allen et al., 1998), local measurements are still required to estimate K_c to account for local crop varieties and for year-to-year variation in weather conditions (Pereira et al., 2015).

Although the K_c method has been widely used for estimating AET for crops, it has not been widely used for natural ecosystems for the purpose of estimating AET due to limited continuous measurements in these systems (Zhang et al., 2001). However, as discussed earlier, ecologists and hydrologists have started to use K_c to quantify ecosystem stress and have considered K_c as a variable rather than a constant. Past studies found that K_c was influenced by the growing stages and leaf area index for maize (Kang et al., 2003; Ding et al., 2015), winter wheat (Allen et al., 1998; Kang et al., 2003), watermelon (Shukla et al., 2014b), and fruit trees (Marsal et al., 2014b; Taylor et al., 2015). The K_c values are tabulated for each and every growth stage for many more crops all over the world (Allen et al., 1998). Variations of mid-season crop coefficients for a mixed riparian vegetation dominated by common reed (*Phragmites australis*) could be predicted by growing degree days in central Nebraska, USA (Irmak et al., 2013a). K_c ranged from 0.50 to 0.85 for small, open-grown shrubs, and from 0.85 to 0.95 for well-developed shrubland. The K_c values had a close logarithmic relationship with the canopy cover fraction in the highlands of northern Ethiopia (Descheemaeker et al., 2011). Overall, the nonagricultural ecosystems, such as forests, grasslands and shrublands, are heterogeneous in nature and have high soil water variability. Thus, K_c values for natural ecosystems have high variability (Allen and Pereira, 2009; Allen et al., 2011).

Therefore, the goal of this study was to explore how K_c varies among multiple ecosystems with various vegetation types over multiple seasons. Another goal was to determine the key biophysical and environmental factors such as latitude, precipitation, and leaf area index that could be used to estimate K_c, and if K_c can be modeled with a reasonable accuracy at a larger spatial scale. We examined the K_c variations for seven land cover types by analyzing the FLUXNET eddy flux data (Baldocchi et al., 2001; Fang et al., 2015). Specifically, our objectives were to (1) understand the variation of monthly K_c for seven distinct land covers by analyzing the influences of environmental factors (e.g., precipitation, site latitude) on K_c and (2) to develop simple land-cover-specific regression models for estimating K_c with key environmental factors as independent variables. Specifically, we developed quantitative relationships between environmental factors and K_c by land cover types using data from FLUXNET sites for 8 croplands (CRO), 13 deciduous broad leaf forests (DBF), 5 evergreen broad leaf forests (EBF), 34 evergreen needle leaf forests (ENF), 9 grasslands (GRA), 10 mixed forests (MF), and 2 open shrublands (OS). In-depth understanding of the biophysical controls on K_c for different ecosystems is important for accurately estimating AET and anticipating the impacts of climate change on ecosystem water stress and water balances.

2 Methods

This synthesis study used the LaThuile eddy flux dataset that was developed by FLUXNET (http://fluxnet.ornl.gov/; Fig. 1), a global network that measures the exchanges of carbon dioxide, water vapor, and energy between the biosphere and atmosphere (Baldocchi et al., 2001). The FLUXNET data (Baldocchi et al., 2001; Baldocchi and Ryu, 2011) have been widely used to understand the evapotranspiration processes and trend (Jung et al., 2010; Fang et al., 2015), develop AET and ecosystem models (Sun et al., 2011b; Zhang

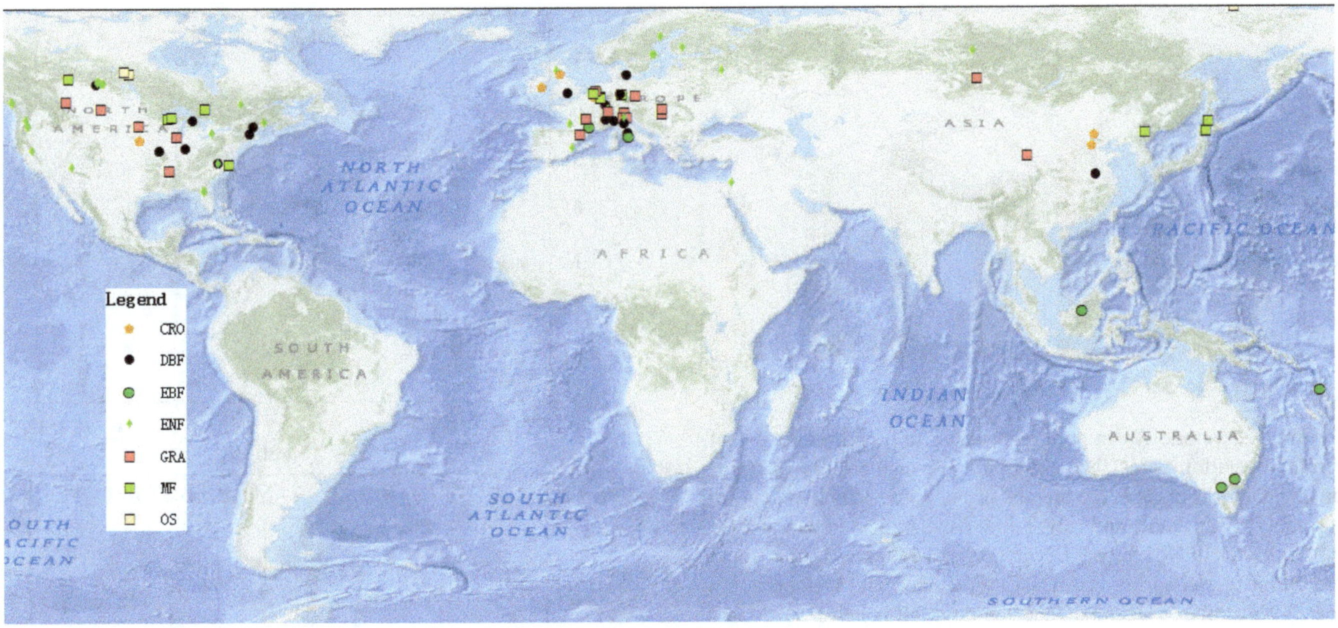

Figure 1. Location of eddy flux sites from which climate and evapotranspiration data are collected.

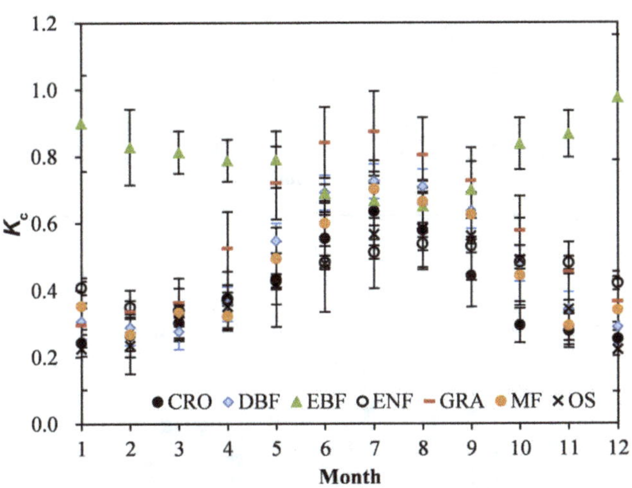

Figure 2. The variation of K_c for the different IGBP codes. The error bars are standard errors among different sites. The seven vegetation covers are open shrubland (OS), cropland (CRO), grassland (GRA), deciduous broad leaf forest (DBF), evergreen needle leaf forest (ENF), evergreen broad leaf forest (EBF), and mixed forest (MF). For sites in the Southern Hemisphere, July data were plotted as in January.

et al., 2016), and map continental-scale ecosystem productivity (Xiao et al., 2014; Zhang et al., 2016).

We used an existing database that was developed from the eddy flux measurements from 111 sites (Fang et al., 2015). A total of 81 sites were selected to calculate monthly K_c for multiple years and develop K_c models for different ecosystems, and 30 sites with 1 or 2 years of data were used for validating the models. According to the International Geosphere-Biosphere Program (IGBP) land cover classification system, these eddy flux sites represent seven land cover types: open shrubland (OS), cropland (CRO), grassland (GRA), deciduous broad leaf forest (DBF), evergreen needle leaf forest (ENF), evergreen broad leaf forest (EBF), and mixed forest (MF). For each eddy flux tower site (Fig. 1), we acquired AET and associated micrometeorological data, such as vapor pressure deficit, precipitation (P), winds speed, and net radiation at a daily timescale during 2000–2006. Based on the hypothesis that the soil surface closely maintains a uniform height, as it is actively growing grass and completely shading the ground, PET was calculated by the FAO Penman–Monteith equation as follows (Allen et al., 1998):

$$\mathrm{PET}=\frac{0.408\Delta\left(R_{\mathrm{n}}-G\right)+\gamma\frac{900}{T+273}u_2\left(e_{\mathrm{s}}-e_{\mathrm{a}}\right)}{\Delta+\gamma\left(1+0.34u_2\right)}, \tag{1}$$

where R_n is net radiation at the cover surface (MJ m^{-2} d^{-1}), G is soil heat flux (MJ m^{-2} d^{-1}), T is mean air temperature (°C), u_2 is wind speed (m s^{-1}), e_s is saturation vapor pressure (kP_a), e_a is actual vapor pressure (kP_a), $e_s - e_a$ is the saturation vapor pressure deficit (kP_a), Δ is slope of saturation vapor pressure curve (kP_a °C^{-1}), and γ is the psychrometric constant (kP_a °C^{-1}). Most sites are in the Northern Hemisphere, except three EBF sites.

The monthly K_c, which is defined as the ratio of the measured total monthly AET and the total monthly PET calculated by Eq. (1), varies by month and vegetation types (Eq. 2). The average annual K_c values were calculated by averaging monthly K_c from January to December for each

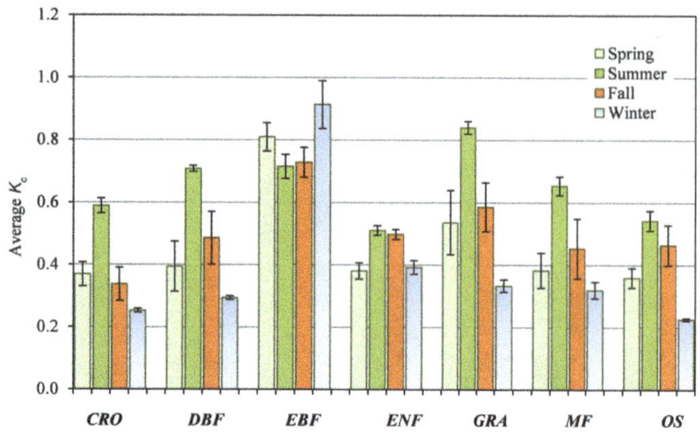

Figure 3. Average K_c in spring, summer, fall, and winter in different vegetation types. The error bars are standard errors among different sites. Spring is the months of February, March, and April; summer is May, June, and July; fall is August, September, and October; winter is November, December, and January. In the Southern Hemisphere, spring is August, September, and October; summer is November, December, and January; fall is February, March, and April; and winter is May, June, and July.

site.

$$K_c = \frac{\text{AET}}{\text{PET}} \tag{2}$$

The leaf area index (LAI) time series data for each tower site were downloaded from the Oak Ridge National Laboratory Distributed Active Archive Center (http://daac.ornl.gov/cgi-bin/MODIS/GR_col5_1/mod_viz.html). Moderate Resolution Imaging Spectroradiometer (MODIS) LAI data were derived from the fraction of absorbed photosynthetically active radiation (FPAR) that a plant canopy absorbs for photosynthesis and growth in the 0.4–0.7 nm spectral range. The MODIS LAI/FPAR algorithm exploits the spectral information of MODIS surface reflectance at up to seven spectral bands. We extracted monthly LAI data for the periods from 2000 to 2006 across 111 sites using 8-day GeoTIFF data from the MODIS land subsets' 1 km LAI global fields. We estimated monthly LAI for each flux tower by computing the mean of the 8-day daily values for each month (Fang et al., 2015).

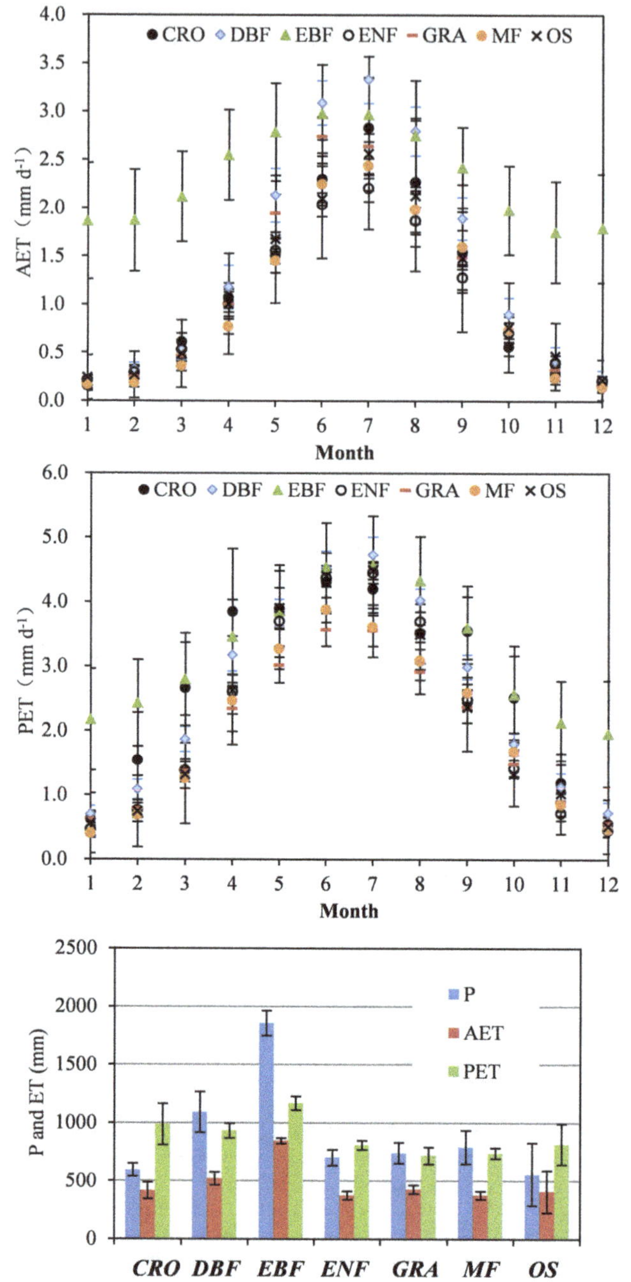

Figure 4. Monthly AET and PET, and annual total precipitation (P), AET, and PET for different vegetation types. The error bars are standard errors among different sites.

3 Results

3.1 Seasonal variations and long-term means of K_c by land cover

The average monthly K_c based on eddy flux data from 2000 to 2007 increased gradually from January to July and then decreased (Fig. 2). Evergreen broad leaf forest (EBF) had the highest mean monthly K_c (0.97 ± 0.19) (mean $\pm$ standard error) in December (June for sites in the Southern Hemisphere). K_c for both EBF and ENF varied less seasonally than other forest types (Fig. 2). Standard errors for grassland (GRA), evergreen needle leaf forest (ENF), and open shrubland (OS) (0.10–0.17) were larger than for other land cover types (0.03–0.10) for April to August. EBF had higher K_c for all seasons than other land covers with a peak value of 0.91 ($\pm$0.08) in the winter season (Fig. 3). In winter seasons, cropland (CRO) and OS had the lowest K_c: 0.25 ($\pm$0.006) and 0.22 ($\pm$0.004), respectively.

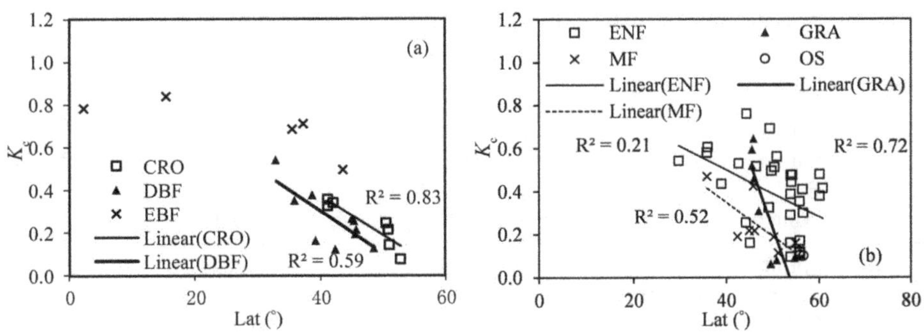

Figure 5. Variation of annual K_c at different latitudes (lat): **(a)** cropland (CRO), deciduous broad leaf forest (DBF), and evergreen broad leaf forest (EBF); **(b)** evergreen needle leaf forest (ENF), grassland (GRA), mixed forest (MF), and open shrubland (OS). The absolute values of the latitude were used in EBF for sites in the Southern Hemisphere, and all the determination coefficients (R^2) listed in the figure were significant ($p<0.05$).

The mean annual K_c was 0.39 (±0.04), 0.47 (±0.05), 0.75 (±0.03), 0.45 (±0.02), 0.57 (±0.06), 0.45 (±0.05), and 0.40 (±0.04) for CRO, DBF, EBF, ENF, GRA, MF, and OS, respectively. Yearly average precipitation was higher in EBF and DBF than other land covers (Fig. 4). The precipitation ranking by land cover type was EBF > DBF > MF > GRA> ENF > CRO > OS. Consequently, OS, MF, GRA, CRO, and ENF had relatively lower yearly AET (376–425 mm) than EBF and DBF. Moreover, DBF, EBF, and CRO had higher PET than other vegetation surfaces. The variations for monthly AET and PET were presented in Fig. 4 to the contrasting patterns of these two variables. The AET and PET reached maximum values 2.2–3.3 and 3.6–4.7 mm d^{-1} in June or July (December or January for the Southern Hemisphere), respectively.

3.2 Environmental controls on K_c

As indicated in Eq. (1), factors such as temperature and solar radiation were used for PET calculations, and were not independent of K_c. Therefore, we chose other independent factors to simulate K_c. Site latitude is a readily available variable for a particular location, but is crucial to the day length and incoming radiation over the year.

The results showed that annual K_c was negatively ($p<0.05$) correlated with latitude (Fig. 5) for CRO, DBF, ENF, GRA, and MF with a determination coefficient (R^2) of 0.83, 0.59, 0.21, 0.72, and 0.52, respectively. For OS, annual mean K_c also decreased with the increase in site latitude. Most of the study site fell between 30 and 60° N in latitude.

At the seasonal scale, the linear relationships between monthly K_c and total monthly precipitation differed among different land cover types (Fig. 6). Monthly K_c increased with monthly precipitation in the same ecosystem type with the R^2 ranking from high to low: OS > MF > GRA > ENF > CRO > DBF. The monthly K_c for OS was especially sensitive to precipitation ($R^2=0.69$, $p<0.001$). The monthly K_c for EBF was not as sensitive to precipitation as other ecosystems because EBF was generally found in a wet environment with a peak monthly precipitation of 468 mm. Moreover, K_c for OS, GRA, and MF in relatively drier environments had lower values (Fig. 2). Therefore, K_c was closely related to the monthly precipitation.

In addition to growing season, site latitude, and monthly precipitation, leaf area index also affected the monthly K_c (Fig. 7). K_c was obviously influenced by LAI for all land covers except EBF. The determination coefficients for different land covers were OS > MF = GRA > ENF > DBF > CRO. The LAI range was up to 6 $m^2\,m^{-2}$ in most land covers, while it only reached 3–4 $m^2\,m^{-2}$ in OS and CRO.

3.3 K_c models

A series of empirical K_c models have been developed using a multiple linear regression approach with precipitation, LAI, and site latitude as independent variables (Table 1). The monthly precipitation, LAI, and site latitude influence K_c ($p<0.1$) for most ecosystems studied in different seasons except at EBF in summer and fall, and for OS in the spring. As annual precipitation increases, total leaf area increases; therefore, K_c increases for ENF in all seasons and most of the time for DBF and MF. As site latitude increases, K_c values are found to decrease in some periods at CRO, DBF, and MF sites. In addition, K_c is closely correlated to LAI, site latitude, and monthly precipitation at ENF in fall and OS in winter with an R^2 of 0.55 and 0.99, respectively. All land covers have peak values (0.53 ± 0.04–1.01 ± 0.17) in the summer months. Except for EBF and GRA, K_c values have a close relationship with the monthly precipitation in the summer with R^2 ranging from 0.21 to 0.90. The linear relationships are significant for most vegetation types, suggesting that the regression models (Table 1) can be used to estimate monthly K_c if LAI and precipitation for a specific ecosystem are available.

Table 1. Multiple linear regression relationships among crop coefficient and LAI, precipitation, and site latitude in different seasons.

IGBP	Season	N	R^2	K_c	b	a_1	a_2	a_3
CRO	Spring	24	0.16	0.31	0.242***	0.141*		
	Summer	24	0.21	0.57	0.331**			0.0033*
	Fall	23	0.78	0.48	0.036	0.472***		
	Winter	21	0.36	0.26	0.920***		−0.0141**	
DBF	Spring	39	0.49	0.30	0.479**		−0.0076*	0.0022***
	Summer	39	0.42	0.65	0.536***			0.0011***
	Fall	39	0.13	0.60	0.462***			0.0014*
	Winter	39	0.15	0.30	0.713***		−0.0094*	
EBF	Spring	15	0.25	0.74	0.875***		−0.0050*	
	Summer	15	–	0.91	0.911***			
	Fall	15	–	0.80	0.798***			
	Winter	15	0.42	0.72	0.676***	0.050*	−0.0050**	
ENF	Spring	96	0.39	0.37	0.225***	0.060***		0.0017***
	Summer	99	0.59	0.49	0.211***	0.053***		0.0020***
	Fall	98	0.55	0.52	−0.040	0.066***	0.0049*	0.0025***
	Winter	92	0.21	0.44	0.293***	0.084*		0.0010*
GRA	Spring	27	0.48	0.45	0.237***			0.0052***
	Summer	27	0.23	0.86	0.572***	0.110*		
	Fall	27	0.30	0.76	0.499***	0.123**		
	Winter	27	0.26	0.41	0.256**			0.0038**
MF	Spring	30	0.67	0.31	0.099**	0.188***		0.0012***
	Summer	30	0.40	0.61	0.372***			0.0029***
	Fall	30	0.54	0.58	0.250***	0.071***		0.0018***
	Winter	30	0.13	0.33	0.961**		−0.0136*	
OS	Spring	6	–	0.23	0.230***			
	Summer	6	0.90	0.35	−5.419*		0.1005*	0.0026*
	Fall	6	0.88	0.42	−9.921*	0.051*	0.1828*	
	Winter	6	0.99	0.14	−4.919*	0.629*	0.0882*	0.0032*

Note: N is the number of observations used, R^2 the determination coefficient, K_c is the average K_c for seasons. b is the intercept of the multiple linear equation, a_1 the coefficient of LAI, a_2 the coefficient of site latitude (absolute values), a_3 the coefficient of precipitation. IGBP is the International Geosphere-Biosphere Program land cover classification system: cropland (CRO), deciduous broad leaf forest (DBF), evergreen broad leaf forest (EBF), evergreen needle leaf forest (ENF), grassland (GRA), mixed forest (MF), and open shrubland (OS). ***, **, * stand for $p<0.001$, $p<0.01$, $p<0.1$, and the blank spaces indicate nonsignificant values. In the Northern Hemisphere, spring is the months of February, March, and April; summer is May, June, and July; fall is August, September, and October; winter is November, December, and January. In the Southern Hemisphere, spring is August, September, and October; summer is November, December, and January; fall is February, March, and April; and winter is May, June, and July.

3.4 The validation of the regression models of K_c

All K_c multiple regression models for different seasons were validated by ecosystem type (Fig. 8). The model validation was carried out for 30 sites at a monthly scale. The results showed that the modeled AET calculated from the multiple K_c models compared well to measurements with R^2 ranging 0.28–0.56. Among the ecosystems, the model for DBF appeared to be the most accurate one, with an R^2 of 0.56. However, model validation results for CRO, EBF, and OS were not as satisfactory as indicated by the slopes (< 1.0 or > 1.0) of the regression equations.

4 Discussion

Our study estimated annual and seasonal crop coefficient (K_c) for seven land cover types using measured global eddy flux data. We comprehensively evaluated environmental controls (i.e., precipitation, LAI, and site latitude) on annual and growing season K_c and developed a series of multiple linear regression models that can be used for estimating monthly AET over time and space for some vegetation types.

4.1 Crop coefficient variation in different seasons

Several recent studies had shown that K_c reached the maximum value in the middle of the growing season in many

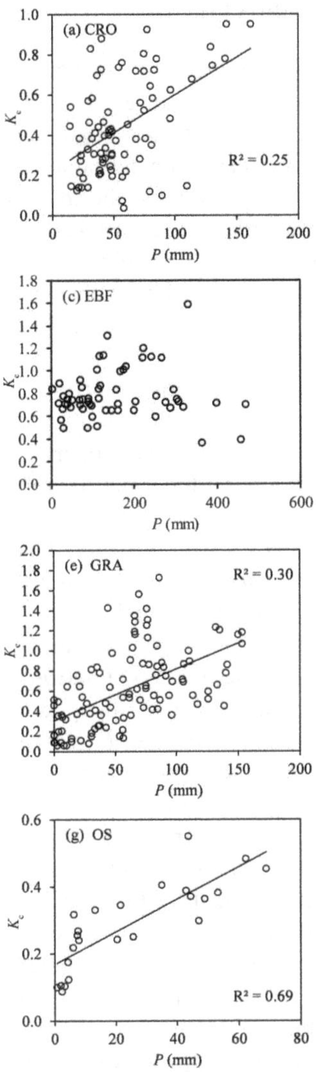

Figure 6. Relationships between the average monthly K_c and monthly precipitation (P, mm) for different vegetation surfaces. Panels **(a–g)** represent cropland (CRO), deciduous broad leaf forest (DBF), evergreen broad leaf forest (EBF), evergreen needle leaf forest (ENF), grassland (GRA), mixed forest (MF), and open shrubland (OS), respectively. All the determination coefficients (R^2) listed in the figure were significant ($p<0.001$).

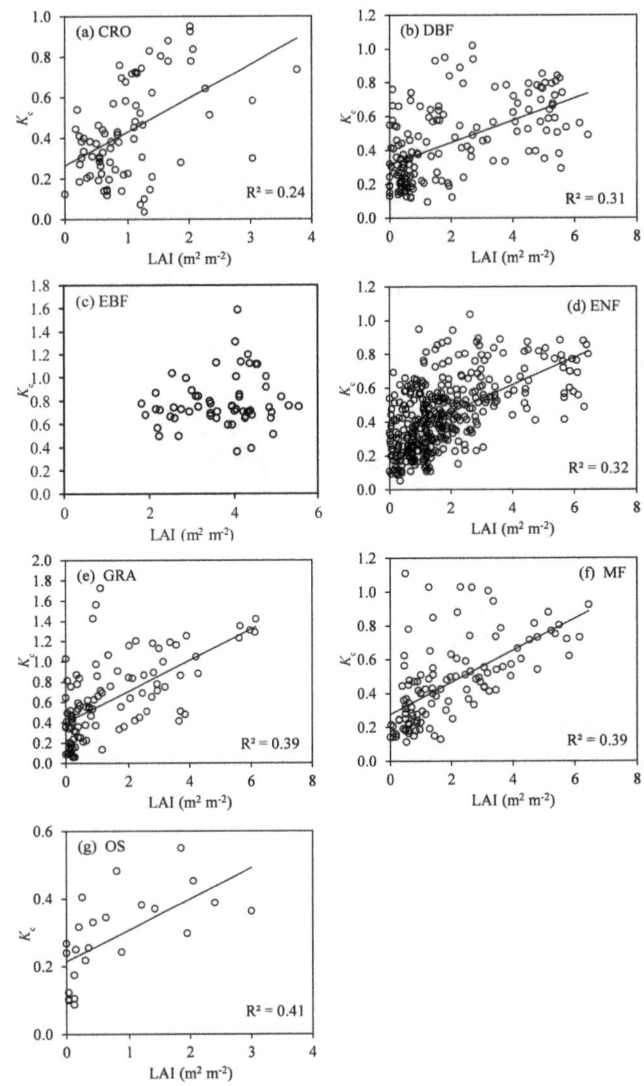

Figure 7. Relationships between the average monthly K_c and leaf area index for different vegetation surfaces. Panels **(a–g)** stand for cropland (CRO), deciduous broad leaf forest (DBF), evergreen broad leaf forest (EBF), evergreen needle leaf forest (ENF), grassland (GRA), mixed forest (MF), and open shrubland (OS). All the determination coefficients (R^2) listed in the figure were significant ($p<0.05$).

ecosystems, such as a *P. euphratica* forest in the riparian area (Hou et al., 2010) in a desert environment, a watermelon crop covered with plastic mulch in Florida (Shukla et al., 2014a, b), soybean in Nebraska (Irmak et al., 2013b), and a temperate desert steppe in Inner Mongolia (Zhang et al., 2012). As Fig. 2 shows, most of the land covers have peak K_c during June to August (in the Northern Hemisphere), while the seasonal patterns of ENF and EBF vary less than other surfaces. Vegetation growth for both the ENF and EBF sites is active throughout the year. The mean crop coefficient for medium-density fruit trees in the early growing season is about 0.5 (Allen et al., 1998; Allen and Pereira, 2009), which is similar to those found for DBF or MF during April and May. In addition, the middle season K_c values for apple and peach trees with active ground cover were higher than K_c for DBF sites during the summer. It is likely that the orchards had higher evapotranspiration rates than natural forests due to irrigation. We also find that the CRO has relatively low precipitation with a high PET because of irrigation. The irrigation has been proven to be a determining factor for AET at the local and even at the global scale (Jaramillo and Destouni, 2015). Thus, the K_c for CRO mainly depends on the irrigation schedule and the primary crops. The loss of leaves on DBF and MF lead to an larger obvious standard error for

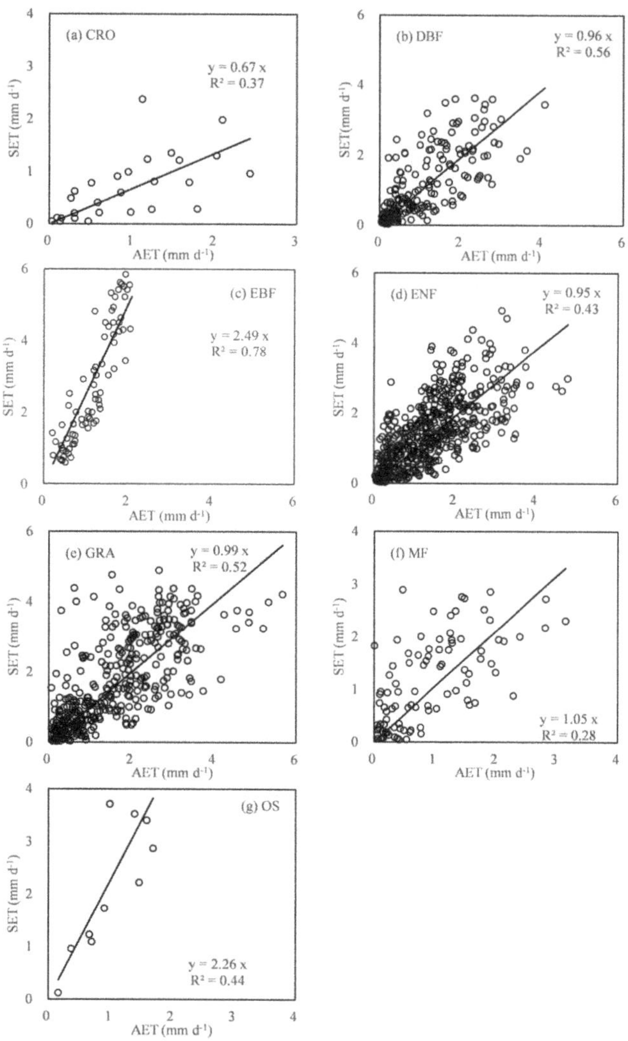

Figure 8. Relationships between the simulated ET using K_c from Table 1 (SET) and the measured ET (AET) for different vegetation surfaces. Panels **(a–f)** stand for cropland (CRO), deciduous broad leaf forest (DBF), evergreen broad leaf forest (EBF), evergreen needle leaf forest (ENF), grassland (GRA), mixed forest (MF), and open shrubland (OS). All the determination coefficients (R^2) listed in the figure were significant ($p < 0.001$).

K_c in fall (Fig. 3). The soil water evaporation represents the main water loss, which is thus a key component of K_c when the ecosystems lack leaves or plants in winter (Allen et al., 1998). Moreover, the K_c is biologically meaningful in vegetation type distribution (Stephenson, 1998); thus, when LAI becomes small for DBF during winter, the K_c reflects the characteristics of evaporation capacity for the ground surface.

4.2 Environmental control factors for K_c

The ecosystem covers and the distributions of the vegetation classes are determined by the latitude (Potter et al., 1993). Crop coefficient varies predominately by ecosystems, and K_c increases as the site latitude decreases for the same land cover type (Fig. 5). As the latitude decreases, the increasing temperature and the solar radiation results of PET are increasing; thus, the acceleration for AET should be faster than PET. The reason may be that the vegetation characteristics are different for the same land cover type in different latitudes. Models developed from the FLUXNET data may be best used on flat areas for a specific latitude given that eddy covariance towers were generally installed on flat lands (Baldocchi et al., 2001). For areas with complex topography, the relationship between K_c and site latitude may be more complicated.

Spatial variations of K_c are characteristic of ecosystems, but K_c is also affected by climate factors such as rainfall. For example, K_c was highly correlated with precipitation for most land covers (Fig. 6). The rainfall is the major source of soil water and AET in natural ecosystems (Parent and Anctil, 2012). During dry years or periods, a lack of precipitation may cause a reduction of the leaf area index, and K_c will decrease. During rainy seasons, as leaf area index and stomatal conductance of trees and rain-fed crops increases, so does K_c (Kar et al., 2006; Zeppel et al., 2008). Irrigation of cropland is a primary mechanism for increasing yield (Fereres and Soriano, 2007; Du et al., 2015), so the CRO may have a high monthly K_c even at sites with a low precipitation. In contrast, K_c does not have a close relationship with precipitation under a wet environment. For example, the EBF site had a monthly precipitation as high as 468 mm month^{-1} and generally exceeded monthly AET. In an opposite case for the OS sites, monthly precipitation values were between 0.7 and 69 mm, and K_c was highly correlated with monthly precipitation. Moreover, the time lag between precipitation and soil moisture might cause errors in calculating AET and modeling K_c in the long dry or wet season. However, at the monthly scale, previous modeling work (Fang et al., 2015) suggests that considering a time lag does not increase the prediction power dramatically (G. Sun, personal communication, 2015).

Besides precipitation, LAI also affects K_c in dry and semihumid areas (Kang et al., 2003; Zhang et al., 2012). Unlike precipitation, LAI directly affects K_c in AET calculations (Tolk and Howell, 2001; Novák, 2012). Interannual K_c values are stable at the GRA and OS sites due to the steady seasonal LAI between years while the plantation forest sites had a more dynamic LAI pattern (Marsal et al., 2014a). As the growth rate of the perennial plants could have large effects on the relationship between K_c and LAI, long-term data are needed to estimate K_c as a function of all environmental factors.

4.3 Modeling the dynamics of K_c

Our study results are consistent with previous studies that show that the growing stage is a key factor for estimating K_c in agricultural crops (Allen et al., 1998; Zhang et al., 2013; Alberto et al., 2014; Wei et al., 2015), fruit trees (Abrisqueta et al., 2013; Marsal et al., 2014b), salt grass (Bawazir et al., 2014), and *Populus euphratica Oliv.* forest (Hou et al., 2010). Additionally, our study showed that K_c fluctuated more dramatically in DBF, GRA, and MF than other land covers in different seasons (Table 1). Studies also show that monthly leaf stomatal resistance that varies over time is important in estimating the seasonal crop coefficient for a citrus orchard (Taylor et al., 2015). The LAI and total monthly precipitation were considered as independent factors (Bond-Lamberty and Thomson, 2010) and both of them varied in both time and space while the site latitude only represented spatial influences on K_c. The modeled AET was acceptable for DBF, ENF, GRA, and MF (Fig. 8), and could be used for monthly AET calculation for large-spatial-scale and homogeneous ecosystems. The slope of CRO modeling ET to AET was 50 % below the 1 : 1 line which may be because the crops were irrigated when the soil lacked water content. Meanwhile, the OS having a large proportion of bare soil with low soil water content may be the result of an overestimate in modeling ET. The lack of site samples may cause a low accuracy of validation in OS and EBF modeling ET. Thus, the multiple linear regression equations developed from this study take into account both spatial and temporal changes in land surface characteristics and offer a powerful tool for estimating seasonal dynamics of K_c for most ecosystems (Table 1).

5 Conclusions

In seeking a convenient method to calculate monthly AET at large spatial scales, we comprehensively examined the relations between K_c and environmental factors using eddy flux data from 81 sites (mainly in the Northern Hemisphere) with different land covers. We found that K_c values varied largely among CRO, DBF, EBF, GRA, and MF, and across seasons. Besides EBF, precipitation determined K_c in the growing seasons (such as summer) and was chosen as a key variable to calculate K_c. We established multiple linear equations for different land covers and seasons to model the dynamics of K_c as function of LAI, site latitude, and monthly precipitation. These empirical models could be helpful in calculating monthly AET at the regional scale with readily available climatic data and vegetation structure information. Our study extended the applications of the traditional K_c method for estimating crop water use to estimating AET rates and evaporative stress for natural ecosystems. Future studies should further test the applicability of the empirical K_c models under extreme climatic conditions and for those ecosystems that are underrepresented by the FLUXNET.

6 Data availability

The FLUXNET data are not publicly accessible; however, the data can be downloaded for free by the user at the following website: http://fluxnet.fluxdata.org/data/la-thuile-dataset/.

Acknowledgement. We are grateful for grants from the National Natural Science Foundation of China (no. 51309132), for supporting this collaborative work between Nanjing University of Information Science and Technology and the Eastern Forest Environmental Threat Assessment Center at the USDA Forest Service Southern Research Station. This work used eddy covariance data acquired by the FLUXNET community and in particular by the following networks: AmeriFlux (US Department of Energy, Biological and Environmental Research, Terrestrial Carbon Program (DEFG02-04ER63917 and DE-FG02-04ER63911)), AfriFlux, AsiaFlux, CarboAfrica, CarboEuropeIP, CarboItaly, CarboMont, ChinaFlux, Fluxnet-Canada (supported by CFCAS, NSERC, BIOCAP, Environment Canada, and NRCan), GreenGrass, KoFlux, LBA, NECC, OzFlux, TCOS-Siberia, and the United States China Carbon Consortium (USCCC). We acknowledge the financial support to the eddy covariance data harmonization provided by CarboEuropeIP, FAO-GTOS-TCO, iLEAPS, Max Planck Institute for Biogeochemistry, National Science Foundation, University of Tuscia, Université Laval and Environment Canada, and the US Department of Energy, and the database development and technical support from Berkeley Water Center, Lawrence Berkeley National Laboratory, Microsoft Research eScience, Oak Ridge National Laboratory, University of California, Berkeley, and University of Virginia. This work also used MODIS land subset (Oak Ridge National Laboratory Distributed Active Archive Center (ORNL DAAC) 2011 MODIS subsetted land products, collection 5). We also thank the reviewers and associate editor for their constructive comments on the manuscript.

Edited by: K. Bishop

References

Abrisqueta, I., Abrisqueta, J. M., Tapia, L. M., Munguía, J. P., Conejero, W., Vera, J., and Ruiz-Sánchez, M. C.: Basal crop coefficients for early-season peach trees, Agr. Water Manage., 121, 158–163, doi:10.1016/j.agwat.2013.02.001, 2013.

Alberto, M. C. R., Quilty, J. R., Buresh, R. J., Wassmann, R., Haidar, S., Correa, T. Q., and Sandro, J. M.: Actual evapotranspiration and dual crop coefficients for dry-seeded rice and hybrid maize grown with overhead sprinkler irrigation, Agr. Water Manage., 136, 1–12, doi:10.1016/j.agwat.2014.01.005, 2014.

Allen, R. G. and Pereira, L. S.: Estimating crop coefficients from fraction of ground cover and height, Irrigation Sci., 28, 17–34, doi:10.1007/s00271-009-0182-z, 2009.

Allen, R. G., Pereira, L. S., Raes, D., and Smith, M.: Crop evapotranspiration, FAO irrigation and drainage paper, No. 56, 1998.

Allen, R. G., Pruitt, W. O., Wright, J. L., Howell, T. A., Ventura, F., Snyder, R., Itenfisu, D., Steduto, P., Berengena, J., Yrisarry, J. B., Smith, M., Pereira, L. S., Raes, D., Perrier, A., Alves, I., Walter, I., and Elliott, R.: A recommendation on standardized surface resistance for hourly calculation of reference ETo by the FAO56 Penman-Monteith method, Agr. Water Manage., 81, 1–22, doi:10.1016/j.agwat.2005.03.007, 2006.

Allen, R. G., Pereira, L. S., Howell, T. A., and Jensen, M. E.: Evapotranspiration information reporting: I. Factors governing measurement accuracy, Agr. Water Manage., 98, 899–920, doi:10.1016/j.agwat.2010.12.015, 2011.

Anda, A., Silva, J. A. T. D., and Soos, G.: Evapotranspiration and crop coefficient of common reed at the surroundings of Lake Balaton, Hungary, Aquat. Botany, 116, 53–59, doi:10.1016/j.aquabot.2014.01.008, 2014.

Anderson, M. C., Allen, R. G., Morse, A., and Kustas, W. P.: Use of Landsat thermal imagery in monitoring evapotranspiration and managing water resources, Remote Sens. Environ., 122, 50–65, 2012.

Baldocchi, D., Falge, E., Gu, L., Olson, R., Hollinger, D., Running, S., Anthoni, P., Bernhofer, C., Davis, K., and Evans, R.: FLUXNET: A new tool to study the temporal and spatial variability of ecosystem-scale carbon dioxide, water vapor, and energy flux densities, B. Am. Meteorol. Soc., 82, 2415–2434, 2001.

Baldocchi, D. D. and Ryu, Y.: A synthesis of forest evaporation fluxes – from days to years – as measured with eddy covariance, in: Forest Hydrology and Biogeochemistry, Springer, 101–116, 2011.

Bawazir, A. S., Luthy, R., King, J. P., Tanzy, B. F., and Solis, J.: Assessment of the crop coefficient for saltgrass under native riparian field conditions in the desert southwest, Hydrol. Process., 28, 6163–6171, doi:10.1002/Hyp.10100, 2014.

Bond-Lamberty, B. and Thomson, A.: Temperature-associated increases in the global soil respiration record, Nature, 464, 579–582, 2010.

Budyko, M.: Climate and Life, Academic Press, New York, 1974.

Consoli, S. and Vanella, D.: Mapping crop evapotranspiration by integrating vegetation indices into a soil water balance model, Agr. Water Manage., 143, 71–81, doi:10.1016/j.agwat.2014.06.012, 2014.

Descheemaeker, K., Raes, D., Allen, R., Nyssen, J., Poesen, J., Muys, B., Haile, M., and Deckers, J.: Two rapid appraisals of FAO-56 crop coefficients for semiarid natural vegetation of the northern Ethiopian highlands, J. Arid Environ., 75, 353–359, doi:10.1016/j.jaridenv.2010.12.002, 2011.

Ding, R. S., Tong, L., Li, F. S., Zhang, Y. Q., Hao, X. M., and Kang, S. Z.: Variations of crop coefficient and its influencing factors in an arid advective cropland of northwest China, Hydrol. Process., 29, 239–249, doi:10.1002/Hyp.10146, 2015.

Donohue, R. J., Roderick, M. L., and McVicar, T. R.: On the importance of including vegetation dynamics in Budyko's hydrological model, Hydrol. Earth Syst. Sci., 11, 983–995, doi:10.5194/hess-11-983-2007, 2007.

Du, T., Kang, S., Zhang, J., and Davies, W. J.: Deficit irrigation and sustainable water-resource strategies in agriculture for China's food security, J. Exp. Bot., 66, 2253–2269, doi:10.1093/jxb/erv034, 2015.

Fang, Y., Sun, G., Caldwell, P., McNulty, S. G., Noormets, A., Domec, J. C., King, J., Zhang, Z., Zhang, X., and Lin, G.: Monthly land cover – evapotranspiration models derived from global eddy flux measurements and remote sensing data, Ecohydrology, 9, 248–266, doi:10.1002/eco.1629, 2015.

Fereres, E. and Soriano, M. A.: Deficit irrigation for reducing agricultural water use, J. Exp. Bot., 58, 147–159, 2007.

Hao, L., Sun, G., Liu, Y., Gao, Z., He, J., Shi, T., and Wu, B.: Effects of precipitation on grassland ecosystem restoration under grazing exclusion in Inner Mongolia, China, Landscape Ecol., 29, 1657–1673, doi:10.1007/s10980-014-0092-1, 2014.

Hao, L., Sun, G., Liu, Y., and Qian, H.: Integrated Modeling of Water Supply and Demand under Management Options and Climate Change Scenarios in Chifeng City, China, J. Am. Water Resour. As., 51, 655–671, 2015a.

Hao, L., Sun, G., Liu, Y., Wan, J., Qin, M., Qian, H., Liu, C., Zheng, J., John, R., Fan, P., and Chen, J.: Urbanization dramatically altered the water balances of a paddy field-dominated basin in southern China, Hydrol. Earth Syst. Sci., 19, 3319–3331, doi:10.5194/hess-19-3319-2015, 2015b.

Hasper, T. B., Wallin, G., Lamba, S., Hall, M., Jaramillo, F., Laudon, H., Linder, S., Medhurst, J. L., Rantfors, M., Sigurdsson, B. D., and Uddling, J.: Water use by Swedish boreal forests in a changing climate, Funct. Ecol., 30, 690–699, doi:10.1111/1365-2435.12546, 2016.

Hou, L. G., Xiao, H. L., Si, J. H., Xiao, S. C., Zhou, M. X., and Yang, Y. G.: Evapotranspiration and crop coefficient of Populus euphratica Oliv forest during the growing season in the extreme arid region northwest China, Agr. Water Manage., 97, 351–356, 2010.

Irmak, S., Kabenge, I., Rudnick, D., Knezevic, S., Woodward, D., and Moravek, M.: Evapotranspiration crop coefficients for mixed riparian plant community and transpiration crop coefficients for Common reed, Cottonwood and Peach-leaf willow in the Platte River Basin, Nebraska-USA, J. Hydrol., 481, 177–190, doi:10.1016/j.jhydrol.2012.12.032, 2013a.

Irmak, S., Odhiambo, L. O., Specht, J. E., and Djaman, K.: Hourly And Daily Single And Basal Evapotranspiration Crop Coefficients as a Function Of Growing Degree Days, Days after Emergence, Leaf Area Index, Fractional Green Canopy Cover, And Plant Phenology for Soybean, T. ASABE, 56, 1785–1803, 2013b.

Jaramillo, F., Prieto, C., Lyon, S. W., and Destouni, G.: Multimethod assessment of evapotranspiration shifts due to non-irrigated agricultural development in Sweden, J. Hydrol., 484, 55–62, doi:10.1016/j.jhydrol.2013.01.010, 2013.

Jaramillo, F. and Destouni, G.: Local flow regulation and irrigation raise global human water consumption and footprint, Science, 350, 1248–1251, doi:10.1126/science.aad1010, 2015.

Jung, M., Reichstein, M., Ciais, P., Seneviratne, S. I., Sheffield, J., Goulden, M. L., Bonan, G., Cescatti, A., Chen, J., and De Jeu, R.: Recent decline in the global land evapotranspiration trend due to limited moisture supply, Nature, 467, 951–954, 2010.

Kang, S., Gu, B., Du, T., and Zhang, J.: Crop coefficient and ratio of transpiration to evapotranspiration of winter wheat and maize in a semi-humid region, Agr. Water Manage., 59, 239–254, 2003.

Kar, G., Verma, H. N., and Singh, R.: Effects of winter crop and supplemental irrigation on crop yield, water use efficiency and profitability in rainfed rice based cropping system of eastern India, Agr. Water Manage., 79, 280–292, doi:10.1016/j.agwat.2005.03.001, 2006.

Lu, J., Sun, G., McNulty, S. G., and Amatya, D.: A comparison of six potential evapotranspiration methods for regional use in the Southeastern United States, J. Am. Water Resour. As., 41, 621–633, 2005.

Marsal, J., Casadesus, J., Lopez, G., Girona, J., and Stöckle, C.: Disagreement between tree size and crop coefficient in "conference" pear: comparing measurements by a weighing Lysimeter and prediction by Cropsyst, Acta Hortic., 1038, 303–310, doi:10.17660/ActaHortic.2014.1038.36, 2014a.

Marsal, J., Johnson, S., Casadesus, J., Lopez, G., Girona, J., and Stöckle, C.: Fraction of canopy intercepted radiation relates differently with crop coefficient depending on the season and the fruit tree species, Agr. Forest Meteorol., 184, 1–11, doi:10.1016/j.agrformet.2013.08.008, 2014b.

Mu, Q., Zhao, M., Kimball, J., McDowell, N., and Running, S.: A remotely sensed global terrestrial drought severity index, in: Evapotranspiration in the Soil-plant-atmosphere System, AGU Fall Meeting Abstracts, L02, 2012.

Novák, V.: Evapotranspiration in the Soil-plant-atmosphere System, Springer Science & Business Media, 2012.

Parent, A. C. and Anctil, F.: Quantifying evapotranspiration of a rainfed potato crop in South-eastern Canada using eddy covariance techniques, Agr. Water Manage., 113, 45–56, doi:10.1016/j.agwat.2012.06.014, 2012.

Pereira, L. S., Allen, R. G., Smith, M., and Raes, D.: Crop evapotranspiration estimation with FAO56: Past and future, Agr. Water Manage., 147, 4–20, 2015.

Potter, C. S., Randerson, J. T., Field, C. B., Matson, P. A., Vitousek, P. M., Mooney, H. A., and Klooster, S. A.: Terrestrial ecosystem production: a process model based on global satellite and surface data, Global Biogeochem. Cy., 7, 811–841, 1993.

Rao, L., Sun, G., Ford, C., and Vose, J.: Modeling potential evapotranspiration of two forested watersheds in the southern Appalachians, T. ASABE, 54, 2067–2078, 2011.

Shukla, S., Shrestha, N. K., and Goswami, D.: Evapotranspiration And Crop Coefficients for Seepage-Irrigated Watermelon with Plastic Mulch In a Sub-Tropical Region, T. ASABE, 57, 1017–1028, 2014a.

Shukla, S., Shrestha, N. K., Jaber, F. H., Srivastava, S., Obreza, T. A., and Boman, B. J.: Evapotranspiration and crop coefficient for watermelon grown under plastic mulched conditions in sub-tropical Florida, Agr. Water Manage., 132, 1–9, doi:10.1016/j.agwat.2013.09.019, 2014b.

Stephenson, N.: Actual evapotranspiration and deficit: biologically meaningful correlates of vegetation distribution across spatial scales, J. Biogeogr., 25, 855–870, 1998.

Sun, G., Noormets, A., Gavazzi, M. J., McNulty, S. G., Chen, J., Domec, J. C., King, J. S., Amatya, D. M., and Skaggs, R. W.: Energy and water balance of two contrasting loblolly pine plantations on the lower coastal plain of North Carolina, USA, Forest Ecol. Manag., 259, 1299–1310, doi:10.1016/j.foreco.2009.09.016, 2010.

Sun, G., Alstad, K., Chen, J. Q., Chen, S. P., Ford, C. R., Lin, G. H., Liu, C. F., Lu, N., McNulty, S. G., Miao, H. X., Noormets, A., Vose, J. M., Wilske, B., Zeppel, M., Zhang, Y., and Zhang, Z. Q.: A general predictive model for estimating monthly ecosystem evapotranspiration, Ecohydrology, 4, 245–255, doi:10.1002/Eco.194, 2011a.

Sun, G., Caldwell, P., Noormets, A., McNulty, S. G., Cohen, E., Moore Myers, J., Domec, J. C., Treasure, E., Mu, Q., and Xiao, J.: Upscaling key ecosystem functions across the conterminous United States by a water – ecosystem model, J. Geophys. Res.-Biogeo., 116, G00J05, doi:10.1029/2010JG001573, 2011b.

Sun, S., Chen, H., Ju, W., Yu, M., Hua, W., and Yin, Y.: On the attribution of the changing hydrological cycle in Poyang Lake Basin, China, J. Hydrol., 514, 214–225, 2014.

Sun, S., Sun, G., Caldwell, P., McNulty, S., Cohen, E., Xiao, J., and Zhang, Y.: Drought impacts on ecosystem functions of the US National Forests and Grasslands: Part II assessment results and management implications, Forest Ecol. Manag., 353, 269–279, 2015a.

Sun, S., Sun, G., Caldwell, P., McNulty, S. G., Cohen, E., Xiao, J., and Zhang, Y.: Drought impacts on ecosystem functions of the US National Forests and Grasslands: Part I evaluation of a water and carbon balance model, Forest Ecol. Manag., 353, 260–268, 2015b.

Tabari, H., Grismer, M. E., and Trajkovic, S.: Comparative analysis of 31 reference evapotranspiration methods under humid conditions, Irrigation Sci., 31, 107–117, 2013.

Taylor, N., Mahohoma, W., Vahrmeijer, J., Gush, M., Allen, R. G., and Annandale, J. G.: Crop coefficient approaches based on fixed estimates of leaf resistance are not appropriate for estimating water use of citrus, Irrigation Sci., 33, 153–166, 2015.

Tolk, J. A. and Howell, T. A.: Measured and simulated evapotranspiration of grain sorghum grown with full and limited irrigation in three high plains soils, T. ASAE, 44, 1553–1558, 2001.

Vose, J. M., Sun, G., Ford, C. R., Bredemeier, M., Otsuki, K., Wei, X., Zhang, Z., and Zhang, L.: Forest ecohydrological research in the 21st century: what are the critical needs?, Ecohydrology, 4, 146–158, 2011.

Wei, Z., Paredes, P., Liu, Y., Chi, W. W., and Pereira, L. S.: Modelling transpiration, soil evaporation and yield prediction of soybean in North China Plain, Agr. Water Manage., 147, 43–53, doi:10.1016/j.agwat.2014.05.004, 2015.

Xiao, J., Ollinger, S. V., Frolking, S., Hurtt, G. C., Hollinger, D. Y., Davis, K. J., Pan, Y., Zhang, X., Deng, F., and Chen, J.: Data-driven diagnostics of terrestrial carbon dynamics over North America, Agr. Forest Meteorol., 197, 142–157, 2014.

Zeppel, M. J. B., Macinnis-Ng, C. M. O., Yunusa, I. A. M., Whitley, R. J., and Earnus, D.: Long term trends of stand transpiration in a remnant forest during wet and dry years, J. Hydrol., 349, 200–213, doi:10.1016/j.jhydrol.2007.11.001, 2008.

Zhang, B., Liu, Y., Xu, D., Zhao, N., Lei, B., Rosa, R. D., Paredes, P., Paço, T. A., and Pereira, L. S.: The dual crop coefficient approach to estimate and partitioning evapotranspiration of the winter wheat–summer maize crop sequence in North China Plain, Irrigation Sci., 31, 1303–1316, 2013.

Zhang, F., Zhou, G. S., Wang, Y., Yang, F. L., and Nilsson, C.: Evapotranspiration and crop coefficient for a temperate desert steppe ecosystem using eddy covariance in Inner Mongolia, China, Hydrol. Process., 26, 379–386, 2012.

Zhang, L., Dawes, W. R., and Walker, G. R.: Response of mean annual evapotranspiration to vegetation changes at catchment scale, Water Resour. Res., 37, 701–708, 2001.

Zhang, Y., Song, C., Sun, G., Band, L. E., McNulty, S., Noormets, A., Zhang, Q., and Zhang, Z.: Development of a coupled carbon and water model for estimating global gross primary productivity and evapotranspiration based on eddy flux and remote sensing data, Agr. Forest Meteorol., 223, 116–131, 2016.

Zhou, L., Zhou, G. S., Liu, S. H., and Sui, X. H.: Seasonal contribution and interannual variation of evapotranspiration over a reed marsh (Phragmites australis) in Northeast China from 3-year eddy covariance data, Hydrol. Process., 24, 1039–1047, 2010.

6

Time series analysis of the long-term hydrologic impacts of afforestation in the Águeda watershed of north-central Portugal

D. Hawtree[1], **J. P. Nunes**[2], **J. J. Keizer**[2], **R. Jacinto**[2], **J. Santos**[2], **M. E. Rial-Rivas**[2], **A.-K. Boulet**[2], **F. Tavares-Wahren**[1], **and K.-H. Feger**[1]

[1]Technische Universität Dresden, Institute of Soil Science & Site Ecology, Pienner Str. 19, 01737 Tharandt, Germany
[2]University of Aveiro, Centre for Environmental and Marine Studies (CESAM), Department of Environment and Planning, 3810-193 Aveiro, Portugal

Correspondence to: D. Hawtree (daniel.hawtree@mailbox.tu-dresden.de)

Abstract. The north-central region of Portugal has undergone significant land cover change since the early 1900s, with large-scale replacement of natural vegetation types with plantation forests. This transition consisted of an initial conversion primarily to *Pinus pinaster*, followed by a secondary transition to *Eucalyptus globulus*. This land cover change is likely to have altered the hydrologic functioning of this region; however, these potential impacts are not fully understood. To contribute to a better understanding of the potential hydrologic impacts of this land cover change, this study examines the temporal trends in 75 years of data from the Águeda watershed (part of the Vouga Basin) over the period of 1936–2010. A number of hydrometeorological variables were analyzed using a combined Thiel–Sen/Mann–Kendall trend-testing approach, to assess the magnitude and significance of patterns in the observed data. These trend tests indicated that there have been no significant reductions in streamflow over either the entire test period, or during sub-record periods, despite the large-scale afforestation which has occurred. This lack of change in streamflow is attributed to the specific characteristics of the watershed and land cover change. By contrast, a number of significant trends were found for baseflow index, with positive trends in the early data record (primarily during *Pinus pinaster* afforestation), followed by negative trends later in the data record (primarily during *Eucalyptus globulus* afforestation). These trends are attributed to land use and vegetation impacts on streamflow generating processes, both due to species differences and to alterations in soil properties (i.e., infiltration capacity, soil water repellency). These results highlight the importance of considering both vegetation types/dynamics and watershed characteristic when assessing hydrologic impacts, in particular with respect to soil properties.

1 Introduction

Water resource management is inherently tied to watershed-scale land dynamics, and proper management requires understanding how changes in land cover/use will impact hydrological processes (Calder, 2005). A key land cover type are forests, as changes in forest cover have the potential to significantly affect watershed-scale hydrologic processes, particularly by altering interception, evaporation, and streamflow. Changes in water availability due to afforestation/deforestation are driven by several factors controlling the water consumption of different vegetation species, in particular canopy interception and evapotranspiration rates, which are typically higher in tree species than in shrub and herbaceous species (Calder, 1998).

Meta-analyses of paired catchments studies have found that afforestation typically results in decreased streamflow while deforestation typically leads to increased streamflow (e.g., Bosch and Hewlett, 1982; Brown et al., 2005). However, the hydrologic response to deforestation is in general more consistent than the response to afforestation. This difference may be due to higher variability in land cover following afforestation compared to deforestation, and the effects of different transitional species and/or changes in forest physiology (Andréassian, 2004). In a global synthesis of afforestation studies, Farley et al. (2005) found that afforestation of grasslands or shrublands will lead, on average, to reductions of one-third to two-thirds of streamflow, with these reductions occurring rapidly after planting (i.e., within the first 5 years) and reaching their maximum reduction 15 to 20 years following planting.

Changes in forest cover can also impact hydrologic processes by altering physical soil conditions, for example by reducing soil bulk density, increasing macroporosity, or changing soil water repellency. Forested areas tend to have higher infiltration and groundwater recharge rates than alternate land cover types (e.g., Bruijnzeel, 2004). Higher infiltration rates will increase soil moisture levels, and therefore increase water availability as well as streamflow during dry periods (e.g., Scott and Lesch, 1997). The increased infiltration capacity of forested areas may also help mitigate storm-driven peak flows, and therefore reduce potential flood damage; however, this effect may be subordinate to other watershed characteristics, particularly during severe flooding events (Calder, 2005; Wahren et al., 2012).

While the hydrologic impacts of forests at the watershed scale are generally well understood, predicting the effects of forest land cover change for a specific watershed requires consideration of both the physical site conditions and vegetation types involved. In this respect, Andréassian (2004) identified several prerequisite conditions that need to be met in order to observe hydrologic impacts at the watershed scale. These include climatic (i.e., periods of hydrologic surplus/deficit), pedological (i.e., soil depth), and ecophysiological (i.e., forest age-dependence) conditions.

The European Mediterranean region has undergone significant land cover changes over its long history of human habitation, which has left only an estimated 4.7 % of primary vegetation unaltered (Geri et al., 2010). These land cover changes are likely to have altered hydrologic processes at multiple scales, and the impacts of these changes are often not well understood. Gaining a better understanding of these past changes is critical for predicting the impact of future land cover changes, particularly given widespread concerns over potential water shortages in this region due to changing temperature and rainfall regimes (Giorgi and Lionello, 2008). Some of the most significant land cover/use changes observed in the European Mediterranean region in recent decades have been increased rural abandonment, a decrease in traditional agricultural/pastoral activities, and widespread planting of fast-growing tree species (Geri et al., 2010; Serra et al., 2008).

These regional trends are representative of the changes which have taken place in north-central Portugal, where traditional rural agrosilvopastoral activities have been widely replaced by plantations of the tree species *Pinus pinaster* and *Eucalyptus globulus* (Jones et al., 2012; Moreira et al., 2001). Both of these tree species have relatively high consumptive water demand and the potential to substantially reduce local water availability. Bosch and Hewlett (1982) estimated that pine and eucalypt forests cause an average decrease of over 40 mm yr^{-1} in water yield per 10 % change in land cover, while Farley et al. (2005) reported that afforestation with pines and eucalypti lead to reductions in streamflow of 40 % (±3 %) and 75 % (±10 %), respectively. Rodríguez-Suárez et al. (2011) found that afforestation with *Eucalyptus globulus* caused a drop in water table depth as well as a decrease in streamflow during the summer period, which they attributed to the higher transpiration capacity of the eucalypt plantations compared to the original crop lands.

In addition to consumptive water use through transpiration, evaporation from canopy interception is an important component of water use by Mediterranean forests. Interception rates have been found to vary widely in this region, depending on the tree species, canopy density, and climatic conditions. With respect to *Pinus pinaster*, Ferreira (1996) reported interception rates of 15–18 % in the Águeda watershed of north-central Portugal (mean precipitation ≈ 1700 mm yr^{-1}), while Valente et al. (1997) found similar rates of 17 % in a drier region of central Portugal (mean precipitation ≈ 600 mm yr^{-1}). For *Eucalyptus globulus*, both Ferreira (1996) and Valente et al. (1997) observed lower rates amounting to 10–14 and 11 %, respectively. By contrast, much higher interception rates have been found for other tree species in different parts of the Mediterranean, with values near and even exceeding 50 %. For example, Scarascia-Mugnozza et al. (1988) found canopy interception rates of 68 % for a mature *Quercus cerris* forest in central Italy (mean precipitation 1006 mm yr^{-1}), Iovino et al. (1998) found rates of 58 % for a mature *Pinus negra* forest in southern Italy (mean precipitation 1179 mm yr^{-1}, and Tarazona et al. (1996) observed rates of 48 % for a mature *Pinus sylvestris* forest in northern Spain (long-term mean precipitation of 895 mm yr^{-1}; 1253 mm yr^{-1} during the study period).

A further hydrologic factor relevant to afforestation in north-central Portugal is the potential for impacts on soil water repellency (SWR). Both pine and eucalyptus tree species can promote SWR in the topsoil due to the considerable amount of resins, waxes, and aromatic oils contained in their organic matter (Benito and Santiago, 2003; Doerr and Thomas, 2000; Ferreira et al., 2000; Keizer et al., 2005a, b). SWR is a key factor in triggering land degradation processes due to reductions in infiltration capacity and increased overland flow (Benito and Santiago, 2003; Doerr and Thomas, 2000; Keizer et al., 2005b; Shakesby et al., 2000). While in many regions SWR is associated primarily with post-fire soil conditions, Doerr et al. (1996) demonstrated that SWR is a widespread characteristic of both burned and unburned soils in the Águeda watershed during dry periods, in particular for stands of *Eucalyptus globulus*. Santos et al. (2013) examined temporal patterns in topsoil hydrophobicity in the Águeda watershed between July 2011 and June 2012 in unburnt pine and eucalypt plantations. Their findings suggested that the breakdown of SWR following dry summer conditions occurs through different mechanisms in the pine and eucalypt stands. In the pine stands, SWR breakdown occurred from the top-down (i.e., vertically downwards), while in the eucalypt stands, breakdown occurred from the bottom-up (i.e., vertically upwards). Unpublished results indicated that this contrast reflected varying infiltration patterns, with infiltration occurring relatively slowly (i.e., matrix flow) in pine stands, as opposed to much faster (i.e., macropore flow) in eucalypt stands. This contrast in infiltration patterns appeared to be a product of SWR-induced alterations in flow pathways.

Despite the well-documented potential for hydrologic impacts from afforestation in the Mediterranean region, there has been little investigation into the long-term effects in north-central Portugal. This is in part due to a lack of long-term streamflow records that allow for historical analyses. A notable exception to this lack of data is the Águeda watershed in the Caramulo Mountains, where streamflow data records are available from 1936 until the present.

Afforestation/deforestation studies typically focus on small paired watersheds, of which one has undergone fairly abrupt and well-recorded changes in land cover (e.g., Bosch and Hewlett, 1982). By contrast, this study is conducted on a mesoscale watershed (404 km^2), where afforestation has occurred progressively over an extended period of time. Furthermore, the present study case lacks a nearby watershed to serve as a paired site, which has a similarly long data record, similar physical–environmental characteristics, or a land use history without similar land cover changes (to serve as a control site).

To assess the hydrologic impacts of afforestation in the Águeda watershed, this study therefore adopts a data-driven and exploratory approach, which conducts multiple trend analyses on the 75-years of hydrometeorological data available from 1936 to 2010. This assessment is conducted over the entire data record as well as over multiple (overlapping) sub-periods for both annual and seasonal trends. The significant trends detected through this analysis are then considered with respect to the regional afforestation trends, and discussed in the context of previous field studies conducted in this watershed. Therefore, the objective of this study is to apply a trend-testing methodology to a long-term data set in a watershed which has undergone progressive afforestation over a 75-year period, to assess what significant trends can be detected, and to relate these changes to the afforestation which has occurred there.

2 Methods

2.1 Watershed description

The Águeda watershed is located in the Caramulo Mountains of north-central Portugal, east of the coastal city of Aveiro (Fig. 1). From the streamflow gauging point of Ponte

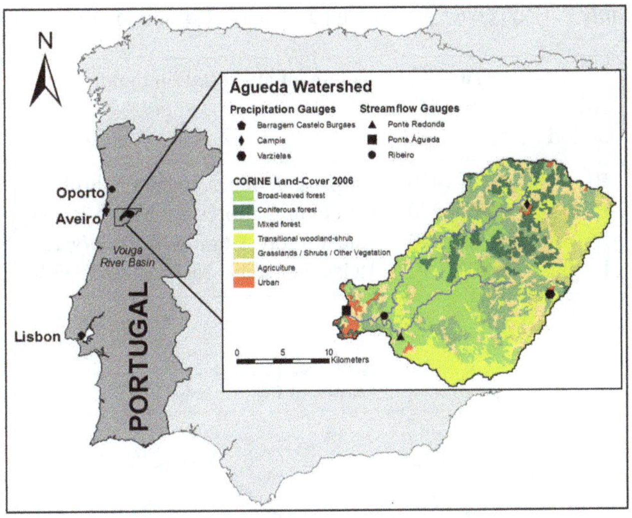

Figure 1. Location and land cover of the Águeda watershed.

Águeda, the watershed area is approximately 404 km^2. The Águeda River is a left bank tributary to the Vouga River, which terminates at the coastal wetland of the Ria de Aveiro lagoon. This region of Portugal is categorized as a wet Mediterranean climate zone, with pronounced seasonal differences in temperature and precipitation between dry summer and wet winter seasons (Fig. 2). The Serra do Caramulo Mountains, which forms the source area of the Águeda River network, receive a substantial amount of annual rainfall, which can range from 1000 to 2500 $mm\,yr^{-1}$. Topographically the landscape is dominated by steep hill slopes with stony and shallow soils (< 0.5 m), which have a long history of anthropogenic impacts. These shallow soils were characterized by Ferreira et al. (2000) as stony, sandy loam, weakly structured Umbric Leptosols.

North-central Portugal has undergone substantial land cover/use changes over the past centuries, which have fundamentally altered the vegetative landscape of this region. From the 1800s until the 1980s, the region had a general trend towards both increased agricultural and forest land cover, with reductions in natural vegetation types (e.g., *matos* (wooded) shrublands and mixed forests). This trend was primarily driven by the adoption of fertilizers and mechanization, as well as the abolition of feudal land systems (Estêvão, 1983; Jones et al., 2011; Silva et al., 2004). The period between 1930 and 1980 saw particularly rapid afforestation, due to incentives from the establishment of related government regulations and subsidies.

A key driver was the enactment of legislation in 1938 which encouraged afforestation of areas classified as "uncultivated/wasteland", which often consisted of areas of *matos* (shrublands), mountain ranges, and sand dunes (Coelho et al., 1995; Estêvão, 1983; Ferreira et al., 2010; Jones et al., 2011; Silva et al., 2004). The primary species planted during this earlier period was *Pinus pinaster*, and beginning in

Table 1. Land cover periods and dominant afforestation trends in Águeda watershed from 1935 to 2010.

Land cover period	Time period	Dominant afforestation trend
P1	1935–1950	Large-scale replacement of shrubland with *Pinus pinaster*.
P2	1950–1970	Continuing afforestation with *Pinus pinaster*, but at a slower rate.
E1	1970–1990	Rapid reforestation with *Eucalyptus globulus* (particularly post 1986 wildfire), replacement of *Pinus pinaster*.
E2	1990–2010	Relatively stable forested area, with continued replacement of *Pinus pinaster* with *Eucalyptus globulus*.

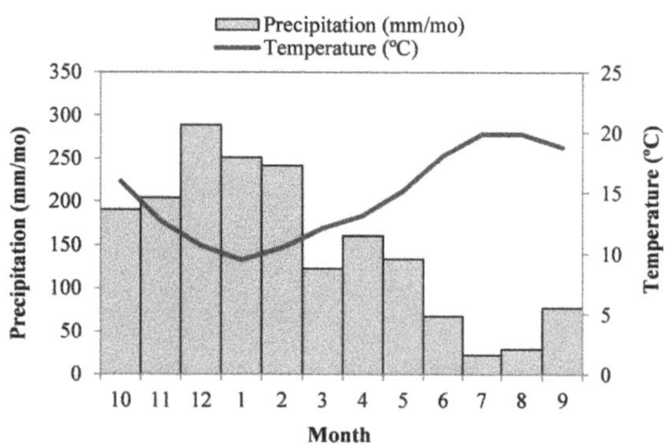

Figure 2. Average monthly precipitation and temperature in the Águeda watershed from 1971 to 2000.

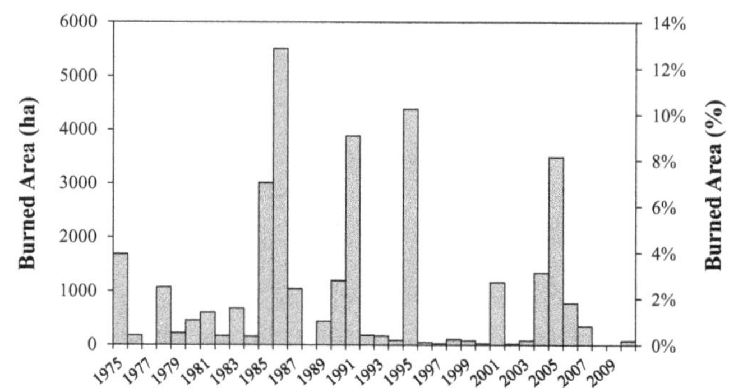

Figure 3. Burned area in the Águeda watershed from 1975 to 2010; total watershed area is 404 km^2.

the 1970s *Eucalyptus globulus* became the preferred species due to its faster growth and higher profitability for use in the paper pulp industry. During this period, eucalypt plantations began to replace pine forests as these were harvested, as well as being widely introduced into remaining areas of shrublands and in recently burned areas (Jones et al., 2011).

Wildfire is another important factor in land cover/use change in Portugal, which has some of the highest rates of wildfire in Europe. Figure 3 shows the burned area of the Águeda watershed from 1975 to 2010, during which a total of 30 790 ha burned, with some single years having wildfires over more than 10 % of the watershed (i.e., 1986 and 1995; Instituto da Conservaçao da Natureza e das Florestas, 2014). Wildfire can have significant short-term impacts on hydrologic functions in the study region, such as decreased infiltration and increased surface runoff/erosion (Malvar et al., 2011; Prats et al., 2012; Shakesby et al., 1993). In addition to these short-term impacts, wildfire can have potential long-term impacts by promoting changes in vegetation type. Wildfire has been a major driver of land cover change in north-central Portugal in this respect, by allowing land owners to convert from pine to eucalyptus plantations in the post-fire period.

This region-wide trend of the afforestation of shrubland with *Pinus pinaster*, followed by a secondary transition from *Pinus pinaster* to *Eucalyptus globulus* plantations, is representative of the land cover changes in the Vouga Basin as a whole, and for the Águeda watershed in particular. From this regional pattern, and from afforestation maps of the Serra do Caramulo Mountains (Rego, 2001), a general timeline of land cover change in the Águeda watershed during the period of investigation can be approximated, which is summarized in Table 1.

The current land cover in the Águeda watershed reflects this large-scale transition towards eucalyptus forests. According to the Corine Land Cover classification of 2006, approximately 46 % of the watershed was covered by broadleaf forest, which is predominantly eucalyptus (Corine Land Cover, 2010). Other land cover types with significant areal coverage in 2006 include 22 % mixed forest (mostly mixed stands of eucalypt and pine), 14 % agriculture, 10 % pine forest, 6 % mato shrubland, 2 % urban, and 1% grasslands (Fig. 1).

2.2 Hydrometeorological data

Hydrometeorological records for the Águeda watershed were compiled from hydrological year 1935–1936 (i.e., 01 October 1935 to 30 September 1936) until hydrological year 2009–2010, for the variables: precipitation, temperature, potential evapotranspiration, streamflow quantity, streamflow yield, baseflow quantity, and baseflow index. Table 2 provides an overview of the hydrometeorological variables used in this study.

Precipitation data were obtained from the rain gauge "Campia", of the "Sistema Nacional de Informação de Re-

Table 2. Summary of hydrometeorological variables.

Hydrometeorological variables			
Variable	Description	Data source	Unit
P	Precipitation	SNIRH gauge data	mm
T	Temperature	IPMA gauge data	°C
PET	Potential evapotranspiration	Thornthwaite equation	mm
Q	Streamflow quantity	SNIRH gauge data	mm
Q_{yld}	Streamflow yield	$\sum Q_{mm}/\sum P$	%
BF	Baseflow quantity	Recursive digital filter	mm
BFI	Baseflow index	$\sum BF_{mm}/\sum Q_{mm}$	%

cursos Hídricos" (SNIRH, 2013), which consists of 24 h rainfall totals collected at 09:00 LT each day. The SNIRH provides a reliability ranking for the data in the range of 5–15, for which Campia is ranked as 14 (highly reliable). Data gaps occurred with the greatest frequency between 1997 and mid-2003, which were filled by linear regression with the nearby rain gauges "Varzielas" ($r^2 = 0.82$) and "Barragem de Castelo Burgães" ($r^2 = 0.79$).

Temperature data were compiled using data from the gauge Campia of the Instituto Portugues do Mar e Atmosfera (IPMA, 2014). When data for "Campia" were not available, the time series gaps were filled using linear regression with the temperature gauge Coimbra ($r^2 = 0.93$) which is part of the Global Historical Climate Network available at the National Climatic Data Center (NCDC).

Potential evapotranspiration (PET) was estimated using the Thornthwaite (1948) equation, using the temperature data from the gauge Campia. The Thornthwaite equation was selected for PET, as opposed to more sophisticated equations (e.g., Hargreaves, Penman–Monteith), as there was insufficient data available over the entire time series to support calculations from these equations. Another option considered for long-term PET values is the gridded Penman–Monteith-based data set available from the Climatic Research Unit (CRU) at the University of East Anglia (Harris et al., 2014). To assess the suitability of this data set to the study's watershed, the monthly PET values of the CRU data and the locally derived Thornwaite values were compared against locally derived Penman–Monteith values, over the period of January 2002 to September 2010. This assessment indicated that the locally derived Thornwaite values are better correlated than the gridded CRU data set with local Penman–Monteith values. This may be due to the mountainous terrain of the study watershed, and the relatively large grid size of the CRU data set (0.5°) being unable to capture smaller scale impacts on PET. Based on this assessment, the locally derived Thornwaite PET values were identified as the most reliable and representative data source available for assessing long-term trends.

Streamflow data consist of daily average-discharge measurements from the gauging station "Ponte Águeda" of the SNIRH (2013). This station was operational from June 1935 to the end of September 1990, and was then reactivated in October 1999. Streamflow for the interim period (1990–1991 to 1998–1999) was estimated by linear regression with the upstream gauges "Ribeiro" ($r^2 = 0.76$) and "Ponte Redonda" ($r^2 = 0.75$). However, the streamflow estimates from the hydrologic years of 1999–2000 to 2002–2003 were eliminated from the data set due to low data quality, owing to the absence of an adequate stage-discharge curve during this period.

In addition, a number of smaller streamflow gaps occurred throughout the daily streamflow data set. When they occurred during periods with little or no precipitation, the gaps were filled by fitting a logarithmic decay curve (traditional linear reservoir with a semi-log fitting) to the streamflow recession. If gaps occurred during a precipitation event, then this approach was not applied and the gaps were left unfilled. If the number of gaps was greater than 5 % of the total record, then the entire period was removed from analysis, which was the case for the hydrologic years 1954–1955 and 1975–1976. Finally, data for the driest months of the year (i.e., June to September) during the period from before 1963 and after 2004 have very high uncertainty, due to unreported and variably occurring impoundments of streamflow during these months. Therefore, this 4 month period had to be removed from the streamflow analysis for the entire data record, to keep the inter-annual comparisons consistent. After the streamflow gaps were filled, the ratio of precipitation which becomes streamflow was calculated, to allow potential changes in the streamflow–precipitation relationship to be assessed. This ratio is defined as the "streamflow yield", which is the total streamflow divided by total precipitation, with the period of summation determined by the period being considered (i.e., the annual or the seasonal ratio).

The final data set utilized in this study is a baseflow time series calculated with the Eckhardt digital filter (Eckhardt, 2005) using the daily streamflow data set. Baseflow corresponds to the portion of streamflow which does not come directly from a precipitation event, and can be used as a proxy of the sustained streamflow contribution from slow flow. The relative proportion of baseflow from each day of streamflow was estimated, which was then aggregated to the time periods used for analysis. To assess the baseflow time series calculated using the Eckhardt digital filter, a supplementary data set from 2001 to 2009 was also utilized, which calculates baseflow contribution using conductivity data from the SNIRH streamflow data using the "conductivity mass balance method" (Stewart et al., 2007).

2.3 Thiel–Sen/Mann–Kendall trend-testing approach

To examine the magnitude and significance of potential trends in the hydrometeorological time series, a multi-step trend-testing approach was applied, following the general approach presented in Yue et al. (2002). This approach first determines the magnitude (i.e., slope) of any potential trend in

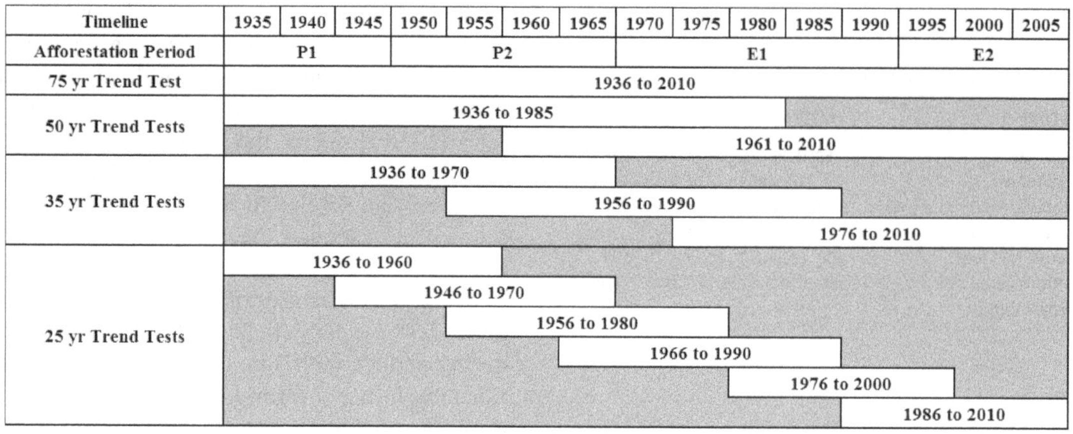

Figure 4. Timeline of the trend-testing periods and their correspondence with the different afforestation periods.

the data using the non-parametric Thiel–Sen slope estimator (Sen, 1968). This value is determined by calculating the median slope among the set generated between all sample points. This method also estimates the 95 % confidence intervals of the true slope, based on the set of slopes from the sample points, which provides a measure of uncertainty of the median Thiel–Sen value. If a potential trend is detected by the Thiel–Sen test (i.e., a non-zero slope), then the data are processed using the "trend-free pre-whitening" procedure of Yue et al. (2002). This step reduces the overestimation of significance which can occur in time series data that exhibits positive serial correlation, as is typically the case for streamflow time series data.

After the trend-free pre-whitening, a Mann–Kendall test was applied to assess the statistical significance of any non-zero slope identified by the Thiel–Sen test. The Mann–Kendall test is a widely used, rank-based significance test, where the null hypothesis is that there is no trend in the observed data (Helsel, 1993). For this study, statistical significance was determined using an α value of 0.05.

For each hydrometeorological variable, this trend-testing procedure was applied over 12 different time periods with varying start/end dates and lengths (Fig. 4). The longest period tested contains the entire 75-year data record (hydrologic year 1936–2010), followed by two periods of 50 years, three periods of 35 years, and six periods of 25 years. These overlapping periods of different lengths aim to thoroughly sample the potential range of years, while still allowing for enough years of data to produce a robust significance test within each test period (i.e., a minimum of 25 years). Figure 4 provides an overview of the testing periods, and their temporal correspondence with the afforestation periods listed in Table 1.

When conducting multiple simultaneous hypothesis tests, it is necessary to correct for the false discovery rate (FDR). FDR corresponds to the expected proportion of incorrectly rejected null hypotheses, and therefore a method is needed to reduce the chance of receiving false positive results (i.e., type I errors). A number of different methods can be applied to control for FDR; however, given the overlapping time periods examined in this study, a method is needed which can deal with FDR under the assumption of positive dependence. Therefore, the Benjamini–Hochberg–Yekutieli procedure was applied to the trend-testing output from each individual "analysis set" (Benjamini and Yekutieli, 2001). An analysis set corresponds to a group of tests which are expected to exhibit mutual positive dependence, which in this case are the 12 overlapping periods over which each hydrometeorological variable was tested for the different annual and seasonal periods (i.e., Fig. 4 for a given variable and period).

Over the time periods shown in Fig. 4, the trend testing was conducted over both annual and seasonal time periods. The seasonal breakdown corresponds to the prevailing precipitation patterns of the study site, which consists of a "wet season" from October to January when the largest amount of precipitation occurs, a "transitional season" from February to May when precipitation rates are reduced, and a "dry season" from June to September when precipitation is lowest. Due to gaps in the streamflow record (discussed in Sect. 2.2), the hydrologic years 1999–2000 through 2002–2003 were unavailable for the trend testing for both the annual and seasonal time periods, and the hydrologic years 1954–1955 and 1975–1976 were unavailable for the annual and transitional season. In addition, the trend tests were not conducted during the dry period for streamflow (and therefore also baseflow), due to the uncertain data quality during these months.

3 Results

3.1 Summary of the seasonal breakdown

To characterize the hydrometeorological conditions of the three seasons used in this study, the median values of the hydrometeorological variables during the study period are

Table 3. Seasonal and annual median values of the hydrometeorological variables in Águeda watershed from 1936 to 2010.

Season	Months	Median values: 1936–2010						
		P (mm)	T (°)	PET (mm)	Q (mm)	Q_{yld} (%)	BF (mm)	BFI (%)
Wet	Oct–Jan	965	11.7	145	301	30 %	149	55 %
Transitional	Feb–May	626	12.6	198	281	43 %	184	63 %
Dry	Jun–Sep	193	19.3	390	NA	NA	NA	NA
Annual	All*	1 787	14.7	732	565	36 %	320	59 %

* The months of June to September are not included for Q (mm), Q_{yld} (%), BF (mm), and BFI (%). NA = not available.

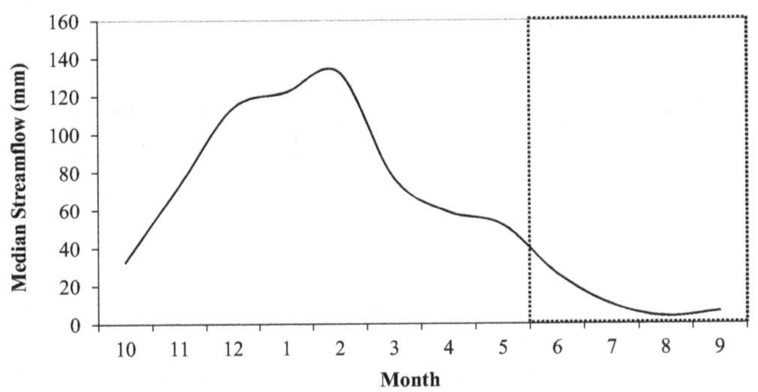

Figure 5. Monthly mean streamflow during the years without seasonal impoundments; the boxed off period (June–September) indicates the period removed from the streamflow and baseflow analysis.

presented in Table 3. This summary shows the strong climatic pattern in the watershed, with distinctly contrasting precipitation, temperature, and potential evapotranspiration values between seasons. With respect to streamflow, the values are similar during the wet and transitional seasons; however, both streamflow yield and baseflow index are higher during the transitional season, which reflects the sustained streamflow carried over from the wet season precipitation, and the lower proportion of streamflow coming directly from precipitation events.

3.2 Analysis of the elimination of the dry season streamflow

As discussed in the data section, the months of June to September had to be removed from all streamflow analyses, due to uncertainty related to unrecorded seasonal impoundments during this part of the year. To quantify the percentage of streamflow that this excluded from the analysis, an assessment was made over the years when streamflow impoundments did not occur (45 % of years). During these years, approximately 6.5 % of streamflow occurred between the months of June to September (Fig. 5; monthly mean values presented).

3.3 Assessment of the baseflow calculations

The baseflow index used in this study is estimated by the Eckhardt digital filter, which contains two parameters: BFI_{max} (maximum value of the baseflow index) and a (a filter parameter). These parameters were set at $BFI_{max} = 0.80$, and $a = 0.98$, based on an examination of different recommended values provided by Eckhardt (2005). To provide a check on the baseflow values estimated using this method, the results were then compared against baseflow values calculated using conductivity data from 2001 to 2009 with the conductivity mass balance method (Stewart et al., 2007).

At the monthly timescale, the two compared baseflow data sets have a Pearson's correlation coefficient of 0.96 for all months (Fig. 6a), and 0.83 for months with less than 100 mm of baseflow (Fig. 6b), which indicates that the Eckhardt method agreed well with the more empirical conductivity mass balance method. This in itself does not confirm the accuracy of the baseflow values utilized, but it does indicate their consistency over the study period, and thus their suitability for the time series trend analysis.

3.4 Thiel–Sen/Mann–Kendall trend-testing results

The results for the Thiel–Sen/Mann–Kendall trend tests for the variables with the most noteworthy results (i.e., precipitation, temperature, potential evapotranspiration, streamflow yield, and baseflow index) are presented by Fig. 7. The full test results for all hydrometeorological variables and test periods are provided in the supplementary material.

For the precipitation data, three significant trends were identified during the transitional season. All trends corresponded to decreases in precipitation: a $-7.9\ mm\ yr^{-1}$ trend over the 50 years from 1961 to 2010, a $-11.3\ mm\ yr^{-1}$ trend over the 35 years from 1976 to 2010, and a $-14.3\ mm\ yr^{-1}$ trend over the 25-year period from 1976 to 2000. These trends indicate that there was a pattern of decreasing precipitation totals during the transitional season (February to May) starting during the P2 land cover period, and this pattern continued through the E1 and E2 land cover periods (cf. Table 1).

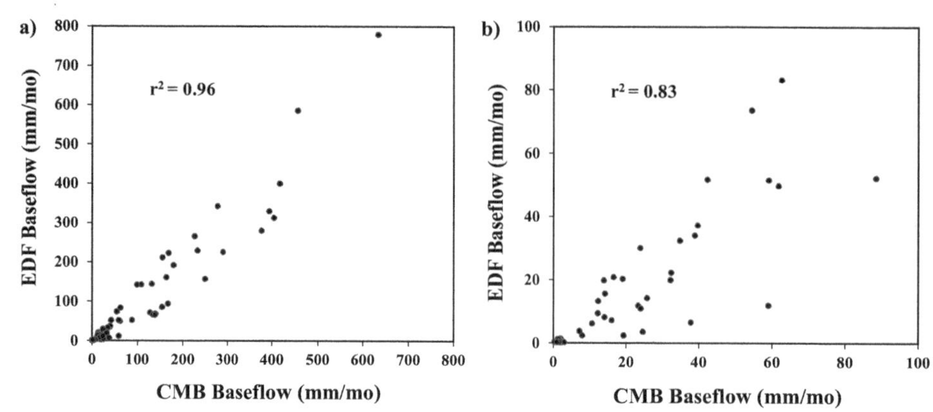

Figure 6. Monthly plots of baseflow from the conductivity mass balance (CMB) and Eckhardt digital filter calculations; (**a**) includes all months ($r^2 = 0.96$) and (**b**) includes months with less than 100 mm of baseflow ($r^2 = 0.83$).

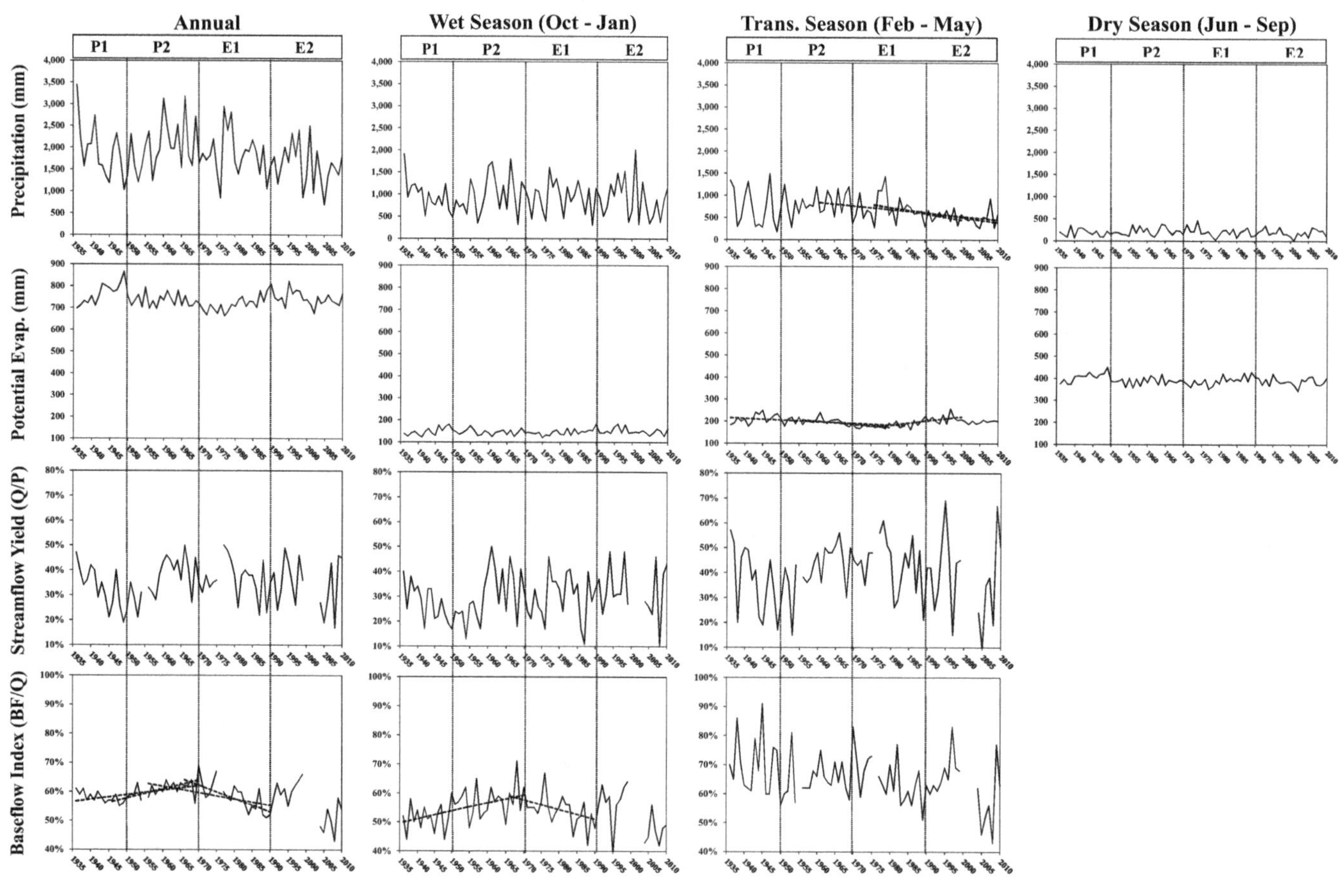

Figure 7. Summary of the trend-testing results, with the afforestation periods (P1, P2, E1, E2; cf. Table 1) overlain for comparison. Significant trends are indicated with dashes lines.

Three significant trends were also found for PET during the transitional season: a $-0.8\ \mathrm{mm\,yr^{-1}}$ trend over the 50 years from 1936 to 1985, a $-1.3\ \mathrm{mm\,yr^{-1}}$ trend over the 25 years from 1956 to 1980, and a $1.7\ \mathrm{mm\,yr^{-1}}$ trend over the 25-year period from 1976 to 2000. Therefore, the PET data show a pattern of negative trends throughout the P1, P2, and into the E1 land cover periods, which reverses and becomes positive during the E1 period and into the E2 land cover period (cf. Table 1).

For the streamflow data record, no significant trends were found for either streamflow quantity or streamflow yield. No significant trends were found for baseflow quantity either; however, a number of significant trends were found for baseflow index (BFI). For the annual test period, four significant trends were found in total: including significant positive trends of $0.16\,\%\,yr^{-1}$ for the 35 year period from 1936 to 1970 and of $0.31\,\%\,yr^{-1}$ for the 25 year period from 1946 to 1970, and negative trends of $-0.22\,\%\,yr^{-1}$ for the 35 year period from 1956 to 1990 and a $-0.46\,\%\,yr^{-1}$ trend for the 25 year period from 1966 to 1990. Two significant trends were found for BFI during the wet season: a $0.28\,\%\,yr^{-1}$ trend for the 35 year period from 1936 to 1970 and a $-0.33\,\%\,yr^{-1}$ trend for the 25 year period from 1966 to 1990. Therefore, the BFI data showed an overall pattern of positive trends during the P1 and P2 land cover periods, which reverse to negative trends during the P2 period and throughout the E1 land cover period (cf. Table 1).

4 Discussion

4.1 Streamflow trends

The streamflow trend tests revealed that there were no significant trends for either quantity or yield over any of the periods tested (Fig. 7). These results therefore contrast with the overall pattern found in meta-analysis studies dealing with the hydrologic impacts of afforestation/deforestation, which indicate that afforestation tends to reduce streamflow (e.g., Bosch and Hewlett, 1982; Brown et al., 2005; Farley et al., 2005). However, there are a number of individual cases within these meta-analyses studies which show contrasting trends to the overall pattern. These cases are difficult to directly compare to the current study, however, as most were conducted at the plot to micro-catchment scale, which underwent relatively rapid land cover change. By contrast, this study was conducted on a $404\,km^2$ watershed, which underwent relatively gradual land cover change over a 75 year period. In this case, any potential changes in hydrologic processes are likely to be far more diffuse and difficult to detect, when compared to the paired catchment studies.

Despite this limitation, some comparisons can be made to sites with similar site conditions, in terms of having winter-dominant precipitation and shallow soils. Across a number of catchments with winter-dominant rainfall, Brown et al. (2005) found that afforestation led to much larger proportional reductions in summer flows compared to winter flows, which they attributed to the afforestation-induced changes in interception and evapotranspiration. Among these catchments, those of Gallart et al. (2001) and Lewis et al. (2000) demonstrated the importance of soil depth in controlling the hydrological response of Mediterranean mountain catchments in the Pyrenees and California, respectively. Other studies with somewhat similar site conditions (i.e., Bari et al., 1996; Van Lill et al., 1980) were conducted at very different temporal and spatial scales than the present study, making comparisons to their findings difficult. In spite of the lack of comparable studies for direct comparison, the absence of a marked reduction in streamflow was an unexpected finding, given the scale of afforestation in the Águeda watershed.

A potential explanation for this lack of observed impact could be the presence of offsetting climatic trends over the same period. Either an increase in water availability due to higher precipitation (P) and/or a reduction in atmospheric demand due to lower PET could compensate for any land cover induced changes. While no significant trends were found for either P or PET at the annual timescale, or during the wet or dry seasons, significant trends were found during the transitional season, which may have impacted water availability.

With respect to increasing water availability during the transitional season, negative trends in PET were found from 1936 to 1985 and from 1956 to 1980 (Fig. 7). These trends occur primarily during the periods of pine afforestation (P1, P2) and partially during the transition to eucalyptus (E1; cf. Table 1). The trends in PET would lead to a reduction in atmospheric demand during this period, and therefore could be responsible for offsetting an increase in consumptive demand that occurred from afforestation.

With respect to reductions in water availability during the transitional season, negative trends in P were found from 1961 to 2010, 1976 to 2010, and 1976 to 2000, and a positive trend in PET was found from 1976 to 2000 (Fig. 7). These trends indicate movement toward a relatively more arid environment, which could therefore lead to a reduction in water availability. However, no corresponding trends in streamflow were found during this period. This lack of change is particularly noteworthy given that these trends occurred during the eucalyptus afforestation periods (E1, E2; cf. Table 1), which would also be expected to increase consumptive demand, and would therefore amplify, rather than offset, an increase in atmospheric demand.

Given the lack of significant climate trends at the annual timescale, and the contrasting findings during the transitional season, offsetting climatic trends do not appear to be an adequate explanation for the overall lack of observed streamflow changes in the Águeda watershed. However, given that the observed climate trends occurred during the transitional season, there may have been streamflow impacts during the (following) dry season. This can only be speculated on however, since no assessment can be made on streamflow during the dry season, due to the limitations in the streamflow data (i.e., the summer streamflow impoundments). Therefore, no comparison could be made with the findings of Rodríguez-Suárez et al. (2011), who found dry season reductions in the water table and streamflow discharge following afforestation with eucalyptus, or to Brown et al. (2005) whose work found that afforestation led to much larger proportional reductions in summer flows compared to winter flows.

An alternate explanation for the lack of streamflow change could relate to the specific characteristics of the watershed, which may make it less responsive to changes in forest land cover than is typical. With respect to watershed characteristics, Andréassian (2004) identifies several prerequisite conditions necessary to observe hydrologic impacts, including soil, climatic, and eco-physiological factors.

With respect to soil conditions, the characteristics of the soils of the Águeda watershed may be a key factor in the lack of a reduction in streamflow. Under conditions of well-developed soils, the deeper rooting depths of trees will give greater access to soil moisture, allowing for more transpiration, resulting in higher water consumption. However, the soils of the Águeda watershed tend to be fairly shallow, being typically less than 1 m deep and often as shallow as 20–30 cm (Santos et al., 2013). These depths are less than the maximum rooting depth of pine and eucalypt trees, and therefore are likely to be a constraint to deep rooting for both species (Canadell et al., 1996). In addition, the schist and granite bedrock in this watershed is relatively impermeable and not easily penetrated by tree roots, which restricts the access of tree species to groundwater reserves as well. Therefore, the capability of the fast-growing pine and eucalypt trees to access deeper sources of soil moisture than the original shrub and slow-growing tree species is likely much less relevant in this watershed than it would be in a location with deeper soils. In the case of the Águeda watershed, the most important soil related factor in water consumption appears to be the low moisture storage capacity of the soils, severely offsetting the potential impact of widespread planting of trees with higher water consumptive capacity.

A second factor which could contribute to the lack of reductions in streamflow is the Mediterranean climate regime of the study area. In all Mediterranean-type climates, the period of peak sunlight and temperature, and therefore potential evapotranspiration, is out of phase with the maximum precipitation period (Brown et al., 2005). Given the low amount of summer precipitation, and the shallowness of soils in this watershed, there will typically be little soil water available for summer evapotranspiration (David et al., 1997; Doerr and Thomas, 2000). In this regard, the climatic conditions of the Águeda catchment may have an amplifying effect on the impacts of the shallow soils, by further reducing the higher evapotranspiration potential of fast-growing trees species.

With respect to eco-physiological conditions, the specific land cover changes in the Águeda watershed may also be a factor in the lack of an observed reduction in streamflow. One of the primary drivers of increased consumptive water use by tree species is their typically high canopy interception capacity (Domingo et al., 1994; Scarascia-Mugnozza et al., 1988; Tarazona et al., 1996). In the Águeda watershed, however, the interception rates appear to be comparatively low for pine and eucalypt species (Coelho et al., 2008; Ferreira, 1996; Valente et al., 1997), while the interception capacity of Mediterranean shrublands can be relatively high. Garcia-Estringana et al. (2010) found that Mediterranean shrub species can have interception capacities similar to those of forests. In addition, interception rates are particularly high in shrublands growing in dense stands (Llorens and Domingo, 2007). These characteristics apply to the *matos* shrubland that was the most common vegetation type in the Águeda watershed prior to pine afforestation, as it has a relatively high leaf area index and the tendency to grow in very dense stands (Asner et al., 2003). By contrast, given the poor soil conditions of the study site, the densities of the tree plantations are not as high as they could be on well-developed soils. Average tree density from unpublished plot assessments put the density of unevenly spaced eucalyptus stands (< 15 years old) at 1600 trees ha^{-1}, of evenly spaces eucalyptus stands on terraces (< 5 years old) at 1,500 trees/ha, of eucalyptus on flat terrain (< 5 years old) at 2600 trees ha^{-1}, and of unevenly spaced pines (< 30 years old) at 500 trees ha^{-1}. Therefore, the land cover/use change from shrubland to pine/eucalypt forest might not have resulted in large changes in either transpiration rates or canopy interception rates.

Therefore, the Águeda watershed does not meet the prerequisite conditions identified by Andréassian (2004) for observing afforestation-driven streamflow changes at the watershed scale. Given this lack of prerequisites conditions, and the absence of offsetting climate trends as an alternative explanation, the streamflow findings of this study appear to be primarily a function of watershed characteristics, with soil properties as the most important factor.

4.2 Baseflow trends

No significant trends were found for baseflow quantity (BF) over any of the periods or seasons tested. However, a number of trends were found for BFI, for both the annual data and the wet season data, which includes both positive and negative trends over different parts of the data record.

Positive trends in BFI were found from 1936 to 1970 for the annual data and the wet season, and from 1946 to 1970 for the annual data (Fig. 7). These trends correspond with the pine afforestation land cover periods P1 and P2 (cf. Table 1). These trends could be an indication that the pine afforestation promoted slower flow pathways, by increasing the amount of water entering the soil matrix via infiltration, and reducing surface flow and fast sub-surface flow (i.e., via macropores). However, given that previous studies in Águeda watershed have found SWR conditions at pine stands (during dry periods), pine afforestation would not normally be expected to increase matrix infiltration in this location (Keizer et al., 2005a, b; Santos et al., 2013). However, the land cover state during the initial conversion to pine forests were significantly different from the state during these studies, which may have led to a more positive impact on infiltration rates. This is due to the ground preparation and planting operations used, which would have the effect of breaking up the repellent topsoil layer and creating sinks for overland flow, both of which would promote infiltration. This effect would be reduced over time, and eventually SWR would recover in established stands.

Negative BFI trends were found from 1956 to 1990 for the annual data, and from 1966 to 1990 for the wet season (Fig. 7). This corresponds with the early part of the P2 land cover period, and the entirety of the first eucalyptus afforestation period (E1; cf. Table 1). Therefore, the negative BFI trends occur during the period when *Pinus pinaster* plantations reached greater maturity and (after logging) were being rapidly replaced with *Eucalyptus globulus*. The reductions in baseflow during this period may therefore be related to high rates of SWR in the established pine stands and the newly established eucalypt stands. An increase in SWR could lead to an increase in quick flow, particularly via fast sub-surface flow from macropore infiltration, and lead to more rapid conversion of precipitation into streamflow.

The temporal correspondence between the significant trends in BFI and land cover changes which could affect hydrologic flow pathways indicate there may be a relationship between afforestation and changes in baseflow index in the Águeda watershed. These findings are further supported by field studies conducted in the watershed, which show the strong impact of SWR in pine and (particularly) eucalyptus stands on hydrologic flow pathways (Santos et al., 2013). However, given that there is no field data available to verify the site conditions during the time of the observed trends, the attribution of the changes in BFI to land cover change is necessarily speculative. To test this hypothesis, further field studies would be needed to examine baseflow dynamics under land cover conditions which replicate the historic conditions.

5 Conclusions

This study did not detect statistically significant – negative or positive – trends in streamflow quantity or yield in the Águeda watershed of north-central Portugal over the 75 year period examined, despite the large-scale afforestation with *Pinus pinaster* and later *Eucalyptus globulus* which has taken place there. While these findings differ from the general conclusion of afforestation/deforestation meta-analysis studies, such as Bosch and Hewlett (1982), Brown et al. (2005), and Farley et al. (2005), they do support the assertion of Andréassian (2004) that there are prerequisite climatic, pedological, and eco-physiological watershed conditions that are necessary to observe hydrologic impacts at the watershed scale. These conditions are not present in the Águeda watershed, and the lack of soil moisture holding capacity is likely the primary controlling factor.

With respect to baseflow trends, the initial conversion from more natural land cover types (i.e., *matos* shrublands, mixed forests) to pine plantations appears to have had a significant – initial – positive impact on baseflow index, while the substitution of pine plantations by eucalypt plantations had a negative impact on baseflow index. The positive trends are attributed to the impact of the site preparation methods applied during the initial pine planting on soil infiltration capacity, while the negative baseflow trends are attributed to the onset of soil water repellency (SWR) under the mature pine and eucalypt stands. Therefore, from the standpoint of promoting well-regulated streamflow (i.e., higher baseflow), the impacts of the afforestation with pine appear generally positive, while those of re-/afforestation with eucalypti were generally negative.

However, it is important to stress that the pine and eucalypt planting in the study catchment took place on dissimilar types of land cover. Pines were primarily replacing naturally occurring shrublands, which was followed by the replacement of the planted pines by eucalypti. Therefore, a direct comparison between the impacts of widespread planting with pine or with eucalypt cannot be drawn from this study. In addition, these baseflow findings are based on a statistical/historical analysis, with no field data available for validation. To further test this hypothesis, field studies would be needed to examine baseflow dynamics under different land cover conditions replicating the historic conditions.

Acknowledgements. The authors would like to express their gratitude to Professor Celeste Coelho at the University of Aveiro, Centre for Environmental and Marine Studies (CESAM) for providing her expertise on the background and land use history of the study region. Thanks are also given to Ameur Manceur and Robert Schlicht, for providing assistance on the statistical methods. Funding for this research was provided by the "Erasmus Mundus – Forest and Nature for Society (FONASO) Joint Doctoral Program"; the European Regional Development Fund (through COMPETE), the European Social Fund and the Portuguese Republic (through FCT), both by individual fellowships (references: SFRH/BPD/87571/2012; SFRH/BPD/64425/2009; SFRH/BD/61451/2009) and research projects (references: FCOMP-01-0124-FEDER-009308; FCOMP-01-0124-FEDER-008534; PTDC/CTE-ATM/111508/2009); and the German Academic Exchange Service (DAAD) for sponsoring a bilateral collaboration action between Universidade de Aveiro and TU Dresden.

Edited by: C. Harman

References

Andréassian, V.: Waters and forests: from historical controversy to scientific debate, J. Hydrol., 291, 1–27, doi:10.1016/j.jhydrol.2003.12.015, 2004.

Asner, G. P., Scurlock, J. M. O., and Hicke, J. A.: Global synthesis of leaf area index observations: implications for ecological and remote sensing studies, Global Ecol. Biogeogr., 12, 191–205, doi:10.1046/j.1466-822X.2003.00026.x, 2003.

Bari, M. A., Smith, N., Ruprecht, J. K., and Boyd, B. W.: Changes in streamflow components following logging and regeneration in the southern forest of Western Australia, Hydrol. Process., 10, 447–461, doi:10.1002/(SICI)1099-1085(199603)10:3<447::AID-HYP431>3.0.CO;2-1, 1996.

Benito, E. and Santiago, J. L.: Deforestation of water repellent soils in Galicia (NW Spain): effects on surface runoff and erosion under simulated rainfall, Earth Surf. Proc. Land., 28, 145–155, doi:10.1002/esp.431, 2003.

Benjamini, Y. and Yekutieli, D.: The control of the false discovery rate in multiple testing under dependency, Ann. Stat., 29, 1165–1188, 2001.

Bosch, J. M. and Hewlett, J. D.: A review of catchment experiments to determine the effect of vegetation changes on water yield and evapotranspiration, J. Hydrol., 55, 3–23, doi:10.1016/0022-1694(82)90117-2, 1982.

Brown, A. E., Zhang, L., McMahon, T. A., Western, A. W., and Vertessy, R. A.: A review of paired catchment studies for determining changes in water yield resulting from alterations in vegetation, J. Hydrol., 310, 28–61, doi:10.1016/j.jhydrol.2004.12.010, 2005.

Bruijnzeel, L. A.: Hydrological functions of tropical forests: not seeing the soil for the trees?, Agric. Ecosyst. Environ., 104, 185–228, doi:10.1016/j.agee.2004.01.015, 2004.

Calder, I. R.: Water use by forests, limits and controls, Tree Physiol., 18, 625–631, doi:10.1093/treephys/18.8-9.625, 1998.

Calder, I. R.: Blue revolution: integrated land and water resource management, Earthscan, London, 2005.

Canadell, J., Jackson, R. B., Ehleringer, J. B., Mooney, H. A., Sala, O. E., and Schulze, E.-D.: Maximum rooting depth of vegetation types at the global scale, Oecologia, 108, 583–595, doi:10.1007/BF00329030, 1996.

Coelho, C. O. A., Shakesby, R. A., and Walsh, R. P. D.: Effects of Forest Fires and Post-fire Land Management Practice on Soil Erosion and Stream Dynamics, Águeda Basin, Portugal, Soil and groundwater research report V, European Commision, Luxembourg, 1995.

Coelho, C. O. A., Ferreira, A. J. D., Prats, S. A., Tomé, M., Soares, P., Cortiçada, A., Tomé, J. A., Salas, G. R., Páscoa, F., and Amaral, A.: Assessment of climatic change impact on water resources and CO_2 fixation in fast growing stand in Portugal, Final Report Silvaqua Project POCTI/MGS/49210/2002, University of Aveiro, 2008.

Corine Land Cover: Corine Land Cover 2006 raster data, European Environment Agency (EEA), available at: http://www.eea.europa.eu/data-and-maps/data/ds_resolveuid/a645109f7a11d43f5d7e275d81f35c61 3 November 2014, 2010.

David, T. S., Ferreira, M. I., David, J. S., and Pereira, J. S.: Transpiration from a mature Eucalyptus globulus plantation in Portugal during a spring–summer period of progressively higher water deficit, Oecologia, 110, 153–159, doi:10.1007/PL00008812, 1997.

Doerr, S. H. and Thomas, A. D.: The role of soil moisture in controlling water repellency: new evidence from forest soils in Portugal, J. Hydrol., 231–232, 134–147, doi:10.1016/S0022-1694(00)00190-6, 2000.

Doerr, S. H., Shakesby, R. A., and Walsh, R. P. D.: Soil hydrophobicity variations with depth and particle size fraction in burned and unburned Eucalyptus globulus and Pinus pinaster forest terrain in the Águeda Basin, Portugal, Catena, 27, 25–47, doi:10.1016/0341-8162(96)00007-0, 1996.

Domingo, F., Puigdefabregas, J., Moro, M. J., and Bellot, J.: Role of vegetation cover in the biogeochemical balances of a small afforested catchment in southeastern Spain, J. Hydrol., 159, 275–289, doi:10.1016/0022-1694(94)90261-5, 1994.

Eckhardt, K.: How to construct recursive digital filters for baseflow separation, Hydrol. Process., 19, 507–515, doi:10.1002/hyp.5675, 2005.

Estêvão, J. A.: A florestação dos baldios, Análise Soc., XIX, 77–79, 1983.

Farley, K. A., Jobbágy, E. G., and Jackson, R. B.: Effects of afforestation on water yield: a global synthesis with implications for policy, Global Change Biol., 11, 1565–1576, doi:10.1111/j.1365-2486.2005.01011.x, 2005.

Ferreira, A. J. D., Coelho, C. O. A., Walsh, R. P. D., Shakesby, R. A., Ceballos, A., and Doerr, S. H.: Hydrological implications of soil water-repellency in Eucalyptus globulus forests, north-central Portugal, J. Hydrol., 231–232, 165–177, doi:10.1016/S0022-1694(00)00192-X, 2000.

Ferreira, C. G.: Erosão hídrica em solos florestais: estudo em povoamentos de Pinus Pinaster e Eucalyptus Globulus em macieira de Alcôba-Águeda, Rev. Fac. Let. – Geogr. Sér., XII/XIII, 145–244, 1996.

Ferreira, R. V., Cerqueira, M. A., de Melo, M. T. C., de Figueiredo, D. R., and Keizer, J. J.: Spatial patterns of surface water quality in the Cértima River basin, central Portugal, J. Environ. Monit., 12, 189–199, doi:10.1039/b914409a, 2010.

Gallart, F., Llorens, P., Latron, J., and Regüés, D.: Hydrological processes and their seasonal controls in a small Mediterranean mountain catchment in the Pyrenees, Hydrol. Earth Syst. Sci., 6, 527–537, doi:10.5194/hess-6-527-2002, 2001.

Garcia-Estringana, P., Alonso-Blázquez, N., and Alegre, J.: Water storage capacity, stemflow and water funneling in Mediterranean shrubs, J. Hydrol., 389, 363–372, doi:10.1016/j.jhydrol.2010.06.017, 2010.

Geri, F., Amici, V., and Rocchini, D.: Human activity impact on the heterogeneity of a Mediterranean landscape, Appl. Geogr., 30, 370–379, doi:10.1016/j.apgeog.2009.10.006, 2010.

Giorgi, F. and Lionello, P.: Climate change projections for the Mediterranean region, Global Planet. Change, 63, 90–104, doi:10.1016/j.gloplacha.2007.09.005, 2008.

Harris, I., Jones, P. D., Osborn, T. J., and Lister, D. H.: Updated high-resolution grids of monthly climatic observations – the CRU TS3.10 Dataset , Int. J. Climatol., 34, 623–642, doi:10.1002/joc.3711, 2014.

Helsel, D. R. and Hirsch, R. M.: Statistical Methods in Water Resources Techniques of Water Resources Investigations, Book 4, chapter A3, US Geological Survey, 522 pp., 2002.

Instituto da Conservaçao da Natureza e das Florestas: Mapas de áreas ardidas entre 1990–1996, 1997–2004, 2010, 2011, 2012 e 2013, available at: http://www.icnf.pt/portal/florestas/dfci/inc/mapas, last access: 1 July 2014.

Iovino, F., Cinnirella, S., Veltri, A., and Callegari, G.: Processus hydriques dans des écosystèmes forestiers, Ecologie, 29, 369–375, 1998.

IPMA: Instituto Portugues do Mar e Atmosfera, IPMA, available at: http://snirh.pt (last access: 22 July 2014), 2014.

Jones, J. A., Creed, I. F., Hatcher, K. L., Warren, R. J., Adams, M. B., Benson, M. H., Boose, E., Brown, W. A., Campbell, J. L., Covich, A., Clow, D. W., Dahm, C. N., Elder, K., Ford, C. R., Grimm, N. B., Henshaw, D. L., Larson, K. L., Miles, E. S., Miles, K. M., Sebestyen, S. D., Spargo, A. T., Stone, A. B., Vose, J. M., and Williams, M. W.: Ecosystem Processes and Human Influences Regulate Streamflow Response to Climate Change at Long-Term Ecological Research Sites, BioScience, 62, 390–404, doi:10.1525/bio.2012.62.4.10, 2012.

Jones, N., de Graaff, J., Rodrigo, I., and Duarte, F.: Historical review of land use changes in Portugal (before and after EU integration in 1986) and their implications for land degradation and conservation, with a focus on Centro and Alentejo regions, Appl. Geogr., 31, 1036–1048, doi:10.1016/j.apgeog.2011.01.024, 2011.

Keizer, J. J., Coelho, C. O. A., Matias, M. J. S., Domingues, C. S. P., and Ferreira, A. J. D.: Soil water repellency under dry and wet antecedent weather conditions for selected land-cover types in the coastal zone of central Portugal, Soil Res., 43, 297–308, 2005a.

Keizer, J. J., Coelho, C. O. A., Shakesby, R. A., Domingues, C. S. P., Malvar, M. C., Perez, I. M. B., Matias, M. J. S., and Ferreira, A. J. D.: The role of soil water repellency in overland flow generation in pine and eucalypt forest stands in coastal Portugal, Soil Res., 43, 337–349, 2005b.

Lewis, D., Singer, M. J., Dahlgren, R. A., and Tate, K. W.: Hydrology in a California oak woodland watershed: a 17-year study, J. Hydrol., 240, 106–117, doi:10.1016/S0022-1694(00)00337-1, 2000.

Llorens, P. and Domingo, F.: Rainfall partitioning by vegetation under Mediterranean conditions. A review of studies in Europe, J. Hydrol., 335, 37–54, doi:10.1016/j.jhydrol.2006.10.032, 2007.

Malvar, M. C., Prats, S. A., Nunes, J. P., and Keizer, J. J.: Post-fire overland flow generation and inter-rill erosion under simulated rainfall in two eucalypt stands in north-central Portugal, Environ. Res., 111, 222–236, doi:10.1016/j.envres.2010.09.003, 2011.

Moreira, F., Rego, F. C., and Ferreira, P. G.: Temporal (1958–1995) pattern of change in a cultural landscape of northwestern Portugal: implications for fire occurrence, Landsc. Ecol., 16, 557–567, 2001.

Prats, S. A., MacDonald, L. H., Monteiro, M., Ferreira, A. J. D., Coelho, C. O. A., and Keizer, J. J.: Effectiveness of forest residue mulching in reducing post-fire runoff and erosion in a pine and a eucalypt plantation in north-central Portugal, Geoderma, 191, 115–124, doi:10.1016/j.geoderma.2012.02.009, 2012.

Rego, F. C.: Florestas Públicas, Ministério da Agricultura, Desenvolvimento Rural e Pescas, Lisbon, Portugal, 2001.

Rodríguez-Suárez, J. A., Soto, B., Perez, R., and Diaz-Fierros, F.: Influence of Eucalyptus globulus plantation growth on water table levels and low flows in a small catchment, J. Hydrol., 396, 321–326, doi:10.1016/j.jhydrol.2010.11.027, 2011.

Santos, J. M., Verheijen, F. G. A., Tavares Wahren, F., Wahren, A., Feger, K.-H., Bernard-Jannin, L., Rial-Rivas, M. E., Keizer, J. J., and Nunes, J. P.: Soil Water Repellency Dynamics in Pine and Eucalypt Plantations in Portugal – a High-Resolution Time Series, Land Degrad. Dev., doi:10.1002/ldr.2251, 2013.

Scarascia-Mugnozza, G., Valentini, R., and Spinelli, R. E. G.: Osservazioni sul ciclo dell'acqua in un bosco ceduo di Quercus cerris L., Ann. Dell'Accademia Ital. Sci. For., XXXVII, 3–21, 1988.

Scott, D. F. and Lesch, W.: Streamflow responses to afforestation with Eucalyptus grandis and Pinus patula and to felling in the Mokobulaan experimental catchments, South Africa, J. Hydrol., 199, 360–377, doi:10.1016/S0022-1694(96)03336-7, 1997.

Sen, P. K.: Estimates of the Regression Coefficient Based on Kendall's Tau, J. Am. Stat. Assoc., 63, 1379–1389, doi:10.1080/01621459.1968.10480934, 1968.

Serra, P., Pons, X., and Saurí, D.: Land-cover and land-use change in a Mediterranean landscape: A spatial analysis of driving forces integrating biophysical and human factors, Appl. Geogr., 28, 189–209, doi:10.1016/j.apgeog.2008.02.001, 2008.

Shakesby, R. A. Shakesby, Coelho, C., Ferreira, A., Terry, J., and Walsh, R.: Wildfire Impacts on Soil-Erosion and Hydrology in Wet Mediterranean Forest, Portugal, Int. J. Wildland Fire, 3, 95–110, 1993.

Shakesby, R. A., Doerr, S. H., and Walsh, R. P. D.: The erosional impact of soil hydrophobicity: current problems and future research directions, J. Hydrol., 231–232, 178–191, doi:10.1016/S0022-1694(00)00193-1, 2000.

Silva, G., Antunes, A., Escada, A., and Marques, J.: Relações Agricultura/Floresta Ambiente, Gabinete de Planeamento e Política Agro-Alimentar (GPPAA), Ministério da Agricultura, do Desenvolvimento Rural e das Pescas, Lisbon, Portugal, 2004.

SNIRH: Sistema Nacional de Informação de Recursos Hídricos, Agência Portuguesa do Ambiente, Lisbon, available at: http://snirh.pt (last access: 11 February 2015), 2013.

Stewart, M., Cimino, J., and Ross, M.: Calibration of base flow separation methods with streamflow conductivity, Ground Water, 45, 17–27, doi:10.1111/j.1745-6584.2006.00263.x, 2007.

Tarazona, T., Santa Regina, I., and Calvo, R.: Interception, throughfall and stemflow in two forest of the "Sierra de la Demanda" in the province of Burgos [Spain], Pirin. Espana, available at: http://agris.fao.org/agris-search/search.do?f=1998_FES_FES98017.xml%3BES1998000049 (last access: 04 May 2014), 1996.

Thornthwaite, C. W.: An Approach toward a Rational Classification of Climate, Geogr. Rev., 38, 55–94, doi:10.2307/210739, 1948.

Valente, F., David, J. S., and Gash, J. H. C.: Modelling interception loss for two sparse eucalypt and pine forests in central Portugal using reformulated Rutter and Gash analytical models, J. Hydrol., 190, 141–162, doi:10.1016/S0022-1694(96)03066-1, 1997.

Van Lill, W. S., Kruger, F. J., and Van Wyk, D. B.: The effect of afforestation with Eucalyptus grandis Hill ex Maiden and Pinus patula Schlecht. et Cham. on streamflow from experimental catchments at Mokobulaan, Transvaal, J. Hydrol., 48, 107–118, doi:10.1016/0022-1694(80)90069-4, 1980.

Wahren, A., Schwärzel, K., and Feger, K.-H.: Potentials and limitations of natural flood retention by forested land in headwater catchments: evidence from experimental and model studies, J. Flood Risk Manage., 5, 321–335, doi:10.1111/j.1753-318X.2012.01152.x, 2012.

Yue, S., Pilon, P., Phinney, B., and Cavadias, G.: The influence of autocorrelation on the ability to detect trend in hydrological series, Hydrol. Process., 16, 1807–1829, doi:10.1002/hyp.1095, 2002.

7

A regional-scale ecological risk framework for environmental flow evaluations

Gordon C. O'Brien[1], **Chris Dickens**[2], **Eleanor Hines**[3], **Victor Wepener**[4], **Retha Stassen**[1], **Leo Quayle**[5], **Kelly Fouchy**[6], **James MacKenzie**[1], **P. Mark Graham**[7], **and Wayne G. Landis**[3]

[1]University of KwaZulu-Natal, College of Agriculture, Engineering and Science, School of Life Sciences, Private Bag X01, Scottsville, South Africa
[2]International Water Management Institute, Private Bag X813, Silverton, 0127, South Africa
[3]Western Washington University, Institute of Environmental Toxicology, Bellingham, Washington, USA
[4]Water Research Group (Ecology), Unit for Environmental Sciences and Management, North-West University, Private Bag x6001, Potchefstroom, 2520, South Africa
[5]Institute of Natural Resources NPC, P.O. Box 100 396, Scottsville, 3209, South Africa
[6]IHE Delft Institute for Water Education, P.O. Box 3015, 2601 DA Delft, the Netherlands
[7]University of KwaZulu-Natal, College of Agriculture, Engineering and Science, School of Hydrology, Centre for Water Resources Research, Private Bag X01, Scottsville, South Africa

Correspondence: Gordon C. O'Brien (obrieng@ukzn.ac.za)

Abstract. Environmental flow (E-flow) frameworks advocate holistic, regional-scale, probabilistic E-flow assessments that consider flow and non-flow drivers of change in a socio-ecological context as best practice. Regional-scale ecological risk assessments of multiple stressors to social and ecological endpoints, which address ecosystem dynamism, have been undertaken internationally at different spatial scales using the relative-risk model since the mid-1990s. With the recent incorporation of Bayesian belief networks into the relative-risk model, a robust regional-scale ecological risk assessment approach is available that can contribute to achieving the best practice recommendations of E-flow frameworks. PROBFLO is a holistic E-flow assessment method that incorporates the relative-risk model and Bayesian belief networks (BN-RRM) into a transparent probabilistic modelling tool that addresses uncertainty explicitly. PROBFLO has been developed to evaluate the socio-ecological consequences of historical, current and future water resource use scenarios and generate E-flow requirements on regional spatial scales. The approach has been implemented in two regional-scale case studies in Africa where its flexibility and functionality has been demonstrated. In both case studies the evidence-based outcomes facilitated informed environmental management decision making, with trade-off considerations in the context of social and ecological aspirations. This paper presents the PROBFLO approach as applied to the Senqu River catchment in Lesotho and further developments and application in the Mara River catchment in Kenya and Tanzania. The 10 BN-RRM procedural steps incorporated in PROBFLO are demonstrated with examples from both case studies. PROBFLO can contribute to the adaptive management of water resources and contribute to the allocation of resources for sustainable use of resources and address protection requirements.

1 Introduction

The global use of water resources has altered the wellbeing of aquatic ecosystems and the benefits that people derive from them (Acreman and Dunbar, 2004; Dudgeon et al., 2006; Growns, 2008; Vörösmarty, 2010; Isaak et al., 2012; Murray et al., 2012; Grafton et al., 2013; Dudgeon, 2014). Environmental flows (E-flows), according to the Brisbane Declaration (2007), are defined as the "quantity, timing and quality of water flows required to sustain freshwater and estuarine ecosystems and the human livelihoods and wellbeing that depend on these ecosystems". In an effort to determine E-flows, the international community has developed a plethora of E-flow assessment methods, which have been applied on numerous spatial scales in a wide range of ecosystem types across the globe (Tharme, 2003; Acreman and Dunbar, 2004; Pahl-Wostl et al., 2013; Poff and Matthews, 2013). These methods have evolved during three distinct periods according to Poff and Matthews (2013) including an emergence and synthesis period, consolidation and expansion period and the current globalization period. During this globalization period a range of best practice E-flow management and assessment principles, and associated frameworks to undertake E-flow on multiple spatial scales in multiple political and/or legislative contexts, have been developed (Poff et al., 2010; Pahl-Wostl et al., 2013). These principles promote the use of holistic assessment tools that consider both social and ecological features of ecosystems on regional spatial scales, are adaptive and incorporate risk evaluation and address uncertainty (Poff et al., 2010; Acreman et al., 2014).

Ecological risk assessments have been undertaken internationally at different spatial scales using the relative-risk model (RRM) established since the mid-1990s (Hunsaker et al., 1990; Landis and Weigers, 1997, 2007; Wiegers et al., 1998; Landis, 2004, 2016). The RRM has been applied to evaluate a range of natural and anthropogenic stressors including water pollution, diseases, alien species and a range of altered environmental states (Walker et al., 2001; Moraes et al., 2002; Hayes and Landis, 2004; Colnar and Landis, 2007; Anderson and Landis, 2012; Ayre and Landis, 2012; Bartolo et al., 2012; O'Brien et al., 2012; Hines and Landis, 2014; Ayre et al., 2014). This tool can be used to carry out holistic, probabilistic assessments of the risk to the availability and conditions of ecosystem service and ecological endpoints, and facilitate socio-ecological trade-offs. For more information on the application of the RRM consider Colnar and Landis (2007), Anderson and Landis (2012) or O'Brien and Wepener (2012). Recent developments to the RRM incorporate the use of Bayesian networks (BNs) that have been established as a powerful tool for ecological risk assessment, ecosystem management and E-flow assessment (Pollino et al., 2007; Hart and Pollino, 2008; Shenton et al., 2011; Chan et al., 2012; Pang and Sun, 2014; Liu et al., 2016; McDonald et al., 2016). In 2012, Ayre and Landis combined both approaches and incorporated BNs into RRMs, which was then formalized into a BN-RRM approach (Hines and Landis, 2014; Herring et al., 2015; Landis et al., 2016).

Between 2013 and 2016 a BN-RRM-based holistic E-flow assessment approach was developed that adheres to the principles of best E-flow management practice and can easily be incorporated into regional E-flow frameworks such as the Ecological Limits of Hydrologic Alteration framework (Poff et al., 2010). This BN-RRM approach, which we have called PROBFLO, is a transparent and adaptable, evidence-based probabilistic modelling approach that can also incorporate expert solicitations and explicitly address uncertainty. PROBFLO is a scenario-based E-flow assessment tool that allows for the evaluation of the socio-ecological consequences of altered flows with consideration of the synergistic effects of non-flow drivers of ecosystem impairment. It adheres to the regional-scale ecological risk assessment exposure and effects, or sources of multiple stressors, habitats and ranked ecological impacts relationship (Wiegers et al., 1998). This paper presents the PROBFLO BN-RRM approach that was used to establish E-flows for the Senqu River in Lesotho, and it evaluates the socio-ecological effects of altered flow and non-flow stressors and developments made in the Mara River in the Kenya and Tanzania case study.

2 Study area

The Lesotho Highlands Water Project (LHWP) is a USD multi-billion water transfer and hydropower project implemented by the governments of Lesotho and South Africa (LHWP, 1986, 2011). Phase I of the LHWP involved the application of the Downstream Response to Imposed Flow Transformations (DRIFT) approach to establish the E-flows associated with the construction of the Katse and Mohale dams on the Malibamats'o and Senqunyane rivers in Lesotho respectively (Arthington et al., 2003; King et al., 2003). Phase II involves the augmentation of the LHWP by construction of the Polihali Dam to divert water directly from the upper Senqu River to the existing Phase I infrastructure of the LHWP (Fig. 1). For Phase II, the custodians of the project, the Lesotho Highlands Development Authority (LHDA), required the service provider awarded with the E-flow determination project to review and implement current best E-flow practice. This included the requirements to implement a probabilistic, regional-scale modelling approach that is transparent and holistic, addressing socio-ecological components and endpoints, and one that considers uncertainty explicitly. The PROBFLO approach has, as a result, been selected for Senqu River in Lesotho as a part of Phase II of LHWP between the proposed Polihali Dam site (29.289593° S; 28.863890° E) and the border of South Africa (30.413231° S; 27.564090° E) (LHWP, 2016).

The entire Mara River in Kenya and Tanzania upstream of the mouth into Lake Victoria (1.518178° S; 33.943497° E) was considered in this regional-scale PROBFLO case study (NBI, 2016) (Fig. 2). The Mara River and its tributaries are an essential source of water for domestic needs, agriculture, pastoralism and wildlife, including tourism, in Kenya and Tanzania (Mati et al., 2008; Defersha and Melesse, 2012). Although extensive research has been undertaken into the management of the terrestrial ecosystems of the Serengeti National Park and Maasai Mara National Reserve, there are limited studies on the effects of land use threats to the wellbeing of the Mara River, with only site-specific evaluations of the environmental flows of the river (Broten and Said, 1995;

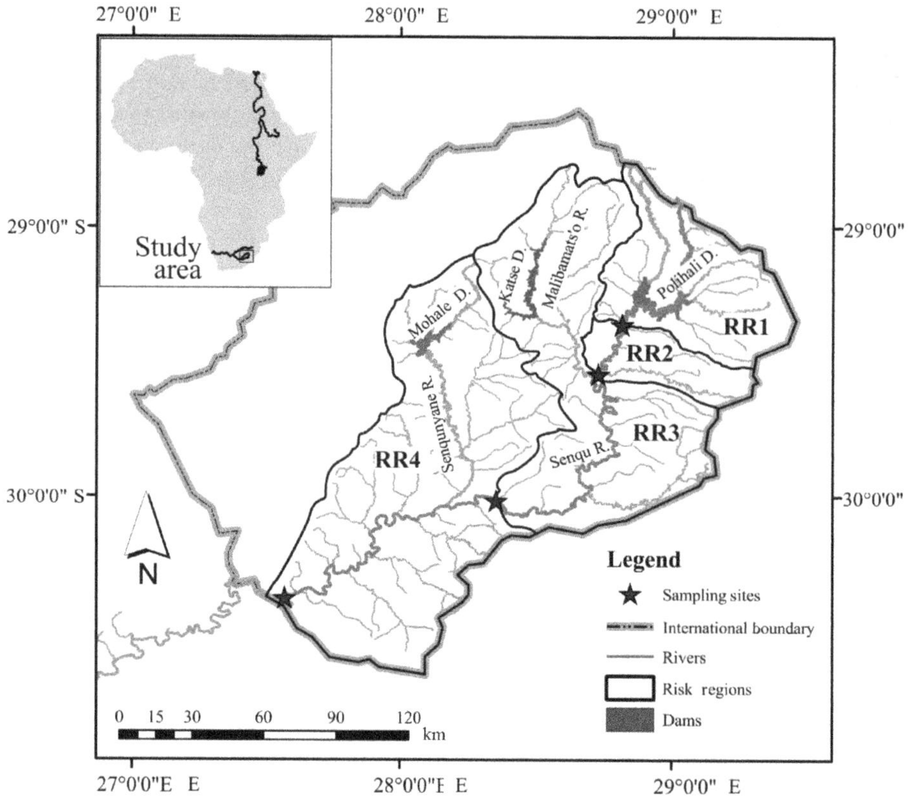

Figure 1. The upper Senqu River study area with risk regions (RRs) established for the study including dams associated with Phase I of the Lesotho Highlands Water Project and the location of the new Polihali Dam planned to be built in Phase II.

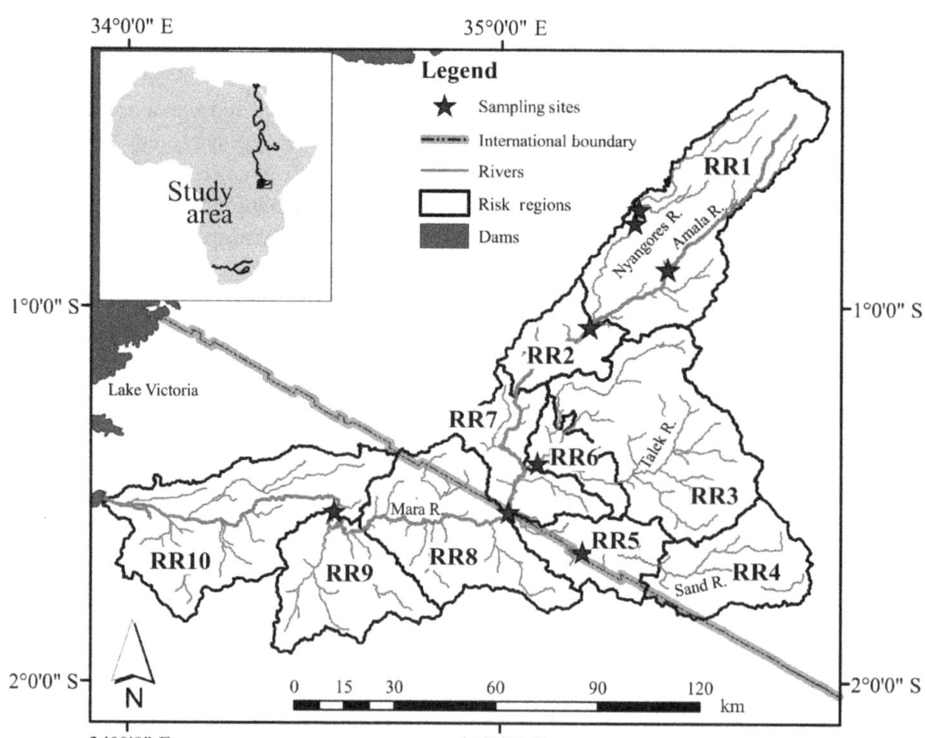

Figure 2. The Mara River basin considered in the study with risk regions (RRs) and sampling sites.

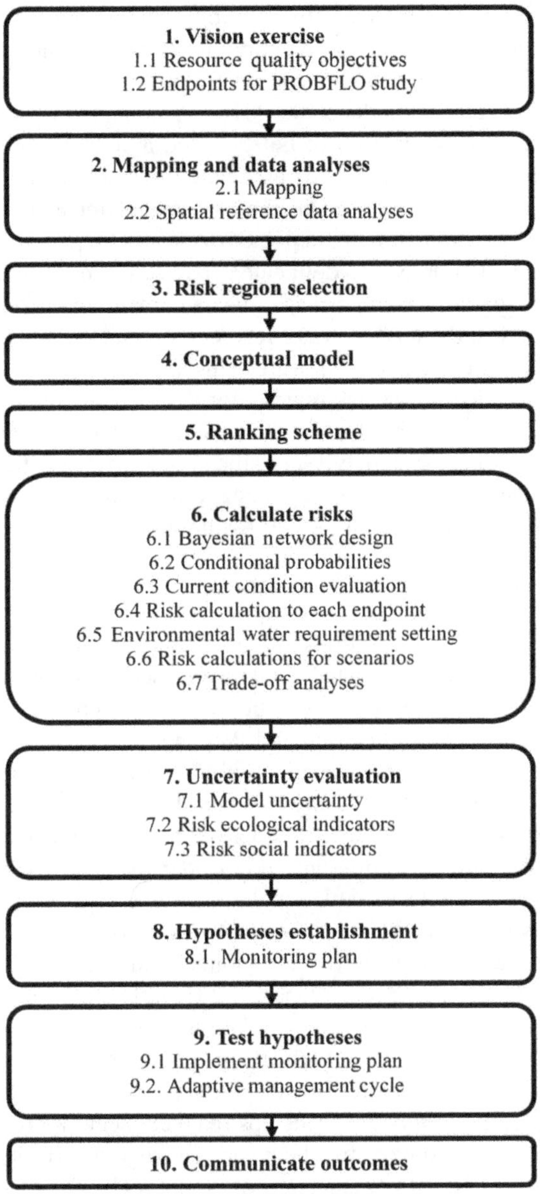

Figure 3. The 10 procedural steps of PROBFLO.

Gereta et al., 2002; Onjala, 2002; Karanja, 2003; Lamprey and Reid, 2004; Hoffman, 2007; Mati et al., 2008; Atisa, 2009; GLOWS-FIU, 2012; Majule, 2010; Hoffman et al., 2011; Ogutu et al., 2011; Defersha and Melesse, 2012; Kiambi et al., 2012; Dessu et al., 2014).

3 PROBFLO framework for E-flows

The PROBFLO framework is based on 10 procedural RRM steps (Landis, 2004), and it incorporates BN development and evaluation procedures (Marcot et al., 2006; Ayre and Landis, 2012) into a robust E-flow assessment method that gives emphasis to adaptive management for holistic E-flow management (Fig. 3). The PROBFLO approach has been implemented in the Senqu and Mara River case studies to evaluate the socio-ecological consequences of altered flows and determine E-flows, which is demonstrated through application of the following 10 procedural steps.

3.1 Step 1: Vision exercise

The importance of having clear water resource management objectives cannot be overemphasized. Numerous Integrated Water Resource Management strategies, regional management plans and frameworks, national legislation and established E-flow assessment tools advocate the establishment of clear goals or visions to direct the use and protection of water resources (Biswas, 2004; Mitchell, 2005; Dudgeon et al., 2006; Richter et al., 2006; Poff et al., 2010; King and Pienaar, 2011; NBI, 2016). Although many vision development approaches are available, the initial application of PROBFLO involved the application of the Resource Quality Objectives (RQOs) determination procedure (DWA, 2011) to describe and document the water quality, water quantity, habitat and biota objectives for the water resource being evaluated (NBI, 2016; DWA, 2011). The RQO process results in narrative and numerical descriptions of various ecosystem features required to achieve a balance between the use and protection of water resources and hence to achieve a documented vision. As part of the initial development of the RRM approach, multiple social and ecological endpoints were evaluated in a relative manner. Social endpoints were limited to the availability and quality of ecosystem services and ecological endpoints, including the requirements to maintain selected ecological indicators in an acceptable integrity state or wellbeing. In addition, for these environmental flow assessment (EFA) case studies, endpoints associated with socio-ecological impacts of the rivers resulting directly or indirectly with altered flows were considered. Findings resulted in relative risks to endpoints that could be compared and used to consider cost-benefit trade-offs between social and ecological endpoints by adjusting water resource use and protection scenarios.

The treaties for Phase I and Phase II of the LHWP entered into by the Kingdom of Lesotho and the Republic of South Africa provided the requirements for the establishment of a vision for water resource use and protection for the Senqu River case study (LHWP, 1986, 2011). The Treaty gives emphasis to protection of the existing quality of the environment and, in particular, requires maintenance of the well-being of persons and communities immediately affected by the project, including those downstream of the dam. Accordingly, the vision states that there should be no change to the existing quality of the downstream environment and that the net effect of the dam should not be negative to the people living downstream of the dam. For the PROBFLO assessment, RQOs describing the desired quality and quantity of water, habitat and biota for the study area were established. The endpoints selected to represent the social and ecological management objectives for the PROBFLO assessment were based on the vision represented by the RQOs in this case study, including the maintenance of the following ecosystem services and ecological objectives affected by the river: (1) the supply of building sand from the Senqu River, (2) water for domestic use, (3) recreation/spiritual use of the river, (4) fish stocks as food for people, (5) edible plants from the riparian zone as food for people, (6) medicinal plants for people, (7) floodplain non-woody plants (for grazing), (8) woody plants for fuel and construction, (9) reeds for construction and (10) fish stocks, (11) aquatic macro-invertebrate and (12) riparian ecosystem integrity or wellbeing.

The vision for the Mara River case study was based on existing regional Mara River management objectives (WRMA, 2014). In 2014, a Catchment Management Strategy (CMS) for the Mara Basin in Kenya was developed to facilitate the management of the water resources, environment and human behaviour in ways that achieve equitable, efficient and sustainable use of water for the benefit of all users (WRMA, 2014). The aims of the Mara River basin as part of the Strategic Environmental Assessment (EAC, 2003) to maintain "the people living in harmony with nature while achieving human wellbeing and sustainable economic development in perpetuity" were also considered. Also considered were the objectives for the Mara River basin as described by the Biodiversity, and Strategy Action Plan, which describes "a region rich in biodiversity which benefits the present and future generations and ecosystem functions" (GLOWS-FIU, 2012). The Kenyan Water Resources Management Authority (WRMA) established a high ecological importance, high livelihoods value and low commercial value vision for the upper Mara River basin (Nyangores and Amala rivers) and a high ecological importance, moderate livelihoods value and moderate commercial value vision for the mainstream Mara River. In this context the endpoints selected for the study included (1) the provision of water for basic human needs according to the national legislation of Kenya and Tanzania, (2) the maintenance of the ecological integrity of the riverine ecosystem (instream and riparian ecosystems), (3) the provision of flows for the commercial production of crops, (4) the maintenance of existing livestock industry, (5) the maintenance and viability of the ecotourism industry and (6) the maintenance of the ecological integrity of the Mara Wetland in the lower reaches of the basin.

3.2 Step 2: Mapping and data analyses

The BN-RRM approach that forms the basis of PROBFLO includes the relative evaluation of multiple sources of stressors to endpoints on a regional scale, which should be spatially and temporally referenced for regional comparisons/evaluations in a PROBFLO assessment (Landis, 2004; Landis and Wiegers, 2007). For this the spatial extent of the study area must be defined and described, and the locations of potential sources, habitats and impacts must be identified and spatially referenced. In addition, source-stressor exposure and habitat/receptor to endpoint pathways/relationships should be spatially referenced where possible (O'Brien and Wepener, 2012; Landis et al., 2016). Available data describing the ecosystem need to be reviewed and spatially referenced and the uncertainties associated with the availability and quality of data used in the assessment must be documented for evaluation in Step 7. O'Brien and Wepener (2012) provide an approach to delineate ecosystem types, the topological features of importance, the catchment and ecoregion boundaries, the land or water resource use scenarios and the pathways of stressors' exposure. This approach is used to direct the selection of risk regions for assessment (Smit et al., 2016). Best practice E-flow frameworks accentuate the importance of ecosystem type classification as part of E-flow assessments to improve our understanding of flow–ecosystem relationships (Poff et al., 2010; Arthington, 2012).

3.3 Step 3: Risk region selection

In this step, combinations of the management objectives, source information and habitat data are used to establish geographical risk regions that can be assessed in a relative manner (Landis, 2004; O'Brien and Wepener, 2012). In the end, the outcomes of the assessment will be available at the spatial scale established during this step for multiple temporal scenarios associated with alternative management options. In this regard it is important to consider the spatial connectivity of multiple variables including flows and other variables within the study area so that risk regions incorporate appropriate sources, stressors, habitats and endpoints (Landis, 2004; O'Brien and Wepener, 2012). The approach can address spatial and temporal relationships of variables between risk regions, such as the downstream effect of a source on multiple risk regions, in the context of the assimilative capacity of the ecosystem or the upstream connectivity requirements of a migratory fish between risk regions. To demonstrate that PROBFLO can conform to the regional E-flow assessment frameworks such as ELOHA (Poff et al., 2010), the selection of RRs should include explicit hydrological and geomorphological classification. The relative risk outcomes of the assessment can later be directly related to the system classification as proposed by ELOHA. With additional E-flow information for a range of hydrological and geomorphological ecosystem types, the outcomes can be used to establish regional E-flows.

The selection of risk regions for the Senqu River E-flow assessment was based on the proposed location of the Polihali Dam and catchment boundaries of the Senqu River and large tributaries (Malibamatso and Senqunyane rivers) for this E-flow assessment. Physical access to sampling sites within Lesotho to conduct biophysical field surveys was extremely difficult and this also contributed to risk region selection. Four broad risk regions were selected for the Senqu River PROBFLO study (Fig. 1).

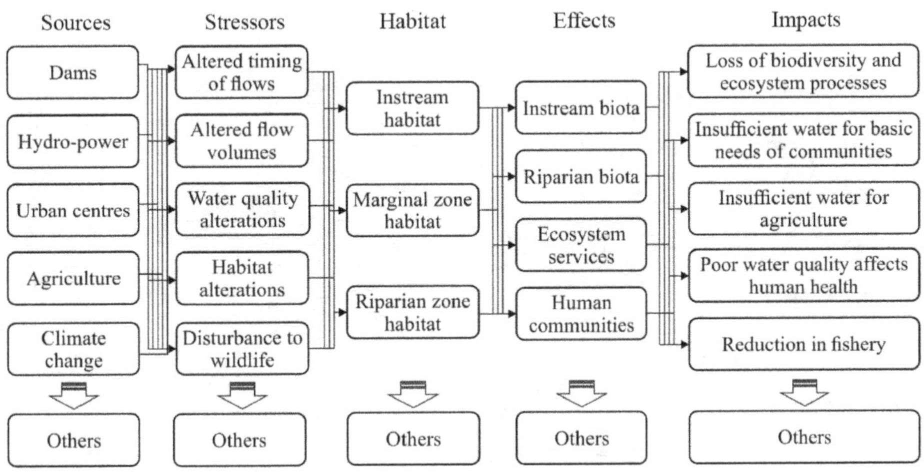

Figure 4. Example of a holistic conceptual model established in a PROBFLO case study that describes causal risk relationships (arrows) between sources, stressors, habitats, effects and impacts to endpoints considered in an assessment.

In the Mara River case study, a review of land use and land cover (Mati et al., 2008; Atisa et al., 2014), hydrology (Mango et al., 2011; McClain et al., 2014), the vision for the case study, current and future land and water resource use options and socio-ecological importance (Karanja, 2002; LVBC and WWF-ESARPO, 2010; Mango et al., 2011; Defersha and Melesse, 2012; GLOWS-FIU, 2012; Dessu et al., 2014; Dutton et al., 2013) was used to select risk regions. Ten risk regions were selected for the Mara River case study which conformed to catchment boundaries, ecoregions, land use practices and the international boundary (Fig. 2).

3.4 Step 4: Conceptual model

In this step conceptual models that describe hypothesized relationships between multiple sources, stressors, habitats and impacts to endpoints selected for the study are generated (Wiegers et al., 1998) (Fig. 4). This includes the holistic (consider flow- and non-flow-related variables in a spatial–temporal context), best practice characterization of flow–ecosystem and flow–ecosystem service relationships in the context of a regional-scale E-flows framework (Poff et al., 2010), with relevant non-flow (water quality and habitat) relationships in the models. Conceptual models should be constructed by expert stakeholders usually including hydrologists, geomorphologists, ecologists and ecosystem services, including social and resource economics scientists. These experts should be familiar with socio-ecological system processes and be able to describe probable cause and effect variables and relationships of sources to stressors to multiple receptors in relation to their impacts on the endpoints, selected for the study. The conceptual models for the case studies presented addressed requirements of the ELOHA and the Nile Basin regional-scale E-flow frameworks to conform to these frameworks (Poff et al., 2010; NBI, 2016). The Nile Basin regional-scale E-flow framework expands on the ELOHA framework to include an initial situation assessment, data review and alignment phase and a governance and Resource Quality Objectives setting phase. The PROBFLO conceptual model thus conforms to the regional-scale E-flow framework procedures in (1) the selection of socio-ecological endpoints, to direct the hydrologic foundations for the study including the selection of hydrological statistics required; (2) the classification of ecosystem types based on geomorphic, water quality, quantity and ecoregion considerations; and with these data, (3) the incorporation of evidence-based flow–ecosystem relationships and flow–ecosystem service relationships, with relevant non-flow variable relationships upon which the assessment is based. Initial conceptual model development considers all relevant sources, stressors, habitat, effects and impact relationships with spatial and temporal considerations.

3.5 Step 5: Ranking scheme

Ranking schemes are used to represent the state of variables, with unique measures and units to be comparable as non-dimensional ranks and combined in BN-RRMs (Landis, 2004; Landis et al., 2016). Four states, designated as zero, low, moderate and high as traditionally used in RRMs (Colnar and Landis, 2007; O'Brien and Wepener, 2012; Hines and Landis, 2014; Landis et al., 2016), have been incorporated into the PROBFLO process. The states represent the range of wellbeing conditions, levels of impacts and management ideals as follows.

- *Zero.* This describes a pristine state, with no impact/risk, comparable to the pre-anthropogenic source establishment, baseline or reference state.
- *Low.* This is a largely natural state with low impact/risk, and it describes an ideal range for sustainable ecosystem use.

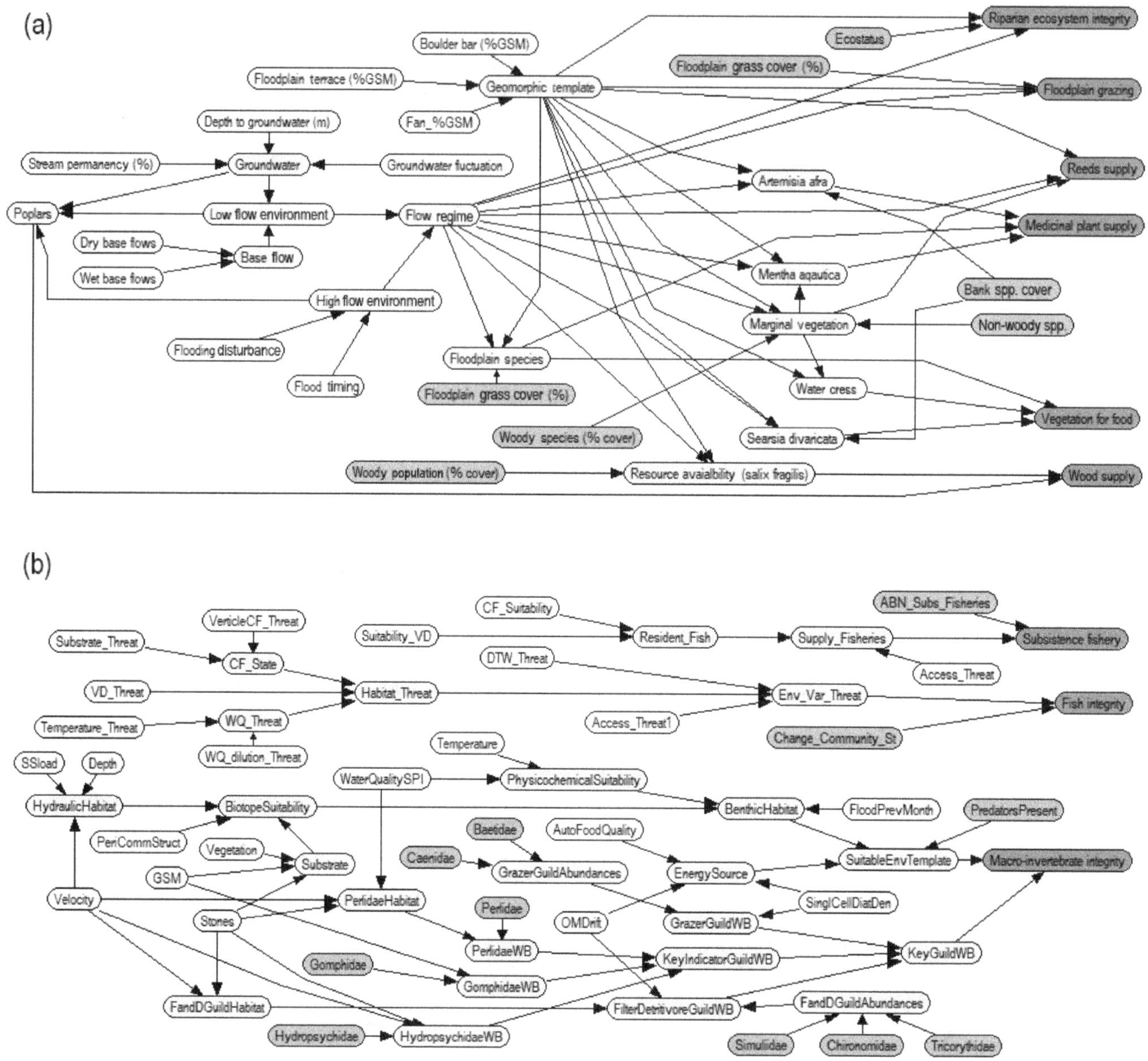

Figure 5. Bayesian network models used in the Senqu River case study to evaluate the risk of flow alterations associated with Phase II of the Lesotho Highlands Water Project to riparian ecosystem integrity and ecosystem service endpoints **(a)** and instream ecological endpoints and fisheries supply endpoints **(b)**. White nodes represent input exposure variables, light grey nodes complete the expose leg of the risk assessment, grey nodes represent effect leg of models and dark grey nodes represent endpoints. Working Netica™ models are provided in the Supplement.

- *Moderate.* This state refers to moderate use or a modified state, with moderate impact/risk, representing the threshold of potential concern or alert range.
- *High.* This describes a significantly altered or impaired state, with unacceptably high impact/risk.

This ranking scheme selected for PROBFLO represents the full range of potential risk to the ecosystem and ecosystem services with management options. Low risk states usually represent management targets with little impact, and moderate risk states represent partially suitable ecosystem conditions that usually warrant management/mitigation measures to avoid high risk conditions. The incorporation of BN modelling into PROBFLO allows the approach to incorporate the variability between ranks for each model variable, represented as a percentage for each rank. Indicator flow and non-flow variables representing the socio-ecological system being evaluated in a PROBFLO assessment are selected (linked to endpoints – step 1), and unique measures and units of measurement are converted into, and represented by, ranks

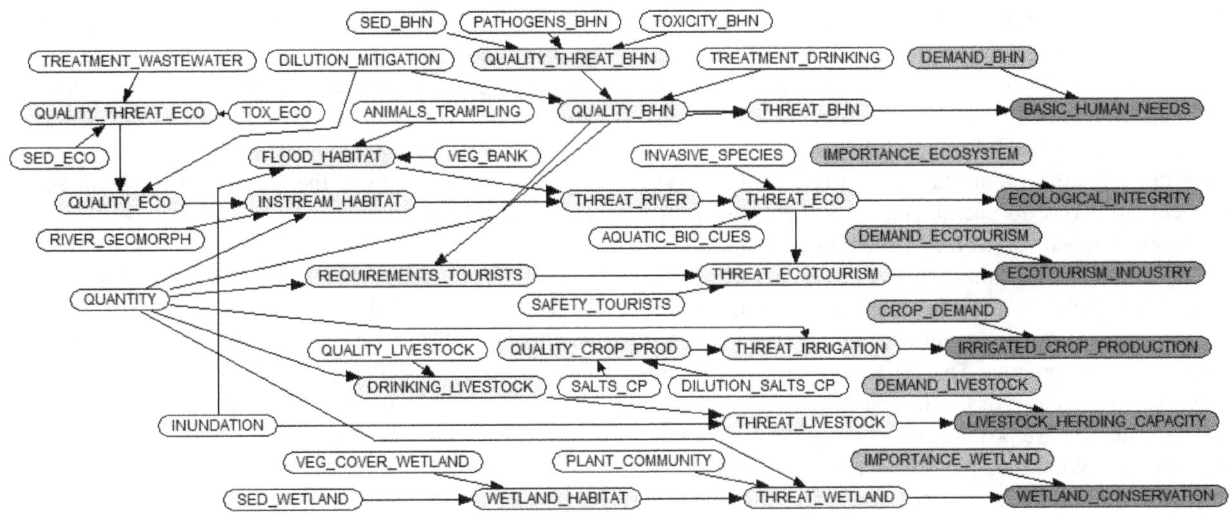

Figure 6. Bayesian network models used in the Mara River case study to evaluate the risk of water resource use to a range of socio-ecological endpoints. White nodes represent input exposure variables, light grey nodes complete the expose leg of the risk assessment, grey nodes represent effect leg of models and dark grey nodes represent endpoints. Working Netica™ models are provided in the Supplement.

for integration in BN assessments. For the BN assessment, ranks are assigned scores along a percentage continuum representing the state of the variables using natural breaks of 0.25 (zero), 0.5 (low), 0.75 (moderate) and 1 (high) in the calculation.

3.6 Step 6: Calculate risks

From the general inclusive conceptual models (step 4), with the principle of requisite simplicity (Stirzaker et al., 2010), smaller social and ecological endpoint specific models that represent the system being assessed are unpacked and converted into Bayesian network models (Figs. 5 and 6) for analyses. These models can be analysed individually or integrated using a range of BN modelling tools, using nodes representing variables that share the same indicators and measures. Bayesian networks are probabilistic modelling networks that graphically represent joint probability distributions over a set of statistical values (Pollino et al., 2007; Korb and Nicholson, 2010). They include parent or input nodes and child or conditional nodes with links that represent causal relationships between nodes combined by Conditional Probability Tables (CPTs) (McCann et al., 2006; Landis et al., 2016). Conditional Probability Tables describe conditional probabilities between the occurrence of states in the parent nodes and the resulting probabilities of states in the child nodes (Landis et al., 2016). The two PROBFLO case studies presented here made use of the Netica™ BN software by Norsys Software (http://www.norsys.com/).

The BNs are initially used to evaluate the risk of anthropogenic/natural hazards to endpoints per risk region, in relative manner for comparisons, for multiple temporal periods (high or low flow months and wet or drought phases etc.) which can also be compared relative to each other. Bayesian networks also make use of available data and expert solicitations as evidence to represents risks to current or present scenarios. Present projections of risk to the endpoints can generally be easily validated using available data, knowledge of existing relationships between variables and by carrying out directed field survey campaigns to describe/test risk relationships. Present risk projections are then calibrated by evaluating benchmark or historical scenario risk projections using the established models, which can often be validated with historical data (see data in the Supplement). An example of how a BN can represent a component of the socio-ecological ecosystem being evaluated is presented in Figs. 5 and 6. Within Fig. 5 we for example hypothesize that the ecological integrity of fish in the Senqu can be selected as an indicator of the Senqu River ecosystem as a suitable ecological endpoint. In this example the ecological integrity of fish is hypothesized to be a function of the conditions of the Senqu River environment for fish, representing the exposure leg of the risk assessment and the potential for fish to occur within the reach of the river being considered as an effects component of the study (refer to Table S1 in the Supplement). The conditions of the Senqu River environment itself are hypothesized in this study to be a function of the potential of communities to be a "disturbance to wildlife". This describes the instream habitat conditions and migration access for fish as source-stressor relationships with local communities, multiple barriers and other source/stressors that affect instream habitat wellbeing, selected indicators representing ecosystem components with associated measures in the study. Water resource use scenarios were used to describe the state of source/stressor and effect nodes as inputs into the model. These variables were integrated using CPTs to represent other system variables, which ultimately result in risk described to endpoints (refer to Table S2). These models are then used to determine E-flow requirements according to acceptable trade-off of risk to endpoints selected for the study, and the consequences of alternative water resource use, management and/or climatic condition scenarios.

To determine E-flow requirements in PROBFLO, trade-offs of acceptable risk to social and ecological endpoints are initially established for each risk region by stakeholders. This usually takes place within a legislative context where regional legislation/policies/agreements are addressed, such as requirements to maintain ecosystem sustainability, which affects trade-off developments. These trade-offs of acceptable risk are represented in the BNs as forced endpoint risk distributions or profiles. These profiles usually range between low and moderate risk, with usually no high risk probabilities. In relation to the definitions of the ranks used in PROBFLO, trade-offs of acceptable risk for E-flow determination should only dominate the "moderate" risk range when there is certainty that the E-flow requirements can be provided, such as in the case of E-flow releases from a dam. In case studies where there is high uncertainty associated with the ability to provide E-flow requirements, such as the management of multiple water resource users to cumulatively maintain E-flows, then a buffer should be provided according to the definition of ranks and the "low" risk range should be selected. After the selection of trade-offs of acceptable risk are established, the calibrated BNs are forced to generate the state (rank distributions) of input flow variables used in the assessments. These flow-related variable state requirements that are spatially and temporally referenced are provided to a hydrologist or geomorphologist, for example, to describe the E-flow requirements which can be presented in various formats, such as daily or monthly water (usually $m^3\,s^{-1}$) and sediment (usually $kg\,m^{-3}\,s^{-1}$) discharge duration percentiles. During E-flow determination procedures, the state of non-flow variable nodes, which contribute to the risk to endpoints, associated with flow variables can either be maintained in their current state, and described as such or amended with available water resource use information. This can include the increased requirement of water for basic human needs, for increases in growths of human populations depending on the resource, for example. Following the establishment of E-flows, the socio-ecological consequences of altered flows, associated with alternative water resource management options or climate change variability, for example, can be evaluated in a relative manner by generating and evaluating a range of future scenarios in PROBFLO.

3.6.1 Senqu River risk calculation

In the Lesotho case study the nine social endpoints and three ecological endpoints were used to represent the socio-ecological endpoints of interest in the study. The 12 BN models established for the study included cause and effect linkages used to estimate risk (refer to BN models provided in the Supplement). These BNs were used to evaluate the risk of multiple sources and stressors with flow-related stressors for base winter (low), summer (high) and drought flows. Where appropriate, CPTs of the BNs for endpoints were adjusted between RRs to represent the subtle changes in ecosystem process dynamics down the length of the Senqu River (refer to Tables S1 and S2).

Evidence used includes the historical understanding of flow–ecosystem and flow–ecosystem service relationships established during Phase I of the LHWP and data derived from a series of biophysical surveys of the study area (refer to Tables S1 and S2). The surveys included hydrology, hydraulic, water quality and geomorphology components grouped as ecosystem driver components, and fish, macroinvertebrates and riparian vegetation grouped as ecosystem responder components. Hydrological data used in the case study included recently updated basin-wide Integrated Water Resources Management Plan outcomes (ORASECOM, 2014). This database was updated in the study with latest available rainfall records, obtained from the Lesotho Meteorological Services, and regional evaporation information. Observed monthly flow data with actual discharge measurements determined in the study were used to calibrate the monthly WRSM2000 model against rainfall to obtain catchment-specific rainfall-runoff parameters (Pitman et al., 2006). A range of hydrological statistics were used in the RRM-BN model. Findings were used to identify a range of indicators to represent hypothesized causal relationships of the socio-ecological system being evaluated and to identify measures for indicators, with units of measurement and node rank thresholds and relationships between variables in the form of CPTs (refer to Table S2). Netica™ was used to carry out the assessment (Ayre et al., 2014, for example). The tool is versatile and incorporates a range of features used to optimize the assessment. This includes equation features to weight the relative importance of parent variables and generate initial CPTs that were easily refined and applied to the daughter nodes for the assessment. The tool includes case file generation options, which allow the BNs to be linked to Microsoft® Excel, where data can be rapidly analysed and used to populate BNs for the analyses. Risk outcome distributions were also linked to Excel, where scenarios and social and ecological endpoints could be integrated using Monte Carlo randomization approaches that are part of the Oracle Crystal Ball software (Landis, 2004). After establishing BN models for each RR, input parameters were changed using RR-specific data for a range of scenarios including the following.

- Scenario 1 represents the present-day scenario based on present state hydrology and associated source to endpoint variable state relationships that represent observable conditions. This scenario is based on existing data and additional data collected during the field surveys.

- Scenario 2 represents a pre-anthropogenic water resource development scenario, considered to represent "natural" hydrology, which was modelled using historical and modelled hydrology and rainfall data, and hypothesized state distributions for non-flow variables. This scenario was selected to calibrate the PROBFLO model for the study.
- Scenario 3 includes the presence of the new proposed Polihali Dam with full modelled interbasin transfer (IBT) supply. Only large floods overtopping the dam have been considered to be available downstream of the dam, with the existing E-Flows from the downstream lateral tributaries bringing water from Katse and Mohale dams available in RR3 and 4. Non-flow source/stressor catchment conditions were based on the present-day scenario.
- Scenario 4 is based on scenario 3 but includes E-Flow releases established as 36 % of the natural mean annual runoff (MAR) from the Polihali Dam, with suitable freshet and flood flows. The range of percentages of the MAR considered in these scenarios were selected by stakeholders to evaluate different levels of water resource use for the development.
- Scenario 5 is based on scenario 3, but with only 25 % of the natural MAR available to contribute towards E-flows, with all floods retained in the Polihali Dam for transfer into the IBT.
- Scenario 6 is based on scenario 5 with one additional $40\,m^3\,s^{-1}$ freshet (small spring flood) released from the dam in addition to the 25 % of the natural MAR to contribute towards E-flows.
- Scenario 7 is based on scenario 3 with only 18 % of the natural MAR available to contribute towards E-flows with one single $40\,m^3\,s^{-1}$ freshet (small spring flood).
- Scenario 8 is based on scenario 6 but with additional stress imposed by further reduction of available flows to 12 % of the natural MAR, released for maintenance but including the single $40\,m^3\,s^{-1}$ freshet (small spring flood).

In this assessment, risk was calculated for 12 endpoints, for three temporal periods and for eight scenarios, thus representing 312 BN models that were relatively comparable. The results include the mean relative risk rank scores with associated standard deviation for each endpoint including the maintenance of riparian vegetation, macro-invertebrates and fish wellbeing as ecological endpoints and the maintenance of wood for fuel, marginal vegetation for livestock grazing and fish for food as social endpoints (Figs. 7 and 8). These initial relative mean risk scores allow for the comparison of alternative spatial and temporal socio-ecological risk projections to the endpoints used in the assessment. Initial risk to ecological endpoints compared between the natural (SC2) and present (SC1) scenarios demonstrates that the number of sources and stressors with associated risk to endpoints has increased in the study area, particularly in RR2 to RR4. These changes can largely be attributed to the consequences of Phase I of the LHWP (Fig. 7). These findings include the synergistic effect of non-flow stressors (such as water quality and habitat condition) to the wellbeing of the Senqu River ecosystem in the study area. Effects of the altered hydrology between natural and present-day scenarios to the social endpoints were less obvious (Fig. 8). Spatial trends in the risk results associated with SC3 to SC8 generally include elevated risk to RR1, directly downstream of the proposed dam in particular. These results demonstrate that the impact on socio-ecological endpoints considered will be highest directly below the dam. Thereafter scenarios that exclude floods and freshets (SC3 and SC5) resulted in excessive risk, demonstrating the importance of flood and freshet flows to the socio-ecological endpoints. Outcomes for scenarios 6 to 7 for riparian vegetation and invertebrates include consistent increases in risk spatially from the proposed new dam towards the lower reaches of the study area, which is ascribed to accumulative effects of the existing Phase I dams on the lateral tributaries. The relative risk to the fish community endpoint includes an opposite trend where a reduction in risk from RR1 to RR4 was observed for all scenarios. These results are indicative of the increased relative resilience of the resident and seasonal migratory fish communities to flow alterations in the Senqu River associated with dam developments, due to the increasing size of the river and associated increases in habitat diversity towards the lower reaches of the study area. In addition, reductions in river connectivity (barrier formation) associated with existing impacts from Phase I and the synergistic new stressors associated with Phase II of the LHWP were also shown to contribute to the increase in risk from the lower reaches of the study area in RR4 for fish migrating upstream to RR1. Interestingly the outcomes included improvements to the condition of, or reductions in risk to the wellbeing of social endpoints for scenarios 6 to 8. This included potential improvements in the availability of and/or condition of wood for fuel in RR2 and grazing for livestock for RR1 and RR4 in particular. These projections allow for trade-off considerations where the wellbeing of some social endpoints will decrease in some parts of the study area and increase in others. From our current understanding of the socio-ecological system represented in this study, the results evidence some additional sustainable development opportunities that may be available for stakeholders to consider (Fig. 8). These opportunities may improve the viability of the Phase 2 development and contribute to the establishment of a suitable balance between the use and protection of the ecosystem being evaluated. These relative risk projections to the multiple socio-ecological endpoints considered are based on flow alterations associated with devel-

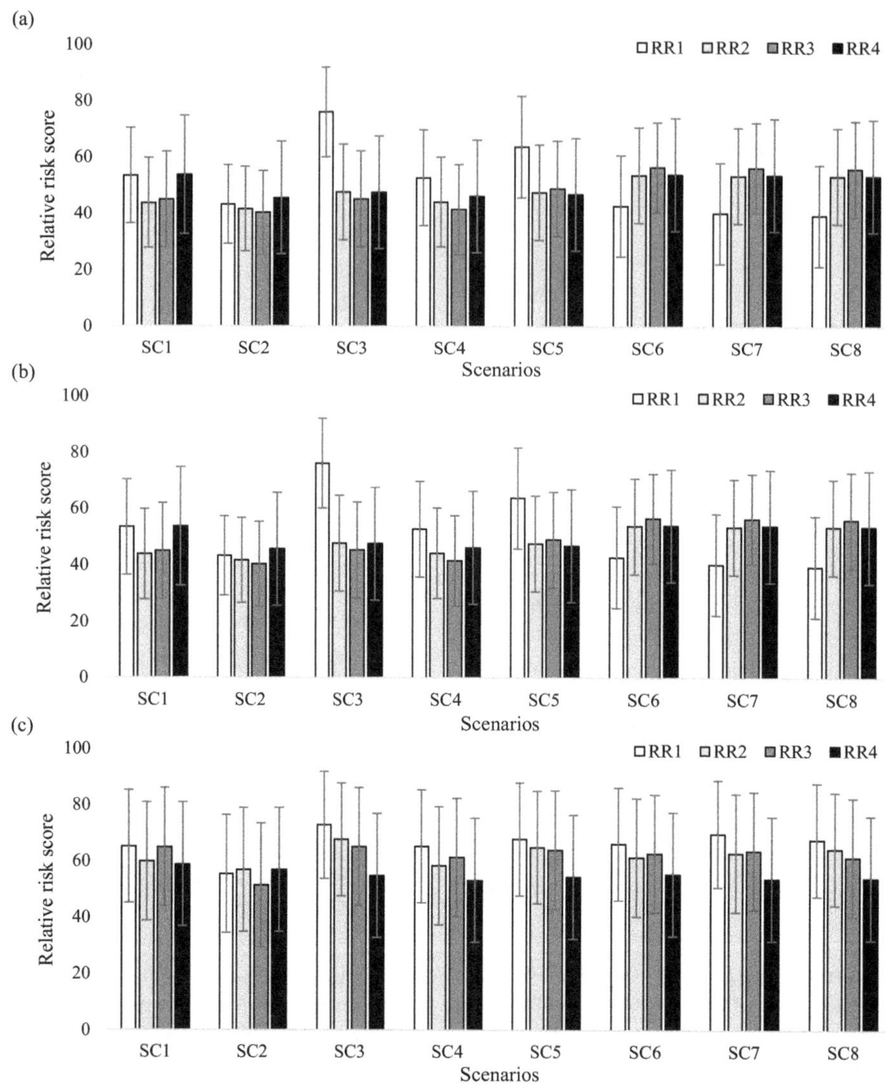

Figure 7. Senqu River case study mean relative risk scores (with SD) considered in the assessment including relative risk to: riparian vegetation **(a)**, macro-invertebrates **(b)** and fish **(c)** wellbeing endpoints for the four risk regions per scenario (SC).

opment scenarios, in the context of the exacerbating of non-flow variable determinants on regional scales.

The cumulative risk of all ecological and social endpoints for each RR, for each temporal period, per scenario, were evaluated using Monte Carlo simulations (5000 trials, Oracle Crystal Ball software, Oregon) (Ayre et al., 2014). The outcomes included relative risk projections, displayed as relative profiles to single endpoints from multiple RRs, and multiple social and ecological endpoints or all endpoints per RR in the study for comparisons and evaluation. These profiles were generated for multiple scenarios to evaluate the potential social and ecological consequences of alternative water resource development scenarios. This is demonstrated by considering the cumulative risk projections to the fish wellbeing endpoint, which demonstrates that relative to the "natural" hydrology scenario (Scenario 2), in which there is a 83 % probability that risk to the fish endpoint occurred in a zero to low risk range, for the present scenario (Scenario 1), Phase II with the dam and no E-flows scenario (Scenario 3) and Scenario 7 (Phase II with the dam, 18 % release of natural MAR and 40 $m^3\,s^{-1}$ freshets) all range between the moderate and high risk range (Fig. 9). The risk outcomes of all future management options suggest that objectives of the stakeholders to maintain the existing wellbeing of the ecosystem could not be achieved because flow scenarios considered would not satisfactorily mitigate the effects of fish migration barriers, which themselves reduce ecosystem wellbeing. To address this, an additional, amended scenario (Scenario 7) was then modelled, which included successful mitigation measures for the existing man-made barriers in the Senqu River as amendments. The outcomes included a reduction in risk in the low to moderate risk ranges, demonstrating that sce-

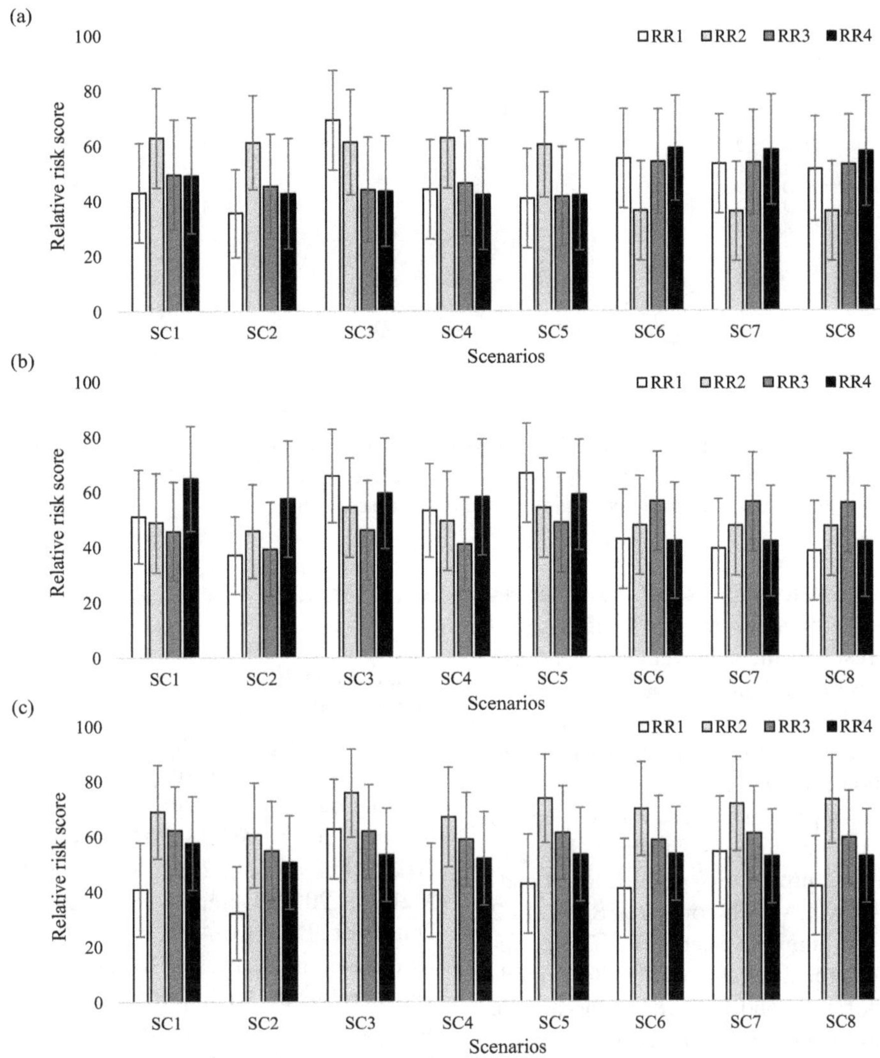

Figure 8. Senqu River mean relative risk scores (with SD) for wood for fuel **(a)**, marginal vegetation for livestock grazing **(b)** and fish for food **(c)** social endpoints for the four risk regions per scenario (SC) considered in the study.

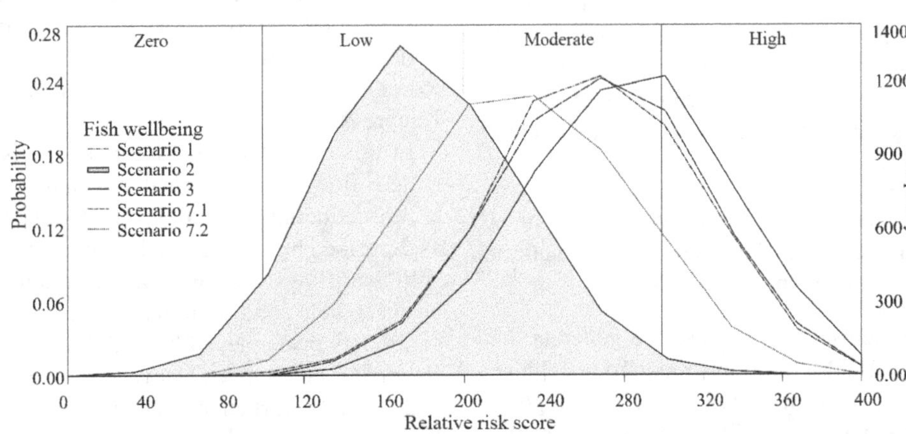

Figure 9. Probability profiles generated during a PROBFLO assessment, overlaid onto risk rank range (Zero, Low, Moderate and High) to describe the relative risk of the multiple sources and stressors, including altered flows, associated with alternative management scenarios considered in the Lesotho case study to the fish wellbeing endpoint.

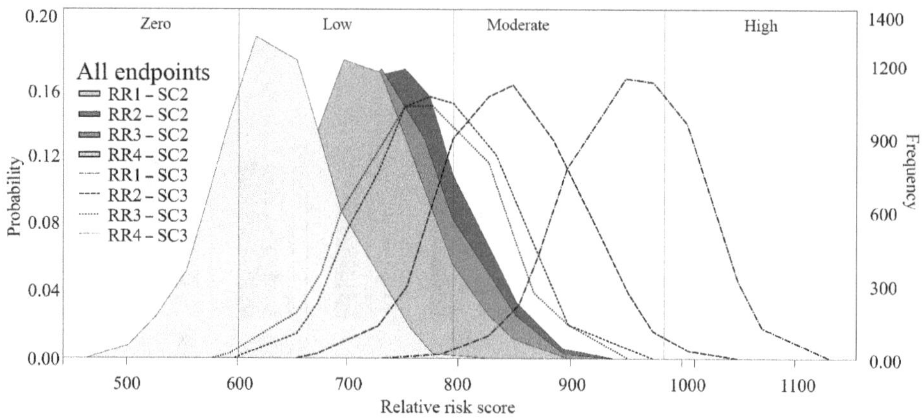

Figure 10. Probability profiles generated during a PROBFLO assessment, overlaid onto risk rank range (Zero, Low, Moderate and High) to describe the relative risk of the multiple sources and stressors, including altered flows, associated with alternative management scenarios considered in the Lesotho case study to all of the endpoints integrated in the assessment.

narios that promote moderate to high use of the water resources, with barrier mitigation measures (such as construction of fishways) could result in the achievement of the fish wellbeing endpoints in the study. This approach established for this case study allows for the relative comparison of the integrated social and/or ecological consequences of altered flows in the context of non-flow variables for each scenario for each endpoint used to represent the use and protection management objectives of the study as shown in Fig. 10. In Fig. 10 the integrated risk probability profiles to all endpoints for each RR are shown, which compares Scenario 2 (reference scenario) to the high use Scenario 3. These results include elevated risk probabilities for RR1 (84 % moderate and 15 % high rank range) and RR2 (81 % moderate), while existing E-flows from Phase I dams reduce the risk posed for this scenario in RR3 and RR4. The relative risk results to endpoints and integrated risk profiles were presented to stakeholders who used these outcomes to select E-flows and associated water resource use mitigation measures (such as barrier mitigation measures) to be implemented for Phase II of the LHWP. In this case study E-flows were established by initially evaluating the individual flow tolerance exceedance thresholds for numerous social and ecological in-dicators used in the BNs.

This information resulted in the de-velopment of a hydrological scenario matching the shape of the natural hydrology that was evaluated in the BNs.

Thereafter the BNs themselves were used to generate the hydrolog-ical statistic variable conditions required to achieve a suitable risk profile for each social and ecological endpoint that represents the Eflows for the study in the context of the vision for RRs in the study. With this information, a range of alterative water resource use scenarios were generated and considered to explore risk tradeoffs between social and ecological end-points.

3.6.2 Mara River risk calculation

In the Mara River case study, the relative risk of stressors and the E-flows were established according to the four social and two ecological endpoints considered in the assessment. The Mara River case study (refer to Table S1) was based on existing data from historical surveys (Mati et al., 2008; McCartney, 2010; Majule, 2010; LVBC and WWF-ESARPO, 2010; Mango et al., 2011; Kanga et al., 2011; Defersha and Melesse, 2012; Defersha et al., 2012; GLOWS-FIU, 2012; Dutton et al., 2013; Atisa et al., 2014; Gichana et al., 2014; Kilonzo et al., 2014; McClain et al., 2014) and a single site visit to refine the CPTs (NBI, 2016). During this survey seven sites were selected to represent the variability of all of the RRs in the study area. After establishing BN models for each RR (refer to BN models provided in the Supplement), input parameters were changed using RR-specific data for two scenarios, including the present condition and alternately the E-flow requirement to achieve the basic human needs and ecological wellbeing of the Mara River known as the Ecological Reserve (United Republic of Tanzania, 2009; Government of Kenya, 2002).

In this case study relative risk results were used to generate E-flow requirements that would not pose excessive risk to the wellbeing of ecological endpoints and social endpoints as described by the RQOs. The assessment hypothesizes that sufficient flows currently exist to maintain the endpoints in an acceptable condition. In addition, in the context of the precautionary principle, additional flows can be allocated before risk to the endpoints exceeds acceptable, sustainable thresholds. Results further demonstrate that sustainable water allocations would reduce risk to selected social endpoints selected in the study and meet the desired balance between the use and protection of the resource (Fig. 11). The approach highlighted the probable effect of non-flow-related stressors that are affecting the ecological wellbeing of Mara River, in-

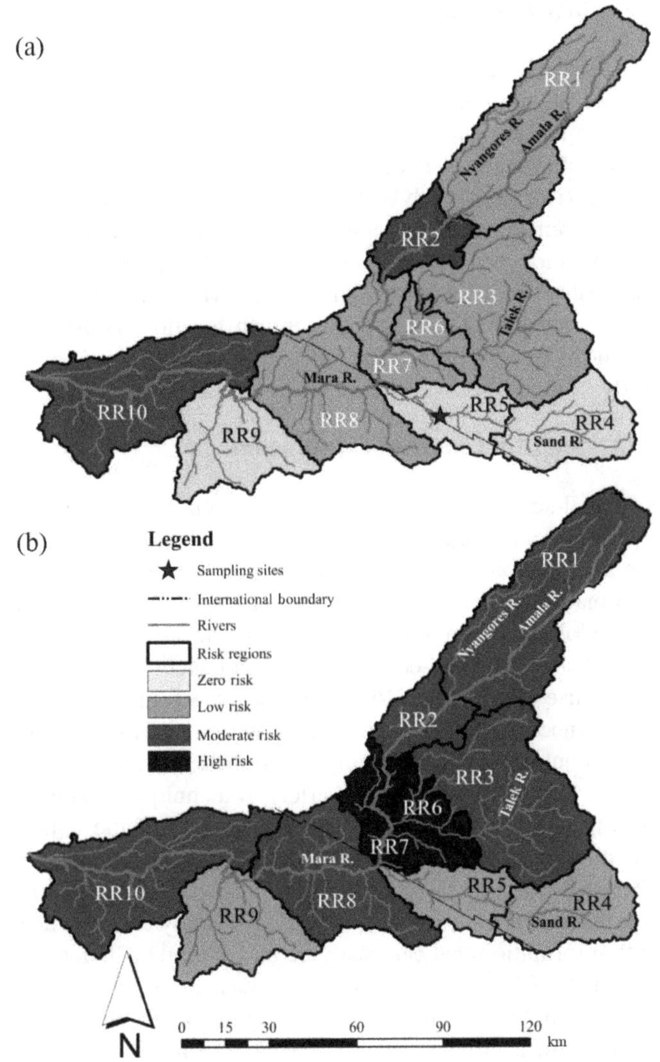

Figure 11. Relative spatial risk of the multiple sources and stressors associated with current **(a)** and planned **(b)** water resource use to risk regions (RRs) selected for the Mara River basin case study.

cluding water physiochemical impacts and habitat alteration stressors associated with urban and rural communities, livestock grazing and watering and the effect of the recent exponential increase in local *Hippopotamus amphibius* populations in the tributaries of the Mara River in particular that are affecting water quality in the system (Kanga et al., 2011; Subalusky et al., 2015). These results were used to demonstrate the relative risk of sources or water resource activities that affect flows relative to other sources at the risk of ecosystem wellbeing (Fig. 11). The approach successfully demonstrated how the BN-RRM approach in PROBFLO can be used to generate acceptable risk profiles for endpoints to evaluate the socio-ecological consequences of altered flows, and it is also demonstrated how these models can be used to determine E-flows and associated information for water resource use.

3.7 Step 7: Uncertainty evaluation

Best ecological risk assessment practice requires the explicit evaluation of uncertainty, or confidence assessment, (O'Brien and Wepener, 2012; Landis, 2004), which has been incorporated into the PROBFLO approach. Any and all aspects of uncertainty associated with the entire BN-RRM process, including objectives and endpoint selection for the assessment, availability and use of evidence, expert solicitations and model uncertainty, for example, must be addressed. In an effort to reduce uncertainty, the BN-RRM approach adopted by PROBFLO inherently considers uncertainty associated with cause and effect relationships and the use of real data with expert solicitations (Uusitalo, 2006; Landis et al., 2016). The additional incorporation of entropy reduction analysis in relative risk calculations using Monte Carlo simulations also contributes to uncertainty reduction in PROBFLO. Additional analyses of the sensitivity of the BN-RRM should be addressed within the uncertainty evaluation section (Pollino et al., 2007; Hines and Landis, 2014), where the relative influence of input nodes on the endpoints can be evaluated as part of the PROBFLO assessment.

The results of the uncertainty assessment are used to provide context to the stakeholders of a PROBFLO assessment and contribute to the decision making process in E-flow assessment studies.

For all of the BNs created in the PROBFLO assessments of the Senqu and Mara River case studies, the sensitivity of the input variables was evaluated in Netica™ using the "Sensitivity to Findings" tool (Marcot, 2012). This approach allows for the relative contribution of each variable to be evaluated. These assessments are used to evaluate model structure and interpret risk result outcomes with the stakeholders of the assessment (Marcot, 2012; Landis et al., 2016). This test demonstrates to both PROBFLO operators and stakeholders where models and associated assessments are sensitive to input data. Evidence to justify these sensitive determinants are imperative to a robust assessment. Additional sources of uncertainty include the comparative availability of evidence and expert knowledge pertaining to the socio-ecological systems considered in the assessments. The Senqu River case study addressed the second phase of a water resource use development that already has two substantial flow-altering developments with more than 15 years of pre- and post-development E-flow assessment (using holistic EFA methods; Arthington et al., 2003) monitoring and evaluations. Additional field surveys of the study area were carried out to generate additional information and to test existing hypotheses for the assessment. The Mara River case study was based largely on available historical information and existing EFA results for parts of the study area (McClain and Kashaigili, 2013; Dessu et al., 2014).

3.8 Step 8: Hypotheses establishment

In the hypotheses establishment step of PROBFLO, suitable hypotheses for field and laboratory experiments are established to test flow–ecosystem and flow–ecosystem service relationships (Landis, 2004; O'Brien and Wepener, 2012). In PROBFLO the fundamental adaptive management approach to improving our understanding of socio-ecological risk relationships, while revisiting outcomes and re-evaluating approaches, is formalized in the hypotheses establishment and testing phase. This process is based on a similar process in the RRM approach, established to reduce uncertainties and to confirm the risk rankings in risk assessments (Landis, 2004). In PROBFLO these adaptive management principles acknowledge that socio-ecological systems are dynamic and that our limited understanding of these processes necessitates the incorporation of many assumptions. In many case studies, uncertainties associated with the outcomes need to be mitigated before they can be used to inform decision making. To reduce uncertainty, assumptions can be tested rigorously and early. The adaptive management processes should (1) be informed by iterative learning about the flow–ecosystem and flow–ecosystem service relationships, (2) consider and respond to earlier management successes and failures and (3) increase present-day socio-ecological system resilience that can improve the ability of E-flows management to respond to the threats of increasing resource use (Lee, 2004).

In the Senqu River case study, many hypotheses associated with the flow–ecosystem and flow–ecosystem service relationships, largely established on data associated with Phase I of the LHWP, were established and tested during the field surveys. These hypotheses included the following. (1) Woody vegetation communities, sustainably harvested by local communities for fuel, respond to reduced average flows by increasing in abundance due to reduced flow variability, reduced stream power and through the colonization of new lower marginal zones. (2) Migratory cyprinid fish respond to ecological cue flows, which include increased discharges associated with reduced salinity, and which initiate fish migration. (3) Grazing for livestock of local communities depends on freshet flows lifting water onto the river banks and floodplains to stimulate vegetation growth. Data were collected from the study area to address these hypotheses and improve the understanding of the flow–ecosystem and flow–ecosystem service relationships considered in the study. In the Mara River case study available flow–ecosystem and flow–ecosystem service information was used in the PROBFLO assessment. A range of hypotheses associated with our understanding of the relationships were generated to refine and improve E-flow assessments of the study area.

3.9 Step 9: Test hypotheses

The two PROBFLO case studies included the design of long-term monitoring programmes to test the accuracy of risk projections and improve the understanding of the flow–ecosystem and flow–ecosystem service relationships. In the Senqu River case study a data management system (DMS) with automated data evaluation components was established. In the Mara River case study a range of hypotheses were established and used to design a monitoring plan and associated research programme to confirm the flow–ecosystem and flow–ecosystem service relationships considered in the study.

3.10 Step 10: Communicate outcomes

Regional-scale ecological risk assessments of water resources are carried out on behalf of stakeholders of the use and/or protection of those resources. Stakeholders need information generated with robust, best scientific practice methodologies in a transparent, clear and concise format, to evaluate the socio-ecological consequences of water resource use options. The PROBFLO approach highlights the importance of communicating the outcomes of assessments in the context of the uncertainty identified in an assessment (Hayes and Landis, 2004). A variety of techniques and tools are available to assist in the communication of the E-flow outcomes and associated socio-ecological consequences of altered flows and careful attention must be paid to ensure that the relevant stakeholders of any case study are presented with information that can easily be understood (O'Brien and Wepener, 2012). In the Senqu River case study, the LHDA with South African and Lesotho governmental delegates participated in a project outcomes workshop in 2014. During this workshop the PROBFLO approach adopted for the study, results and outcomes were discussed. Risk results of sources and stressors to social and ecological endpoints were compared in a relative manner, facilitating water resource use and protection trade-off considerations for the LHWP Phase II. In the Mara River case study the PROBFLO assessment successfully formed a part of the Nile E-flows framework development (NBI, 2016) and the ongoing Mau Mara Serengeti (MaMaSe) Sustainable Water Initiative (http://mamase.org). The application of the PROBFLO and associated uncertainty assessment was used to establish a monitoring plan that should be implemented with water resource use scenarios selected from the case studies. These plans were designed to validate the model by testing the response of the receiving environment to observed ecosystem driver conditions, associated with implemented scenarios, and to improve the understanding of the causal relationships hypothesized in the original assessment with real data.

4 Conclusion

The regional-scale ecological risk assessment approach was established in 1997 in response to the need to apply ecological risk assessments that consider multiple sources, stressors and receptors in the context of spatial and temporal ecosystem dynamics, on multiple spatial scales (Landis and Wiegers, 1997, 2007). The approach, which includes the RRM, has been widely implemented, reviewed and proven to be a robust probabilistic modelling tool to contribute to the sustainable management of ecological resources (Landis and Wiegers, 2007). Recent developments in E-flow frameworks (Poff et al., 2010; NBI, 2016) now also call for holistic, regional-scale, probabilistic E-flow assessments that consider flow and non-flow drivers of change in socio-ecological contexts. We have established a regional-scale ecological risk assessment method, incorporating the BN-RRM approach, to evaluate the socio-ecological consequences of altered flows and establish E-flows called PROBFLO. PROBFLO is a robust method for E-flow assessments that meets current standards of best scientific practice and can make a positive contribution to the sustainable management of water resources. The approach provides true transparency and adaptability options for holistic E-flow management. PROBFLO has already been successfully implemented in two major case studies where its flexibility and functionality has been demonstrated. In both case studies the evidence-based outcomes facilitated informed environmental management decision making, in the context of social and ecological aspirations. From these outcomes, stakeholders have, in addition, been able to consider sustainable social and ecological trade-offs in an attempt to balance the use and protection of water resources. The PROBFLO outcomes used to direct the sustainable use of water resources in the case studies are probabilistic and need to be validated with monitoring data during implementation phases. PROBFLO is an adaptable tool that allows for the rapid incorporation of new information, which will inform adaptive management and reduce uncertainty associated with the accuracy of the projections. In the case studies, stakeholders were presented with evidence-based probabilistic projections of PROBFLO and used the risk projections to consider water resource use trade-off options. Both of these case studies are being used by stakeholders to make water resource use decisions that are currently being undertaken. PROBFLO is a holistic, evidence-based probability modelling E-flow assessment tool that is transparent and adaptable, and it is suitable for application on multiple spatial scales. PROBFLO has the potential to contribute to the sustainable management of water resources for the benefit of social and ecological components of these systems.

Code availability. The code used in this paper is available in the Supplement.

Information about the Supplement

The Supplement includes justification tables for the exposure and effects input nodes, and the conditional probability justification tables for the daughter nodes use in the Bayesian Network models in the study. In addition two Netica™ working Bayesian Network models used in the risk assessment for the Mara River and Senqu River case study is provided.

Author contributions. The BN-RRM approach established for use in E-flow assessments was co-developed by GCO'B, CD and VW. The approach was implemented by this team including LQ, KF, JM, MG and RS. The paper was written by GCO'B and CD and edited by WGL, VW, EH and RS.

Competing interests. The authors declare that they have no conflict of interest.

Acknowledgements. The PROBFLO E-flow assessment approach was established through the Institute of Natural Resources NPC (Pietermaritzburg, South Africa) as part of the LHWP Phase II study funded by LHDA. We acknowledge the contributions made to the initial Senqu River case study by the extended team of ecological and social scientists and engineers who worked on the project. Contributions were also made through the NBI Guidance Document on Environmental Flows study, prepared by HYDROC GmbH in collaboration with Mau Mara Serengeti Sustainable Water Initiative study on behalf of the Nile Basin Initiative and Deutsche Gesellschaft für Internationale Zusammenarbeit (GIZ). In particular, the contributions made by Michael McClain and John Conallin are acknowledged. The authors also acknowledge the advice provided by fellow E-flow assessment scientists that have contributed to the development of PROBFLO including Jacqueline King, Cate Brown, Rebecca Tharme and Jay O'Keeffe.

Finally, we appreciate the comments on the manuscript provided during the discussion document phase of the HESS publication process and the highly valuable comments and suggestions provided by the reviewers of the manuscript.

Edited by: Stacey Archfield

References

Acreman, M. C. and Dunbar, M. J.: Defining environmental river flow requirements – a review, Hydrol. Earth Syst. Sci., 8, 861–876, https://doi.org/10.5194/hess-8-861-2004, 2004.

Acreman, M. C., Arthington, A. H., Colloff, M. J., Couch, C., Crossman, N. D., Dyer, F., Overton, I., Pollino, C. A.,

Stewardson, M. J., and Young, W.: Environmental flows for natural, hybrid, and novel riverine ecosystems in a changing world In a nutshell?, Front. Ecol. Environ., 12, 466–473, https://doi.org/10.1890/130134, 2014.

Anderson, S. and Landis, W.: A pilot application of regional scale risk assessment to the forestry management of the Upper Grande Ronde watershed, Oregon, Hum. Ecol. Risk Assess., 18, 705–732, https://doi.org/10.1080/10807039.2012.688696, 2012.

Arthington, A.: "Environmental Flows: Saving Rivers in the Third Millennium", University of California Press, Berkeley, CA, 406 pp., 2012.

Arthington, A., Rall, J., and Kennard, M.: Environmental flow requirements of fish in Lesotho Rivers using the DRIFT methodology, River Res, available at: http://onlinelibrary.wiley.com/doi/10.1002/rra.728/abstract (last access: 19 August 2016), 2003.

Atisa, G.: Economic Assessment of Best Management Practices in the Mara River Basin: Toward Implementing Payment forWatershed Services. MSc Thesis Florida International University, Miami, Florida, 135 pp., 2009.

Atisa, G., Bhat, M., and McClain, M.: Economic assessment of best management practices in the Mara River Basin: toward implementing payment for watershed services, Water Resour. Manag., 28, 1751–1766, https://doi.org/10.1007/s11269-014-0585-3, 2014.

Ayre, K. and Landis, W.: A Bayesian approach to landscape ecological risk assessment applied to the Upper Grande Ronde Watershed, Oregon, Hum. Ecol. Risk Assess., 18, 946–970, https://doi.org/10.1080/10807039.2012.707925, 2012.

Ayre, K., Caldwell, C., Stinson, J., and Landis, W.: Analysis of regional scale risk of whirling disease in populations of colorado and rio grande cutthroat trout using a bayesian belief network model, Risk Anal., 34, 1589–1605, 2014.

Bartolo, R., van Dam, R. A., and Bayliss, P.: Regional ecological risk assessment for Australia's tropical rivers: Application of the relative risk model, Hum. Ecol. Risk Assess., 18, 16–46, https://doi.org/10.1080/10807039.2012.631467, 2012.

Biswas, A.: Integrated water resources management: a reassessment: a water forum contribution, Water Int., 29, 248–256, https://doi.org/10.1080/02508060408691775, 2004.

Colnar, A. M. and Landis, W. G.: Conceptual model development for invasive species and a regional risk assessment case study: the European green crab, Carcinus maenas, at Cherry Point, Hum. Ecol. Risk Assess., 13, 120–155, https://doi.org/10.1080/10807030601105076, 2007.

Defersha, M. and Melesse, A. M.: Field-scale investigation of the effect of land use on sediment yield and runoff using runoff plot data and models in the Mara River basin, Kenya, Catena, 89, 54–64, 2012.

Defersha, M., Melesse, A. M., and McClain, M. E.: Watershed scale application of WEPP and EROSION 3D models for assessment of potential sediment source areas and runoff flux in the Mara River Basin, Kenya, Catena, 95, 63–72, https://doi.org/10.1016/j.catena.2012.03.004, 2012.

Department of Water Affairs (DWA): Procedures to Develop and Implement Resource Quality Objectives, Report, available at: http://inr.org.za/wp-content/uploads/2014/09/dwa-rqo-report_march-2011.pdf, 2011.

Dessu, S., Melesse, A., Bhat, M., and McClain, M.: Assessment of water resources availability and demand in the Mara River Basin, Catena, 115, 104–114, https://doi.org/10.1016/j.catena.2013.11.017, 2014.

Dudgeon, D.: Threats to freshwater biodiversity in a changing world, Global Environmental Change, 1, 243–253, https://doi.org/10.1007/978-94-007-5784-4_108, 2014.

Dudgeon, D., Arthington, A. H., Gessner, M. O., Kawabata, Z.-I., Knowler, D. J., Lévêque, C., Naiman, R. J., Prieur-Richard, A.-H., Soto, D., Stiassny, M. L. J., and Sullivan, C. A.: Freshwater biodiversity: importance, threats, status and conservation challenges, Biol. Rev. Camb. Philos. Soc., 81, 163–82, https://doi.org/10.1017/S1464793105006950, 2006.

Dutton, C., Anisfeld, S., and Ernstberger, H.: A novel sediment fingerprinting method using filtration: application to the Mara River, East Africa, J. Soils Sediments, 13, 1708–1723, 2013.

East African Community (EAC): Memorandum of Understanding (MoU) with the World Wide Fund for Nature Eastern Africa Regional Programme Office (WWF EARPO) for WWF-EARPO to support the EAC in programmes contributing to the vision of the Lake Victoria Development Programme and Established Lake Victoria Basin Commission, Kampala, Uganda, 2003.

Gichana, Z., Njiru, M., Raburu, P., and Masese, F.: Effects Of Human Activities On Microbial Water Quality In Nyangores Stream, Mara River Basin, Int. J. Sci. Technol. Res., 3, 153–157, 2014.

GLOWS-FIU: Environmental Flow Recommendations for the Mara River, Kenya and Tanzania, Global Water for Sustainability Program (GLOWS), Miami, FL, 2012.

Government of Kenya: Kenya Water Act No. 8 of 2002, Government of Kenya, 2002.

Grafton, R., Pittock, J., Davis, R., Williams, J., and Fu, G.: Global insights into water resources, climate change and governance, Nat. Clim. Change, 3, 315–321, https://doi.org/10.1038/nclimate1746, 2013.

Growns, I.: The influence of changes to river hydrology on freshwater fish in regulated rivers of the Murray–Darling basin, Hydrobiologia, 596, 203–211, https://doi.org/10.1007/s10750-007-9097-y, 2008.

Hart, B. and Pollino, C.: Increased use of Bayesian network models will improve ecological risk assessments, Hum. Ecol. Risk Assess., 14, 851–853, https://doi.org/10.1080/10807030802235037, 2008.

Hayes, E. and Landis, W.: Regional ecological risk assessment of a near shore marine environment: Cherry Point, WA, Hum. Ecol. Risk Assess., 10, 299–325, https://doi.org/10.1080/10807030490438256, 2004.

Herring, C. E., Stinson, J., and Landis, W. G.: Evaluating nonindigenous species management in a Bayesian networks derived relative risk framework for Padilla Bay, Washington, Integr. Environ. Assess. Manag., 11, 640–652, 2015.

Hines, E. and Landis, W.: Regional risk assessment of the Puyallup River Watershed and the evaluation of low impact development in meeting management goals, Integr. Environ. Assess., 10, 269–278, https://doi.org/10.1002/ieam.1509, 2014.

Hoffman, C. M.: Geospatial Mapping and Analysis of Water Availability-Demand-Use Within the Mara River Basin, Florida International University, Miami, 2007.

Hoffman, C. M., Melesse, A. M., and McClain, M. E.: Geospatial mapping and analysis of water availability-demand-use within the Mara River Basin, in: Melesse, A. M., Nile River Basin: Hydrology, Climate and Water Use, Springer, Dordrecht, New York, 359–382, 2011.

Hunsaker, C. T., Graham, R. L., Suter II, G. W., O'Neill, R. V., Barnthouse, L. W., and Gardner, R. H.: Assessing ecological risk on a regional scale, Environ. Manage., 14, 325–332, 1990.

Isaak, D., Wollrab, S., Horan, D., and Chandler, G.: Climate change effects on stream and river temperatures across the northwest US from 1980–2009 and implications for salmonid fishes, Clim. Change, 113, 499–524, 2012.

Kanga, E., Ogutu, J., Olff, H., and Santema, P.: Population trend and distribution of the Vulnerable common hippopotamus Hippopotamus amphibius in the Mara Region of Kenya, Oryx, 45, 20–27, 2011.

Karanja, G.: Tourist Impacts in Masai Mara National Reserve, University of Kent at Canterbury, ISNI: 0000 0001 3594 9345, 2002.

Karanja, G.: Tourism impacts in Masai Mara National Reserve, in: Walpole, edited by: Karanja, M. J., Sitati, G., and Leader-Williams, N., Wildlife and People: Conflict and Conservation in Masai Mara, IIED, London, 5–16, 2003.

Kiambi, S., Kuloba, B., Kenana, L., Muteti, D., and Mwenda, E.: Wet Season Aerial Count of Large Herbivores inMasai Mara National Reserve and the Adjacent Community Areas (2010), Mara Research Station, Kenya Wildlife Service, Narok, Kenya, 2012.

Kilonzo, F., Masese, F. O., Van Griensven, A., Bauwens, W., Obando, J., and Lens, P. N. L.: Spatial–temporal variability in water quality and macro-invertebrate assemblages in the Upper Mara River basin, Kenya, Phys. Chem. Earth, 67–69, 93–104, 2014.

King, J. and Pienaar, H. (Eds.): Sustainable use of South Africa's inland waters, Water Research Commission, Technical report, available at: http://www.wrc.org.za/KnowledgeHubDocuments/ResearchReports/TT491-11.pdf, 2011.

King, J., Brown, C., and Sabet, H. A scenario-based holistic approach to environmental flow assessments for rivers, River Res. Appl., 19, 619–639, 2003.

Korb, K. and Nicholson, A.: Bayesian artificial intelligence, 2nd edn., CRC Press, Boca Raton, FL, USA, 451 pp., 2010.

Lamprey, R. H. and Reid, R. S.: Expansion of human settlement in Kenya's Maasai Mara: what future for pastoralism and wildlife?, J. Biogeogr., 31, 997–1032, 2004.

Landis, W. (Ed.): Regional scale ecological risk assessment: using the relative risk model, CRC Press, Boca Raton, FL, USA, 309 pp., 2004.

Landis, W.: Comment on ET&C perspectives, November 2015 – A holistic view, Environ. Toxicol. Chem., 35, 1337–1339, https://doi.org/10.1002/etc.3378, 2016.

Landis, W. and Wiegers, J.: Design considerations and a suggested approach for regional and comparative ecological risk assessment, Hum. Ecol. Risk Assess., 3, 287–297, 1997.

Landis, W. and Wiegers, J.: Ten years of the relative risk model and regional scale ecological risk assessment, Hum. Ecol. Risk Assess., 13, 25–38, 2007.

Landis, W. G., Ayre, K. K., Johns, A. F., Summers, H. M., Stinson, J., Harris, M. J., Herring, C. E., and Markiewicz, A. J.: The multiple stressor ecological risk assessment for the mercury contaminated South River and upper Shenandoah River using the Bayesian Network-Relative Risk Model, Integr. Environ. Assess. Manag., 13, 85–99, https://doi.org/10.1002/ieam.1758, 2016.

Lee, K. N.: Appraising Adaptive Management, Conserv. Ecol., 3, 1–18, 2004.

Lesotho Highlands Water Project (LHWP): Treaty on the Lesotho highlands water project between the government of the Kingdom of Lesotho and the government of the Republic of South Africa signed at Maseru, 24 October 1986.

Lesotho Highlands Water Project (LHWP): Agreement on Phase II of the Lesotho highlands water project between the government of the Kingdom of Lesotho and the government of the Republic of South Africa signed at Maseru, 11 August 2011.

Lesotho Highlands Water Project (LHWP): Specialist Consultants to Undertake Baseline Studies (Flow, Water Quality and Geomorphology) and Instream Flow Requirement (IFR) Assessment for Phase 2: Instream Flow Requirements for the Senqu River – Final report No. 6001/2/e, Lesotho Highlands Development Authority, Maseru, 2016.

LVBC & WWF-ESARPO: Biodiversity Strategy and Action Plan for Sustainable Management of the Mara River Basin, Nairobi and Kisumu, Kenya, 2010 Lake Victoria Basin Commission and WWF-ESARPO, 2010.

Majule, A.: Towards sustainable management of natural resources in the Mara river basin in Northeast Tanzania, J. Ecol. Nat. Environ., 2, 213–224, 2010.

Mango, L. M., Melesse, A. M., McClain, M. E., Gann, D., and Setegn, S. G.: Land use and climate change impacts on the hydrology of the upper Mara River Basin, Kenya: results of a modeling study to support better resource management, Hydrol. Earth Syst. Sci., 15, 2245–2258, https://doi.org/10.5194/hess-15-2245-2011, 2011.

Marcot, B.: Metrics for evaluating performance and uncertainty of Bayesian network models, Ecol. Modell., 230, 50–62, 2012.

Marcot, B. G., Steventon, J. D., Sutherland, G. D., and Mccann, R. K.: Guidelines for developing and updating Bayesian belief networks applied to ecological modeling and conservation, Can. J. For. Res., 36, 3063–3074, 2006.

Mati, B., Mutie, S., Gadain, H., and Home, P.: Impacts of land-use/cover changes on the hydrology of the transboundary Mara River, Kenya/Tanzania, Lakes & Reservoirs: Research & Management, 13, 169–177, 2008.

McCann, R. K., Marcot, B. G., and Ellis, R.: Bayesian belief networks: applications in ecology and natural resource management, 36, 3053–3062, 2006.

McCartney, B.: Evaluation of Water Quality and Aquatic Ecosystem Health in the Mara River Basin East Africa, Masters thesis, Florida International University, available at: http://dpanther.fiu.edu/sobek/content/FI/MA/00/00/17/00001/FIMA000017.pdf, 2010.

McClain, M. and Kashaigili, J.: Environmental flow assessment as a tool for achieving environmental objectives of African water policy, with examples from East Africa, Int. J. Water Resour. D, 29, 650–665, 2013.

McClain, M. E., Subalusky, A. L., Anderson, E. P., Dessu, B., Melesse, A. M., Ndomba, P. M., Joseph, O. D., Tamatamah, R. A., Mligo, C., Mcclain, M. E., Subalusky, A. L., Anderson, E. P., Dessu, B., Melesse, A. M., Ndomba, P. M., Mtamba, J. O. D., and Rashid, A.: Comparing flow regime, channel hydraulics, and biological communities to infer flow–ecology relationships in the Mara River of Kenya and Tanzania, Hydrol. Sci. J., 59, 801–819, https://doi.org/10.1080/02626667.2013.853121, 2014.

Mitchell, B.: Integrated water resource management, institutional arrangements, and land-use planning, Environ. Plann. A, 37, 1335-1352, https://doi.org/10.1068/a37224, 2005.

Moraes, R., Landis, W., and Molander, S.: Regional risk assessment of a Brazilian rain forest reserve, Hum. Ecol. Risk Assess., 8, 1779–1803, 2002.

Murray, S., Foster, P., and Prentice, I.: Future global water resources with respect to climate change and water withdrawals as estimated by a dynamic global vegetation model, J. Hydrol., 448–449, 14–29, https://doi.org/10.1016/j.jhydrol.2012.02.044, 2012.

NBI: Preparation of NBI Guidance Document on Environmental Flows: Nile E-flows Framework Technical Implementation Manual, HYDROC GmbH on behalf of the Nile Basin Initiative and Deutsche Gesellschaft für Internationale Zusammenarbeit, 2016.

O'Brien, G. and Wepener, V.: Regional-scale risk assessment methodology using the Relative Risk Model (RRM) for surface freshwater aquatic ecosystems in South Africa, Water SA, 38, 153–166, https://doi.org/10.4314/wsa.v38i2.1, 2012.

Ogutu, J. O., Owen-Smith, N., Piepho, H. P., and Said, M. Y.: Continuing wildlife population declines and range contraction in the Mara region of Kenya during 1977–2009, J. Zool., 285, 99–109, 2011.

Onjala, J. O.: Managing Water Scarcity in Kenya: Industrial Response to Tariffs and Regulatory Enforcement, PhD Dissertation Thesis Roskilde University, Copenhagen, Denmark, 329 pp., 2002.

Orange-Senqu River Commission (ORASECOM): Integrated Water Resources Management Plan For The Orange-Senqu River Basin: Main Report, Report No. ORASECOM 019/2014, Orange-Senqu River Basin Commission, 2014.

Pahl-Wostl, C., Arthington, A., Bogardi, J., Bunn, S. E., Hoff, H., Lebel, L., Nikitina, E., Palmer, M., Poff, L. N., Richards, K., Schlüter, M., Schulze, R., St-Hilaire, A., Tharme, R., Tockner, K., and Tsegai, D.: Environmental flows and water governance: Managing sustainable water uses, Curr. Opin. Environ. Sustain., 5, 341–351, https://doi.org/10.1016/j.cosust.2013.06.009, 2013.

Pitman, W. V., Bailiey, A. K., and Kakebeeke, J. P.: WRSM2000 Water Resources Simulation Model for Windows: Users Guide, Department of Water Affairs and the Water Research Commission, 2006.

Poff, N. and Matthews, J.: Environmental flows in the Anthropocence: past progress and future prospects, Curr. Opin. Environ. Sustain., 5, 667–675, 2013.

Poff, N. L., Richter, B. D., Arthington, A. H., Bunn, S. E., Naiman, R. J., Kendy, E., Acreman, M., Apse, C., Bledsoe, B. P., Freeman, M. C., Henriksen, J., Jacobson, R. B., Kennen, J. G., Merritt, D. M., O'Keeffe, J. H., Olden, J. D., Rogers, K., Tharme, R. E., and Warner, A.: The ecological limits of hydrologic alteration (ELOHA): A new framework for developing regional environmental flow standards, Freshw. Biol., 55, 147–170, https://doi.org/10.1111/j.1365-2427.2009.02204.x, 2010.

Pollino, C. A., Woodberry, O., Nicholson, A., Korb, K., and Hart, B. T.: Parameterisation and evaluation of a Bayesian network for use in an ecological risk assessment, Environ. Modell. Softw., 22, 1140–1152, https://doi.org/10.1016/j.envsoft.2006.03.006, 2007.

Richter, B. D., Warner, A. T., Meyer, J. L., and Lutz, K.: A collaborative and adaptive process for developing environmental flow recommendations, River Res. Appl., 22, 297–318, https://doi.org/10.1002/rra.892, 2006.

Smit, N. J., Vlok, W., van Vuren, J. H. J., Du Preez, L. H., Van Eeden, E., O'Brien, G. C., and Wepener, V.: Socio-ecological System Management of the Lower Phongolo River and Floodplain Using Relative Risk Methodology, Water Resarch Commission, ISBN: 978-1-4312-0811-1, 2016.

Stirzaker, R., Biggs, H., Roux, D., and Cilliers, P.: Requisite simplicities to help negotiate complex problems, Ambio, 39, 600–607, 2010.

Subalusky, A. L., Dutton, C. L., Rosi-Marshall, E. J., and Post, D. M.: The hippopotamus conveyor belt: vectors of carbon and nutrients from terrestrial grasslands to aquatic systems in sub-Saharan Africa, Freshw. Biol., 60, 512–525, 2015.

Tharme, R.: A global perspective on environmental flow assessment: emerging trends in the development and application of environmental flow methodologies for rivers, River Res. Appl., 19, 397–441, 2003.

United Republic of Tanzania: Water Resources Management Act, ISSN: 0856-033IX, 73 pp., 2009.

Uusitalo, L.: Advantages and challenges of Bayesian networks in environmental modelling, Ecol. Model., 203, 312–318, https://doi.org/10.1016/j.ecolmodel.2006.11.033, 2006.

Vörösmarty, C. J., McIntyre, P. B., Gessner, M. O., Dudgeon, D., Prusevich, A., Green, P., Glidden, S., Bunn, S. E., Sullivan, C. A., Reidy Liermann, C., and Davies, P. M.: Global Threats to Human Water Security and River Biodiversity, Nature, 467, 555–561, 2010.

Walker, R., Landis, W., and Brown, P.: Developing a regional ecological risk assessment: A case study of a Tasmanian agricultural catchment, Hum. Ecol. Risk Assess., 7, 417–439, doi:10.1080/20018091094439, 2001.

Wiegers, J., Feder, H., and Mortensen, L.: A Regional Multiple-Stressor Rank-Based Ecological Risk Assessment for the Fjord of Port Valdez, Alaska, Hum. Ecol. Risk Assess., 4, 1125–1173, 1998.

WRMA: Catchment Management Strategy: Lake Victoria South Catchment Area, Kenya, Water Resources Management Authority of Kenya, 2014.

8

Hydrologic controls on aperiodic spatial organization of the ridge–slough patterned landscape

Stephen T. Casey[1], **Matthew J. Cohen**[1], **Subodh Acharya**[1], **David A. Kaplan**[2], **and James W. Jawitz**[3]

[1]School of Forest Resources and Conservation, University of Florida, Gainesville, FL, USA
[2]Engineering School of Sustainable Infrastructure and Environment, Environmental Engineering Sciences Department, University of Florida, Gainesville, FL, USA
[3]Soil and Water Science Department, University of Florida, Gainesville, FL, USA

Correspondence to: Matthew J. Cohen (mjc@ufl.edu)

Abstract. A century of hydrologic modification has altered the physical and biological drivers of landscape processes in the Everglades (Florida, USA). Restoring the ridge–slough patterned landscape, a dominant feature of the historical system, is a priority but requires an understanding of pattern genesis and degradation mechanisms. Physical experiments to evaluate alternative pattern formation mechanisms are limited by the long timescales of peat accumulation and loss, necessitating model-based comparisons, where support for a particular mechanism is based on model replication of extant patterning and trajectories of degradation. However, multiple mechanisms yield a central feature of ridge-slough patterning (patch elongation in the direction of historical flow), limiting the utility of that characteristic for discriminating among alternatives. Using data from vegetation maps, we investigated the statistical features of ridge-slough spatial patterning (ridge density, patch perimeter, elongation, patch size distributions, and spatial periodicity) to establish more rigorous criteria for evaluating model performance and to inform controls on pattern variation across the contemporary system. Mean water depth explained significant variation in ridge density, total perimeter, and length : width ratios, illustrating an important pattern response to existing hydrologic gradients. Two independent analyses (2-D periodograms and patch size distributions) provide strong evidence against regular patterning, with the landscape exhibiting neither a characteristic wavelength nor a characteristic patch size, both of which are expected under conditions that produce regular patterns. Rather, landscape properties suggest robust scale-free patterning, indicating genesis from the coupled effects of local facilitation and a global negative feedback operating uniformly at the landscape scale. Critically, this challenges widespread invocation of scale-dependent negative feedbacks for explaining ridge-slough pattern origins. These results help discern among genesis mechanisms and provide an improved statistical description of the landscape that can be used to compare among model outputs, as well as to assess the success of future restoration projects.

1 Introduction

The coupling of ecosystem processes operating at different scales can cause vegetation communities to form a wide variety of spatial patterns (Borgogno et al., 2009), ranging from highly regular striping, stippling, or maze-like patterns in woodland landscapes (Ludwig et al., 1999), tidal mud flats (Weerman et al., 2012), and boreal peatlands (Eppinga et al., 2010) to scale-free patterning in semi-arid landscapes (Kéfi et al., 2007; Scanlon et al., 2007). The mechanisms that produce these patterns are integral to understanding landscape origins and, thus, for predicting appropriate remedies where patterns and underlying processes have been degraded and require restoration. The spatial arrangement of vegetation on the landscape has long been viewed as a manifestation of the dominant interactions and drivers (Hutchinson, 1957; Levin, 1992) and the scales at which they operate. By quantifying this spatial arrangement we can make process-based inferences about the underlying mechanisms (Gardner et al., 1987; Turner, 2005).

The ridge-slough landscape comprised ~55 % of the pre-development Everglades in southern Florida (McVoy et al., 2011). However, processes that created, and in some places still maintain, the characteristic ridge-slough patterning are only partially understood (SCT, 2003; Larsen et al., 2011; Cohen et al., 2011). The landscape pattern consists of flow-parallel bands of higher-elevation ridges dominated by emergent sedge saw grass (*Cladium jamaicense*), interspersed within a matrix of lower-elevation sloughs (~25 cm lower in the best conserved portions of the landscape; Watts et al., 2010), which contain a variety of submerged and emergent herbaceous macrophytes. The Everglades has undergone massive hydrologic modification through the construction of a system of levees and canals over the past century (Light and Dineen, 1994), and ensuing ecological degradation has prompted a complex, expensive, and ambitious restoration effort. Because the ridge-slough landscape was so prevalent in the pre-development system, pattern restoration is a central priority (SCT, 2003; McVoy et al., 2011). The mechanisms that control the emergence of patterning and explain variation in pattern geometry are thus integral to specifying hydrologic restoration objectives.

To understand the landscape processes that produce patterning, and by extension gain insight into how to restore them (Pickett and Cadenasso, 1995), requires a testable mechanistic framework for pattern genesis and maintenance. However, experiments to test alternative mechanisms are constrained by the spatial extent and timescales of peat accumulation responses. Paradoxically, compartmentalization by the extensive canal and levee system has created artificial gradients that are informative for assessing trajectories of landscape pattern degradation. Here we focus on Water Conservation Area 3 (WCA-3), located in the central Everglades, an area historically dominated by the ridge-slough landscape (Fig. 1), and where the best conserved patterning is found. The hydrologic gradient in WCA-3 spans from relatively dry (i.e., short hydroperiod) conditions in the north due to major canals that drain water to the southeast, to extended inundation (i.e., long hydroperiod) in the south and southeast due to impoundment caused by US41/Tamiami Trail (which runs orthogonal to flow) and the L-67 levee. The best conserved patterning (SCT, 2003; Watts et al., 2010) is found between these hydrologic extremes.

Several alternative hypotheses have been proposed to explain ridge–slough patterns, and all have been evaluated using process-based models. The mechanisms invoked vary and include evaporative nutrient redistribution (Ross et al., 2006; Cheng et al., 2011), flow-driven sediment redistribution from sloughs to ridges (Larsen et al., 2007; Larsen and Harvey, 2011; Lago et al., 2010), self-optimization of patterning for discharge and hydroperiod (Cohen et al., 2011; Kaplan et al., 2012; Heffernan et al., 2013), and a suite of mechanisms that couple pattern-hydroperiod effects with directional local facilitation processes (i.e., where patches expand more rapidly in one direction than another; Acharya et al., 2015). Clearly, these mechanisms are not mutually exclusive, so process models have sought to explore the sufficiency of each alternative, while acknowledging the potential that multiple processes may overlap. One central criterion used to evaluate the models has been whether simulations can produce morphologies qualitatively consistent with the extant landscape (principally replicating the elongation of patches in the flow direction). To date, however, almost all models either accomplish (Ross et al., 2006; Larsen and Harvey, 2010; Lago et al., 2010; Cheng et al., 2011; Acharya et al., 2015) or strongly imply (Heffernan et al., 2013) patch elongation (albeit sometimes under conditions markedly different than those observed in the Everglades), limiting discrimination among pattern genesis mechanisms and highlighting the need for a more rigorous and quantitative characterization of landscape pattern.

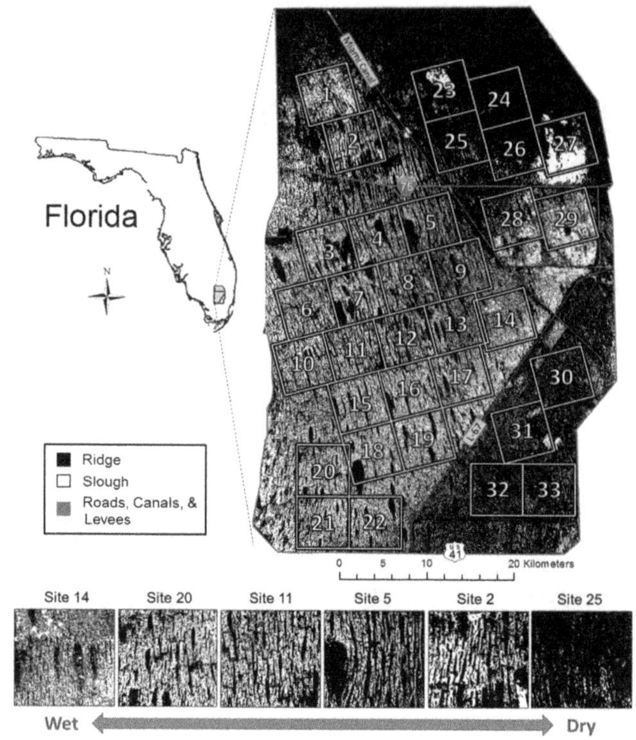

Figure 1. Study area and site locations, including major roads, canals, and levees for the primary map (M1). Sites spanning the pattern gradient in WCA3 are shown in the bottom panel. Two additional maps (Supplement) were used to corroborate the primary results.

To better characterize patterns in both the best conserved state and spanning a gradient of degradation requires spatial analyses that yield quantitative properties against which model outputs can be compared. Although numerous metrics have been developed to quantify different pattern attributes (Wu et al., 2006; Yuan et al., 2015), significant gaps in our understanding of how to interpret these metrics remain (Turner, 2001; Remmel and Csillag, 2003). Real landscapes clearly depart from regular Euclidean geometry, making characterization problematic in some cases (Mandelbrot, 1983). Likewise, changes in mapping procedures (e.g., grain size, extent, classification schemes) can yield significantly different metric values for the same landscape (Li and Wu, 2004). To remedy some of these issues, we focused on a set of relatively direct and easily interpreted metrics of fundamental aspects of the pattern and used multiple maps produced with varying methods to rule out mapping-related artifacts. We were interested in three aspects of landscape patterning: density and shape statistics, patch size distributions, and spectral (i.e., pattern wavelength) characteristics. For each aspect, we explored the magnitude of site-to-site variation and the support for hydrologic control of that variation.

Density and shape statistics focus on the most basic and intuitive geometric properties of the landscape: areal coverage of the patch types (density), landscape pattern complexity (perimeter), and the degree of elongation. While inundation has been shown to control species composition (Givnish et al., 2008; Zweig and Kitchens, 2008; Todd et al., 2010), the relationship between hydrologic drivers and other aspects of landscape pattern remain relatively unknown, so this effort also serves as an inventory of hydrologic controls on pattern geometry.

Patch size distributions (i.e., frequency of different patch sizes) have been used in many systems to identify underlying landscape processes (e.g., Manor and Shnerb, 2008a; Kéfi et al., 2011; Bowker and Maestre, 2012; Weerman et al., 2012). For example, regular patterning is associated with a characteristic patch size (Rietkerk and van de Koppel, 2008; von Hardenberg et al., 2010), arising in response to an inhibitory feedback operating at a particular spatial scale (van de Koppel and Crain, 2006) that limits patch expansion. Under these conditions, there should be a distinct mode in patch area distribution, or at least the absence of very large patches (Manor and Shnerb, 2008a; von Hardenberg et al., 2010; Kéfi et al., 2014). In contrast, patch size distributions that follow a power law (i.e., $y = x^{\alpha}$, where α is a scaling parameter) lack a characteristic spatial scale (e.g., Scanlon et al., 2007) and may suggest genesis mechanisms that operate equally across scales. Correspondingly, power law distributions are often referred to as scale-free, in that the distribution form remains the same regardless of the measurement scale.

Scale-free distributions can arise via a number of mechanisms (Newman, 2005). In a landscape where grid cells are randomly occupied, patch distributions show relatively few large patches, up to a critical density (~ 0.59; known as the percolation threshold) at which patches span the domain, yielding power-law area scaling. At densities slightly above and below the percolation threshold, area distributions depart from power laws. The narrow range of density space over which scale-free area distributions emerge would seem to suggest that this mechanism is rare. However, some systems can endogenously maintain themselves near this critical point in a phenomenon referred to as self-organized criticality (Bak et al., 1989). This is accomplished through disturbance processes that propagate via patch contiguity (e.g., forest fires, see Drossel and Schwabl, 1992), maintaining patterns near the percolation threshold through a cycle of large-scale disturbance and slow recovery (Pascual and Guichard, 2005).

Alternatively, power-law scaling of patch areas can arise from the coupled action of local facilitation, which causes patches to expand, and competition for a global resource (Pascual et al., 2002; Scanlon et al., 2007) that ultimately limits the density of that patch type at the landscape scale. In contrast to regular patterning mechanisms, these feedback processes limit landscape-level patch density but not the size of individual patches, leading to the creation, via local facilitation, of very large patches. This is known as robust criticality because power-law scaling in response can occur over a wide range of external conditions and patch densities, including densities well below the percolation threshold. Robust criticality has been noted in Everglades vegetation distributions (Foti et al., 2012), as well as in a variety of dryland vegetation patterns (Kéfi et al., 2011). Widespread occurrence of both local facilitation and global resource competition in ecological systems suggests this process may operate in a multitude of landscapes.

Finally, spectral characteristics provide insights on the presence and wavelength of regular landscape pattern. Useful information about the scale at which spatial feedbacks operate in self-organized systems has been obtained by evaluating two-dimensional pattern periodicity (Couteron, 2002; Kéfi et al., 2014). This is particularly important in the Everglades because the prevailing conceptual model for ridge-slough pattern genesis invokes interactions between spatial feedbacks operating on different characteristic scales, resulting in a pattern wavelength of approximately 150 m in the direction perpendicular to historical flow (SCT, 2003; Watts et al., 2010). Several models (e.g., Ross et al., 2006; Lago et al., 2010; Cheng et al., 2011) produce distinctly periodic landscapes, which arise from the action of local facilitation feedbacks and, crucially, negative feedbacks on patch expansion that operate at a characteristic scale. In contrast, the feedback between hydroperiod and landscape geometry suggested by Cohen et al. (2011), enumerated by Heffernan et al. (2013), and tested at the landscape scale in Kaplan et al. (2012) operates at the global scale, implying no characteristic spatial scale. To that end, we tested the hypothesis that the ridge-slough landscape is regularly patterned (i.e., exhibits a characteristic wavelength), consistent with scale-specific negative feedbacks, or whether the landscape lacks periodicity, consistent with scale-free feedbacks.

Together, these spatial analyses encompass a novel and rigorous set of metrics for improved quantification of observed and modeled landscape pattern. While developed to improve descriptions of the ridge-slough pattern, these metrics may also be useful for identifying pattern and discriminating genesis mechanisms in other patterned landscapes.

2 Methods

2.1 Vegetation and hydrologic data

We used multiple vegetation maps of the central Everglades, which vary in scale, extent, mapping schemes, and time frame. For all maps, we aggregated vegetation types into binary classes (reclassification scheme in Table S2 in the Supplement) of ridges (value = 1) and sloughs (value = 0). Our primary map (M1) was produced by the South Florida Water Management District (SFWMD) using 1 : 24 000-scale color infrared photos from September 1994 (Rutchey et al., 2005). This map was chosen due to its large, continuous spatial extent and fine mapping detail. The presence of small ($< 25\,\mathrm{m}^2$) landscape features permitted us to select raster representation of dominant vegetation at high (i.e., 1×1 m cells) resolution. While the presence of small features does not imply map accuracy at that fine scale, it does imply loss of patch geometric detail with larger cells. Features at this scale can be subject to mapping error and artifacts, likely under-representing their prevalence. As such, patches below $100\,\mathrm{m}^2$ were omitted from patch-level analyses.

We selected 33 6×6 km sites to span the range of current hydrological conditions (i.e., dry in northern areas to wet in southern areas; Fig. 1). We sought to maximize the number of sites with minimal overlap while avoiding roads and canals. All sites except 20–22 and 32–33 were rotated to align with the prevailing direction of patch elongation (15° counterclockwise). Ridge cells were grouped into patches if they shared at least one edge with an adjacent ridge (i.e., a von Neumann neighborhood).

Within each site, *point-specific* daily average water depths at a grid spacing of 200 m were obtained from the Everglades Depth Estimation Network (EDEN) xyLocator (http://sofia.usgs.gov/eden/edenapps/xylocator.php). We note these water depths are spatially interpolated from a network of water elevation monitoring stations and, as such, represent only an estimate of actual conditions. *Site-specific* mean water depth (MWD) values were obtained by averaging all point-specific values in each site over the period of record from 1991 to 2010.

We used two additional maps (M2 and M3), which vary in spatial extent, resolution, and sampling date, to corroborate M1 analyses and test map resolution effects and temporal changes. M2 was generated from 1 : 24 000 scale aerial photographs taken in 2004 (RECOVER, 2014) and rasterized at 50 m resolution. M3 was generated from 1 m resolution digital orthophotos and rasterized at 1 m (Nungesser, 2011). Methodological details for both M2 and M3 are given as Supplement.

2.2 Shape and density

We compared ridge density, edge density, and elongation across sites. Ridge density is the proportion of ridge area to site area, while edge density is total patch perimeter divided by site area. In order to measure elongation, E, we first identify individual lengths and widths (l and w, respectively) as any group of contiguous ridge cells (i.e., unbroken by slough cells) along a row or column. Elongation is the ratio of the mean of these contiguous row and column sections:

$$E = \frac{\frac{1}{n_\mathrm{c}}\sum l}{\frac{1}{n_\mathrm{r}}\sum w} = \frac{n_\mathrm{r}}{n_\mathrm{c}}, \tag{1}$$

where n_r and n_c represent the number of contiguous rows and columns. Elongation simplifies to their ratio since the summation terms both yield the total number of ridge cells. Elongation metrics are sensitive to orientation differences between the grid and landscape features. Sites with tortuous flow paths or a poorly aligned grid will underestimate E. We provide estimates of grid alignment with feature orientation as a mean patch angle, $\overline{A}_\mathrm{p}$, where A_p is the angle between the grid y axis and the major axis of an ellipse with the same second moment as the patch.

Hydrologic trends were identified by regressing MWD against site-level metrics and were considered statistically significant at $p < 0.05$. For analyses that are highly dependent on mapping resolution (i.e., edge density), we omit M1 sites north of Interstate 75, as these were mapped using significantly lower resolution than those to the south (Rutchey et al., 2005). Because elongation values are dominated by the domain shape at very high ridge densities, we omitted sites where ridge density exceeded 0.8.

2.3 Patch size distributions

Patch size scaling properties were evaluated by comparing empirical distributions to several candidate models. Patch size distributions can be described in terms of their complementary cumulative distribution function (CCDF), which gives the probability that the area of an observed patch is greater than or equal to a given area, x. Preliminary analyses showed that empirical CCDFs exhibited extremely heavy tails consistent with power laws, but only above a minimum cutoff, below which patches were less abundant and the CCDFs were rounded. This form is in relative agreement with both the generalized Pareto (GP) and truncated lognormal distributions. The GP is given by its CCDF as

$$P(x) = \begin{cases} \left(1 + \dfrac{k\,(x - x_{\min})}{\delta}\right)^{-\frac{1}{k}} & \text{for } k \neq 0 \\ \exp\left(-\dfrac{x - x_{\min}}{\delta}\right) & \text{for } k = 0, \end{cases} \tag{2}$$

for $x \geq x_{\min}$ when $k \geq 0$, and for $x_{\min} \leq x \leq (x_{\min} - \delta/k)$ when $k < 0$. The GP reduces to the exponential distribution when $k = 0$ and $x_{\min} = 0$ and reduces to a power function when $k > 0$ and $x_{\min} = \delta/k$. For $k > 0$ and $x_{\min} < \delta/k$ the GP shows exponential-like behavior for low values of x, while

the tail asymptotically approaches a power law for $x \gg x_{\min}$. Within this range of parameters, δ indicates the curvature in the upper end of the distribution (higher values correspond to greater curvature and, hence, relatively fewer small patches), while k indicates the scaling properties of the tail, such that for $x \gg x_{\min}$, the power-law scaling exponent α approaches $\alpha^* = (1 + 1/k)$ (Pisarenko and Sornette, 2003). Where the GP fits the data well, we can use the estimated parameters as general information about patch size scaling properties. The CCDF for a truncated lognormal distribution uses the mean ($\mu_{\ln_x}$) and standard deviation ($\sigma_{\ln_x}$) of $\ln(x)$.

$$P(x) = \frac{\operatorname{erf}\left(\frac{\sqrt{2}[\mu_{\ln_x} - \ln(x)]}{2\sigma_{\ln_x}}\right) + 1}{\operatorname{erf}\left(\frac{\sqrt{2}[\mu_{\ln_x} - \ln(x_{\min})]}{2\sigma_{\ln_x}}\right) + 1} \quad x \geq x_{\min} \tag{3}$$

We compared empirical distributions to synthetic data sets from Monte Carlo simulations ($n = 20\,000$ per model) and compared candidate distributions based on log-likelihood ratios and significance values (Clauset et al., 2009). Distribution testing details are given in SI.

2.4 Spectral characteristics

Spectral characteristics of the ridge-slough landscape were evaluated from 2-D periodograms generated following the methods of Mugglestone and Renshaw (1998). In brief, we constructed a discrete 2-D Fourier transform (available in most computational software packages) for each binary vegetation map (Kéfi et al., 2014) and then took the absolute value to obtain the real number component. The resulting 2-D periodogram (i.e., spectral density) is a grid representing the magnitude of cosine and sine waves of possible wavenumbers (i.e., spatial frequencies) and orientations to the spectrum. Values were averaged across all orientations in equally spaced wavenumber bins to generate radial spectra (r spectra), which indicate the relative spectral density for each corresponding wavenumber bin. Local maxima indicate dominant wavelengths and, thus, suggest the presence of spatial periodicity (Couteron, 2002; Kéfi et al., 2014) or regular patterning. The absence of local maxima indicates an aperiodic landscape. Because the ridge-slough pattern has been described as regular in the direction orthogonal to flow, we generated both lateral and longitudinal r spectra derived from the spectral densities observed within $\pm 10°$ perpendicular and parallel to the main axis of pattern elongation. For both directions, we noted the wavelength at which either clear spectral peaks (i.e., for periodic patterns) or locations of spectral shouldering (i.e., slope breaks), which may indicate a secondary scale-dependent feedback mechanism, were evident. Since smaller features are underrepresented in low-resolution maps, we omitted wavelengths < 10 m from our analyses.

3 Results

3.1 Visual comparisons

Visual inspection of the vegetation maps reveals a remarkable range of pattern morphology (Fig. 1). Ridges in northwestern sites (1–5) show pronounced striping, which is less apparent in southern sites (18–22), where ridges appear more elliptical. Eastern sites located below I-75 (5, 9, 13, 14, 17, 28–33) show fine-scale speckling and disaggregation, with sites 14, 28, and 29 appearing random, with faint outlines of historic pattern.

Individual ridges exhibit numerous connections between adjacent elongated portions, with larger patches forming complex webs composed of multiple individual elements. Although this behavior is apparent in all sites, it appears to be density dependent, with most of the landscape spanned by one large patch in denser sites (e.g., 2, 5, 8, 9, 11–13, 23–28, 30–33). Within sites, large patches are always more web-like than smaller ones, which appear more distinctly separated.

3.2 Density and shape

Ridge density was negatively correlated to MWD (Fig. 2a; $R^2 = 0.38$, $p = 0.0002$). Deviation from this association was similar across maps and related to geographic position. Specifically, ridge densities in the eastern half of the domain (sites 9, 13, 14, 17, 23–33) were consistently higher than in the west, suggesting a strong east–west control on density. The correlation between MWD and ridge density increased markedly when sites were partitioned into east and west blocks (east: $R^2 = 0.81$, $p < 0.0001$; west: $R^2 = 0.61$, $p = 0.0004$). Based on recent aerial imagery, low ridge density in site 1 is a misclassification of sparse saw-grass prairies as slough; that site was omitted from regression analyses.

Site-level elongation was also strongly correlated to MWD (Fig. 2b; $R^2 = 0.65$, $p < 0.0001$). Sites with ridge densities greater than 0.8 showed elongation values much lower than this trend. Average patch orientations ($\overline{A}_{\mathrm{p}}$) indicate consistency between the grid and feature elongation (i.e., $\overline{A}_{\mathrm{p}}$ values close to zero; Table S1). Sites with values of $|\overline{A}_p| \geq 5°$ (e.g., 1, 22) may be underestimated due to mismatch between patch orientation and map orientation. Finally, edge density was strongly correlated to MWD, indicating greater perimeter at deeper sites (Fig. 2c; $R^2 = 0.79$, $p < 0.0001$).

3.3 Patch size distributions

Patch area distributions were consistent with the GP distribution (Fig. 3c), with 16 of 25 sites passing GP Monte Carlo tests for M1 and 4 of 9 passing for M3 (Table S1). The majority of sites that were not significant contained extremely large patches but had little deviation in the rest of the distribution; in some cases (e.g., sites 2, 5, 8, 9, 11, 12, 13, 28, 31) the largest patch was over an order of magnitude larger than

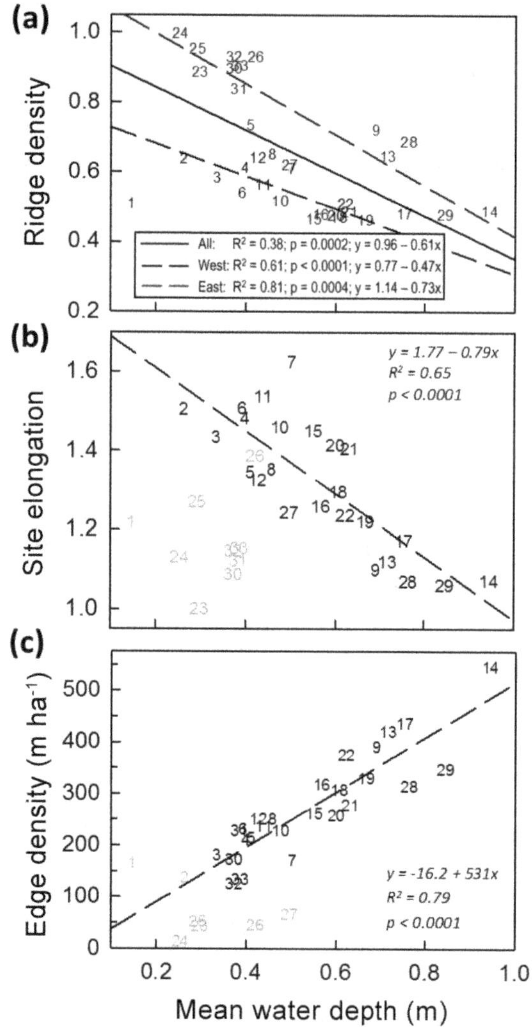

Figure 2. (a) Ridge density is negatively correlated with mean water depth. Eastern sites (blue) show consistently higher ridge densities than those in the west (black). Trends associated with east–west segregation (dashed lines) show much stronger relationships than the composite trend (solid line). Site 1 was omitted due to possible misclassification. **(b)** Site elongation shows a strong negative relationship with mean water depth. Sites with ridge densities greater than 0.8 (indicated in grey) were omitted from regressions and show elongation values lower than expected from this trend. **(c)** Edge density is positively correlated to mean water depth indicating higher perimeters in deeper sites. Sites indicated in grey were mapped at lower resolution and were omitted from regressions. The relationships observed for site elongation and edge density are both consistent with patches becoming disaggregated with increased water depth.

predicted based on the GP distribution. All these sites with extremely large patches have ridge densities above or very close to the percolation threshold of a square lattice (~ 0.59, Stauffer and Aharony, 1991). Above this percolation threshold, the largest patch becomes "over-connected", suggesting that failure of Monte Carlo tests within this group may be density driven rather than a result of an underlying patterning mechanism. Note that these sites are largely located in the north and eastern sections of the study area, a region typified by high ridge densities. The presence of tree islands, a third landform modality distinct from ridges and sloughs, may also affect patch scaling relationships; however, we neglect these effects here because across all blocks tree islands represent less than 10 % of the area, and often much less. The log-normal distribution was significant in only 4 of 25 sites for M1 and 2 of 9 sites for M3. Although these sites (15, 16, 19, 21) showed slight rounding in the extreme tail, log-likelihood ratios were not different enough to distinguish between the two candidate distributions (Table S1).

Within each map, GP parameters were remarkably consistent across sites, with almost constant estimates of k and δ for sites that passed Monte Carlo tests (Table S1). Area scaling in the tail of the distribution is illustrated by α^* (analogous to the scaling exponent of a power-law distribution) $= 1.77 \pm 0.06$ for M1 and 1.87 ± 0.13 for M3. The δ parameter indicates how sharply the distribution head deviates from a power law, with larger values indicating that smaller patch areas are exceedingly rare. For M1 and M3, $\delta = 474 \pm 88$ and 1490 ± 219; these differences are likely due to map resolution, with M3 under-representing smaller patches.

3.4 Spectral characteristics

We found no evidence of periodicity in either the lateral or longitudinal r spectra. The absence of peak values other than the smallest wavenumber indicates that no dominant pattern wavelength exists, a finding consistent across hydrologic conditions and pattern morphologies (Fig. 3a, Fig. S7 in the Supplement). Spearman correlations, ρ, show the r spectra are nearly perfectly approximated by a monotonic function across all sites (Table S1), with $\rho < -0.99$ for both lateral and longitudinal r spectra. As with patch-scaling relationships, tree islands may introduce some noise in the observed r spectra; however, this effect is likely to be small, given that they constitute less than 10 % of the landscape.

For both lateral and longitudinal directions, the form of the r spectra appeared to contain a mix of both power law and exponential scaling. Lateral r spectra largely appear linear in log–log space (i.e., power law form) at higher wavenumbers while rounding towards an exponential (i.e., curved) at lower wavenumbers. This curvature appears over a wider range of wavenumbers for longitudinal spectra, but the morphology and mean transition location are same laterally and longitudinally. Sites with the best-conserved patterning (e.g., sites 2, 5, 11, 20) show more localized curvature in lateral r spectra compared to degraded sites, potentially signifying the action of a secondary, scale-dependent patterning mechanism. We note, however, that this finding is inconsistent with proposed patterning mechanisms that invoke a characteristic wavelength in the lateral direction but include no mecha-

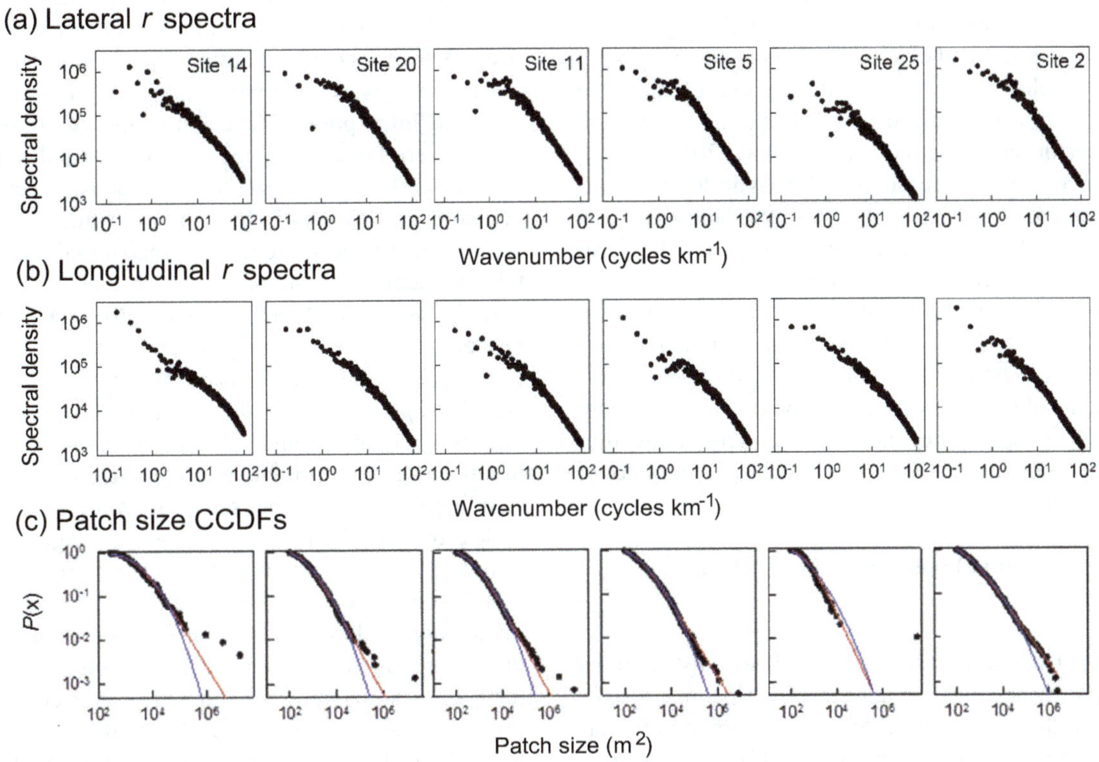

Figure 3. (a) Lateral r spectra (limited to $\pm 10°$ perpendicular to the pattern) monotonically decreased with no evidence of peaks, indicating aperiodic behavior in the direction of presumed regularity. **(b)** Longitudinal r spectra (limited to $\pm 10°$ in the direction parallel to the pattern) show similar monotonic behavior. The form for both lateral and longitudinal directions is similar, with both exhibiting a mixture of power-law and exponential behavior. The location of the exponential-like curvature appears to be influenced by both orientation and pattern condition, suggesting a weak-acting scale-dependent mechanism. **(c)** Patch size distributions across sites are well described by the generalized Pareto distribution (red lines). Sites with high ridge densities (e.g., sites 2, 5 and 25) have maximum patch sizes much greater than expected from the GP distribution. Conversely, sites in excessively inundated sections (e.g., site 20) show slightly steeper tails, consistent with a lognormal distribution (blue lines), though not enough to rule out the GP.

nism to generate regular patterning in the longitudinal direction. Alternatively, this shouldering may result from undersampling large features at low wavenumbers due to a limited domain size.

4 Discussion and conclusions

4.1 Water depth controls pattern attributes

Our results provide strong observational support for water depth as a dominant control on several key shape and density properties of the ridge-slough landscape. Although these findings are correlative and not necessarily mechanistic, they align with current understanding about the mechanisms that create, maintain, and degrade the landscape. The observed decline in ridge abundance with MWD is consistent with conceptual models that predict that changes in water levels precipitate transitions between ridge and slough by modifying production and respiration dynamics (Givnish et al., 2008; Watts et al., 2010) and inducing state changes in vegetation composition (Zweig et al., 2008). The implication that these dynamics differ in eastern and western sections of the study area was unexpected and points to unexplained controls on ridge expansion. The largest difference between the east and west trends occurs at low water depths, indicating that this control is most pronounced in drier sites. In short, the deviation seen in eastern sites represents a shifting of the relationship to favor saw-grass expansion in extremely dry sites rather than a general reduction of the hydrologic limitation (since deep sites remain the least affected).

Mean water depth also exerted strong control on ridge-slough pattern shape. The most salient features of the pattern, elongation and perimeter, both showed strong dependence on MWD, with maximum elongation observed at low to intermediate water depths and minimum edge density values at low water depths. This is consistent with ridge features fragmenting into smaller, less elongated patches under deeper water conditions, a finding previously observed anecdotally (McVoy et al., 2011) and in the spatial statistics of soil elevation (Watts et al., 2010). Likewise, sites with very low MWD show a significant loss of pattern, with ridge densities approaching unity and elongation values that are largely isotropic. The coherent response of these pattern features to hydrologic modification suggests promise for their use as restoration performance measures (Yuan et al., 2015).

In this work we provide support for hydrological controls on ridge-slough pattern shape; however landscape patterning (specifically ridge density and elongation) has also been shown to exert reciprocal control on regional hydrology (Kaplan et al., 2012). Loss of sloughs in sites with very low MWD alters drainage characteristics. Coupled to observations of patch fragmentation in sites with higher water depths, these results strongly reinforce the commanding role of hydrology in maintaining landscape pattern, indicating that reversal of modern hydrologic modification is paramount for ongoing restoration.

4.2 The ridge-slough landscape is aperiodic and scale-free

Both spatial periodogram results and patch size distributions strongly suggest the ridge-slough landscape pattern is aperiodic, a marked departure from extensive literature qualitatively describing the pattern as periodic (SCT, 2003; Wetzel et al., 2005; Ross et al., 2006; Larsen et al., 2007; Givnish et al., 2008; Larsen and Harvey, 2010; Lago et al., 2010; Watts et al., 2010; Cheng et al., 2011; Nungesser, 2011; Sullivan et al., 2014). Because negative feedbacks operating at a characteristic spatial scale result in regular patterning (Rietkerk and Van de Koppel, 2008), aperiodic patterning in the ridge-slough landscape implies the absence, or least secondary importance, of such feedbacks, ruling them out as the dominant control on patterning many of the mechanisms invoked to explain pattern formation (Borgogno et al., 2009).

While our results clearly support the primacy of aperiodic patterning mechanisms, the r spectra in both lateral and longitudinal directions do exhibit persistent curvature, whose location and degree appears dependant on both orientation and pattern condition. This suggests ridge-slough patterning is secondarily influenced by scale-dependent (but omnidirectional) feedbacks, possibly suggesting links with vegetative propagation or fire behavior. Additional investigation and modeling, requiring higher-resolution mapping, would be necessary to better understand the mode and scale of these secondary feedbacks.

The observation that patch size distributions uniformly follow power-law scaling suggests a scale-free patterning process. While power-law scaling can be produced via several mechanisms (Newman, 2005), our results can be used to rule out some alternatives. For example, power-law scaling of patch areas can arise in systems near the percolation threshold (i.e., at criticality), which occurs within a relatively narrow region of patch density. Observed patch area scaling in our study occurs across a wide range of patch densities, suggesting robust criticality that comports with Foti et al. (2012), who observed similar power-law scaling behavior over a wide range of vegetation types and densities.

Caution is warranted when using contemporary aerial imagery to infer pre-drainage landscape conditions; the first aerials were taken ~ 65 years after Everglades drainage began. Several pattern attributes (e.g., density, perimeter) may adjust readily with hydrologic modification, and while some areas remain largely unchanged since initial imagery was obtained, pattern in many other areas has degraded, sometimes entirely (Wu et al., 2006; Nungesser 2011). However, pattern properties that are relatively invariant with hydrologic modification (e.g., the general forms of the r spectrum and patch area distributions) are more likely to reflect pre-drainage conditions. In contrast, while measures that vary with hydrologic modification are correlative, they remain useful for understanding landscape responses to hydrologic forcing but may be less informative for inferring pre-drainage conditions and long-term processes such as landscape formation.

Self-organized criticality can also produce power-law scaling at varying densities (i.e., far from the percolation threshold) but requires large temporal variation in ridge density as the system endogenously readjusts towards criticality following disturbances (Pascual and Guichard, 2005). Recent paleoecological evidence (Bernhardt and Willard, 2009) suggests that ridge-slough configurations and densities have remained relatively stable since initial formation 2700 years before present, though temporal variation in density (e.g., during the Medieval Warm Period) may have been sufficient to modestly alter landscape pattern metrics. Moreover, no documented disturbance regime exhibits the characteristic separation of timescales between growth and disturbance associated with self-organized criticality. While peat fires could be invoked, there is little evidence for widespread incidence and large-scale impacts of these prior to modern hydrologic modification (McVoy et al., 2011).

Rather, power-law scaling in patch areas over a range of densities along environmental gradients is consistent with robust criticality, wherein local facilitation induces clustering (i.e., patch growth) while a global limitation maintains landscape heterogeneity (Pascual and Guichard, 2005). Although robust criticality is typically suggested in isotropic landscapes, Acharya et al. (2015) recently showed that anisotropy in the local facilitation kernel of a robust criticality model can produce directional banding without periodicity, yielding simulated ridge-slough patterns with high statistical and visual fidelity to the observed landscape. Local facilitation may take the form of autogenic peat accretion (Larsen et al., 2007), clonal propagation of saw grass (Brewer, 1996), nutrient accumulation dynamics (Cohen et al., 2009; Larsen et al. 2015), or local seed dispersal, although the relative importance and directionality of these mechanisms remains unknown (Acharya et al., 2015). Screening possible mechanisms for anisotropic local facilitation emerges from our analysis as a priority for future investigations.

Several candidate processes could limit patch expansion in the ridge–slough landscape. Each implies a distinct spatial pattern geometry, and we can use the extant scale-free and aperiodic geometry to evaluate their respective plausibilities. A key distinction between limiting processes that produce periodic versus scale-free patterning is the spatial range over which the limiting factor acts (Manor and Shnerb, 2008a; von Hardenberg et al., 2010) the landscape, the effect is considered global or uniform. Conversely, when the limiting effect acts in a more localized manner, limitation gradients can develop and produce periodic patterning.

Phosphorus limitation and sediment transport mechanisms are both potentially important feedbacks on patch expansion. While phosphorus is strongly limiting of primary production in the Everglades (Noe et al., 2001) and can be dramatically enriched in tree islands (Wetzel et al., 2009) and ridges (Ross et al., 2006) via multiple mechanisms, this process of local enrichment and depletion is inconsistent with robust criticality. Indeed, the presence of strong local phosphorus gradients indicates that limitation feedbacks are distinctly local and not spread uniformly across the landscape. If phosphorus limitation were the dominant control, the result would be regular patterning. Similarly, sediment transport mechanisms (Larsen et al., 2007; Lago et al., 2010) yield a balance between entrainment and deposition governed by focused flow in sloughs, the velocity of which is controlled by cross-sectional occlusion of flow by ridges. Because patch geometry is controlled by local heterogeneity in flow velocity, this suggests an inhibitory feedback operating at a limited spatial scale, as the velocity field responds most strongly to local flow occlusion.

Water level (and hydroperiod) is another potential feedback on patch expansion. Our observations of water depth control on ridge density comport with numerous studies (Givnish et al., 2008; Zweig and Kitchens, 2008; Todd et al., 2010) suggesting ridges are significantly impacted by water depths. Moreover, pattern geometry strongly influences landscape hydrology (Kaplan et al., 2012; Acharya et al., 2015). As ridges expand into adjacent sloughs, they displace water and alter landscape flow capacity, causing regional water levels to increase (Kaplan et al., 2012), and creating a negative feedback that likely limits further ridge expansion (Cohen et al., 2011). Indeed, the RASCAL model of ridge-slough development (Larsen and Harvey, 2011) represents this feedback, though in that model velocity-field feedbacks alone could not impose elongation and regular patterning; disentangling sediment transport and water-level feedbacks in that model, and interrogating pattern output, may enable tests of the relative importance of overlapping feedbacks at different scales. We note here that because water depths equilibrate quickly, local patch expansion effects on water level are distributed rapidly and evenly across the landscape. This expansion is consistent with the global limitation necessary to create observed aperiodic and scale-free pattern. Therefore, water depth effects are strong candidates for the requisite global feedback to induce ridge-slough formation.

Our results also indicate that elongated landscape features do not necessarily require pattern periodicity, suggesting that spatial structures in numerous ecosystems may have been misclassified as regularly patterned and that aperiodic banding may be more prevalent than the literature suggests. Invoking robust criticality and directional facilitation, as in Acharya et al. (2015), may be of general value for explaining aperiodic banding in other settings.

The ridge-slough landscape pattern has emerged as a key measure of restoration performance in one of the largest and most ambitious ecosystem management endeavors ever. Enumeration of spatial pattern statistical features is a prerequisite for assessing landscape condition and for comparing models with alternative landscape genesis mechanisms. Our results inform about the metrics for comparison between real and simulated landscape patterns and provide insights into the controls on pattern variation across the contemporary system. Given the potentially significant differences in water management implied by comparative genesis explanations, these metrics of real and simulated landscapes are important for restoration planning and assessment.

5 Data availability

All of the data that we used are publicly available, and references to the EDEN xyLocator can be found at http://sofia.usgs.gov/eden/edenapps/xylocator.php. The primary map M1 can be obtained by contacting Ken Rutchey at the South Florida Water Management District: krutchey@sfwmd.gov or by contacting the authors.

Acknowledgements. We gratefully acknowledge funding support from the US Army Corps of Engineers through the Everglades Monitoring and Assessment Program. We also acknowledge the contributions of Danielle Watts, Jim Heffernan and Mark Brown in this work. We also benefited greatly from deep and constructive feedback from Laurel Larsen on an earlier version of the manuscript.

Edited by: R. Moussa

References

Acharya, S., Kaplan, D. A., Casey, S., Cohen, M. J., and Jawitz, J. W.: Coupled local facilitation and global hydrologic inhibition drive landscape geometry in a patterned peatland, Hydrol. Earth Syst. Sci., 19, 2133–2144, doi:10.5194/hess-19-2133-2015, 2015.

Bak, P., Tang, C., and Wiesenfeld, K.: Self-organized criticality, Phys. Rev. A, 38, 364–374, 1989.

Bernhardt, C. E. and Willard, D. A.: Response of the Everglades ridge and slough landscape to climate variability and 20th-century water management, Ecol. Appl., 19, 1723–1738, doi:10.1890/08-0779.1, 2009.

Borgogno, F., D'Odorico, P., Laio, F., and Ridolfi, L.: Mathematical models of vegetation pattern formation in ecohydrology, Rev. Geophys., 47, RG1005, doi:10.1029/2007RG000256, 2009.

Bowker, M. A. and Maestre, F. T.: Inferring local competition intensity from patch size distributions: a test using biological soil crusts, Oikos, 121, 1914–1922, doi:10.1111/j.1600-0706.2012.20192.x, 2012

Brewer, J. S.: Site differences in the clone structure of an emergent sedge, *Cladium jamaicense*, Aquat. Bot., 55, 79–91, 1996.

Cheng, Y., Stieglitz, M., Turk, G., and Engel, V.: Effects of anisotropy on pattern formation in wetland ecosystems, Geophys. Res. Lett., 38, L04402, doi:10.1029/2010GL046091, 2011.

Clauset, A., Shalizi, C. R., and Newman, M. E.: Power-law distributions in empirical data, SIAM Rev., 51, 661–703, 2009.

Cohen, M. J., Osborne, T. Z., Lamsal, S. J., and Clark, M. W.: Regional Distribution of Soil Nutrients-Hierarchical Soil Nutrient Mapping for Improved Ecosystem Change Detection, South Florida Water Management District, West Palm Beach, Florida, USA, 91 pp., 2009.

Cohen, M. J., Watts, D. L., Heffernan, J. B., and Osborne, T. Z.: Reciprocal biotic control on hydrology, nutrient gradients and landform in the Greater Everglades, Crit. Rev. Environ. Sci. Technol., 41, 395–429, 2011.

Couteron, P.: Quantifying change in patterned semi-arid vegetation by Fourier analysis of digitized aerial photographs, Int. J. Remote Sens., 23, 3407–3425, 2002.

Drossel, B. and Schwabl, F.: Self-organized critical forest-fire model, Phys. Rev. Lett., 69, 1629, doi:10.1103/PhysRevLett.69.1629, 1992.

Eppinga, M. B., Rietkerk, M., Belyea, L., Nilsson, M., Ruiter, P., and Wassen, M.: Resource contrast in patterned peatlands increases along a climatic gradient, Ecology, 91, 2344–2355, 2010.

Foti, R., del Jesus, M., Rinaldo, A., and Rodriguez-Iturbe, I.: Hydroperiod regime controls the organization of plant species in wetlands, P. Natl. Acad. Sci. USA, 109, 19596–19600, 2012.

Gardner, R. H., Milne, B. T., Turnei, M. G., and O'Neill, R. V.: Neutral models for the analysis of broad-scale landscape pattern, Landscape Ecol., 1, 19–28, doi:10.1007/BF02275262, 1987.

Givnish, T. J., Volin, J. C., Owen, V. D., Volin, V. C., Muss, J. D., and Glaser, P. H.: Vegetation differentiation in the patterned landscape of the central Everglades: importance of local and landscape drivers, Global Ecol. Biogeogr., 17, 384–402, 2008.

Heffernan, J. B., Watts, D. L., and Cohen, M. J.: Discharge competence and pattern formation in peatlands: a meta-ecosystem model of the everglades ridge-slough landscape, PloS One, 8, e64174, doi:10.1371/journal.pone.0064174, 2013.

Hutchinson, G. E.: A treatise on Limnology, vol. 1, Geography, Physics, and Chemistry, Wiley, New York, 1957

Kaplan, D. A., Paudel, R., Cohen, M. J., and Jawitz, J. W.: Orientation matters: patch anisotropy controls discharge competence and hydroperiod in a patterned peatland, Geophys. Res. Lett., 39, L17401, doi:10.1029/2012GL052754, 2012.

Kéfi, S., Rietkerk, M., Alados, C. L., Pueyo, Y., Papanastasis, V. P., ElAich, A., and De Ruiter, P. C.: Spatial vegetation patterns and imminent desertification in Mediterranean arid ecosystems, Nature, 449, 213–217, 2007.

Kéfi, S., Rietkerk, M., Roy, M., Franc, A., De Ruiter, P. C., and Pascual, M.: Robust scaling in ecosystems and the meltdown of patch size distributions before extinction, Ecol. Lett., 14, 29–35, 2011.

Kéfi, S., Guttal, V., Brock, W. A., Carpenter, S. R., Ellison, A. M., Livina, V. N., Seekell, D. A., Scheffer, M., van Nes, E. H., and Dakos, V.: Early warning signals of ecological transitions: methods for spatial patterns, PloS One, 9, e92097, doi:10.1371/journal.pone.0092097, 2014.

Lago, M. E., Miralles-Wilhelm, F., Mahmoudi, M., and Engel, V.: Numerical modeling of the effects of water flow, sediment transport and vegetation growth on the spatiotemporal patterning of the ridge and slough landscape of the Everglades wetland, Adv. Water Resour., 33, 1268–1278, 2010.

Larsen, L. G. and Harvey, J. W.: How vegetation and sediment transport feedbacks drive landscape change in the Everglades and wetlands worldwide, Am. Nat., 176, E66–E79, 2010.

Larsen, L. G. and Harvey, J. W.: Modeling of hydroecological feedbacks predicts distinct classes of landscape pattern, process, and restoration potential in shallow aquatic ecosystems, Geomorphology, 126.3, 279–296, 2011.

Larsen, L. G., Harvey, J. W., and Crimaldi, J. P.: A delicate balance: ecohydrological feedbacks governing landscape morphology in a lotic peatland, Ecol. Monogr., 77, 591–614, 2007.

Larsen, L. G., Aumen, N., Bernhardt, C., Engel, V., Givnish, T., Hagerthey, S., Harvey, J., Leonard, L., McCormick, P., McVoy, C., Noe, G., Nungesser, M., Rutchey, K., Sklar, F., Troxler, T., Volin, J., and Willard, D.: Recent and historic drivers of landscape change in the Everglades ridge, slough, and tree island mosaic, Crit. Rev. Environ. Sci. Technol., 41, 344–381, 2011.

Larsen, L. G., Harvey, J. W., and Maglio, M. M.: Mechanisms of nutrient retention and its relation to flow connectivity in river–floodplain corridors, Freshwater Sci., 34, 187–205, 2015.

Levin, S. A.: The Problem of Pattern and Scale in Ecology: The Robert H. MacArthur Award Lecture, Ecology, 73, 1943–1967, doi:10.2307/1941447, 1992.

Li, H. and Wu, J.: Use and misuse of landscape indices, Landscape Ecol., 19, 389–399, 2004.

Light, S. S. and Dineen, J. W.: Water Control in the Everglades: a Historical Perspective, Everglades: the Ecosystem and its Restoration, St. Lucie Press, Delray Beach, Florida, 47–84, 1994.

Ludwig, J. A., Tongway, D. J., and Marsden, S. G.: Stripes, strands, or stipples: modelling the influence of three landscape banding patterns on resource capture and productivity in semi-arid woodlands, Australia, Catena, 37, 257–273, 1999.

Mandelbrot, B. B.: The Fractal Geometry of Nature, Freeman, New York, 1983.

Manor, A. and Shnerb, N. M.: Facilitation, competition, and vegetation patchiness: from scale free distribution to patterns, J. Theor. Biol., 253, 838–842, 2008a.

Manor, A. and Shnerb, N. M.: Origin of Pareto-like spatial distributions in ecosystems, Phys. Rev. Lett., 101, 268104, doi:10.1103/PhysRevLett.101.268104, 2008b.

McVoy, C., Park Said, W., Obeysekera, J., VanArman, J., and Dreschel, T.: Landscapes and Hydrology of the Predrainage Everglades, University Press of Florida, Gainesville, FL, 2011.

Mugglestone, M. A. and Renshaw, E.: Detection of geological lineations on aerial photographs using two-dimensional spectral analysis, Comput. Geosci., 24, 771–784, 1998.

Newman, M. E.: Power laws, Pareto distributions and Zipf's law, Contemp. Phys., 46, 323–351, 2005.

Noe, G. B., Childers, D. L., and Jones, R. D.: Phosphorus biogeochemistry and the impact of phosphorus enrichment: why is the Everglades so unique?, Ecosystems, 4, 603–624, 2001.

Nungesser, M. K.: Reading the landscape: temporal and spatial changes in a patterned peatland, Wetland. Ecol. Manage., 19, 475–493, 2011.

Pascual, M. and Guichard, F.: Criticality and disturbance in spatial ecological systems, Trends Ecol. Evol., 20, 88–95, 2005.

Pascual, M., Roy, M., Guichard, F., and Flierl, G.: Cluster size distributions: signatures of self-organization in spatial ecologies, Philos. T. Roy. Soc. B, 357, 657–666, 2002.

Pickett, S. T. and Cadenasso, M. L.: Landscape ecology: spatial heterogeneity in ecological systems, Science, 269, 331–334, 1995.

Pisarenko, V. F. and Sornette, D.: Characterization of the frequency of extreme earthquake events by the generalized Pareto distribution, Pure Appl. Geophys., 160, 2343–2364, 2003.

RECOVER: 2014 System Status Report, Restoration Coordination and Verification Program, c/o US Army Corps of Engineers, Jacksonville, FL, and South Florida Water Management District, West Palm Beach, FL, 2014.

Remmel, T. K. and Csillag, F.: When are two landscape pattern indices significantly different?, J. Geogr. Syst., 5, 331–351, 2003.

Rietkerk, M. and Van de Koppel, J.: Regular pattern formation in real ecosystems, Trends Ecol. Evol., 23, 169–175, 2008.

Ross, M. S., Mitchell-Bruker, S., Sah, J. P., Stothoff, S., Ruiz, P. L., Reed, D. L., Jayachandran, K., and Coultas, C. L.: Interaction of hydrology and nutrient limitation in the Ridge and Slough landscape of the southern Everglades, Hydrobiologia, 569, 37–59, 2006.

Rutchey, K., Vilchek, L., and Love, M.: Development of a vegetation map for Water Conservation Area 3, Technical Publication ERA Number 421, South Florida Water Management District, West Palm Beach, FL, USA, 2005.

Scanlon, T. M., Caylor, K. K., Levin, S. A., and Rodriguez-Iturbe, I.: Positive feedbacks promote power-law clustering of Kalahari vegetation, Nature, 449, 209–212, 2007.

SCT – Science Coordination Team: The Role of Flow in the Everglades Ridge and Slough Landscape, South Florida Ecosystem Restoration Working Group, West Palm Beach, FL, 2003.

Stauffer, D. and Aharony, A.: Introduction to percolation theory, Taylor and Francis, London, 1991.

Sullivan, P. L., Price, R. M., Miralles-Wilhelm, F., Ross, M. S., Scinto, L. J., Dreschel, T. W., Sklar, F. H., and Cline, E.: The role of recharge and evapotranspiration as hydraulic drivers of ion concentrations in shallow groundwater on Everglades tree islands, Florida (USA), Hydrol. Process., 28, 293–304, 2014.

Todd, M. J., Muneepeerakul, R., Pumo, D., Azaele, S., Miralles-Wilhelm, F., Rinaldo, A., and Rodriguez-Iturbe, I.: Hydrological drivers of wetland vegetation community distribution within Everglades National Park, Florida, Adv. Water Resour., 33, 1279–1289, 2010.

Turner, M. G.: Landscape Ecology in Theory and Practice: Pattern and Process, Springer-Verlag, New York, 2001.

Turner, M. G.: Landscape ecology: what is the state of the science?, Annu. Rev. Ecol. Evol. S., 36, 319–344, 2005.

von Hardenberg, J., Kletter, A. Y., Yizhaq, H., Nathan, J., and Meron, E.: Periodic vs. scale-free patterns in dryland vegetation, Philos. Roy. Soc. B, 277, 1771–1776, 2010.

Watts, D. L., Cohen, M. J., Heffernan, J. B., and Osborne, T. Z.: Hydrologic modification and the loss of self-organized patterning in the ridge-slough mosaic of the Everglades, Ecosystems, 13, 813–827, 2010.

Weerman, E. J., Van Belzen, J., Rietkerk, M., Temmerman, S., Kéfi, S., Herman, P. M. J., and de Koppel, J. V.: Changes in diatom patch-size distribution and degradation in a spatially self-organized intertidal mudflat ecosystem, Ecology, 93, 608–618, 2012.

Wetzel, P. R., van der Valk, A. G., Newman, S., Gawlik, D. E., Troxler Gann, T., Coronado-Molina, C. A., Childers, D. L., and Sklar, F. H.: Maintaining tree islands in the Florida Everglades: nutrient redistribution is the key, Front. Ecol. Environ., 3, 370–376, 2005.

Wetzel, P. R., van der Valk, A. G., Newman, S., Coronado, C. A., Troxler-Gann, T. G., Childers, D. L., Orem, W. H., and Sklar, F. H.: Heterogeneity of phosphorus distribution in a patterned landscape, the Florida Everglades, Plant Ecol., 200, 83–90, 2009.

Wu, Y., Wang, N., and Rutchey, K.: An analysis of spatial complexity of ridge and slough patterns in the Everglades ecosystem, Ecol. Complex., 3, 183–192, 2006.

Yuan, J., Cohen, M. J., Kaplan, D. A., Acharya, S., Larsen, L. G., and Nungesser, M. K.: Linking metrics of landscape pattern to hydrological process in a lotic wetland, Landscape Ecol., 30, 1893–1912, doi:10.1007/s10980-015-0219-z, 2015

Zweig, C. L. and Kitchens, W. M.: Effects of landscape gradients on wetland vegetation communities: information for large-scale restoration, Wetlands, 28, doi:10.1672/08-96.1, 2008.

9

Impacts of changing hydrology on permanent gully growth

Stephanie S. Day[1], **Karen B. Gran**[2], **and Chris Paola**[3]

[1]Department of Geosciences, North Dakota State University, Fargo, ND 58102, USA
[2]Department of Earth and Environmental Sciences, University of Minnesota Duluth, Duluth, MN 55812, USA
[3]Department of Earth Sciences, University of Minnesota, Minneapolis, MN 55455, USA

Correspondence: Stephanie S. Day (stephanie.day@ndsu.edu)

Abstract. Permanent gullies grow through head cut propagation in response to overland flow coupled with incision and widening in the channel bottom leading to hillslope failures. Altered hydrology can impact the rate at which permanent gullies grow by changing head cut propagation, channel incision, and channel widening rates. Using a set of small physical experiments, we tested how changing overland flow rates and flow volumes alter the total volume of erosion and resulting gully morphology. Permanent gullies were modeled as both detachment-limited and transport-limited systems, using two different substrates with varying cohesion. In both cases, the erosion rate varied linearly with water discharge, such that the volume of sediment eroded was a function not of flow rate, but of total water volume. This implies that efforts to reduce peak flow rates alone without addressing flow volumes entering gully systems may not reduce erosion. The documented response in these experiments is not typical when compared to larger preexisting channels where higher flow rates result in greater erosion through nonlinear relationships between water discharge and sediment discharge. Permanent gullies do not respond like preexisting channels because channel slope remains a free parameter and can adjust relatively quickly in response to changing flows.

1 Introduction

Permanent gullies are first-order, deeply incised, ephemeral streams with steep head cuts. Also known as ravines (Poesen et al., 2003a), permanent gullies form part of a continuum between ephemeral gullies and perennial channels. Permanent gullies can be highly erosive, accounting for 10–94 % of the total sediment yield in a watershed (Poesen et al., 2003a, and references therein). Gullies pose a hazard to infrastructure, agricultural productivity, and human safety through rapid propagation of head cuts in both rural and urban environments, and they contribute to the rapid transport of sediment and nutrients from the uplands to main-stem channels (Bull and Kirkby, 2002; Gran et al., 2011; Poesen et al., 2003a).

Permanent gullies initiate in places where concentrated flow can erode and move sediment (Mosley, 1974; Merritt, 1984; Bennett et al., 2000; Bryan, 1990; Knapen and Poesen, 2010). These topographic lows may be subtle, but where they connect with incised river valleys, steep knickpoints can rapidly form and propagate as head cuts. Gully head cut propagation occurs primarily by groundwater sapping or piping, overland flow, or a combination of these processes. Groundwater sapping typically forms amphitheater-type head cuts (Hinds, 1925; Howard, 1988; Lamb et al., 2006) which propagate as the soil is weakened through saturation at the water table, causing failure in the sediment above (Dunne, 1990). Where groundwater follows concentrated flow paths, piping can trigger head cut initiation and drive head cut retreat (Fox and Wilson, 2010; Nichols et al., 2017; Wilson, 2009, 2011). Overland flow erosion can also cause head cut propagation in multiple ways including granular erosion and slab failure. Slab failure relies on the constant wetting and drying experienced at the head cut between events. Tension cracks form as the soil dries, and during wet times water flows into these cracks, dislodging the slab (Dietrich and Dunne, 1993; Istanbulluoglu et al., 2005). Another form of slab failure occurs in layered substrate, where a lower highly erosive layer is scoured away, leaving the upper resistant layer overhanging until it eventually breaks away (Chu-Agor et al., 2008; Holland and Pickup, 1976; Robinson and Hanson, 1994; Stein and LaTray, 2002). In both of these slab failure mechanisms, head cut retreat temporarily ceases after slab failure with the slab protecting the head cut until the slab itself is eroded away (Robinson and Hanson, 1994; Istanbulluoglu et al., 2005). Granular erosion is driven by high shear stress at steep head cut slopes and can be enhanced through focused erosion of plunge pools and impinging jet scour (Alonso et al., 2002; Flores-Cervantes et al., 2006). The focus in this paper is on permanent gullies that grow as a result of granular sediment transport in a homogeneous substrate driven primarily by overland flow.

Permanent gully initiation and growth have been found to be a function of sediment texture and erodibility, slope, land use, vegetation cover, soil moisture, head cut size, climate, and hydrology (Ambers et al., 2006; Chiverrell et al., 2007; Flores-Cervantes et al., 2006; Istanbulluoglu et al., 2005; Knapen et al., 2007; Nachtergaele et al., 2001; Poesen et al., 2003b, 2011; Stankoviansky, 2003; Vandekerckhove et al., 2000; Vanwalleghem et al., 2003, 2005), but understanding the relative importance of each factor remains a challenge. Of the many variables impacting gully erosion, the two humans most readily impact are hydrology and land cover. Here the focus is on gully response to changes in hydrology.

Hydrology is altered in agricultural areas through complex changes to the volume and/or rate at which water enters permanent gully networks as overland flow, groundwater flow, or pipe flow. Agricultural development can involve direct modification of hydrology in the form of irrigation in dry climates and sandy soils or artificial drainage in wetter climates and less permeable soils. While irrigation has little impact on overland flow due to transpiration (Haddeland et al., 2005), combined changes to land cover and drainage associated with conversion from native vegetation to row-crop agriculture tend to increase the rate and volume of water entering a stream network (Blann et al., 2009; Robinson and Rycroft, 1999; Van der Ploeg et al., 1999; Wiskow and van der Ploeg, 2003). Following harvest, fields are left bare for a portion of the year (1–6 months), which can lower evapotranspiration and increase the volume of overland flow, particularly if the ground is frozen and infiltration cannot occur (Pierson et al., 2007). In addition, the removal of roughness elements when fields are bare can increase the rates of overland flow entering gullies (Abrahams and Parsons, 1991; Einstein and Barbarossa, 1951; Eitel et al., 2011; Farres, 1978; Römkens and Wang, 1987).

Although changes to hydrology in urban areas are different from those in agricultural areas, the impacts on permanent gully hydrology can be similar, with increased impervious surface area and rerouting of water through infrastructure leading to more direct hydrologic connections between uplands and gullies. Although the presence of storm sewers and water retention ponds may mitigate the effect of increased impervious area, many urban areas see an increase in overland flow rates and the overall volume of overland flow (Arnold Jr. and Gibbons, 1996; Espey Jr. et al., 1965; Paul and Meyer, 2001; Seaburn, 1969). Increased gullying is particularly of concern in urban areas with rapid population growth, intense rainfall, and infrastructure that promotes flow concentration into erodible areas (Adediji et al., 2013; Ebisemiju, 1989).

Permanent gullies evolve over shorter timescales than larger preexisting channels. Previous numerical and physical experiments of gully evolution identified two stages to gully growth: an early stage of rapid evolution and a later stage of growth within a more static form (Parker, 1977; Kosov et al., 1978; Parker and Schumm, 1982; Sidorchuk, 1999; Bennett et al., 2000). During this early phase, gullies should be more responsive to changes in hydrology as both channel cross-sectional geometry and longitudinal profile geometry are more adaptable. During the latter phase, changes in hydrology would be accommodated through erosion focused on head cuts at the channel tips and channel incision and widening with ensuing adjustments to side slopes (Sidorchuk, 1999, 2006).

Because anthropogenic alterations to the landscape involve both changes in flow rate and volume, we chose to investigate those two aspects of hydrologic change using a physical experimental basin. Two different experimental substrates were used to measure the effects of changing flow rates on both detachment-limited and transport-limited systems. These two types of erosion represent a continuum, where the volume of sediment carried out of the system is controlled by the ability of the flow to dislodge sediment from the substrate vs. the ability of the flow to carry easily eroded sediment. Commonly "detachment-limited" is used to describe bedrock rivers, while "transport-limited" is used to describe alluvial rivers. Given that permanent gullies are incisional systems and may be down-cutting through nonalluvial substrate, the detachment-limited model may be more appropriate for permanent gully systems than classic transport-limited models for alluvial systems.

Physical experiments, like the ones we discuss here, offer a setting where most variables can be controlled and timescales for channel evolution are greatly reduced. Moreover, using experiments allows us to make measurements at a high spatial and temporal density to ensure that much of the variability in the system is captured. Such experiments are not intended as scale models; rather, they are small systems in which scale-independent processes can be studied under controlled conditions (Paola et al., 2009). As such the experiments presented here add to the body of literature focused on channel formation across a continuum of scales from rills over short timescales to head-cutting rivers over much longer timescales (i.e., Stein and LaTray, 2002; Grovers et al., 2007). The discussion of this work focuses on permanent gullies or ravines because these features tend to form over relatively short timescales geologically, but similar timescales over which preexisting channel networks might also evolve, thus allowing for a comparison between the two. Permanent gullies can also cause deep and rapid incision, potentially having a significant effect on infrastructure. Here we utilize physical experiments of permanent gullies to focus on a basic question: how does changing the flow rate of a fixed quantity of water change the total volume of sediment removed and the morphology of the resulting permanent gully?

2 Methods

We performed experiments at the St. Anthony Falls Laboratory at the University of Minnesota, in Minneapolis, Minnesota. These experiments tested how different overland flow rates affect erosion and permanent gully growth on a series of permanent gullies initiated through base level fall, a common feature of post-glacial landscapes in the Upper Midwest, USA (Belmont et al., 2011; Gran et al., 2013; Lenhart et al., 2013). We represented a range of natural permanent gullies using two different substrates. These substrates varied in material and grain size, with one representing a more cohesive detachment-limited system and the other a transport-limited system. The fine-grained substrate was composed of 12 µm ceramic spheres (Zeeospheres) with a density of 2.5 g cm^{-3} (Zeeospheres G series data sheet, 2018). The natural cohesion of this substrate caused it to behave as a detachment-limited system, as empirically shown in the results section of this paper. The coarser substrate was composed of quartz sand with a median grain size of 96 µm and a density of 2.65 g cm^{-3} (AGSCO technical data, 2018). This sand lacked cohesion and behaved as a transport-limited system. The experimental basin size varied for each material type; the basin used with the fine-grained substrate was 1 × 1 m, while the basin filled with the sand substrate was 2 × 4 m, thus allowing room for the more rapid growth of the permanent gullies in the more easily erodible substrate. For both substrates, water flowed out through a 0.076 m wide notch at the downstream end of the basin (Fig. 1). To initiate each run, we dropped the notch to 0.14 m below the surface of the substrate, thus creating a single abrupt base level drop. During each run a knickpoint developed at this notch and propagated upstream, carving a deep channel in the substrate.

To initialize each run, the substrate material was mixed with water and smoothed into the basin to create a flat bed. Experiments began with a fully saturated bed to ensure that water flowed over the surface rather than infiltrating into it. We determined that the bed was at saturation and no longer oversaturated when there was no longer water on the surface of the sediment and water was no longer flowing out the downstream outlet. We analyzed the bed saturation by visual inspection rather than waiting a standard length of time because changes in humidity and temperature affected evaporation rates. Beginning with a saturated substrate allowed us to assume that the net flow was relatively constant in the cross-channel direction and no water was lost to infiltration into dry sediment.

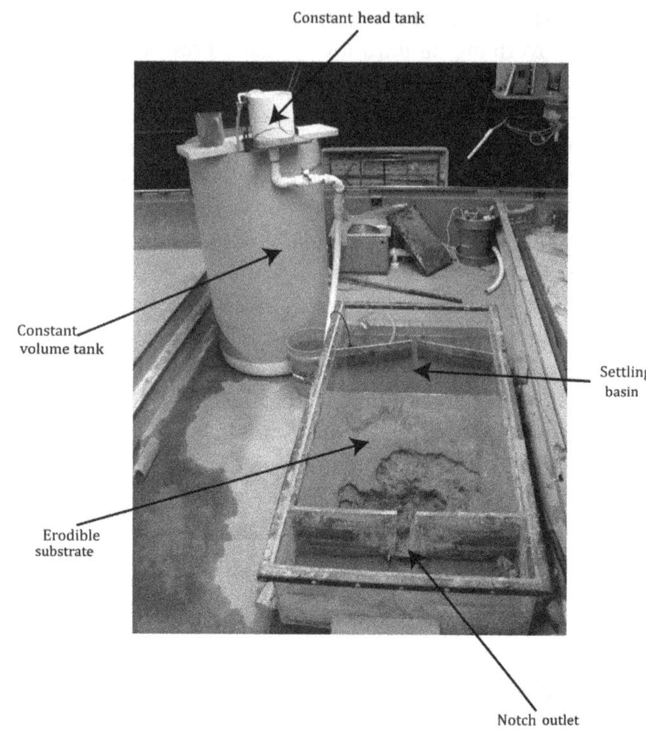

Figure 1. The experimental setup shown here allows water to flow from a settling basin over an erodible substrate and out through a 7.6 × 14 cm notch. The flow rates entering the basin range from 4×10^{-6} to 311×10^{-6} m^3 s^{-1} and are controlled by a constant head tank. For each run a constant volume of water either 190 or 380 L is run over the erodible substrate. This figure shows the setup for the fine-grained substrate. The sand substrate setup was similar, although the erodible substrate was larger.

For each experimental run, water flowed over the uniform substrate as overland flow. We generated an even sheet flow by allowing water to flow over a broad crested weir as water was added to an upstream settling basin (Fig. 1). We controlled the flow rate of the water entering the settling basin with a constant-head tank, which received water from a tank with a predetermined volume of water (190 or 380 L). The flow rate was held constant through the entire run when only 190 L of water were used. When 380 L of water were used, the flow rate was held constant for the first 190 L, then increased and held constant for the second 190 L. Flow rates varied from 4 to 311 mL s^{-1} (4×10^{-6} to 311×10^{-6} m^3 s^{-1}) in the fine-grained substrate and 55 to 262 mL s^{-1} (55×10^{-6} to 262×10^{-6} m^3 s^{-1}) in the sand (Table 1; Fig. 2).

We collected topographic data before, during, and after each experimental run using a fully automated topographic scanner developed by engineers at the St. Anthony Falls Laboratory. The topographic scanner uses a standard laser that collects 2000 points per second in a gridded pattern. Data

Table 1. Experimental run parameters.

Run	Substrate	Water volume (L)	Total time (min)	Flow 1st 190 L ($\times 10^{-6}$ m^3 s^{-1})	Flow 2nd 190 L ($\times 10^{-6}$ m^3 s^{-1})
F-1	Fine	190	60.0	52.58	
F-2	Fine	190	32.0	98.58	
F-3	Fine	190	16.0	197.16	
F-4	Fine	190	21.25	148.45	
F-5	Fine	190	11.0	286.77	
F-6	Fine	190	872.0	3.62	
F-7	Fine	190	10.13	311.40	
F-8	Fine	190	21.5	146.72	
F-9	Fine	190	15.5	203.52	
F-10	Fine	380	91.5	71.69	66.41
F-11	Fine	380	65.0	73.36	143.39
F-12	Fine	380	54.5	76.94	233.67
S-13	Sand	190	25.25	124.93	
S-14	Sand	190	56.5	55.83	
S-15	Sand	190	13.5	233.67	
S-16	Sand	190	12.0	262.88	
S-17	Sand	190	20.25	155.78	
S-18	Sand	380	56.0	80.88	185.56
S-19	Sand	380	71.5	55.34	217.55
S-20	Sand	380	89.0	42.06	225.32

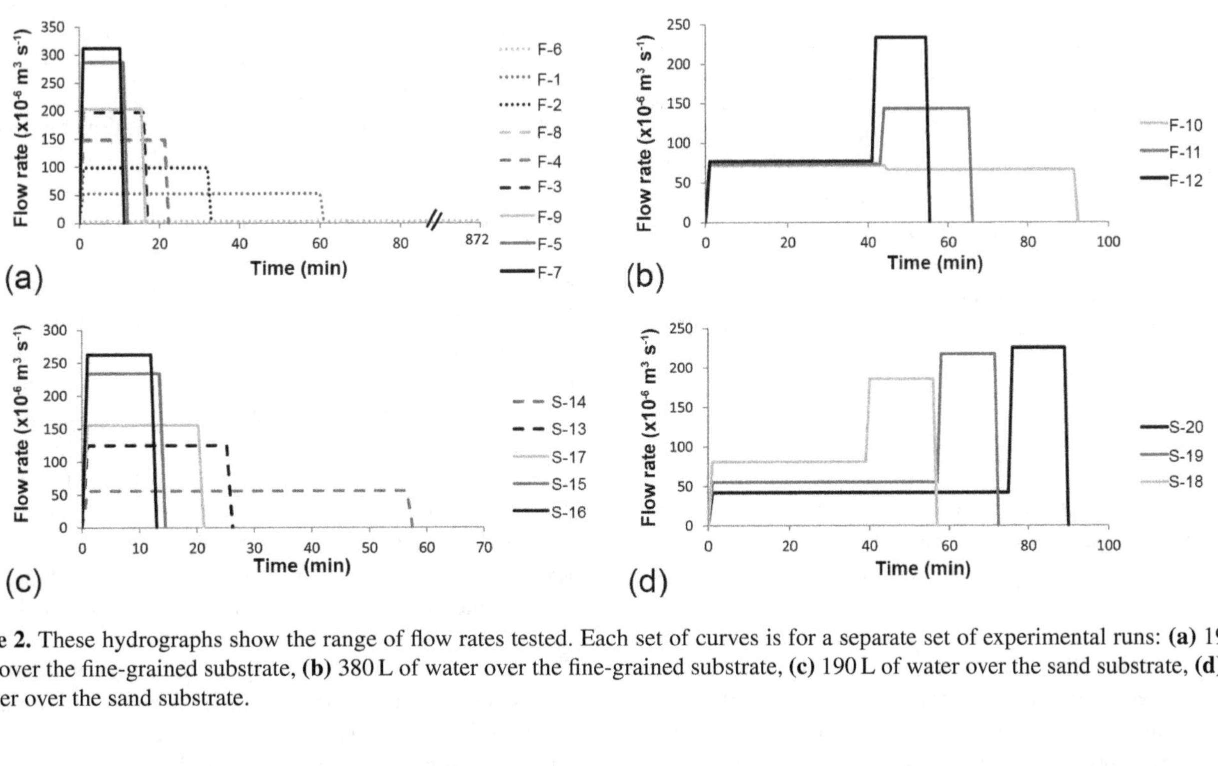

Figure 2. These hydrographs show the range of flow rates tested. Each set of curves is for a separate set of experimental runs: **(a)** 190 L of water over the fine-grained substrate, **(b)** 380 L of water over the fine-grained substrate, **(c)** 190 L of water over the sand substrate, **(d)** 380 L of water over the sand substrate.

were collected with 20 × 25 mm point spacing and approximately 0.5 mm vertical resolution. For each experiment, we collected 2–5 topographic scans per 190 L of water. The experiment was paused at these intervals to complete the scans, ensuring that no data were lost in areas where water was present.

Topographic data were gridded to form a digital elevation model (DEM). To determine the total volume of sediment removed during each experimental run, we subtracted the

DEM of the last scan from the DEM created from the initial scan over the flat initial surface. At each cell the vertical change was multiplied by the area of the cell to give a volume change. The total volume change was summed for each experimental run. We also calculated the volume of sediment eroded at intermediate scans to measure sediment flux over time. Because these data only provide coarse temporal measurements of sediment flux, we sampled sediment flux out of the experiment for 5 s every minute during experimental run F-11 to measure sediment flux at higher time resolution.

We imported the DEMs into ArcGIS to measure channel width and slope using the profile tool in the 3-D analyst toolbox. Channel width is defined as the distance between channel banks on a cross section. Where multiple channels formed, the cross section was selected just below the confluence of these channels. In the fine-grained substrate, channel width was roughly equal to the valley width along the full channel length. In the sand substrate, channels meandered, and therefore we measured channel width in the most upstream portion of the channel where migration had not yet occurred, and the valley width was equal to the channel width. We also measured the channel length and used this measurement to approximate knickpoint retreat rate. In both substrates we measured the channel length from the outlet to the break in slope at the flat upland surface. Slope was measured along the channel profile. In the fine-grained substrate we measured channel slope three separate ways: along the bed, along the knickpoint, and the average channel slope along the complete channel, including both channel bed and knickpoint zones. Where there was more than one knickpoint, we measured the bed slope and knickpoint slope for each section and averaged. Only one slope measurement, the average channel slope, was made on channels formed in the sand substrate because knickpoints were less prominent. In both substrates, average channel slope is a function of channel length, because elevation change for each channel is the same.

3 Results and analysis

These experiments show that the total eroded volume shows no consistent trend with flow rate. For all 190 L experimental runs, the volume of sediment removed ranges from 8.9×10^{-3} to $2.4 \times 10^{-2}\,m^3$ with no clear trend (Fig. 3a). This result was the same for both the sand and the fine-grained substrate. Similarly, there is no trend in the volume of sediment removed in the 380 L runs, yet the volume of eroded sediment is greater (2.8×10^{-2} to $3.9 \times 10^{-2}\,m^3$) than for the 190 L runs (Fig. 3). Put another way, the sediment discharge is linearly related to water discharge (Fig. 3b). For a natural system this implies that in a given storm event the amount of sediment eroded from a permanent gully is proportional to the amount of precipitation rather than the storm intensity, assuming flow rate generates shear stresses that surpass the critical shear stress of the sediment.

In experimental run F-11, the sediment discharge peaked early and rapidly decreased (Fig. 4). This pattern is also seen in several other experimental runs using the coarser resolution erosion data collected from the topographic scans, most prominently in run F-6 (Fig. 5). The experimental runs where this pattern was observed were also those where the channel interacted with a wall (see Supplement). After a channel touches a wall it is no longer able to widen, and flow preferentially remains against that smooth surface. While it was observed that the channels appeared to propagate most rapidly early in the experiment and then slow with each successive time interval, the scan data suggest that the total volume of sediment removed was relatively constant throughout the experiment except where the channel interacted with a wall. For the 380 L runs, where discharge was increased, there was a corresponding increase in sediment discharge, which was then maintained throughout the second half of the experiment.

3.1 Channel morphology

While sediment volumes removed are roughly the same in the fine-grained and sand channels, there are differences in how they erode. Channels in the fine-grained substrate erode primarily via head cut propagation. In contrast channels in the sand substrate erode due to head cut propagation as well as lateral channel migration (see Supplement).

The experiments also demonstrate a relationship between channel width and discharge. Generally, higher flows resulted in shorter, wider channels (Table 2). In particular, the 380 L runs reveal both the importance the total flow volume has for erosion volumes and how increasing flow rate affects channel geometry. In the cohesive fine-grained substrate, higher flows in the second half of the experiment formed a wider channel upstream of the already eroded gully, without altering the preexisting channel during the rest of the experiment (Fig. 6). Channel widths before the increased flow ranged from 0.16 to 0.40 m for a flow rate of 73×10^{-6} to $77 \times 10^{-6}\,m^3\,s^{-1}$ and increased to 0.28 to 0.61 m for flow rates ranging from 143×10^{-6} m to $234 \times 10^{-6}\,m^3\,s^{-1}$. The channels formed in the sand substrate responded differently; these channels widened along the entire channel length when flow was increased (Fig. 6). In the sand channels widths varied from 0.14 to 0.19 m for initial flows which varied from 42×10^{-6} to $81 \times 10^{-6}\,m^3\,s^{-1}$, and the entire channel width increased to 0.24 to 0.43 m for the higher flows ranging from 185×10^{-6} to $225 \times 10^{-6}\,m^3\,s^{-1}$ (Table 2).

The hydraulic geometry equation for width developed by Leopold and Maddock (1953) was used to quantify the relationship between width, W, and discharge, Q, for these experiments to determine how the experimental channels compare with natural channels:

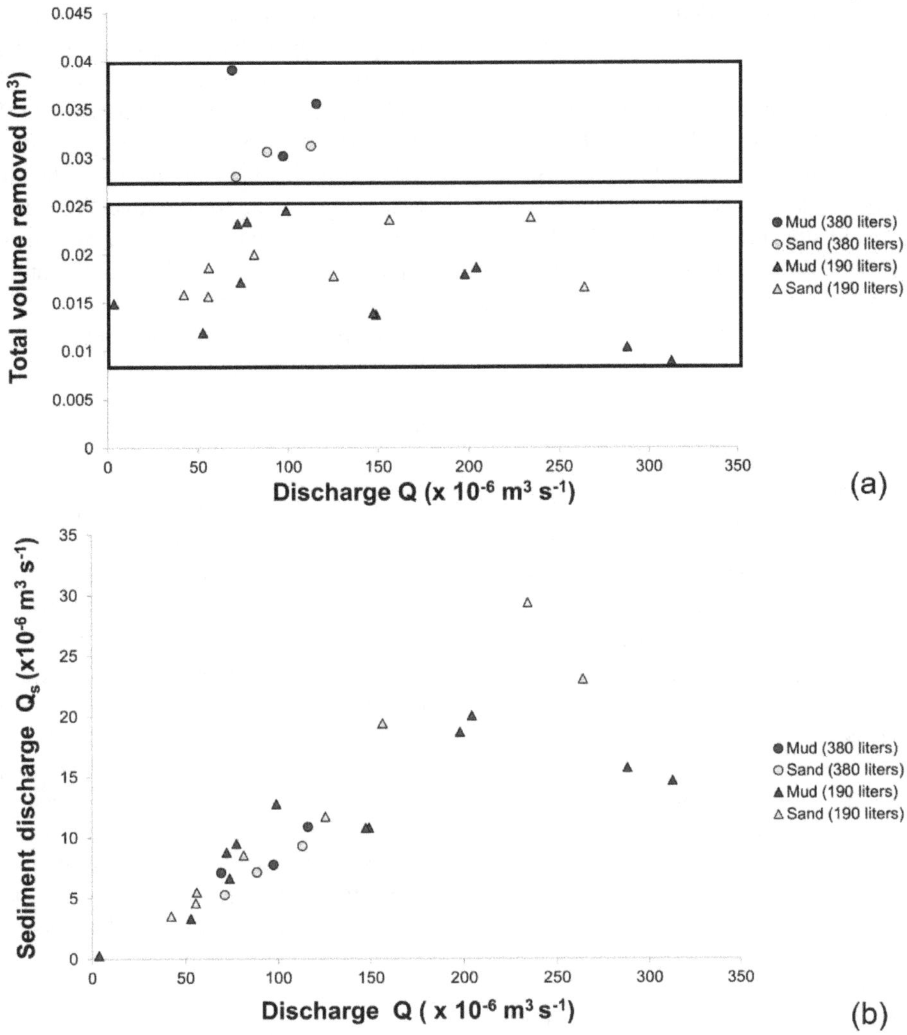

Figure 3. The results from these experiments show that the total volume of sediment removed is not dependent on flow rate **(a)** or that sediment discharge is linearly related to water discharge **(b)**. These charts also show that while sediment discharge is the same for the 380 L runs as it is for the 190 L runs at a given discharge, the total volume of sediment removed is greater in the 380 L runs.

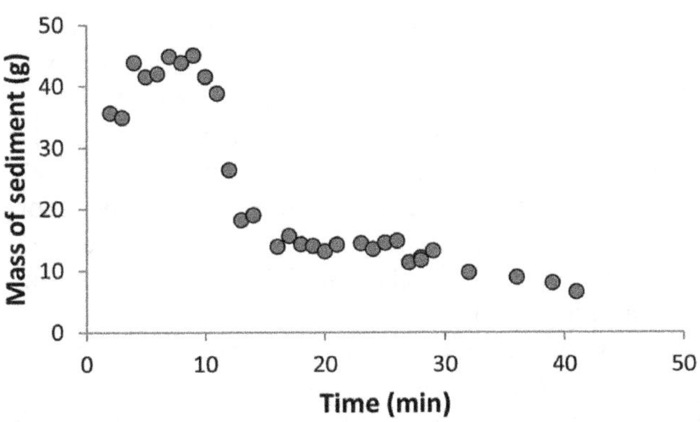

Figure 4. Samples collected for 5 s every minute during the first 190 L of run 11 show a decrease in sediment load with time after an initial peak.

$$W = aQ^b, \tag{1}$$

where a and b are constants. The range of empirically derived exponents for b is between 0.3 and 0.5, derived from field measurements of both bedrock and alluvial streams (Knighton, 1998; Leopold and Maddock, 1953; Montgomery and Gran, 2001; Whipple, 2004). Studies of rills and ephemeral gullies have found that these relationships hold with a b value of 0.4–0.5 (Natchergaele et al., 2002; Torri et al., 2006).

To test how well these field-measured values describe our experimental channels, we calculated width using the reported values for b and iterated to find the best fit for a. In addition, we also iterated to determine what b value most accurately describes the trend in our experimental data. The root mean square error (RMSE) was calculated for each data set and is reported as a percentage of the average measured width value.

Using the field-derived b exponents ranging from 0.3 to 0.5, the RMSE is 29–31 % of the average measured width in the fine-grained substrate. The best fit for the experimental data results in a b value of 0.27, resulting in an RMSE of 28.5 % (Fig. 7). In the sand substrate the best fit for the data falls within the range of field-derived exponents. The error is minimized when b is 0.39, resulting in an RMSE of 9 % (Fig. 7).

3.2 Modeled sediment transport

Results from these experiments appear to conflict with standard sediment transport equations which generally predict a nonlinear increase in sediment flux as discharge increases (e.g., Engelund and Hansen, 1967; Meyer-Peter and Müller, 1948; Parker, 1990; Wilcock and Crowe, 2003), yet these equations typically include an additional factor, often slope, which, if also varied, can account for changes in discharge resulting in a linear increase in sediment flux. Below are examples of commonly used sediment transport models that, when applied to these experiments, account for the linear nature of the sediment flux discharge relationship.

Commonly erosion in detachment-limited systems is modeled by the stream power incision model (Howard and Kirby, 1983):

$$\mathrm{d}z/\mathrm{d}t = kQ^{(m_d)}S^{(n_d)}, \tag{2}$$

where k, m_d and n_d are constants, $\mathrm{d}z\mathrm{d}t^{-1}$ is vertical erosion rate, and S is the slope. In the detachment-limited fine-grained substrate, all erosion took place at the knickpoint and therefore we consider only knickpoint retreat rate and use S_k, knickpoint slope, in place of S. Because this rate is a horizontal retreat rate rather than vertical incision rate we must convert Eq. (2) appropriately:

$$U \cdot S_k = \mathrm{d}z/\mathrm{d}t, \tag{3}$$

$$U = kQ^{(m)_d}S_k^{(n_d^{-1})}, \tag{4}$$

where U is the knickpoint retreat rate. The exponents m_d and n_d have been derived for a variety of natural environments. Typically the $m_d\,/\,n_d^{-1}$ ratio (concavity index) is approximately 0.35–0.6 (Whipple and Tucker, 1999; Baldwin et al., 2003). The exponent n_d ranges between two-thirds and five-thirds depending on the erodibility of the substrate where more easily eroded sediment has a lower value for the exponent n_d (Foley, 1980; Howard and Kirby, 1983; Whipple et al., 2000). The exponents m_d and n_d for uniform detachment-limited landscapes (i.e., badlands) reported by Howard and Kirby (1983) are four-ninths and two-thirds, respectively. While the $m_d\,/\,n_d^{-1}$ ratio is slightly higher than reported elsewhere, we used these values to model erosion in our experiments because, like the badlands, our experimental setup was spatially homogeneous and easily eroded. RMSE was measured between the calculated U and the measured U and minimized by modifying the coefficient k. The RMSE was 25 % of the measured average knickpoint retreat rate for the fine-grained substrate. The detachment-limited equation is not appropriate for the sand substrate because there is both headward and lateral erosion, which is not captured by the equation. In addition, the knickpoint slope cannot be measured in the sand substrate.

Sediment loads from the sand substrate were modeled with the transport-limited equation (Pelletier, 2011):

$$q_\mathrm{s} = kQ^{(m_t)}S^{(n_t)}, \tag{5}$$

where k, m_t and n_t are constants and q_s is the volumetric unit sediment flux. This equation can be derived from the Engelund and Hansen (1967) equation where both m_t and n_t equal five-thirds. The Engelund and Hansen (1967) equation is ideal because it does not have an incipient motion threshold as many other sediment transport equations do. Here again the RMSE was measured between the measured and calculated q_s and minimized using the coefficient k. The RMSE was 41 % of the average volumetric unit sediment discharge for the sand substrate. The transport-limited equation was also tested with the fine-grained substrate. For this test the average slope was used for S and the RMSE calculated was 86 % of the average volumetric unit sediment discharge. This high RMSE value supports the observation that the fine-grained substrate does not behave as a transport-limited system.

4 Discussion

Our experiments demonstrate that, over a range of conditions, the sediment volume eroded during permanent gully growth under application of a fixed volume of water is independent of the rate at which the water is supplied. Thus, sediment discharge is linearly related to water discharge in both detachment-limited and transport-limited systems. These results contrast with data from many preexisting streams where changing flow intensity has resulted in increased erosion volume (i.e., Boateng et al., 2012; Ma et al., 2010; Naik and Jay, 2011), yet these data may not be directly applicable to early-stage gullies. We suggest that because gullies are actively evolving in response to a given hydrology, the channel morphology that develops reflects that hydrology, with erosion balanced by altering channel slope. This is supported by sediment transport Eqs. (4) and (5) which, when applied to our experiments, predict the measured sediment discharges by including the effects of both discharge and slope. In preexisting channels where channel slope may take tens of thousands of years to adjust to changing flows, both equations would predict a nonlinear increase in sediment discharge with increasing water discharge. Gullies evolve more rapidly in re-

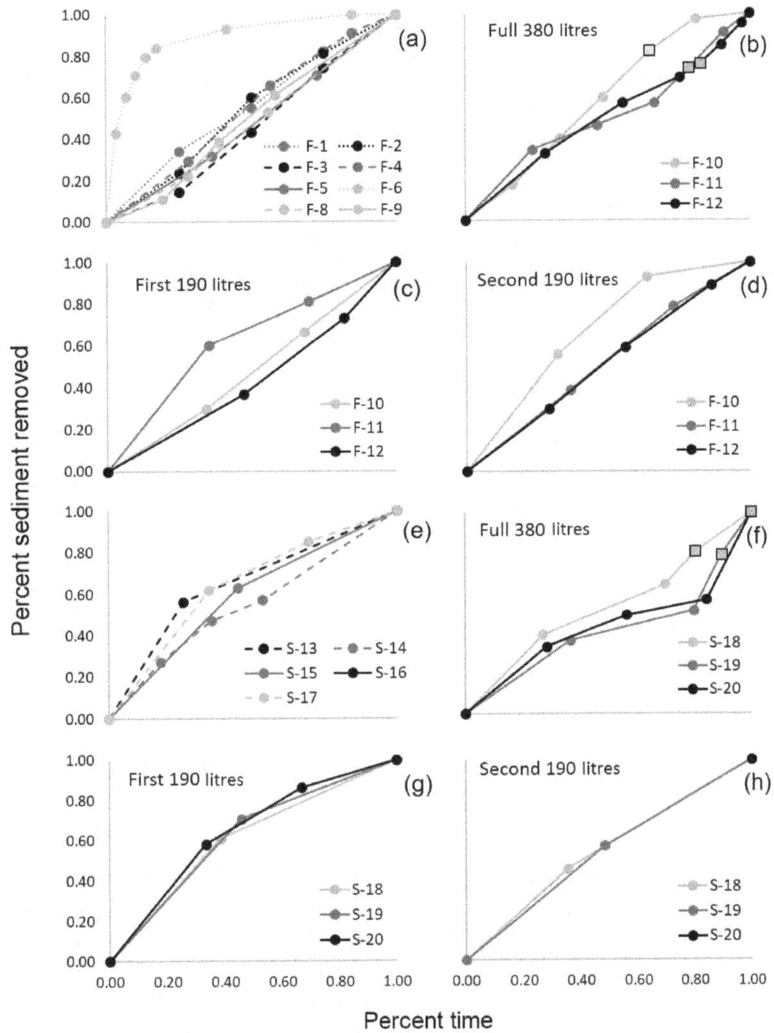

Figure 5. In general, sediment discharge was near constant throughout the experiments, as indicated by the linear relationship between percent time elapsed and percent of total sediment removed in the graphs above. This result was generally consistent between for both fine-grained substrate **(a–d)** and the sand substrate **(e–h)**. Results from the 190 L runs are graphed in **(a)** and **(e)** and the 380 L runs are graphed in **(b)** and **(f)** with the first half broken out in **(c)** and **(g)** and the second half in **(f)** and **(h)**.

sponse to the imposed discharge and can balance erosion by adjusting channel slope in response to a change in the hydrologic regime.

Moreover, our findings suggest that anthropogenic changes to the discharge regime could affect channel morphology (i.e., channel width), without changing sediment output derived from permanent gullies. Where water discharge was increased in the 380 L runs, the channel quickly evolved in response to the new discharge by creating a wider channel. In these experiments the observed response to the increased discharge differed for the fine-grained and sand substrate (Fig. 6), suggesting cohesion may be an important factor in how a channel initially responds to a new discharge regime.

The results of our experiments follow the hydraulic geometry relationship for width in Eq. (1), although with lower exponents than usually measured in field studies for alluvial and bedrock channels (Knighton, 1998; Leopold and Maddock, 1953; Montgomery and Gran, 2001; Whipple, 2004) and rills and ephemeral gullies (Nachtergaele et al., 2002; Torri et al., 2006). While this relationship is typically applied to describing width or discharge changes in a single channel, it works here for these separate channels because each comparable channel is carving through the same substrate. In the fine-grained substrate the empirical exponent b for these data is lower than has been derived for natural channels. This may be a result of the steep channel walls developed in these experiments; in natural permanent gullies where near-vertical channel walls are less common, we expect that this exponent would be closer to the reported values. In the sand substrate where steep channel walls could not develop, the empirical exponent b is 0.39, which is only slightly lower than the

Table 2. Experimental results.

Run	Volume sediment removed (m^3)	Average slope	Bed slope	Knickpoint slope	Width 1st 190 L (m)	Width 2nd 190 L (m)
F-1	0.011920	0.17	0.06	1.19	0.22	
F-2	0.024506	0.20	0.04	0.60	0.48	
F-3	0.017937	0.21	0.04	0.36	0.24	
F-4	0.013776	0.27	0.03	0.62	0.43	
F-5	0.010384	0.29	0.04	1.22	0.26	
F-6	0.014903	0.12	0.10	0.27	0.13	
F-7	0.08913	0.42	0.11	0.71	0.38	
F-8	0.013908	0.37	0.06	0.67	0.43	
F-9	0.018645	0.22	0.07	0.58	0.52	
F-10	0.039053	0.15	0.05	0.81	0.40	0.49
F-11	0.030158	0.14	0.05	0.56	0.16	0.28
F-12	0.035532	0.16	0.03	0.29	0.38	0.61
S-13	0.017762	0.06			0.20	
S-14	0.018649	0.07			0.13	
S-15	0.023796	0.06			0.28	
S-16	0.016590	0.06			0.27	
S-17	0.023569	0.06			0.29	
S-18	0.031174	0.07			0.19	0.39
S-19	0.030587	0.06			0.14	0.43
S-20	0.028028	0.06			0.15	0.24

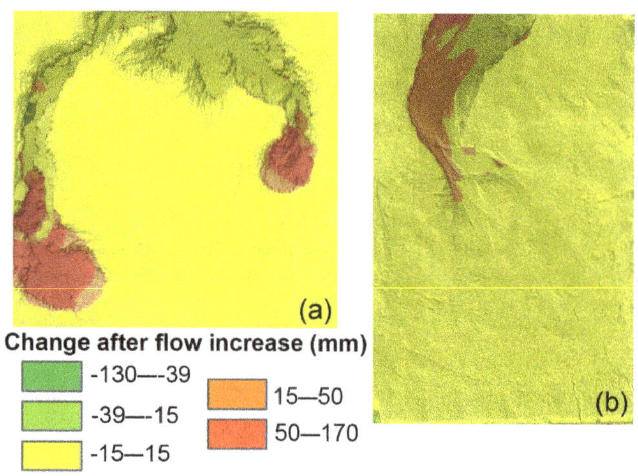

Figure 6. These differenced DEMs show how changes in flow impact channel width. In the fine-grained runs **(a)** the width changed in the newly formed channel, while in the sand runs **(b)** the channel width was altered along the entire channel length. Flow is bottom to top in both images.

range typically considered for alluvial channels (Rodriquez-Iturbe and Rinaldo, 1997), but similar to exponents found in rills and ephemeral gullies (Nachtergaele et al., 2002; Torri et al., 2006).

Channel width has also been modeled by Wells et al. (2013), who found that both slope and discharge play a role in setting channel width. This relationship was tested for the results of these experiments, yet it was not as strong as the relationship with discharge alone. Slope in the Wells et al. (2013) study could not change during the experiment, which contrasts with our experiments where slope is a free parameter that adjusts to discharge. This finding further supports the distinction between rapidly evolving channels like permanent gullies, and more stable systems where slope does not change as quickly, causing channel adjustments to occur primarily through changes in channel width and through a nonlinear erosion response.

Experiments focused on head cut growth completed by Bennett et al. (2000) also reported a linear relationship between water discharge and sediment discharge, yet the water discharge was lower, and the slope of the relationship was much higher in our results. This may suggest that a nonlinear relationship between sediment and water may develop over a wider range of flow rates than tested here, yet more research would be required. In addition, Bennett et al. (2000) noted two dominant processes for head cut migration: surface seal failure, which is similar to slab failure reported in other papers, and plunge pool scour, where head cut migration is driven by undercutting. Although both of these mechanisms could lead to large blocks of sediment collecting at the base of the head cut, creating periods of quiescence in head cut migration, the authors do not indicate that head cut migration stalled in their experiments. The erosion mechanisms described by Bennett et al. (2000) are similar to the

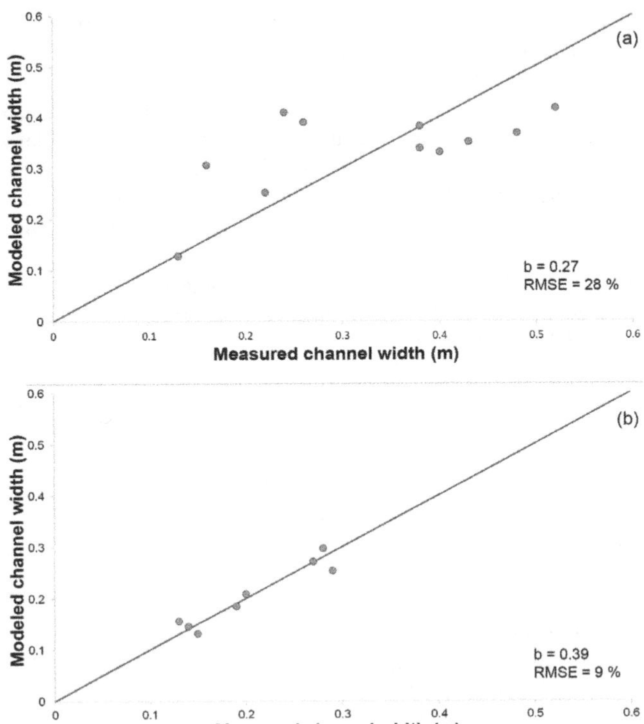

Figure 7. Channel width can be modeled using the hydraulic geometry relationship (Eq. 1). In both plots the solid black line is a 1 : 1 line. **(a)** The relationship measured in the channels formed in the fine-grained substrate has a constant *b* value that is less than the values commonly observed in nature. **(b)** The relationship measured in the channels formed in the sand substrate has a constant *b* value of 0.4, which is in the intermediate range of the commonly proposed constants.

mechanisms observed in our experiments, in that there was a continuous head cut propagation, although we did not observe significant plunge pool development.

In a numerical study, Istanbulluoglu et al. (2005) tested the effect of changing rain intensity while storm volume was held constant. The modeled results of their study showed an increase in the volume of sediment eroded as intensity increased. This result is in direct conflict with our results, yet there are many distinctions between the two studies that may explain this discrepancy. The Istanbulluoglu et al. (2005) model assumed gully erosion due to slab failure in a detachment-limited system. While the fine-grained substrate in our physical experiments was also detachment limited, erosion occurred grain by grain rather than as large blocks. Once these grains were detached, the flow was easily able to carry them through the channel and there was no measurable deposition in any of the experimental runs. In contrast, erosion in the slab failure model occurred in response to pore pressure build up in tension cracks resulting in large failures. This slab failure model does not require that the flow be able to carry the detached sediment and often resulted in deposition at the toe of the knickpoint, which increases resistance to future failure.

It is likely that both slab failure and granular knickpoint propagation occur in permanent gullies throughout the world. The relative importance of each process is dependent on the substrate and the knickpoint slope. Tension cracks develop behind steep slopes where shrinkage occurs due to desiccation and horizontal tensile stresses generated in large part by gravity are greater than the tensile strength of the sediment (Darby and Thorne, 1994). Cracks like this are likely to form on the landscape above steep head cuts in cohesive sediment, between storm events. In neither our study nor the Bennett et al. (2000) did we model individual storm events rather than a constant overland flow. If we had, it is likely we would have also developed tension cracks in the cohesive substrates. Because this was outside of the scope of these experiments it is difficult to form accurate conclusions on how the development of tension cracks and the subsequent failure events would have altered these results, but an analogous study that encouraged erosion by slab failure would be a useful extension.

Our experimental results are not clear with regard to Sidorchuk's (1999) two-stage gully evolution model. For most runs it appears that the second stage of gully evolution was not achieved. In a few specific cases, most notably runs F-6 and F-11, there does appear to be an early peak and later decrease and stabilization in sediment discharge. Surprisingly the results of these runs were not among the lowest total sediment discharges, as might be expected where the second stage was achieved. It is possible that if all the runs were allowed to continue we may have reached a stable system where the second stage of gully evolution was achieved. If this was allowed to occur, we would anticipate that additional water at the same discharge rate would not cause measurable erosion, potentially altering the relationship we have observed. What is not clear is how long it takes for this steady state to occur, and how this may relate to discharge.

Based on the results of this study and the comparison with previous gully studies, it is important to consider a wide range of variables when mitigating permanent gully erosion. The results from our experiments suggest that during early stages of permanent gully growth, increasing overland flow rates will not result in increased sediment yield, if the volumes of water delivered are not changed. Moreover, while sediment loads are not affected by changing flow rates, channel morphology is. Another important variable to consider is the mode of head cut retreat. The experimental results apply in environments with steep slopes where erosion is grain-by-grain. In places where tension cracks develop, the slab failure mechanism highlighted by Istanbulluoglu et al. (2005) might dominate. If the tension crack failure mechanism is the dominant process of head cut retreat, flow rates and storm intensity may become more important than they were in our study because the slabs may require a higher threshold to break up and mobilize, allowing further head cut propagation.

5 Conclusion

These experiments highlight how young incising channels like permanent gullies can respond to changing hydrology differently than higher order channels that are later in their evolution. A relevant future study should investigate how natural gullies, which have a great deal more variability than this experimental system, respond to changing hydrology. The conclusions from this project are outlined below:

- The experiments here suggest that water volume, rather than discharge, controls the total volume of erosion during permanent gully formation. This result holds true for both transport-limited and detachment-limited systems.
- As long as slope is a free parameter in these rapidly evolving systems, changes in flow rate can be accommodated through an adjustment in both cross-sectional and longitudinal channel geometry. Wider channels were typically shorter and thus steeper.
- In both substrates, variations in channel width were described by the hydraulic geometry relationship proposed by Leopold and Maddock (1953), with wider channels forming in response to higher discharge.
- Sediment transport in sand and fine-grained substrates was well described by the transport-limited sediment flux equation and the detachment-limited stream power equation, respectively.

Competing interests. The authors declare that they have no conflict of interest.

Acknowledgements. This work was funded in part by a Geological Society of America Graduate student research grant and by the National Center for Earth-surface Dynamics (NCED), which is funded by the Office of Integrative Activities, National Science Foundation grant EAR-0120914. We appreciate comments and reviews from Robert Wells and two anonymous reviewers.

Edited by: Thom Bogaard

References

Abrahams, A. D. and Parsons, A. J.: Resistance to overland flow on desert pavement and its implications for sediment transport modeling, Water Resour. Res., 27, 1827–1836, 1991.

Adediji, A., Jeje, L. K., and Ibitoye, M. O.: Urban development and informal drainage patterns: Gully dynamics in Southwestern Nigeria, Appl. Geogr., 40, 90–102, 2013.

Agsco technical Data: http://www.agsco.com/assets/pdfs/Silica-Sands-Flours-Tech-Data-Sheet-06-04-2014.pdf, last access: 7 February 2018.

Alonso, C. V., Bennett, S. J., and Stein, O. R.: Predicting head cut erosion and migration in concentrated flows typical of upland areas, Water Resour. Res., 38, 39-1–39-15, 2002.

Ambers, R. K., Druckenbrod, D. L., and Ambers, C. P.: Geomorphic response to historical agriculture at Monument Hill in the Blue Ridge foothills of central Virginia, Catena, 65, 49–60, 2006.

Arnold Jr., C. L. and Gibbons, C. J.: Impervious surface coverage: the emergence of a key environmental indicator, J. Am. Plan. Assoc., 62, 243–258, 1996.

Baldwin, J. A., Whipple, K. X., and Tucker, G. E.: Implications of the shear stress river incision model for the timescale of postorogenic decay of topography, J. Geophys. Res., 108, 2156–2202, 2003.

Belmont, P., Gran, K., Schottler, S., Wilcock, P., Day, S., Jennings, C., Lauer, J., Viparelli, E., Willenbring, J., Engstrom, D., and Parker, G.: Large shift in source of fine sediment in the Upper Mississippi River, Environ. Sci. Technol., 45, 8804–8810, 2011.

Bennett, S. J., Alonso, C. V., Prasad, S. N., and Römkens, M. J. M.: Experiments on headcut growth and migration in concentrated flows typical of upland areas, Water Resour. Res., 36, 1911–1922, 2000.

Blann, K. L., Anderson, J. L., Sands, G. R., and Vondracek, B.: Effects of agricultural drainage on aquatic ecosystems: a review, Crit. Rev. Environ. Sci. Tech., 39, 909–1001, 2009.

Boateng, I., Bray, M., and Hooke, J.: Estimating the fluvial sediment input to the coastal sediment budget: a case study of Ghana, Geomorphology, 138, 100–110, 2012.

Bryan, R. B.: Knickpoint evolution in rillwash, Catena, 17, 111–132, 1990.

Bull, L. J. and Kirkby, M. J.: Gully processes and modelling, Prog. Phys. Geogr., 21, 354– 374, 1997.

Bull, L. J. and Kirkby, M.: Channel heads and channel extension, in: Dryland Rivers: Hydrology and Geomorphology of Semi-Arid Channels, edited by: Bull, L. J. and Kirkby, M. J., Wiley, Chichester, UK, 263–298, 2002.

Chiverrell, R. C., Harvey, A. M., and Foster, G. C.: Hillslope gullying in the Solway Firth–Morecambe Bay region, Great Britain: Responses to human impact and/or climatic deterioration?, Geomorphology, 84, 317–343, 2007.

Chu-Agor, M. L., Fox, G. A., Cancienne, R. M., and Wilson, G. V.: Seepage caused tension failures and erosion undercutting of hillslopes, J. Hydrol., 359, 247–259, 2008.

Darby, S. E. and Thorne, C. R.: Prediction of tension crack location and riverbank erosion hazards along destabilized channels, Earth Surf. Proc. Land., 19, 233–245, 1994.

Dietrich, W. E. and Dunne, T.: The Channel Head, in: Channel Network Hydrology, edited by: Beven, K. J. and Kirkby, M. J., Wiley, Chichester, UK, 175–219, 1993.

Dunne, T.: Hydrology, mechanics, and geomorphic implications or erosion by subsurface flow, in: Groundwater Geomorphology: The Role of Subsurface Water in Earth-Surface Processes and Landforms, edited by: Higgins, C. G. and Coates, D. R., Geological Society of America as Special Paper 252, Geological Society of America, Colorado, Boulder, 1–28, 1990.

Ebisemiju, F. S.: The response of headwater stream channels to urbanization in the humid tropics, Hydrol. Process., 3, 237–253, 1989.

Einstein, H. A. and Barbarossa, N. L.: River channel roughness, T. Am. Soc. Civ. Eng., 117, 1121–1132, 1951.

Eitel, J. U. H., Williams, C. J., Vierling, L. A., Al-Hamdan, O. Z., and Pierson, F. B.: Suitability of terrestrial laser scanning for studying surface roughness effects on concentrated flow erosion processes in rangelands, Catena, 87, 398–407, 2011.

Engelund, F. and Hansen, E., A monograph on sediment transport in alluvial streams, TEKNISKFORLAG, Copenhagen, Denmark, 1967.

Espey Jr., W. H., Morgan, C. W., and Masch, F. D.: A study of some effects of urbanization on storm runoff from a small watershed, Tech. Rep. 44D 07-6501 CRWR-2, Ctr. for Res. in Water Resour., Univ. Texas, Austin, 1965.

Farres, P. J.: The role of time and aggregate size in the crusting process, Earth Surf. Proc. Land., 3, 243–254, 1978.

Flores-Cervantes, J. H., Istanbulluoglu, E., and Bras, R. L.: Development of gullies on the landscape: A model of headcut retreat resulting from plunge pool erosion, J. Geophys. Res., 111, 2156–2202, 2006.

Foley, M. G.: Bed-rock incision by streams, Geol. Soc. Am. Bull., 91, 2189–2213, 1980.

Fox, G. A. and Wilson, G. V.: The role of subsurface flow in hillslope and stream bank erosion: a review, Soil Sci. Soc. Am. J., 74, 717–733, 2010.

Gran, K. B., Finnegan, N., Johnson, A. L., Belmont, P., Wittkop, C., and Rittenour, T.: Landscape evolution, valley excavation, and terrace development following abrupt postglacial base level fall, Geol. Soc. Am. Bull., 125, 1068–1082, 2013.

Gran, K. B., Belmont, P., Day, S., Jennings, C., Lauer, J. W., Viparelli, E., Wilcock, P., and Parker, G.: An integrated sediment budget for the Le Sueur River basin, Final report Presented to the Minnesota Pollution Control Agency, Minnesota, 2011.

Grovers, G., Giménez, R., and Van Oost, K.: Rill erosion: Exploring the relationship between experiments, modelling and field observations, Earth-Sci. Rev., 84, 87–102, 2007.

Haddeland, I., Lettenmaier, D. P., and Skaugen, T.: Effects of irrigation on the water and energy balances of the Colorado and Mekong river basins, J. Hydrol., 324, 210–223, 2005.

Hinds, N. E. A.: Amphitheatre valley heads, J. Geol., 33, 816–818, 1925.

Holland, W. N. and Pickup, G.: Flume study of knickpoint development in stratified sediment, Geol. Soc. Am. Bull., 87, 76–82, 1976.

Howard, A. D.: Groundwater sapping experiments and modeling, in: Sapping Features of the Colorado Plateau: A Comparative Planetary Geology Field Guide, NASA, 491, 71–83, 1988.

Howard, A. D. and Kirby, G.: Channel changes in badlands, Geol. Soc. Am. Bull., 94, 739–752, 1983.

Istanbulluoglu, E., Bras, R. L., Flores-Cervantes, H., and Tucker, G. E.: Implications of bank failures and fluvial erosion for gully development: Field observations and modelling, J. Geophys. Res., 110, 2156–2202, 2005.

Knapen, A. and Poesen, J.: Soil erosion resistance effects on rill and gully initiation points and dimensions, Earth Surf. Proc. Land., 35, 217–228, 2010.

Knapen, A., Poesen, J., Govers, G., Gyssels, G., and Nachtergaele, J.: Resistance of soils to concentrated flow erosion: A review, Earth-Sci. Rev., 80, 75–109, 2007.

Knighton, D.: Fluvial forms and processes, Edward Arnold, London, p. 377, 1998.

Kosov, B. F., Nikol'skaya, I. I., and Zorina, Y. F.: Eksperimental'nyye issledovaniya ovragoobrazovaniya, in: Eksperimental'naya geomorfologiya, edited by: Makkaveev, N. I., Moscow, 113–140, 1978.

Lamb, M. P., Howard, A. D., Johnson, J., Whipple, K. X., Dietrich, W. E., and Perron, J. T.: Can springs cut canyons into rock?, J. Geophys. Res., 111, E07002, https://doi.org/10.1029/2005JE002663, 2006.

Lenhart, C. F., Titov, M. L., Ulrich, J. S., Nieber, J. L., and Suppes, B. J.: The role of hydrologic alteration and riparian vegetation dynamics in channel evolution along the lower Minnesota River, T. ASABE, 56, 549–561, 2013.

Leopold, L. B. and Maddock, T.: The hydraulic geometry of stream channels and some physiographic implications, US Geol. Survey Prof. Paper 252, US Geological Survey, Washington, D.C., 1953.

Ma, H., Yang, D., Tan, S. K., Gao, B., and Hu, Q.: Impact of climate variability and human activity on streamflow decrease in the Miyun Reservoir catchment, J. Hydrol., 389, 317–324, 2010.

Merritt, E.: The identification of four stages during micro-rill development, Earth Surf. Proc. Land., 9, 493–496, 1984.

Meyer-Peter, E. and Müller, R.: Formulas for bed-load transport, IAHR, Stockholm, 1948.

Montgomery, D. R. and Gran, K. B.: Downstream variations in the width of bedrock channels, Water Resour. Res., 37, 1841–1846, 2001.

Mosley, M. P.: Experimental study of rill erosion, T. ASAE, 17, 909–913, 1974.

Nachtergaele, J., Poesen, J., Steegen, A., Takken, I., Beuselinck, L., Vandekerckhove, L., and Grovers, G.: The value of a physically based model versus an empirical approach in the prediction of ephemeral gully erosion for loess derived soils, Geomorphology, 40, 237–252, 2001.

Nachtergaele, J., Poesen, J., Sidorchuk, A., and Torri, D.: Prediction of concentrated flow width in ephemeral gully channels, Hydrol. Process., 16, 1935–1953, 2002.

Naik, P. K. and Jay, D. A.: Distinguishing human and climate influences on the Columbia River: Changes in mean flow and sediment transport, Hydrolog. Sci. J., 404, 259–277, 2011.

Nichols, M. H., Magirl, C., Sayre, N. F., and Shaw, J. R.: The geomorphic legacy of water and erosion control structures in a semiarid rangeland watershed, Earth Surf. Proc. Land., 43, 909–918, 2017.

Paola, C., Straub, K., Mohrig, D., and Reinhardt, L.: The "unreasonable effectiveness" of stratigraphic and geomorphic experiments, Earth-Sci. Rev., 97, 1–43, 2009.

Parker, G.: Surface-based bedload transport relation for gravel rivers, J. Hydraul. Res., 28, 417–436, 1990.

Parker, R. S.: Experimental study of basin evolution and its hydrologic implications, PhD dissertation, Colo. State Univ., Ft. Collins, 1977.

Parker, R. S. and Schumm, S. A.: Experimental study of drainage networks, in: Badland Geomorphology and Piping, edited by: Bryan, R. and Yair, A., Geobooks, Norwich, England, 153–168, 1982.

Paul, M. J. and Meyer, J. L.: Streams in the urban landscape, Ann. Rev. Ecol. Syst., 32, 333–365, 2001.

Pelletier, J. D.: Fluvial and slope-wash erosion of soil-mantled landscapes: detachment- or transport-limited, Earth Surf. Proc. Land., 37, 37–51, 2011.

Pierson, F. B., Bates, J. D., Svejcar, T. J., and Hardegree, S. P.: Runoff and erosion after cutting western Juniper, Rangeland Ecol. Manage., 60, 285–292, 2007.

Poesen, J., Nachtergaele, J., Verstaeten, G., and Valentin, C.: Gully erosion and environmental change: importance and research needs, Catena, 50, 91–133, 2003a.

Poesen, J. W. A., Torri, D. B., and VanWalleghem, T.: Gully erosion: Procedures to adopt when modelling soil erosion in landscapes affected by gullying, in: Handbook of Erosion Modelling, 1st Edn., edited by: Morgan., R. P. C. and Nearing, M. A., Blackwell Publishing Ltd., Hoboken, NJ, 360–386, 2003b.

Robinson, K. M. and Hanson, G. J.: A deterministic headcut advance model, T. ASAE, 37, 1437–1443, 1994.

Robinson, M. and Rycroft, D. W.: Agricultural Drainage, in: Agronomy Monograph 38, edited by: Skaggs, R. W. and van Schifgaarde, J., ASA, CSSA, SSSA, Madison, WI, 767–800, 1999.

Rodriquez-Iturbe, I. and Rinaldo, A.: Fractal River Basins: Chance and Self-Organization, Cambridge Univ. Press, New York, 547 pp., 1997.

Römkens, M. J. M. and Wang, J. Y.: Soil roughness changes from rainfall, T. Am. Soc. Civ. Eng., 31, 408–413, 1987.

Seaburn, G. E.: Effects of urban development on direct runoff to East Meadow Brook, Nassau County, New York, USGS Prof. Paper 627-B, US Geological Survey, Washington, 1969.

Sidorchuk, A.: Dynamic and static models of gully erosion, Catena, 37, 401–414, 1999.

Sidorchuk, A.: Stages in gully evolution and self-organized criticality, Earth Surf. Proc. Land., 31, 1329–1344, 2006.

Stankoviansky, M. I. L. O. Š.: Gully evolution in the Myjava Hill Land in the second half of the last millennium in the context of the central-European area, Geogr. Pol., 76, 89–107, 2003.

Stein, O. R. and LaTray, D. A.: Experiments and modeling of head cut migration in stratified soils, Water Resour. Res., 38, 20-1–20-12, 2002.

Torri, D., Poesen, J., Borselli, L., and Knapen, A.: Channel width–flow discharge relationships for rills and gullies, Geomorphology, 76, 273–279, 2006.

Vandekerckhove, L., Poesen, J., Oostwoud-Wijdenes, D., Nachtergaele, J., Kosmas, D., Roxo, M. J., and Figueiredo, T. D.: Thresholds for gully initiation and sedimentation in Mediterranean Europe, Earth Surf. Proc. Land., 25, 1201–1220, 2000.

Van der Ploeg, R. R., Ehlers, W., and Sieker, F.: Floods and other possible adverse environmental effects of meadowland area decline in former West Germany, Naturwissenschaften, 86, 313–319, 1999.

Vanwalleghem, T., Van Den Eeckhaut, M., Poesen, J., Deckers, J., Nachtergaele, J., Van Oost, K., and Slenters, C.: Characteristics and controlling factors of old gullies under forest in a temperate humid climate: a case study from the Meerdaal Forest (Central Belgium), Geomorphology, 56, 15–29, 2003.

Vanwalleghem, T., Poesen, J., Van Den Eeckhaut, M., Nachtergaelel, J., and Deckers, J.: Reconstructing rainfall and land-use conditions leading to the development of old gullies, Holocene, 15, 378–386, 2005.

Wells, R. R., Momm, H. G., Rigby, J. R., Bennett, S. J., Bingner, R. L., and Dabney, S. M.: An empirical investigation of gully widening rates in upland concentrated flows, Catena, 101, 114–121, 2013.

Whipple, K. X.: Bedrock rivers and the geomorphology of active orogens, Annu. Rev. Earth Planet. Sci., 32, 151–185, 2004.

Whipple, K. X. and Tucker, G. E.: Dynamics of the stream-power river incision model: Implications for height limits of mountain ranges, landscape response timescales, and research needs, J. Geophys. Res.-Solid Ea., 104, 17661–17674, 1999.

Whipple, K. X., Hancock, G. S., and Anderson, R. S.: River incision into bedrock: Mechanics and relative efficacy of plucking, abrasion, and cavitation, Geol. Soc. Am. Bull., 112, 490–503, 2000.

Wilcock, P. R. and Crowe, J. C.: Surface-based transport model for mixed-size sediment, J. Hydraul. Eng., 129, 120–128, 2003.

Wilson, G.: Understanding soil-pipe flow and its role in ephemeral gully erosion, Hydrol. Process., 25, 2354–2364, 2011.

Wilson, G. V.: Mechanisms of ephemeral gully erosion caused by constant flow through a continuous soil-pipe, Earth Surf. Proc. Land., 34, 1858–1866, 2009.

Wiskow, E. and van der Ploeg, R. R.: Calculation of drain spacings for optimal rainstorm flood control, J. Hydrol., 272, 163–174, 2003.

Zeeospheres G Series data sheet: http://www.zeeospheres.com/wp-content/uploads/2015/12/G-Series.pdf, last access: 7 February 2018.

10

Projecting water yield and ecosystem productivity across the United States by linking an ecohydrological model to WRF dynamically downscaled climate data

Shanlei Sun[1,2], Ge Sun[3], Erika Cohen[3], Steven G. McNulty[3], Peter V. Caldwell[4], Kai Duan[1], and Yang Zhang[1]

[1]Department of Marine, Earth, and Atmospheric Sciences, North Carolina State University, Raleigh, NC, USA
[2]Key Laboratory of Meteorological Disaster of Ministry of Education, Nanjing University of Information Science & Technology, Nanjing, Jiangsu Province, China
[3]Eastern Forest Environmental Threat Assessment Center, USDA Forest Service, Raleigh, NC, USA
[4]Coweeta Hydrologic Laboratory, USDA Forest Service, Otto, NC, USA

Correspondence to: Ge Sun (gesun@fs.fed.us)

Abstract. Quantifying the potential impacts of climate change on water yield and ecosystem productivity is essential to developing sound watershed restoration plans, and ecosystem adaptation and mitigation strategies. This study links an ecohydrological model (Water Supply and Stress Index, WaSSI) with WRF (Weather Research and Forecasting Model) using dynamically downscaled climate data of the HadCM3 model under the IPCC SRES A2 emission scenario. We evaluated the future (2031–2060) changes in evapotranspiration (ET), water yield (Q) and gross primary productivity (GPP) from the baseline period of 1979–2007 across the 82 773 watersheds (12-digit Hydrologic Unit Code level) in the coterminous US (CONUS). Across the CONUS, the future multi-year means show increases in annual precipitation (P) of 45 mm yr^{-1} (6 %), 1.8° C increase in temperature (T), 37 mm yr^{-1} (7 %) increase in ET, 9 mm yr^{-1} (3 %) increase in Q, and 106 gC m^{-2} yr^{-1} (9 %) increase in GPP. We found a large spatial variability in response to climate change across the CONUS 12-digit HUC watersheds, but in general, the majority would see consistent increases all variables evaluated. Over half of the watersheds, mostly found in the northeast and the southern part of the southwest, would see an increase in annual Q (> 100 mm yr^{-1} or 20 %). In addition, we also evaluated the future annual and monthly changes of hydrology and ecosystem productivity for the 18 Water Resource Regions (WRRs) or two-digit HUCs. The study provides an integrated method and example for comprehensive assessment of the potential impacts of climate change on watershed water balances and ecosystem productivity at high spatial and temporal resolutions. Results may be useful for policy-makers and land managers to formulate appropriate watershed-specific strategies for sustaining water and carbon sources in the face of climate change.

1 Introduction

Due to human activities, such as emissions of greenhouse gas, aerosol, and land use/cover change (LUCC), the Earth's climate system has been significantly altered over the past 100 years. The Intergovernmental Panel on Climate Change (IPCC, 2014) concludes that global surface temperature has increased 0.85° C during 1880–2012, and increased 0.78° C during 2003–2012 when compared to 1850–1900. Additionally, extreme precipitation and droughts have increased (Tebaldi et al., 2006; Trenberth, 2011; Bony et al., 2013; Hegerl et al., 2014). The global climate is projected to continue to change over this century and beyond (IPCC, 2014). In comparison to the period of 1986–2005, the period 2018–2100 is projected to see 0.3 to 4.8° C increase in global surface temperature (IPCC, 2014). Future changes in precipitation show a small increase in the global average, but a substantial shift in where and how precipitation falls (Noake et al., 2012; Scheff and Frierson, 2012; Liu et al., 2013a).

In response, the hydrological cycle and ecosystems have been markedly changed through various physical, chemical, and biological processes during the past century (Labat et al., 2004; Milly et al., 2005; Dai et al., 2009; Harding et al., 2011; Sedláček and Knutti, 2014). Mounting evidence has suggested that climate and its change has played an important role in controlling the water cycle by changes in evaporation, transpiration, and runoff (McCabe, 2002; Hamlet et al., 2007; Syed et al., 2010; Wang and Hejazi, 2011; Chien et al., 2013; Hegerl et al., 2014; Huntington and Billmire, 2014; McCabe and Wolock, 2014; Sun et al., 2014). Also, climate can exert a dominant control on vegetation structural and phenological characteristics through variations in air temperature, precipitation, solar radiation, wind, and CO_2 concentration (Nemani et al., 2003; Harding et al., 2011; Wang et al., 2014; Zhang, F. et al., 2014; Tang et al., 2014). Climate change affects vegetation dormancy onset date, timing of bud burst, net primary production (NPP), gross primary production (GPP), and ecosystem respiration (Nemani et al., 2003; Scholze et al., 2006; Pennington and Collins, 2007; Anderson-Teixeira et al., 2011; Gang et al., 2013; Peng et al., 2013; Zhang et al., 2013; Williams et al., 2014; Piao et al., 2015; Wang et al., 2015). In addition, future water cycle and ecosystems are affected by the combined forces from natural environment (e.g., climate and land surface properties) and socio-economics (e.g., economic development and population increases) (Cox et al., 2000; Somerville and Briscoe, 2001; Sitch et al., 2008; Alkama et al., 2013; Piontek et al., 2014; Schewe et al., 2014; Zhang, Y. et al., 2014; Aparício et al., 2015).

In the US, average temperature has dramatically increased since the record keeping began in 1895. The most recent decade was believed to be the warmest on record (see the website: http://www.nasa.gov/home/hqnews/2010/jan/HQ_10-017_Warmest_temps.html). Mean precipitation over the US has increased overall since 1900; some areas have increased with a higher rate than the national average, and some areas have decreased (Groisman et al., 2004; Meehl et al., 2005; Anderson et al., 2015). Over the past century, climate change in the US has caused severe water stress, floods and droughts as well as forest morality (Xu et al., 2013), leading to serious economic losses in some regions. Quantifying the impacts on future climate change on water and ecosystem productivity has become a major research area in hydrology and ecosystem sciences (Lettenmaier et al., 1994; Lins and Slack, 1999; Groisman et al., 2001; McCabe and Wolock, 2011; Sagarika et al., 2014).

Because climate change patterns are not uniform across space or time (Sankarasubramanian et al., 2001; Sankarasubramanian, 2003; Wang and Hejazi, 2011; Xu et al., 2013; Brikowski, 2014) climate change impacts on water cycle and ecosystem productivity vary from region to region, and variability will be even bigger across small watersheds. To support future water resource planning, watershed management and to develop sound adaptation strategies over the continental US (CONUS), tools are needed to integrate various climate scenarios from a variety of atmospheric ocean general circulation models (AOGCMs) and community Earth system models (CESMs), and hydrological and vegetation dynamic models (Brown et al., 2013; Blanc et al., 2014; Yu et al., 2014).

Two major research gaps exist in past climate change studies that aim at quantifying the interactions among climate, hydrology and ecosystem productivity. First, few studies have provided projections of future climate change impacts on water and carbon balances at watershed scale using a consistent approach. Various land surface models (LSMs) simulate and predict water fluxes for a large region, but the scale is often too coarse with a spatial resolution ranging from 0.25 to 2.5°. The water budget within each grid cell in LSMs may not be balanced because it is not a closed watershed system. Key hydrological processes (e.g., lateral surface and sub-surface flows among grid boxes) have been rarely considered, potentially resulting in uncertainties in water balance projections (Overgaard et al., 2006; Li et al., 2011). Second, to save computational resource and enhance the computational efficiency, a statistical (or empirical) downscaling method has been mostly used to generate climate forcing to land surface models or watershed ecosystem models. However, this type of method does not consider the effects of atmospheric dynamical processes (Xue et al., 2014) and could introduce uncertainties into the crucial land surface variables.

Therefore, the general goal of this study is to explore how dynamically downscaled climate data can be used to drive a common ecosystem model for climate change assessment at a fine spatial scale (i.e., 12-digit HUC watersheds, whose detailed information can be found in the following text). The specific objectives of this study are to (1) evaluate future climate changes in precipitation and temperature during 1979–2007 and 2031–2060 for one emission scenario over the CONUS using dynamically downscaled climate projections from the WRF (Weather Research and Forecasting) model; (2) project future changes of water yield (Q), ET, and GPP for the study area by linking the WRF dynamically downscaled climate change scenarios and the WaSSI model. The goal is to generate information that can be useful for policy makers to plan for potential shifts in water resources and ecosystem productivity at the watershed to national level.

2 Data and methodology

2.1 Study area

The research area includes the conterminous continental US covering 82 773 12-digit HUC watersheds within the 18 Water Resources Regions (WRRs; Fig. 1a). The size of these HUC12 watersheds ranges from 0.16 to 9238 km^2, with the median and the mean values of 88.2 and 95.0 km^2, respectively. Moreover, the area of the overwhelming majority of the watersheds (> 80 000) is between 50 and

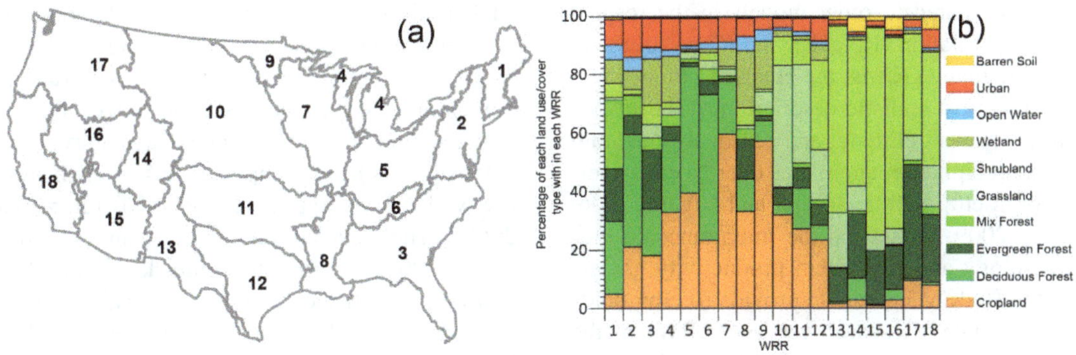

Figure 1. Location of the Water Resource Regions (WRRs) over the CONUS **(a)** with the percentage of each land use/cover type within each WRR. The numeral from 1 to 18 in panel **(a)** represents the number of WRR. In panel **(b)**, the rectangle size notes the percentage of each land use/cover type within each WRR. Note that the percentages of each land use/cover type were calculated based on the 2006 National Land Cover Database (NLCD) of CONUS.

170 km^2. The WRRs vary in size with the maximum of 1.3×10^6 km^2 (WRR10) and the minimum of 1.1×10^5 km^2 (WRR6). In addition, climatology and land surface characteristics (e.g., land cover; Fig. 1b) vary dramatically among these WRRs. From the east to the west CONUS, multi-year mean (1979–2007) annual precipitation as estimated by the Parameter-elevation Regressions on Independent Slopes Model (PRISM) shows longitudinal decreases ranging from 1300 mm yr^{-1} to 341 mm yr^{-1}. For the multi-year mean temperature (1979–2007), the spatial distribution displays the latitudinal characteristic decreasing from the south to the north CONUS, with a range from a high of 18° C to a low of 4° C. The WRRs in the east had the larger percentages (around 10 %) of urban use with WRR2 (13 %) and WRR4 (11 %) ranked as the top two. The wetlands are mainly located in the WRRs in the eastern US, while the western regions had the higher percentages of shrubland (> 30 %). The WRRs in the east generally had higher forest (including mixed, evergreen and deciduous forests) percentages (> 33 %) than the southwest (< 30 %). The deciduous and the evergreen forests were mainly found in the east and the west, respectively. Most of the crop lands were located in the east and central CONUS (Fig. 1b).

2.2 Dynamically downscaled climate by WRF

The IPCC Special Report on Emissions Scenarios (SRES) scenarios were designed to project future global environment with a special reference to the production of greenhouse gases and aerosol precursor emissions (Nakicenvoic and Swart, 2000). The SRES scenarios mainly include four narrative storylines (i.e., A1, A2, B1, and B2), which describe the relationships between the forces affecting greenhouse gas and aerosol emissions and their evolution in the 21st century for large regions and the globe. Each storyline represents a specific and typical demographic, economic, technological, social, and environment progresses with divergence in increasingly irreversible ways. The A2 storyline represents the high end of the SRES emission scenarios (but not the highest) and has been widely used by the scientific communities (Seneviratne et al., 2006; Wi et al., 2012). Therefore, the SRES A2 emission scenario was selected in this study. From an impact and adaptation point of view, if one can adapt to a larger climate change, then the smaller climate changes of the lower-end scenarios can also be adapted to. Moreover, the historic emissions (1990 to present) correspond to a relatively high emission trajectory (http://www.narccap.ucar.edu/about/emissions.html).

The global circulation models (GCMs) have significant issues in representing local climates, mountains in particular, because of their coarse spatial resolution (Leung and Qian, 2003). To downscale the GCMs climate data to a higher spatial resolution for regional and local applications, two types of downscaling method are available: dynamical and statistical (or empirical) downscaling (Huang et al., 2011). Due to better representation of finer-scale physical processes in climate variables (Gao et al., 2011; Xue et al., 2014), dynamical downscaling was used here for generating the current and the future climate.

The HadCM3 (Hadley Centre Coupled Model, Version 3) is a coupled atmosphere–ocean general circulation model (AOGCM) developed by the Hadley Centre in the United Kingdom (Gordon et al., 2000; Pope et al., 2000; Collins et al., 2001), which has been used extensively for climate prediction, detection, and attribution, and other climate sensitivity studies, e.g., the Third, the Fourth, and the Fifth IPCC Assessments reports. For the atmospheric component, the model dynamics and physics are solved on a 3.75° (longitude) × 2.5° (latitude) grid with 19 hybrid vertical levels, and a horizontal resolution of 1.25° (longitude) × 1.25° (latitude) with 20 vertical levels in the oceans. The reader is referred to Pope et al. (2000) for details of the HadCM3 dynamical and physical processes. Generally speaking, although flux adjustments are not utilized by the HadCM3, it still ranks highly compared to other models in the respect of current climate simulation (Reichler and Kim, 2008). In addition, among the many GCMs, the HadCM3 model was believed to have the most realistic description of the ENSO mechanisms in the current climate, and reasonably capture ENSO-associated precipitation anomalies over North America (van Oldenborgh et al., 2005; Joseph and Nigam, 2006; Dominguez et al., 2009). Based on the importance of precipitation in hydrology and ecosystem productivity assessment, we chose the HadCM3 model to provide forcing fields for running the Advanced Research version (ARW) of the Weather Research and Forecasting (WRF) regional climate model (Skamarock et al., 2005).

Data generated from the WRF model are described below. The WRF model was run for the years 1969 to 2079 at a 35 km resolution. HadCM3 inputs with 6 h time resolution were used, and the dynamically downscaled output by the WRF model was also stored at 6 h time intervals. For the model domain, the CONUS and northern Mexico were included (Wi et al., 2012). The model's physical parameterizations mainly included: WRF Single-Moment three-class microphysics (Hong et al., 2004), Kain–Fritsch cumulus parameterization (Kain and Fritsch, 1993), Goddard Shortwave radiation (Chou and Suarez, 1994), Rapid Radiative Transfer Model (RRTM), Longwave (Mlawer et al., 1997), Eta surface layer (Janjic, 2002), Mellor–Yamada–Janjic (MYJ) planetary boundary layer (Janjic, 2002), and the Noah land surface model Version 1.0 (Chen and Dudhia, 2001). To ensure the maintenance of synoptic-scale circulation features, like ridges and troughs, in the RCM (regional climate model), we performed spectral nudging on the zonal and meridional winds, the temperatures and the geo-potential height fields for all pressure levels below 0.36 of the surface pressure (for a surface pressure of 1000 mb this would mean all pressures below 360 mb) effectively nudging only at very high elevations above the surface.

2.3 Climate data bias corrections

The dynamically downscaled precipitation and temperature simulations by WRF were sufficient for a hydrological study (1981–2005) by Wi et al. (2012) in the Colorado River Basin). Our comparison study showed that although downscaled climate simulations agreed well with the observations (PRMS data) in a climatological sense, some large regional biases were found. Therefore, bias correction was performed using a monthly Bias Correction Spatial Disaggregation (BCSD; Wood et al., 2002, 2004) approach. The method has been applied for hydrologic forecasting in the eastern US (Wood et al., 2002). Basically, the bias correction includes the following procedures: (1) scale up the PRISM monthly precipitation and temperature with 4 km × 4 km resolution to match the simulated WRF data (35 km × 35 km) for the time period of 1978–2007; (2) construct cumulative distribution functions (CDFs) for climate variables in each grid cell and month for both historic WRF and upscaled PRISM data sets; (3) the paired CDFs are combined to form a "quantile map", where at each rank probability or percentile, the bias between the WRF and the PRISM (at that location, for that variable, and during that month) is calculated; (4) the computed bias in each month, grid cell and variable are applied to the WRF future outputs (2031–2060). The detailed procedures can be found in Brekke et al. (2013), see; http://gdo-dcp.ucllnl.org/downscaled_cmip_projections/techmemo/downscaled_climate.pdf. Both the corrected WRF monthly precipitation and temperature in historic and future periods were scaled to the 12-digit HUC watershed scale because the WaSSI model operated on the 12-digit HUC watershed level.

2.4 The WaSSI model

The WaSSI model is an integrated, water-centric process-based ecohydrological model designed for modeling water and carbon balance and water supply stress at a broad scale (Sun et al., 2011a; Caldwell et al., 2012; Sun et al., 2015a, b). It operates on a monthly time step at the 8-digit HUC or 12-digit HUC watershed scale for the CONUS. The WaSSI model simulates the full monthly water (ET, Q, and soil moisture storage) and carbon balances (GPP, ecosystem respiration, and net ecosystem productivity) for each land cover class at the given watershed scale. This model has been tested in a variety of geographical regions, and widely used for quantitatively assessing combined or individual effects of climate change, land use/cover change (LUCC), and population dynamics on water supply stress and ecosystem productivity (i.e., carbon dynamic) over the CONUS (Sun et al., 2008, 2011a; Lockaby et al., 2011; Caldwell et al., 2012; Averyt et al., 2013; Tavernia et al., 2013; Marion et al., 2014; Sun et al., 2015a, b). The model has also been applied internationally in Mexico, China (Liu et al., 2013b), and Africa (McNulty et al., 2016).

The key algorithms of the WaSSI model were derived from accumulated knowledge of ecosystem carbon and water cycles gained through the global eddy covariance flux monitoring networks and watershed-based ecohydrological studies across the US. The ecosystem ET sub-module, the core of the WaSSI model, is described as a function of potential ET (PET), LAI, precipitation, and soil water availability by land cover type (Sun et al., 2011a). The snow model embedded within WaSSI (McCabe and Wolock, 1999; McCabe and Markstrom, 2007) estimates snow melt rates, mean monthly snow water equivalent (SWE), mean watershed elevation, and monthly air temperature. Infiltration, surface runoff, soil moisture, and baseflow processes for each watershed are simulated by the Sacramento Soil Moisture Accounting Model (SAC-SMA; Burnash, 1995). The ecosystem productivity module computes carbon dynamics (GPP and respiration) using linear relationships between ET and GPP derived from global eddy covariance flux measurements (Sun et al., 2011a, b). The User Guide of WaSSI Ecosystem Services Model, Version 2.1 (www.forestthreats.org/research/tools/WaSSI) provides detailed description of model algorithms and data requirements (Caldwell et al., 2012).

To run the WaSSI model, the necessary inputs include monthly precipitation, monthly mean air temperature, monthly mean leaf area index (LAI) by land cover, land cover composition within each watershed, and 11 SAC-SMA soil parameters. The historic (1979–1997) climate data (i.e., precipitation and air temperature) derived from the Precipitation Elevation Regression on Independent Slopes Model (Daly et al., 1994; PRISM Climate Group, 2013) at the 4 km × 4 km resolution were scaled to the 12-digit HUC level. The data set of the 2006 National Land Cover Database (NLCD; http://www.mrlc.gov/nlcd06_data.php) with 17 land cover classes was aggregated into 10 classes (Fry et al., 2011): crop, deciduous forest, evergreen forest, mixed forest, grassland, shrubland, wetland, water, urban, and barren. WaSSI The monthly LAI time series data required by WaSSI for each land cover type were derived from the Moderate Resolution Imaging Spectroradiometer (MODIS)–MOD15A2 FPAR/LAI 8-day product (Myneni et al., 2002). The 1 km × 1 km SAC-SMA soil data set provided by the State Soil Geographic Database (STATSGO), based on the Sacramento Soil Moisture Accounting Model Soil Parameters, was aggregated to the 12-digit HUC watershed. No WaSSI model parameters were calibrated during the model evaluation process.

The WaSSI has been evaluated at multiple scales using gaging station data for streamflow and remote sensing products for evapotranspiration across the US (Sun et al., 2011a; Caldwell et al., 2012; Sun et al., 2015a). At the 12-digit HUC scale, the model was validated using monthly and annual water yield data collected at 72 selected USGS watersheds, and ET and GPP data for 170 National Forests over the CONUS (Sun et al., 2015a). Overall, the validation results suggested that this model could capture characteristics of water and carbon balances at the selected spatial levels under various climatic conditions (Sun et al., 2015a, b).

2.5 Impact analysis

We first examined modeled changes in monthly ET and GPP at the 12-digit HUC watershed scale using the WRF dynamically downscaled, bias-corrected historic and future climate data, respectively. Then, we computed future annual changes at three spatial levels: the entire CONUS as a whole, the 12-digit HUC watershed, and the individual WRR. The multi-year means of annual precipitation, temperature, ET, Q, and GPP averaged across the whole CONUS, WRR, or each 12-digit HUC watershed for the 1979–2007 time period were compared to those for the 2031–2060 period. The absolute or percentage (except for temperature) changes for each variable were calculated. Herein, the absolute differences were expressed as the future means minus those in the historical period, while the percentage differences were calculated using the absolute difference divided by baseline mean in the 1979–2007. In addition, the future monthly changes of these ecosystem flux variables were also assessed for the whole CONUS and each WRR.

3 Results

3.1 Baseline characteristics of hydro-climatology and ecosystem productivity (1979–2007)

For the baseline period, multi-year means of annual precipitation (Fig. 2a), ET (Fig. 3a), Q (Fig. 3e), and GPP (Fig. 4a) all generally showed longitudinal decreases from east to west across the CONUS. The Pacific Northwest region has the highest precipitation (> 1800 mm yr^{-1}), followed by the larger values for precipitation in the southeast (> 1200 mm yr^{-1} in Fig. 2a). For ET, the maximum (> 750 mm yr^{-1} in Fig. 3a) mainly appeared in the southeast. The largest Q higher than 600 mm yr^{-1} (Fig. 3e), mainly existed in the Pacific Northwest region, the Rocky Mountains, and the Appalachian Mountains, especially for some 12-digit HUC watersheds in the Pacific Northwest region being greater than 1000 mm yr^{-1}. For GPP (Fig. 4a), the 12-digit HUC watersheds with higher values (> 1000 gC m^{-2} yr^{-1}) were mainly located in the areas of the southeast and the Pacific Northwest. By contrast, the average annual temperature climatology of the CONUS presented a clear latitudinal increase ranging from −0.8° C in the north to 22° C in the south. Because of topographical effects, temperature in the Rocky Mountains was lower than 4° C relative to the surrounding regions.

Taking the CONUS as a whole, the area-weighted average precipitation, temperature, ET, Q, and GPP in the period of 1979–2007 was 801 mm yr^{-1}, 11.2° C, 515 mm yr^{-1}, 290 mm yr^{-1}, and 1232 gC m^{-2} yr^{-1}, respectively (Table 1). Comparing the area-average precipitation among the 18 WRRs, the WRR3, 6, and 8 had the highest precipitation (> 1200 mm yr^{-1}), while the WRR13–16 had the lowest (< 400 mm yr^{-1}). In the WRR3, 8, and 12, the area average temperatures were the highest (> 17° C), while the WRR9 had the lowest temperature (4.2° C). The WRR3, 6, and 8 had the highest ET (> 750 mm yr^{-1}), with the lowest values found in WRR16 (< 300 mm yr^{-1}). The WRR1 had the largest Q of 636 mm yr^{-1}, while the smallest Q was found in the WRR13–16 (< 100 mm yr^{-1}). Similar to the average ET, the highest GPP (> 2100 gC m^{-2} yr^{-1}) were also found in the WRR3, 6, and 8, but the western WRRs (e.g., WRR13–16 and 18) exhibited lowest values (< 800 gC m^{-2} yr^{-1}).

The baseline intra-annual precipitation presented a complicated pattern (Fig. 5). Except in February, precipitation in all the months was more than 65 mm yr^{-1}, and peaked in May with 78 mm yr^{-1}. Overall, temperature (Fig. 5b), ET (Fig. 5c), and GPP (Fig. 5e) all increased gradually

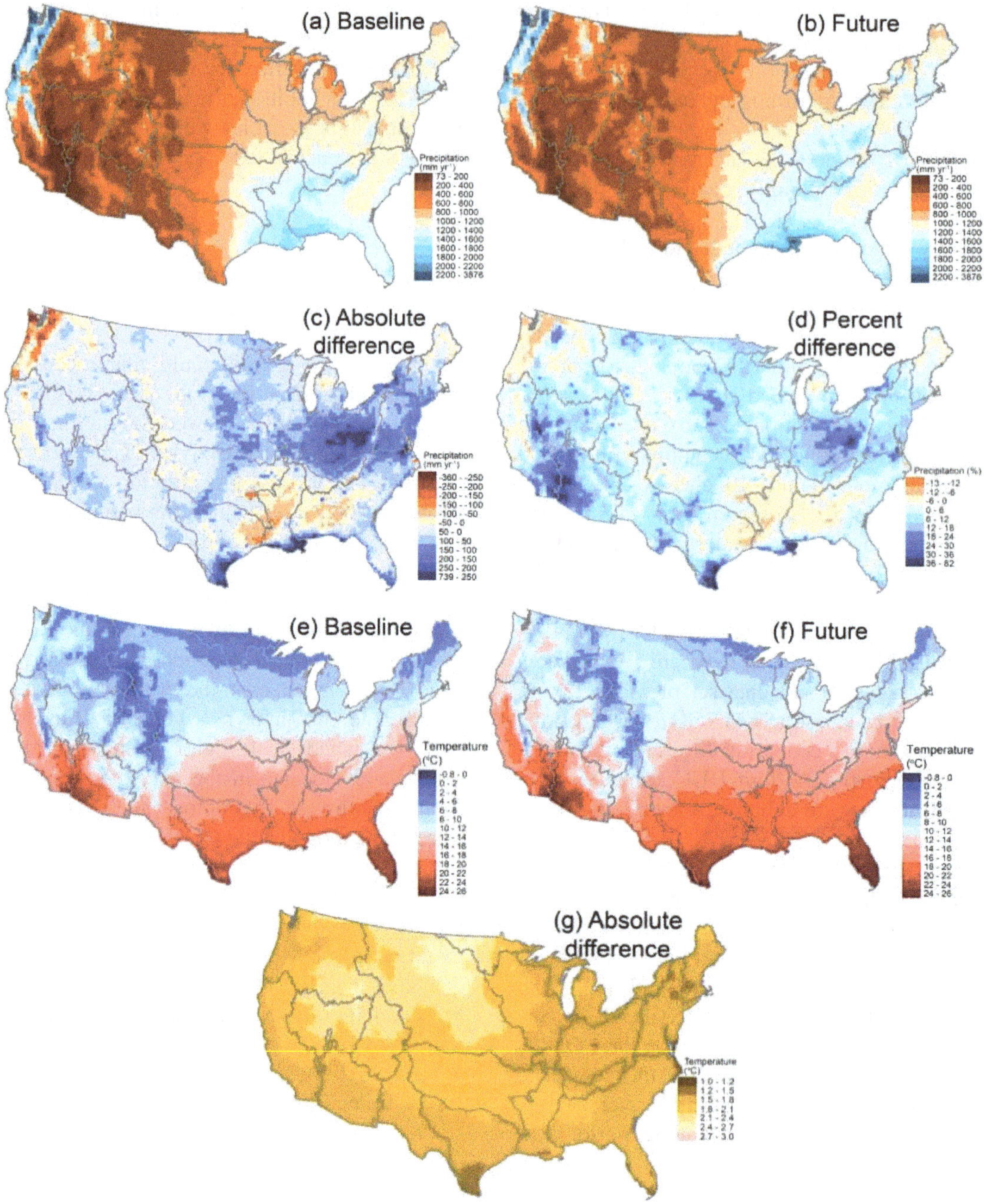

Figure 2. Characteristics of precipitation and temperature during the baseline (1979–2007) and the future periods, and the future changes (future – baseline).

starting from January, peaked (24.8° C, 80 mm yr^{-1} and 205 gC m^{-2} yr^{-1}, respectively) in July, and then decreased sharply. Fluctuations of Q clearly differed from other variables (Fig. 5d) following a pattern similar to a sine function. Q increased in January, peaked in April (36 mm yr^{-1}), decreased to the lowest (15 mm yr^{-1}) in August, and then rose.

We also explored multi-year mean monthly precipitation, temperature, ET, Q, and GPP for each WRR (not shown). Generally, the intra-annual distribution was different (e.g., phases and magnitudes) among the 18 WRRs, due to the complex differences in topography and climate among them. For WRR16–18, most precipitation fell in January–April and October–December, while precipitation in other WRRs mainly concentrated in May–September. In all the WRRs, the intra-annual temperature followed a unimodal curve, with peaks in July or August and the lowest values in January or December. For ET and GPP, the higher values were mainly found from May to November, except for WRR18. Com-

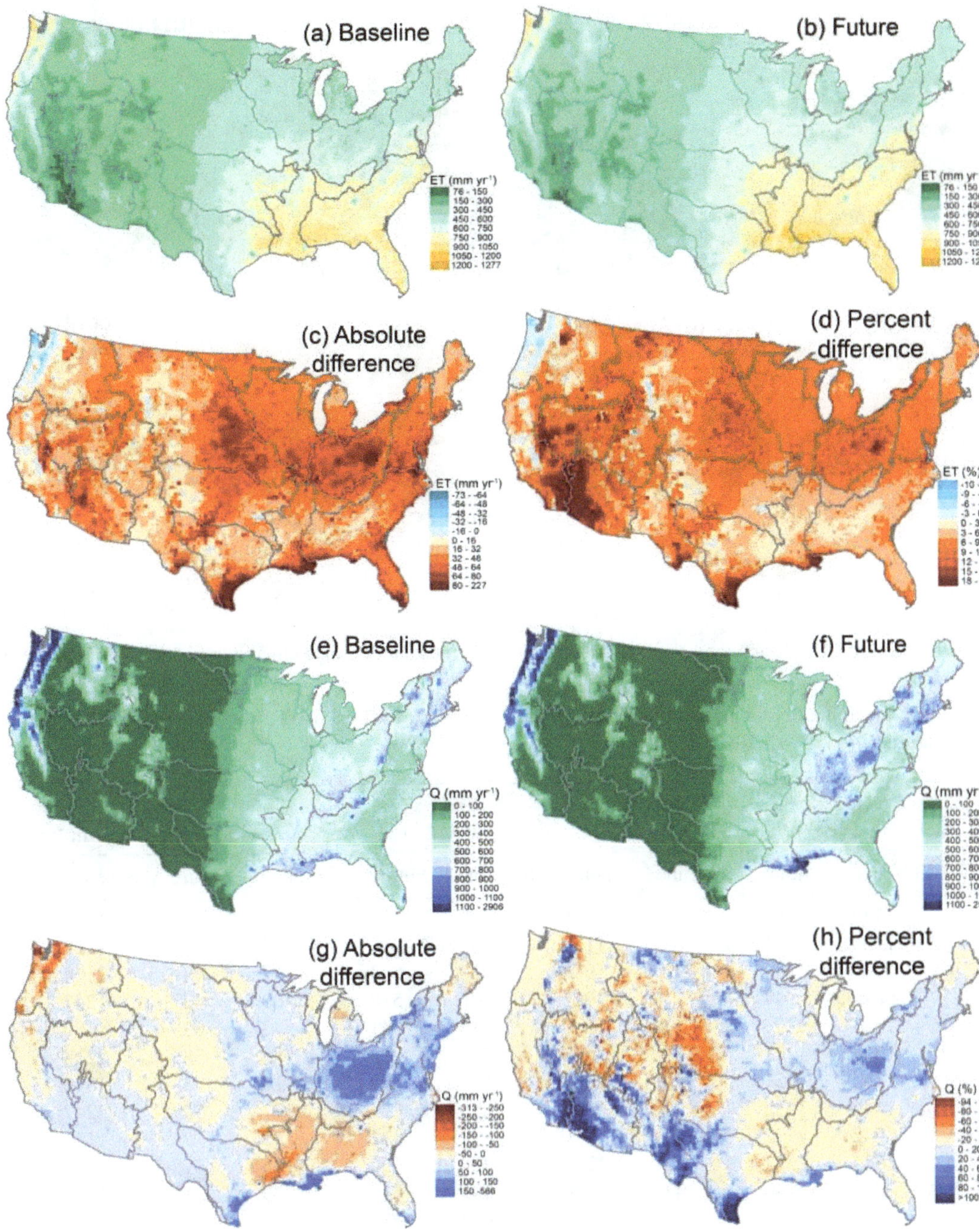

Figure 3. Spatial distribution of ET and Q during the baseline and the future periods, and the future changes.

paring the monthly distributions among the 18 WRRs, they could be divided into three categories: unimodal, sine, and trough curves.

3.2 Future climate change

Future precipitation and temperature followed a similar pattern as the baseline (Fig. 2). Precipitation showed a longitudinal decrease from the east to the west, but temperature presented a clear latitudinal decrease. However, for each 12-digit HUC watershed, these two climate variables would increase or decrease by different magnitudes in the future (Fig. 2c and d for precipitation, and Fig. 2g). During 2031–2060, annual precipitation would increase in 82 % of the CONUS 12-digit HUC watersheds, while decreasing in the rest of the watersheds that were mainly located in the southeast and the west coastal regions. The northeast and the northwest coastal regions would generally have a greater increase (> 150 mm yr^{-1}) or decrease (> 200 mm yr^{-1}), respectively, in P (Fig. 2c). The greater percent increases in

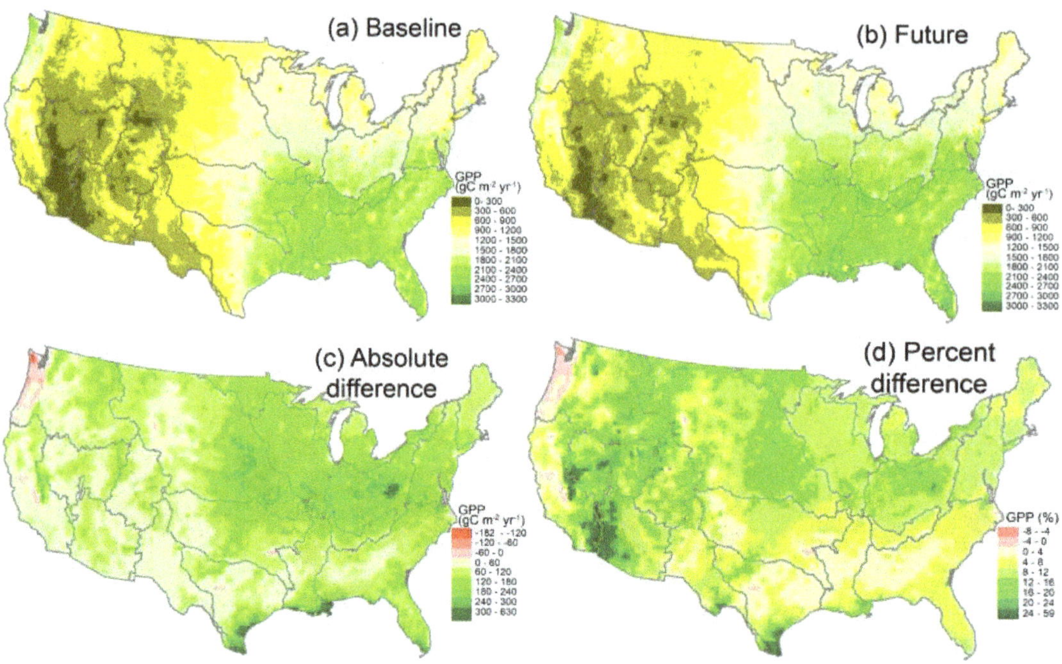

Figure 4. Spatial distribution of GPP during the baseline and the future periods, and climate change impacts (future – baseline).

Table 1. Multi-year mean precipitation, temperature, ET, Q, and GPP averaged over each WRR or the entire CONUS during the baseline (1979–2007) and the future period (2031–2060).

WRR	Precipitation ($mm\,yr^{-1}$)		Temperature (°C)		ET ($mm\,yr^{-1}$)		Q ($mm\,yr^{-1}$)		GPP ($gC\,m^{-2}\,yr^{-1}$)	
	Baseline	Future	Baseline	Future	Baseline	Future	Baseline	Future	Baseline	Future
1	1143	1169	6.3	8.0	506	538	636	632	1218	1316
2	1100	1211	10.2	11.8	582	629	518	583	1564	1712
3	1299	1334	17.5	19.2	823	863	477	471	2104	2207
4	875	944	7.3	9.0	476	518	400	427	1241	1376
5	1123	1297	11.6	13.1	580	641	543	655	1680	1882
6	1354	1395	13.8	15.4	769	810	585	585	2218	2347
7	863	931	8.5	10.3	550	597	314	335	1516	1677
8	1414	1425	17.4	19.2	836	877	577	549	2247	2361
9	542	592	4.2	6.5	429	472	115	123	1120	1256
10	534	572	7.9	10.1	424	462	115	118	985	1104
11	819	840	14.0	15.8	593	626	229	219	1502	1597
12	828	866	18.7	20.3	615	650	215	220	1379	1457
13	392	419	13.9	15.7	368	394	35	35	602	651
14	397	411	7.3	9.4	318	343	86	76	546	614
15	342	387	15.1	16.8	316	354	34	40	522	588
16	339	372	8.6	10.8	298	331	54	50	478	557
17	854	841	7.2	9.2	464	481	395	363	904	972
18	626	647	13.9	15.7	366	391	267	258	740	793
CONUS	801	844	11.2	13.1	515	551	290	297	1232	1339

precipitation (> 18 %) were found in some watersheds in the southwest and the northeast regions (Fig. 2d). Future temperature would increase consistently across watersheds, ranging from 1.0 to 3.0° C. The northwest and the north-central regions would see increases of more than 2.1° C (Fig. 2g).

For the CONUS as a whole, the area-weighted mean annual precipitation and temperature for 2013–2060 would

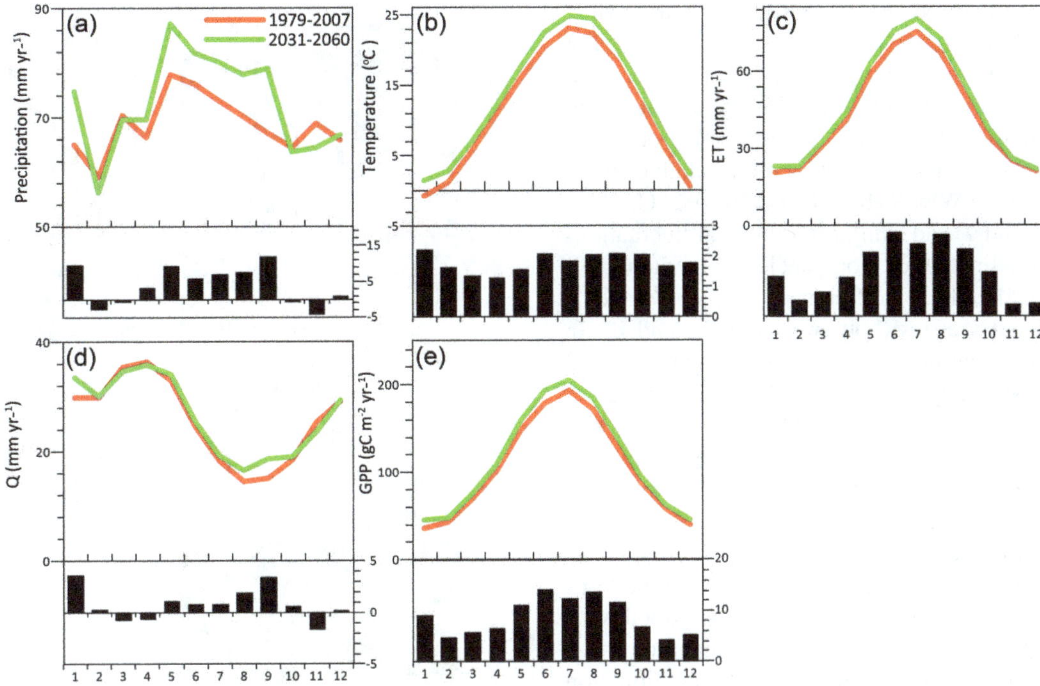

Figure 5. Monthly precipitation **(a)**, temperature **(b)**, ET **(c)**, Q **(d)**, and GPP **(e)** for the whole CONUS during 1979–2007 and 2031–2060 (the top of each panel), and their differences (future – baseline) between the two periods (the bottom of each panel).

Table 2. Future changes in multi-year mean precipitation, temperature, ET, Q, and GPP averaged over each WRR or the entire CONUS relative to the baseline period.

WRR	Precipitation		Temperature	ET		Q		GPP	
	Absolute (mm yr^{-1})	Percent (%)	Absolute (° C)	Absolute (mm yr^{-1})	Percent (%)	Absolute (mm yr^{-1})	Percent (%)	Absolute (gC m^{-2} yr^{-1})	Percent (%)
1	26	2	1.7	32	6	−4	−1	98	8
2	111	10	1.6	46	8	65	13	148	9
3	35	3	1.6	40	5	−7	−1	103	5
4	68	8	1.7	42	9	27	7	135	11
5	174	15	1.6	61	11	113	21	202	12
6	40	3	1.7	41	5	0	0	129	6
7	68	8	1.8	47	9	22	7	160	11
8	11	1	1.8	41	5	−29	−5	114	5
9	50	9	2.2	43	10	8	7	136	12
10	38	7	2.2	39	9	3	3	119	12
11	21	3	1.9	33	6	−10	−4	95	6
12	38	5	1.7	35	6	4	2	78	6
13	26	7	1.8	27	7	1	2	49	8
14	14	4	2.1	25	8	−10	−12	68	13
15	45	13	1.7	39	12	6	16	65	13
16	33	10	2.1	33	11	−3	−6	79	17
17	−13	−1	2.0	18	4	−32	−8	69	8
18	21	3	1.8	25	7	−9	−3	53	7
CONUS	45	6	1.8	37	7	9	3	106	9

be 844 mm yr^{-1} and 13.1° C, respectively (Table 1). The mean annual P for the entire CONUS would increase by 45 mm yr^{-1} (6 %) and T increase by 1.8° C , respectively (Table 2). Except for the WRR17 with a slight decrease in P (13 mm yr^{-1} or 1 %), the other 17 WRRs all exhibited increases. A large absolute increment of precipitation

(> 60 mm yr^{-1}) could be found in WRR2, 4, 5, and 7, while the WRR8 and 14 have lower increases (< 15 mm yr^{-1}). For the percentage increment, the higher increases in precipitation (≥ 10 %) existed in the WRR2, 5, 15, and 16; however, WRR1 and 8 showed lower increases (≤ 1 %). For future temperature, all the 18 WRRs show increase relative to the past period, especially WRR9, 10, 14, and 16 (≥ 2° C).

Both future *P* and *T* had similar intra-annual fluctuations to those of the baseline period (top panels in Fig. 5a and b). However, the magnitudes of differences in both *P* and *T* differed in different seasons (bottom of Fig. 5a and b). In most months, precipitation would increase ranging from 3 to 11 mm yr^{-1}, especially in January, May, and September (> 7 mm yr^{-1}). For February, March, October, and November, *P* would have slight reduction with a range from −5 to −1 mm yr^{-1}. The temperatures for each month would significantly increase by at least 1.5° C, particularly for January and June–October (> 2.0° C) (Fig. 5b).

The comparisons of seasonal climatic change patterns among the 18 WRRs suggested that timings agreed well among WRRs (not shown). However, the magnitudes of changes varied greatly. The future monthly precipitation shows an increase in January and May–October in more than 10 WRRs. The differences are most pronounced for January, July, and September (Fig. 6a). In other months, however, future monthly precipitation would reduce to some extent in most of the WRRs. The future monthly temperature for all the WRRs is shown to increase in a range from 0.5 to 3.0° C. In January and June–October, temperatures in most WRRs show increase at a relatively high rate (> 1.5° C) compared to other months for most WRRs.

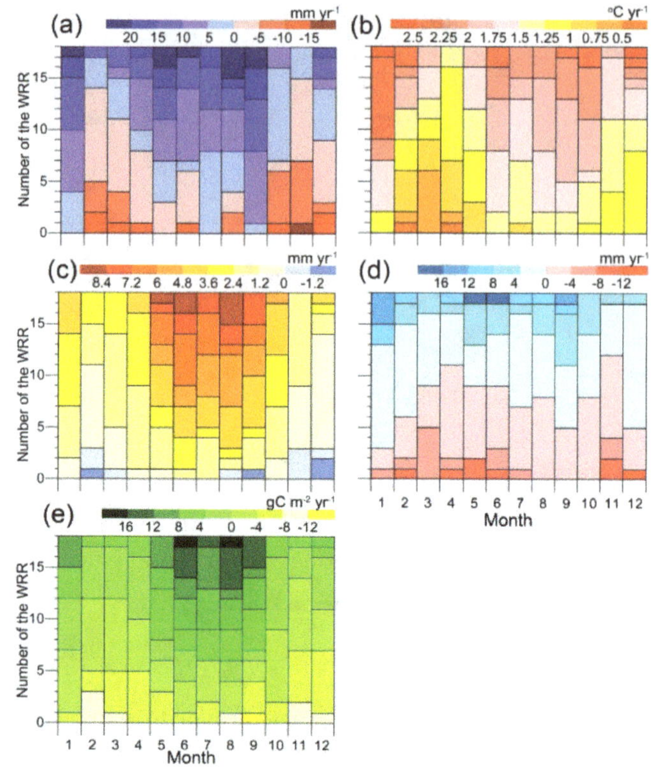

Figure 6. Number of the WRR within a given interval of change (future minus past) for each month: **(a–e)** precipitation (*P*), temperature (*T*), ET, *Q*, and GPP, respectively. The rectangle size for each month represents the number of the WRR that fall in a given interval value.

3.3 Future (2031–2060) changes in ET and *Q*

3.3.1 Annual change

The spatial patterns in ET and *Q* for the baseline are similar to those in the future (Fig. 3). However, the changes of annual ET (Fig. 3c and d) and *Q* (Fig. 3g and h) for each 12-digit HUC watershed vary in magnitude spatially. Overwhelmingly, the majority (98 %) of the CONUS 12-digit HUC watersheds are suggested to increase in annual ET, and the watersheds with annual ET reduction are mainly concentrated in the northwest coastal region. Regarding the absolute difference of ET (Fig. 3c and d), annual ET shows a relatively higher increase (> 32 mm yr^{-1}) in the northeast CONUS, especially in the southeast coastal region and the south part of the northeast CONUS (> 48 mm yr^{-1}). Different from the absolute changes, relative changes (%) in most of the western regions (excluding the west coast) and the northeast had high values (> 6 %) with the highest increments (> 12 %) found in the south of the southwest CONUS.

Across the CONUS, annual *Q* in 52 and 48 % of the CONUS 12-digit HUC watersheds is projected to increase and decrease by 2031–2060, respectively (Fig. 3g and 3h). In general, the northeast and the south part of the south CONUS shows increase in annual *Q*, while other regions show decrease (Fig, 3g and h). The positive (> 100 mm yr^{-1}) and the negative (> 100 mm yr^{-1}) changes in *Q* are mainly in the northeast, and the west coastal and the southeast regions, respectively. *Q* in the south part of the southwest CONUS is projected to significantly increase (> 20 %), while the central part of the west CONUS generally decreases by more than 20 %.

Over the CONUS, projected multi-year mean annual ET and *Q* are 551 and 297 mm yr^{-1} in the future, respectively (Table 1), representing an increase in ET by 37 mm yr^{-1} or 7 %, and in *Q* by 9 mm yr^{-1} or 3 % (Table 2). For each WRR, the future annual ET increases more or less (Table 2). The WRR2, 5, and 7 show the largest absolute increases for ET (> 45 mm yr^{-1}), while WRR17 (18 mm yr^{-1}) had the lowest increases. For the percentage increment, the highest increases of ET (≥ 10 %) are in WRR5, 9, 16, and 17; however, WRR17 showed the lowest increase (4 %). For the future annual *Q*, nine WRRs would increase, eight would reduce and one would have no change compared to the baseline period (Table 2). Among these 18 WRRs, the WRR2 and WRR5 have the largest absolute increase (> 60 mm yr^{-1}), and WRR8 and WRR17 have the largest decline (> 20 mm yr^{-1}). According to the percentage changes of annual Q, the greatest increases (> 10 %) and decreases (> 10 %) could be found in the WRR2, 5, and 15, and the WRR14.

3.3.2 Seasonal change

The variations of future CONUS-wide multi-year mean monthly ET and Q are presented in Fig. 5c and d. Although these two variables had similar intra-annual fluctuations to those of the baseline period, their monthly magnitudes changed to some degree. Overall, the future monthly ET would increase with the largest increments (> 2 mm $month^{-1}$) in January. April–October had higher values than the other 5 months. For monthly Q, most of the months (9 months) would increase, especially in January and September (increase > 3 mm $month^{-1}$).

We also compared the future intra-annual fluctuations of ET and Q to those of the baseline period, and found that each WRR agreed well in their flow timings for the baseline and the future periods (not shown here). Fig. 6c and d present the number of the WRR within a given difference interval for ET or Q by month respectively. Generally, future monthly ET would increase by different rates for each month at each WRR (Fig. 6c). Moreover, ET from May to September (roughly the growing season) would have greater increments (> 2.4 mm yr^{-1}) in most of the 18 WRRs. Q in most of WRRs would increase in January, February, July, September, and December, but would decrease in April and November.

3.4 Future changes in GPP

3.4.1 Annual change

The overall spatial distribution of GPP did not change in the future (Fig. 4b) when compared to the baseline (Fig. 4a). For each 12-digit HUC watershed, GPP would change with great spatial variations (Fig. 4c and d). In the future, the overwhelming majority (98 %) of the CONUS 12-digit HUC watersheds would increase in annual GPP. The watersheds with annual GPP reduction were mainly located in the northwest coastal region. A relatively high increase (> 120 gC m^{-2} yr^{-1}) were found in the northeast, especially in the south part of the region (> 180 gC m^{-2} yr^{-1}; Fig. 4c). In contrast to the absolute difference, most of the west CONUS (excluding the coastal regions) show great increase (> 12 %) in relative change (%) of annual GPP. The highest changes (> 20 %) were mainly located in the south of the southwest region.

Over the CONUS, multi-year mean annual GPP is projected to be 1339 gC m^{-2} yr^{-1} in the future (Table 1), representing an increase of 106 gC m^{-2} yr^{-1} or 9 % (Table 2). Future annual GPP in every WRR would increase ranging from 49 to 202 gC m^{-2} yr^{-1} or from 5 to 12 % (Table 2). The WRR2–WRR10 were found to have the largest absolute increases for GPP (> 100 gC m^{-2} yr^{-1}), especially for the WRR5 with the maximum of 202 gC m^{-2} yr^{-1}, while WRR13 (49 gC m^{-2} yr^{-1}) had the lowest increases. In terms of percentage change, GPP increments ranged from 5 to 17 % among all the WRRs. The higher GPP increases (> 10 %) occurred in WRR4, 5, 7, 9, 10, and WRR14–16, with the largest of 17 % in WRR16, while other WRRs had increments lower than 10 %, particularly WRR3 and 8 with the minimum of 5 %.

3.4.2 Seasonal change

Figure 5e shows the future multi-year mean monthly GPP averaged over the whole CONUS. Despite the similar intra-annual fluctuations of multi-year mean monthly GPP during the baseline and the future periods, the future magnitude in each month is projected to change to some degree (the bottom of Fig. 5e). Overall, the future monthly ET would have larger increments (> 9 gC m^{-2} yr^{-1}) in January and May–October than other months. The future intra-annual fluctuation patterns of GPP for each WRR are similar to the baseline periods (not shown here). As indicated by the number of the WRR within a given GPP difference interval (Fig. 6e), the future monthly GPP generally would increase by different rates for each WRR. Moreover, GPP from May to September would have greater increments (> 4 gC m^{-2} yr^{-1}) in most of the 18 WRRs.

4 Discussions

4.1 Uncertainties

In the present study, we have assumed that the water balance and ecosystems at each 12-digit HUC watershed are unaffected by human activities as represented by a fixed land cover (year 2000), and ecosystem flux changes are attributed to climate change alone. However, one way or another, most catchments in the US have experienced some level of human influence (National Research Council, 2002). Hydrology and ecosystems can be influenced significantly by human activities on various temporal and spatial scales (Foley et al., 2005; Harding et al., 2011). Hydraulic projects such as dam constructions, reservoir management (Hu et al., 2008), groundwater withdrawals for irrigation and domestic use, and land use/cover change all affect watershed balances (Foley et al., 2005; Piao et al., 2007; Wang and Hejazi, 2011; Schilling et al., 2008) and ecosystem productivity (Zhang, Y. et al., 2014).

Similarly, natural disturbances (e.g., wildfire, climate extremes, and pest and pathogen outbreak) also impact water balance and ecosystem productivity – in both the past and the future. For example, the direct effects of wildfire include plant mortality and thus exert adverse impacts on vegetation productivity, consequently leading to a decrease in carbon uptake and stocks (Lenihan et al., 2008; Dore et al., 2010;

Lee et al., 2015). Wildfires alter the watershed hydrologic processes through reducing vegetation canopy interception, transpiration, and infiltration rate (Yao, 2003; Neary et al., 2005; Bond-Lamberty et al., 2009; Brookhouse et al., 2013; Nolan et al., 2014, 2015). As an important natural disturbance, droughts generally increase vapor pressure gradient between leaves and atmosphere and thus cause stress on plant hydraulic systems (Anderegg et al., 2012; Reichstein et al., 2013). As a result, high tension in the xylem can trigger embolism and partial failure of hydraulic transport in the stem, and even tend to result in vegetation mortality, which can aversely impact on water yield and carbon sink capability (Cook et al., 2007; Allen et al., 2010; Guardiola-Claramonte et al., 2011; Adams et al., 2012). Droughts often lead to pest and pathogen outbreaks (Overpeck et al., 1990; Hason and Weltzin, 2000; Marengo et al., 2008; DeRose and Long, 2012; Jactel et al., 2012), and thus predispose an individual plant species to disease or mortality (Schoeneweiss, 1981; Ayres and Lombarder, 2000). Although our modeling approach has considered water stress on productivity, tree mortality has not been dealt with, so the impacts of droughts on GPP might be underestimated, and water yield may be underestimated as well.

Additionally, elevated CO_2 and climate change can also execrate impacts on hydrological and ecosystem productivity through changing water use efficiency (Miller-Rushing et al., 2009; de Kauwe et al., 2013; Zhang, F. et al., 2014; Liu et al., 2015) and vegetation processes (e.g., stomatal conductance and LAI; Sun et al., 2014). However, the WaSSI model did not consider these effects, potentially resulting in errors in estimating ET, GPP, or water yield (Cox et al., 2000; Gedney et al., 2006; Oki and Kanae, 2006; Betts et al., 2007; Piao et al., 2007). Human activities aside, natural disturbances and their couplings may introduce uncertainties into our results. However, the potential errors are largely dependent on specific trajectories of climate change and land cover change (Qi et al., 2009; Thompson et al., 2011; Alkama et al., 2013). The complex interactions of climate, disturbance, and ecohydrological processes require a more mechanistic integrated modeling approach.

4.2 Land management implications

Numerous modeling studies around the world have shown that future climate change could increase or decrease the water availability to certain specific ecosystems and human populations under different climate scenarios (Arnell, 1999; Blanc et al., 2014; Ingjerd et al., 2014; Kundzewicz and Gerten, 2015). Our analyses have shown that, over the whole CONUS, P would increase by 45 mm (6 %) leading to a small increase in Q of 9 mm yr^{-1} (3 %). So, climate change under the SRES A2 scenario has little influence on water shortage for the entire CONUS. However, there are large regional differences in Q responses to future climate change among the 18 WRRs. The magnitude is large, from a decrease of -32 mm yr^{-1} to an increase of 113 mm yr^{-1} or from -12 to 21 %. Despite the increase in annual P, annual Q in the WRR1, 3, 8, 11, 14, 16 and 18 decreased by various degrees, due to the increased ET. Consequently, the climate scenario studied will likely increase stress on the water supply in these WRRs. In addition, it is worth noting that monthly responses of Q to future climate also vary among watersheds. Water yield in about half of the 18 WRRs (mainly located in the west CONUS) decreases and water yield in the WRR2–8 increases. The increased Q in the wet months tends to intensify the flooding risk, while decreased Q in the major dry months would be likely to aggravate the water shortage conditions. Taking California (mostly in the WRR18) as an example, the monthly Q would decrease by around 5 mm during spring through early summer (the major runoff generation season) due to coupling changes in P and ET. The decrease in flow may cause severe water shortage similar to what happened in 2014–2015 in California (Aghakouchak et al., 2014; Mao et al., 2015). Hydrological changes will bring many impacts on water-related economic sectors. For example, droughts would reduce low flows and degrade water quality (high water temperature and nutrient concentrations), thus bringing harmful influence on fisheries (Magoulick and Kobza, 2003; Dolbeth et al., 2008; Gillson et al., 2009), navigation (Theiling et al., 1996), and recreations (Thomas et al., 2013).

The modeling results suggest that GPP over the whole CONUS would increase 106 gC m^{-2} yr^{-1} (9 %) in the future. The increase by WRR ranged from 49 to 202 gC m^{-2} yr^{-1} or from 5 to 17 % among the 18 WRRs. These findings suggest that carbon stock and vegetation capacity to sequester atmospheric CO_2 for the entire CONUS and each WRR tend to be enhanced under the SRES A2 climate scenario. For the intra-annual GPP changes to climate change, most WRRs show GPP increases, particularly during late spring to summer with higher rates, which implies that the capability of ecosystem to sequestrate carbon in these months will be significantly enhanced in future. By contrast, several WRRs would decrease GPP in several months. For example, during August and September, GPP in WRR17 decreased. The ecosystem sequestration carbon capability would be weakened in these months under the SRES A2 climate scenario. For forests, variations of GPP caused by climate change will be ultimately reflected in timber production, soil carbon storage, and other ecosystem such as dissolved carbon loading in aquatic ecosystems. According to this study, under the SRES A2 climate scenario, the forest biomass and timber production is expected to increase, thus climate change may have implications for timber prices in timberland-dominated regions (Sohngen and Mendelsohn, 1998; Irland et al., 2001; Alig et al., 2004). At the same time, densification of forest lands under a warming climate may provide conditions of increased wildfire potential (Liu et al., 2013c).

5 Conclusions

We assessed the impacts of future climate change on the hydrological cycle and GPP over the CONUS by linking an ecohydrology model (i.e., WaSSI) with WRF using dynamically downscaled HadCM3 model climate data under the IPCC SRES A2 emission scenario. The current study represents a coupling of bias-corrected, dynamically downscaled climate data with an ecohydrological model to address regional ecosystem issues. The study provides a potential scenario of likely impacts of future climate change on watershed hydrology and productivity across the CONUS, including 82 773 12-digit HUC watersheds. Although only one future climate scenario (the SRES A2 emission scenario) and one GCM (HadCM3 model) was employed here, the methodology applies to other scenarios when more climate change scenarios generated from the WRF are available.

Future climate change will not likely change the spatial patterns of precipitation, temperature, ET, Q, and GPP. However, a large spatial variability in the hydrological and ecosystem productivity responses is expected among the watersheds at both 12-digit and 2-digit HUC scales. The assessment results provide a benchmark of water yield and ecosystem productivity across the whole CONUS, the 18 WRRs and even the 82 773 12-digit HUC watersheds. This type of information will be useful for prioritizing watershed restoration and developing specific measures to mitigate the negative impacts of future climate to sustain the terrestrial ecosystem on different spatial scales (i.e., 12-digit HUC and WRR).

Acknowledgements. This study was funded by the National Science Foundation EaSM program (AGS-1049200) with a grant award to North Carolina State University, to the Eastern Forest Environment Threat Assessment Center (EFETAC), to the USDA Forest Service, and to the Natural Science Foundation of Jiangsu Province, China (BK20151525 and BK20130987). The authors would like to thank Francina Dominguez (Department of Atmospheric Sciences, University of Illinois, Urbana, Illinois, USA) who provided the WRF climate data.

Edited by: A. Wei

References

Adams, H. D., Luce, C. H., Breshears, D. D., Allen, C. D., Weiler, M., Hale, V. C., Smith, A. M. S., and Huxman, T. E.: Ecohydrological consequences of drought- and infestation-triggered tree die-off: insights and hypotheses, Ecohydrology, 5, 145–159, 2012.

Aghakouchak, A., Cheng, L., Mazdiyasni, O., and Farahmand, A.: Global warming and changes in risk of concurrent climate extremes: Insights from the 2014 California drought, Geophys. Res. Lett., 41, 8847–8852, 2014.

Alig, R. J., Adams, D., Joyce, L., and Sohngen, B.: Climate change impacts and adaptation in forestry: responses by trees and markets, Choices, 19, 1–7, 2004.

Alkama, R., Marchand, L., Ribes, A., and Decharme, B.: Detection of global runoff changes: results from observations and CMIP5 experiments, Hydrol. Earth Syst. Sci., 17, 2967–2979, doi:10.5194/hess-17-2967-2013, 2013.

Allen, C. D., Macalady, A. K., Chenchouni, H., Bachelet, D., McDowell, N., Vennetier, M., Kitzberger, T., Rigling, A., Breshears, D. D., Hogg, E. H., Gonzalez, P., Fensham, R., Zhang, Z., Castro, J., Demidova, N., Lim, J. H., Allard, G., Running, S. W., Semerci, A., and Cobb, N.: A global overview of drought and heat induced tree mortality reveals emerging climate change risks for forests, Forest Ecol. Manag., 259, 660–684, 2010.

Anderegg, W. R. L., Berry, J. A., Smith, D. D., Sperry, J. S., Anderegg, L. D. L., and Field, C. B.: The roles of hydraulic and carbon stress in a widespread climate-induced forest die-off, P. Natl. Acad. Sci. USA, 109, 233–237, 2012.

Anderson, B. T., Gianotti, D., and Salvucci, G.: Detectability of historical trends in station-based precipitation characteristics over the continental United States, J. Geophys. Res., 120, 4842–4859, 2015.

Anderson-Teixeira, K. J., Delong, J. P., Fox, A. M., Brese, D. A., and Litvak, M. E.: Differential responses of production and respiration to temperature and moisture drive the carbon balance across a climatic gradient in New Mexico, Glob. Change Biol., 17, 410–424, 2011.

Aparício, S., Carvalhais, N., and Seixas, J.: Climate change impacts on the vegetation carbon cycle of the Iberian Peninsula-Intercomparison of CMIP5 results, J. Geophys. Res., 120, 641–660, 2015.

Arnell, N. W.: Climate change and global water resources, Global Environ. Chang., 9, S31–S49, 1999.

Averyt, K., Meldrum, J., Caldwell, P. V., Sun, G., McNulty, S. G., Huber-Lee, A., and Madden, N.: Sectoral vulnerabilities to changing water resources in the conterminous U.S., Environ. Res. Lett., 8, 035046, doi:10.1088/1748-9326/8/3/035046, 2013.

Ayres, M. P. and Lombardero, M. J.: Assessing the consequences of climate change for forest herbivore and pathogens, Sci. Total Environ., 262, 263–286, 2000.

Betts, R. A., Boucher, O., Collins, M., Cox, P. M., Falloon, P. D., Gedney, N., Hemming D. L., Huntingford C., Jones C. D., Sexton D. M. H., and Webb, M. J.: Projected increase in continental runoff due to plant responses to increasing carbon dioxide, Nature, 448, 1037–1041, 2007.

Blanc, E., Strzepek, K., Schlosser, A., Jacoby, H., Gueneau, A., Fant, C., Rausch S., and Reilly, J.: Modeling US water resources under climate change, Earth's Future, 2, 197–224, 2014.

Bond-Lamberty, B. E. N., Peckham, S. D., Gower, S. T., and Ewers, B. E.: Effects of fire on regional evapotranspiration in the central Canadian boreal forest, Glob. Change Biol., 15, 1242–1254, 2009.

Bony, S., Bellon, G., Klocke, D., Sherwood, S., Fermepin, S., and Denvil, S.: Robust direct effect of carbon dioxide on tropical circulation and regional precipitation, Nature Geosci., 6, 447–451, 2013.

Brekke, L., Thrasher, B. L., Maurer, E. P., and Pruitt, T.: Downscaled CMIP3 and CMIP5 Climate Projections: Release of Downscaled CMIP5 Climate Projections, Comparison with Preceding Information, and Summary of User Needs. U.S. Department of the Interior, Bureau of Reclamation, Technical Service Center, Denver, Colorado, USA, 116 pp., 2013.

Brikowski, T. H.: Applying multi-parameter runoff elasticity to assess water availability in a changing climate: an example from Texas, USA, Hydrol. Process., 29, 1746–1756, 2014.

Brookhouse, M. T., Farquhar, G. D., and Roderick, M. L.: The impact of bushfires on water yield from south-east Australia's ash forests, Water Resour. Res., 49, 4493–4505, 2013.

Brown, T. C., Foti, R., and Ramirez, J. A.: Projected freshwater withdrawals in the United States under a changing climate, Water Resour. Res., 49, 1259–1276, 2013.

Burnash, R. J. C.: The NWS river forecast system-catchment modeling, in: Computer Models of Watershed Hydrology, edited by: Singh, V. P., Water Resources Publications, Littleton, CO, USA, 311–366, 1995.

Caldwell, P. V., Sun, G., McNulty, S. G., Cohen, E. C., and Moore Myers, J. A.: Impacts of impervious cover, water withdrawals, and climate change on river flows in the conterminous US, Hydrol. Earth Syst. Sci., 16, 2839–2857, doi:10.5194/hess-16-2839-2012, 2012.

Chen, F., and Dudhia, J.: Coupling an advanced land surface–hydrology model with the Penn State–NCAR MM5 modeling system. Part I: Model implementation and sensitivity, Mon. Weather Rev., 129, 569–585, 2001.

Chien, H., Yeh, P. J. F., and Knouft, J. H.: Modeling the potential impacts of climate change on streamflow in agricultural watersheds of the Midwestern United States, J. Hydrol., 491, 73–88, 2013.

Chou, M. D. and Suarez, M. J.: An Efficient Thermal Infrared Radiation Parameterization for Use in General Circulation Models. NASA Technical Memorandum 104606, Technical Report Series on Global Modeling and Data Assimilation, Vol. 3, National Aeronautics and Space Administration, Greenbelt, MD, USA, 85 pp., 1994.

Collins, M., Tett, S. F. B., and Cooper, C.: The internal climate variability of HadCM3, a version of the Hadley Centre coupled model without flux adjustments, Clim. Dynam., 17, 61–81, doi:10.1007/s003820000094, 2001.

Cook, E. R., Seager, R., Cane, M. A., and Stahle, D. W.: North American drought: Reconstructions, causes, and consequences, Earth-Sci. Rev., 81, 93–134, 2007.

Cox, P. M., Betts, R. A., Jones, C. D., Spall, S. A., and Totterdell, I. J.: Acceleration of global warming due to carbon-cycle feedbacks in a coupled climate model, Nature, 408, 184–187, 2000.

Dai, A., Qian, T., Trenberth, K. E., and Milliman, J. D.: Changes in continental freshwater discharge from 1948 to 2004, J. Climate, 22, 2773–2792, 2009.

Daly, C., Neilson, R. P., and Phillips, D. L.: A statistical topographic model for mapping climatological precipitation over mountainous terrain, J. Appl. Meteorol., 33, 140–158, 1994.

de Kauwe, M. G., Medlyn, B. E., Zaehle, S., A. P.. Walker, M. C., Dietze, T., Hickler, Jain, A. K., Luo, Y., Parton, W. J, Prentice, I. C., Smith, B., Thornton, P. E., Wang, S., Wang, Y.-P., Warlind D., Weng, E., Crous, K. Y., Ellsworth, D. S., Hanson, P. J., Kim, H.-S., Warren, J. M., Oren, R., and Norby, R. J.: Forest water use and water use efficiency at elevated CO_2: a model-data intercomparison at two contrasting temperate forest FACE sites, Glob. Change Biol., 19, 1759–1779, 2013.

DeRose, R. J. and Long, J. N.: Drought-driven disturbance history characterizes a southern Rocky Mountain subalpine forest, Can. J. Forest Res., 42, 1649–1660, 2012.

Dolbeth, M., Martinho, F., Viegas, I., Cabral, H., Pardal, M. A.: Estuarine production of resident and nursery fish species: conditioning by drought events?, Estuar. Coast. Shelf S., 78, 51–60, 2008.

Dominguez, F., Cañon, J., and Valdes, J.: IPCC-AR4 climate simulations for the Southwestern US: the importance of future ENSO projections, Climatic Change, 99, 499–514, 2009.

Dore, S., Kolb, T. E., Montes-Helu, M., Eckert, S. E., Sullivan, B. W., Hungate, B. A., Kaye, J.P., Hart, S. C., Koch, G. W., and Finkral, A.: Carbon and water fluxes from ponderosa pine forests disturbed by wildfire and thinning, Ecol. Appl., 20, 663–683, 2010.

Foley, J. A., Defries, R., Asner, G. P., Barford, C., Bonan, G., Carpenter, S. R., Chapin, F. S., Coe, M. T., Daily, G. C., Gibbs, H. K., Helkowski, J. H., Holloway, T., Howard, E. A., Kucharik, C. J., Monfreda, C., Patz, J. A., Prentice, I. C., Ramankutty, N., and Snyder, P. K.: Global consequences of land use, Science, 309, 570–574, 2005.

Fry, J., Xian, G., Jin, S., Dewitz, J., Homer, C., Yang, L., Barnes, C., Herold, N., and Wickham, J.: Completion of the 2006 National Land Cover Database for the conterminous United States, Photogramm. Eng. Rem. S., 77, 858–864, 2011.

Gang, C., Zhou, W., Li, J., Chen, Y., Mu, S., Ren, J., Chen, J., and Groisman, P. Y.: Assessing the spatiotemporal variation in distribution, extent and NPP of terrestrial ecosystems in response to climate change from 1911 to 2000, PLoS One, 8, e80394, doi:10.1371/journal.pone.0080394, 2013.

Gao, Y., Vano, J. A., Zhu, C., and Lettenmaier, D. P.: Evaluating climate change over the Colorado River basin using regional climate models, J. Geophys. Res., 116, D13104, doi:10.1029/2010jd015278, 2011.

Gedney, N., Cox, P. M., Betts, R. A., Boucher, O., Huntingford, C., and Stott, P. A.: Detection of a direct carbon dioxide effect in continental river runoff records, Nature 439, 835–838, 2006.

Gillson, J., Scandol, J., and Suthers, L.: Estuarine gillnet fishery catch rates decline during drought in eastern Australia, Fish. Res., 99, 26–37, 2009.

Gordon, C., Cooper, C., Senior, C.A., Banks, H., Gregory, J. M., Johns, T. C., Mitchell, J. F. B., and Wood, R. A.: The simulation of SST, sea ice extents and ocean heat transports in a version of the Hadley Centre coupled model without flux adjustments, Clim. Dynam., 16, 147–168, doi:10.1007/s003820050010, 2000.

Groisman, P. Y., Knight, R. W., and Karl, T. R.: Heavy precipitation and high streamflow in the contiguous United States: Trends in the twentieth century, B. Am. Meteorol. Soc., 82, 219–246, 2001.

Groisman, P. Y., Knight, R. W., Karl, T. R., Easterling, D. R., Sun, B., and Lawrimore, J. H.: Contemporary changes of the hydrological cycle over the contiguous United States: Trends derived from in situ observations, J. Hydrometeorol., 5, 64–85, 2004.

Guardiola-Claramonte, M., Troch, P. A., Breshears, D. D., Huxman, T. E., Switanek, M. B., Durcik, M., and Cobb, N. S.: Decreased streamflow in semi-arid basins following drought-induced tree die-off: a counter-intuitive and indirect climate impact on hydrology, J. Hydrol., 406, 225–233, 2011.

Hamlet, A. F., Mote, P. W., Clark, M. P., and Lettenmaier, D. P.: Twentieth-century trends in runoff, evapotranspiration, and soil moisture in the western United States, J. Climate, 20, 1468–1486, 2007.

Hanson, P. J. and Weltzin, J. F.: Drought disturbance from climate change: response of United States forests, Sci. Total Environ., 262, 205–220, 2000.

Harding, R., Best, M., Blyth, E., Hagemann, S., Kabat, P., Tallaksen, L. M., Warnaars, T., Wiberg, D., Weedon, G. P., van Lanen, H., Ludwig, F., and Haddeland, I.: WATCH: Current knowledge of the terrestrial global water cycle, J. Hydrometeorol., 12, 1149–1156, 2011.

Hegerl, G. C., Black, E., Allan, R. P., Ingram, W. J., Polson, D., Trenberth, K. E., Chadwick, R. S., Arkin, P. A., Sarojini, B. B., Becker, A., Dai, A., Durack, P. J., Easterling, D., Fowler, H. J., Kendon, E. J., Huffman, G. J., Liu, C., Marsh, R., New, M., Osbornm T. J., Skliris, N., Stotto, P. A., Vidale, P.-L., Wijffels, S. E., Wilcox, L. J., and Zhang, X.: Challenges in quantifying changes in the global water cycle, B. Am. Meteorol. Soc., 96, 1097–1115, doi:10.1175/bams-d-13-00212.1, 2014.

Hong, S.-Y., Dudhia, J., and Chen, S.-H.: A revised approach to ice microphysical processes for the bulk parameterization of clouds and precipitation, Mon. Weather Rev., 132, 103–120, 2004.

Hu, W. W., Wang, G. X., Deng, W., and Li, S. N.: The influence of dams on ecohydrological conditions in the Huaihe River basin, China, Ecol. Eng., 33, 233–241, 2008.

Huang, J., Zhang, J., Zhang, Z., Sun, S., and Yao, J.: Simulation of extreme precipitation indices in the Yangtze River basin by using statistical downscaling method (SDSM), Theor. Appl. Climatol., 108, 325–343, 2011.

Huntington, T. G. and Billmire, M.: Trends in precipitation, runoff, and evapotranspiration for rivers draining to the Gulf of Maine in the United States, J. Hydrometeorol., 15, 726–743, 2014.

Ingjerd, H., Jens, H., Hester, B., Stephanie, E., Martina, F., Naota, H., Markus, K., Fulco, L., Yoshimitsu, M., Jacob, S., Tobias, S., Zachary, D. T., Yoshihide, W., and Dominik W.: Global water resources affected by human interventions and climate change, P. Natl. Acad. Sci. USA, 111, 3251–3256, 2014,

IPCC: Summary for policymakers, in: Climate Change 2014, Impacts, Adaptation, and Vulnerability, Part A: Global and Sectoral Aspects, Contribution of Working Group II to the Fifth Assessment Report of the Intergovernmental Panel on Climate Change, edited by: Field, C. B., Barros, V. R., Dokken, D. J., Mach, K. J., Mastrandrea, M. D., Bilir, T. E., Chatterjee, M., Ebi, K. L., Estrada, Y. O., Genova, R. C., Girma, B., Kissel, E. S., Levy, A. N., MacCracken, S., Mastrandrea, P. R., and White, L. L., Cambridge University Press, Cambridge, UK and New York, NY, USA, 1–32, 2014.

Irland, L., Adams, D., Alig, R., Betz, C., Chen, C., Hutchins, M., McCarl, B., Skog, K., and Sohngen, B.: Assessing socioeconomic impacts of climate change on US forests, wood-product markets, and forest recreation, Bioscience, 51, 753–764, 2001.

Jactel, H., Petit, J., Desprez-Loustau, M.-L., Delzon, S., Piou, D., Battisti, A., and Koricheva, J.: Drought effects on damage by forest insects and pathogens: a meta-analysis, Glob. Change Biol., 18, 267–276, 2012.

Janjic, Z. I.: Nonsingular implementation of the Mellor-Yamada level 2.5 scheme in the NCEP meso-model, NCEP Office Note no. 437, 61 pp., National Centers for Environmental Prediction, Camp Springs, Maryland, USA, 2002.

Joseph, R. and Nigam, S.: ENSO evolution and teleconnections in IPCC's twentieth-century climate simulations: Realistic representation, J. Climate, 19, 4360–4377, 2006.

Kain, J. S. and Fritsch, J. M.: Convective parameterization for mesoscale models: the Kain–Fritsch scheme, in: The Representation of Cumulus Convection in Numerical Models, edited by: Emanuel, K. A. and Raymond, D. J., American Meteorological Society, Boston, USA, 165–170, 1993.

Kundzewicz, Z. W. and Gerten, D. G.: Challenges Related to the Assessment of Climate Change Impacts on Freshwater Resources, J. Hydrol. Eng., 20, A4014011, doi:10.1061/(asce)he.1943-5584.0001012, 2015.

Labat, D., Goddéris, Y., Probst, J. L., and Guyot, J. L.: Evidence for global runoff increase related to climate warming, Adv. Water Res., 27, 631–642, 2004.

Lee, C., Schlemme, C., Murray, J., and Unsworth, R.: The cost of climate change: Ecosystem services and wildland fires, Ecol. Econ., 116, 261–269, 2015.

Lenihan, J. M., Bachelet, D., Neilson, R. P., and Drapek, R.: Simulated response of conterminous United States ecosystems to climate change at different levels of fire suppression, CO_2 emission rate, and growth response to CO_2, Global Planet. Change, 64, 16–25, 2008.

Lettenmaier, D. P., Wood, E. F., and Wallis, J. R.: Hydro-climatological trends in the continental United States, 1948–88, J. Climate, 7, 586–607, 1994.

Leung, L. R. and Qian, Y.: The sensitivity of precipitation and snowpack simulations to model resolution via nesting in regions of complex terrain, J. Hydrometeorol., 4, 1025–1043, 2003.

Li, H., Huang, M., Wigmosta, M. S., Ke, Y., Coleman, A. M., Leung, L. R., Wang, A., and Ricciuto, D. M.: Evaluating runoff simulations from the Community Land Model 4.0 using observations from flux towers and a mountainous watershed, J. Geophys. Res., 116, D24120, doi:10.1029/2011jd016276, 2011.

Lins, H. F. and Slack, J. R.: Streamflow trends in the United States, Geophys. Res. Lett., 26, 227–230, 1999.

Liu, J., Wang, B., Cane, M. A., Yim, S.-Y., and Lee, J.-Y.: Divergent global precipitation changes induced by natural versus anthropogenic forcing, Nature, 493, 656–659, 2013a.

Liu, N., Sun, P., Liu, S., and Sun, G.: Coupling simulation of water–carbon processes for catchment: calibration and validation of the WaSSI-C model, Chinese Journal of Plant Ecology, 37, 492–502, 2013b (in Chinese with English abstract).

Liu, Y., Prestemon, J., Goodrick, S., Holmes, T., Stanturf, J., Vose, J., and Sun, G.: Chapter 5: Future Wildfire Trends, Impacts, and Mitigation Options in the Southern United States, in: Climate Change Adaptation and Mitigation Management Options, edited by: Vose, J. and Klepzig, K. D., CRS Press, Boca Raton, FL, USA, 476 pp., 2013c.

Liu, Y., Xiao, J., Ju, W., Zhou, Y., Wang, S., and Wu, X.: Water use efficiency of China's terrestrial ecosystems and responses to drought, Scientific Report, 5, 13799, doi:10.1038/srep13799, 2015

Lockaby, G., Nagy, C., Vose, J. M., Ford, C. R., Sun, G., McNulty, S. G., Caldwell, P. V., Cohen, E., and Moore Myers, J. A.: Water and forests, in: The Southern Forest Futures Project: Technical Report, edited by: Wear, D. N. and Greis, J. G., General Technical Report, USDA Forest Service, Southern Research Station, Asheville, NC, USA, 2011.

Magoulick, D. D. and Kobza, R. M.: The role of refugia for fishes during drought: a review and synthesis, Freshwater Biol., 48, 1186–1198, 2003.

Mao, Y., Nijssen, B., and Lettenmaier, D. P.: Is climate change implicated in the 2013-2014 California drought? A hydrologic perspective, Geophys. Res. Lett., 42, 2805–2813, 2015.

Marengo, J. A., Tomasella, J., Cardoso, M. F., and Oyama, M. D.: Hydro-climatic and ecological behaviour of the drought of Amazonia in 2005, Philos. T. R. Soc. B, 363, 1773–1778, 2008.

Marion, D. A., Sun, G., Caldwell, P. V., Miniat, C. F., Ouyang, Y., Amatya, D. M., Clinton, B. D., Conrads, P. A., Gull Laird, S., Dai, Z., Clingenpeel, J. A., Liu, Y., Roehl Jr., E. A., Myers, J. A. M., and Trettin, C.: Managing forest water quantity and quality under climate change, in: Climate Change Adaptation and Mitigation Management Options: A Guide for Natural Resource Managers in Southern Forest Ecosystems, edited by: Vose, J. M. and Klepzig, K. D., CRC Press, Boca Raton, FL, USA, 249–305, 2014.

McCabe, G. J.: A step increase in streamflow in the conterminous United States, Geophys. Res. Lett., 29, 2185, doi:10.1029/2002gl015999, 2002.

McCabe, G. J. and Markstrom, S. L.: A monthly water-balance model driven by a graphical user interface, US Geological Survey Open-File report 2007–1088, 6 pp., US Geological Survey, Reston, Virginia, USA, 2007.

McCabe, G. J. and Wolock, D. M.: Future snowpack conditions in the western United States derived from general circulation model climate simulations, J. Am. Water Resour. As., 35, 1473–1484, 1999.

McCabe, G. J. and Wolock, D. M.: Independent effects of temperature and precipitation on modeled runoff in the conterminous United States, Water Resour. Res., 47, W11522, doi:10.1029/2011wr010630, 2011.

McCabe, G. J. and Wolock, D. M.: Spatial and temporal patterns in conterminous United States streamflow characteristics, Geophys. Res. Lett., 41, 6889–6897, 2014.

McNulty, S. G., Cohen, E. C., Sun, G., and Caldwell, P.: Chapter 1.7: Hydrologic Modelling for Water Resource Assessment in a Developing Country: the Rwanda Case Study. in: Watershed Modeling, edited by: Lafforgue M., in press, 2016.

Meehl, G. A., Arblaster, J. M., and Tebaldi, C.: Understanding future patterns of increased precipitation intensity in climate model simulations, Geophys. Res. Lett., 32, L18719, doi:10.1029/2005gl023680, 2005.

Miller-Rushing, A. J., Primack, R. B., Templer, P. H., Rathbone, S., and Mukunda, S.: Long-term relationships among atmospheric CO_2, stomata, and intrinsic water use efficiency in individual trees, Am. J. Bot., 96, 1779–1786, 2009.

Milly, P. C., Dunne, K. A., and Vecchia, A. V.: Global pattern of trends in streamflow and water availability in a changing climate, Nature, 438, 347–350, 2005.

Mlawer, E. J., Taubman, S. J., Brown, P. D., Iacono, M. J., and Clough, S. A.: Radiative transfer for inhomogeneous atmospheres: RRTM, a validated correlated-k model for the longwave, J. Geophys. Res., 102, 16663, doi:10.1029/97jd00237, 1997.

Myneni, R. B., Keeling, C. D., Tucker, C. J., Asrar, G., and Nemani, R. R.: Increased plant growth in the northern high latitudes from 1981 to 1991, Nature, 386, 698–702, 1997.

Nakicenvoic, N. and Swart, R.: Special report on emissions scenarios. A special report of Working Group III of the Intergovernmental Panel on Climate Change. Cambridge University Press, Cambridge, UK, 599 pp., 2000.

National Research Council: Estimating Water Use in the United States – A new paradigm for the national water use information program, The National Academies Press, Washington, DC, USA, 176 pp., 2002.

Nemani, R. R., Keeling, C. D., Hashimoto, H., Jolly, W. M., Piper, S. C., Tucker, C. J., Myneni, R. B., and Running, S. W.: Climate-driven increases in global terrestrial net primary production from 1982 to 1999, Science, 300, 1560–1563, 2003.

Neary, D. G., Ryan, K. C., and DeBano, L. F.: Wildland fire in ecosystems: Effects of fire on soil and water. Gen. Tech. Rep. RMRS-GTR-32-vol. 4. US Department of Agriculture, Forest Service, Rocky Mountain Research Station, Ogden, UT, USA, 250–251, 2005.

Noake, K., Polson, D., Hegerl, G., and Zhang, X.: Changes in seasonal land precipitation during the latter twentieth-century, Geophys. Res. Lett., 39, L03706, doi:10.1029/2011gl050405, 2012.

Nolan, R. H., Lane, P. N. J., Benyon, R. G., Bradstock, R. A., and Mitchell, P. J.: Changes in evapotranspiration following wildfire in resprouting eucalypt forests, Ecohydrology, 7, 1363–1377, 2014.

Nolan, R. H., Lane, P. N. J., Benyon, R. G., Bradstock, R. A., and Mitchell, P. J.: Trends in evapotranspiration and streamflow following wildfire in resprouting eucalypt forests, J. Hydrol., 524, 614–624, 2015.

Oki, T. and Kanae, S.: Global hydrological cycles and world water resources, Science, 313, 1068–1072, 2006.

Overgaard, J., Rosbjerg, D., and Butts, M. B.: Land-surface modelling in hydrological perspective – a review, Biogeosciences, 3, 229–241, doi:10.5194/bg-3-229-2006, 2006.

Overpeck, J. T., Rind, D., and Goldberg, R.: Climate-induced changes in forest disturbance and vegetation, Nature, 343, 51–53, 1990.

Peng, S., Piao, S., Ciais, P., Myneni, R. B., Chen, A., Chevallier, F., Dolman, A. J., Janssens, I. A., Peñuelas, J., Zhang, G., Vicca, S., Wan, S., Wang, S., and Zeng, H.: Asymmetric effects of daytime and night-time warming on Northern Hemisphere vegetation, Nature, 501, 88–92, 2013.

Pennington, D. D. and Collins, S. L.: Response of an arid-land ecosystem to interannual climate variability and prolonged drought, Landscape Ecol., 22, 897–910, 2007.

Piao, S., Friedlingstein, P., Ciais, P., de Noblet-Ducoudre, N., Labat, D., and Zaehle, S.: Changes in climate and land use have a larger direct impact than rising CO_2 on global river runoff trends, P. Natl. Acad. Sci. USA, 104, 15242–15247, 2007.

Piao, S., Yin, G., Tan, J., Cheng, L., Huang, M., Li, Y., Liu, R., Mao, J., Myneni, R. B., Peng, S., Poulter, B., Shi, X., Xiao, Z., Zeng, N., Zeng, Z., and Wang, Y.: Detection and attribution of vegetation greening trend in China over the last 30 years, Glob. Change Biol., 21, 1601–1609, 2015.

Piontek, F., Muller, C., Pugh, T. A., Clark, D. B., Deryng, D., Elliott, J., de Jesus Colón Gonzálezf, F., Flörke, M., Folberth, C., Franssen, W., Frieler, K., Friend, A. D., Gosling, S. N., Hemming, D., Khabarov, N., Kim, H., Lomas, M. R., Masaki, Y., Mengel, M., Morse, A., Neuman, K., Nishina, K., Ostberg, S., Pavlick, R., Ruane, A. C., Schewe, J., Schmid, E., Stacke, T.,

Tang, Q., Tessler, Z. D., Tompkins, A. M., Warszawski, L., Wisser, D., and Schellnhuber, H. J.: Multisectoral climate impact hotspots in a warming world, P. Natl. Acad. Sci. USA, 111, 3233–3238, 2014.

Pope, V. D., Gallani, M. L., Rowntree, P. R., and Stratton, R. A.: The impact of new physical parameterizations in the Hadley Centre climate model-HadAM3, Clim. Dynam., 16, 123–146, doi:10.1007/s003820050009, 2000.

PRISM Climate Group: Descriptions of PRISM spatial climate datasets for the conterminous United States, available at: http://www.prism.oregonstate.edu/documents/PRISM_datasets_aug2013.pdf, last access: 1 May 2013.

Qi, S., Sun, G., Wang, Y., McNulty, S. G., and Moore, M. J. A.: Streamflow response to climate and landuse changes in a coastal watershed in North Carolina, Transactions of the American Society of Agricultural Engineer, 52, 739–749, 2009.

Reichler, T. and Kim, J.: How Well Do Coupled Models Simulate Today's Climate?, B. Am. Meteorol. Soc., 89, 303–311, 2008.

Reichstein, M., Bahn, M., Ciais, P., Frank, D., Mahecha, M. D., Seneviratne, S., Zscheischler, J., Beer, C., Buchmann, N., Frank, D. C., Papale, D., Rammig, A., Smith, P., Thonicke, K., van der Velde, M., Vicca, S., Walz, A., and Wattenbach, M.: Climate extremes and the carbon cycle, Nature, 500, 287–295, 2013.

Sagarika, S., Kalra, A., and Ahmad, S.: Evaluating the effect of persistence on long-term trends and analyzing step changes in streamflows of the continental United States, J. Hydrol., 517, 36–53, 2014.

Sankarasubramanian, A.: Hydroclimatology of the continental United States, Geophys. Res. Lett., 30, 1363, doi:10.1029/2002gl015937, 2003.

Sankarasubramanian, A., Vogel, R. M., and Limbrunner, J. F.: Climate elasticity of streamflow in the United States, Water Resour. Res., 37, 1771–1781, 2001.

Scheff, J. and Frierson, D. M. W.: Robust future precipitation declines in CMIP5 largely reflect the poleward expansion of model subtropical dry zones, Geophys. Res. Lett., 39, L18704, doi:10.1029/2012gl052910, 2012.

Schewe, J., Heinke, J., Gerten, D., Haddeland, I., Arnell, N. W., Clark, D. B., Dankers, R., Eisner, S., Fekete, B. M., Colón-González, F. J., Gosling, S. N., Kim, H., Liu, X., Masaki, Y., Portman, F. T., Satoh, Y., Stacke, T., Tang, Q., Wada, Y., Wisser, D., Albrecht, T., Frieler, K., Piontek, F., Warszawski, L., and Kabat, P.: Multimodel assessment of water scarcity under climate change, P. Natl. Acad. Sci. USA, 111, 3245–3250, 2014.

Schilling, K. E., Jha, M. K., Zhang, Y.-K., Gassman, P. W., and Wolter, C. F.: Impact of land use and land cover change on the water balance of a large agricultural watershed: Historical effects and future directions, Water Resour. Res., 44, W00A09, doi:10.1029/2007wr006644, 2008.

Schoeneweiss, D. F.: The role of environmental stress in diseases of woody plants, Plant Dis., 65, 308–314, 1981.

Scholze, M., Knorr, W., Arnell, N. W., and Prentice, I. C.: A climate-change risk analysis for world ecosystems, P. Natl. Acad. Sci., 103, 13116–13120, 2006.

Sedláček, J. and Knutti, R.: Half of the world's population experience robust changes in the water cycle for a 2° C warmer world, Environ. Res. Lett., 9, 044008, doi:10.1088/1748-9326/9/4/044008, 2014.

Seneviratne, S. I., Luthi, D., Litschi, M., and Schar, C.: Land-atmosphere coupling and climate change in Europe, Nature, 443, 205–209, doi:10.1038/nature05095, 2006.

Sitch, S., Huntingford, C., Gedney, N., Levy, P. E., Lomas, M., Piao, S. L., Betts, R., Ciais, P., Cox, P., Friedlingstein, P., Jones, C. D., Prentice, I. C., and Woodward, F. I.: Evaluation of the terrestrial carbon cycle, future plant geography and climate-carbon cycle feedbacks using five Dynamic Global Vegetation Models (DGVMs), Glob. Change Biol., 14, 2015–2039, 2008.

Skamarock, W. C., Klemp, J. B., Dudhia, J., Gill, G. O., Barker, D. M., Wang, W., and Powers, J. G.: A description of the advanced research WRF version 2, TN-468+STR, NCAR Technical Note, Mesoscale and Microscale Meteorology Division, National Center for Atmospheric Research, Boulder, Colorado, USA, 100 pp., 2005.

Sohngen, B. and Mendelsohn, R.: Valuing the impact of large-scale ecological change in a market: the effect of climate change on US timber, Am. Econ. Rev., 88, 686–710, 1998.

Somerville, C. and Briscoe, J.: Genetic engineering and water, Science, 292, p. 2217, 2001.

Sun, G., Alstad, K., Chen, J., Chen, S., Ford, C. R., Lin, G., Liu, C., Lu, N., McNulty, S. G., Miao, H., Noormets, A., Vose, J. M., Wilske, B., Zeppel, M., Zhang, Y., and Zhang, Z.: A general predictive model for estimating monthly ecosystem evapotranspiration, Ecohydrology, 4, 245–255, 2011a.

Sun, G., Caldwell, P., Noormets, A., Cohen, E., McNulty, S., Treasure, E., Domec, J. C., Mu, Q., Xiao, J., John, R., and Chen, J.: Upscaling key ecosystem functions across the conterminous United States by a water-centric ecosystem model, J. Geophys. Res., 116, G00J05, doi:10.1029/2010JG001573, 2011b.

Sun, G., McNulty, S. G., Myers, J. A. M., and Cohen, E.: Impacts of multiple stresses on water demand and supply across the southeastern United States, J. Am. Water Resour. As., 44, 1441–1457, 2008.

Sun, S., Chen, H., Ju, W., Yu, M., Hua, W., and Yin, Y.: On the attribution of the changing hydrological cycle in Poyang Lake Basin, China, J. Hydrol., 514, 214–225, 2014.

Sun, S., Sun, G., Caldwell, P., McNulty, S., Cohen, E., Xiao, J., and Zhang, Y.: Drought impacts on ecosystem functions of the US National Forests and Grasslands: Part II assessment results and management implications, Forest Ecol. Manag., 353, 269-279, 2015a.

Sun, S., Sun, G., Caldwell, P., McNulty, S. G., Cohen, E., Xiao, J., and Zhang, Y.: Drought impacts on ecosystem functions of the US National Forests and Grasslands: Part I evaluation of a water and carbon balance model, Forest Ecol. Manag., 353, 260–268, 2015b.

Syed, T. H., Famiglietti, J. S., Chambers, D. P., Willis, J. K., and Hilburn, K.: Satellite-based global-ocean mass balance estimates of interannual variability and emerging trends in continental freshwater discharge, P. Natl. Acad. Sci. USA, 107, 17916–17921, 2010.

Tavernia, B. G., Nelson, M. D., Caldwell, P. V., and Sun, G.: Water stress projections for the northeastern and midwestern United States in 2060: Anthropogenic and ecological consequences, J. Am. Water Resour. As., 49, 938–952, 2013.

Tebaldi, C., Hayhoe, K., Arblaster, J. M., and Meehl, G. A.: Going to the extremes: An intercomparison of model-simulated historical and future changes in extreme events, Climatic Change, 79, 185–211, 2006.

Theiling, C. H., Maher, R. J., and Sparks, R. E.: Effects of variable annual hydrology on a river regulated for navigation: Pool 26, Upper Mississippi River System, J. Freshwater Ecol., 11, 101–114, 1996.

Thomas, D. S. K., Wilhelmi, O. V., Finnessey, T. N., and Deheza, V.: A comprehensive framework for tourism and recreation drought vulnerability reduction, Environ. Res. Lett., 8, 575–591, 2013.

Thompson, J. R., Foster, D. R., Scheller, R., and Kittredge, D.: The influence of land use and climate change on forest biomass and composition in Massachusetts, USA, Ecol. Appl., 21, 2425–2444, 2011.

Trenberth, K. E.: Changes in precipitation with climate change, Clim. Res., 47, 123–138, 2011.

van Oldenborgh, G. J., Philip, S. Y., and Collins, M: El Niño in a changing climate: a multi-model study, Ocean Sci., 1, 81–95, doi:10.5194/os-1-81-2005, 2005.

Wang, C., Cao, R., Chen, J., Rao, Y., and Tang, Y.: Temperature sensitivity of spring vegetation phenology correlates to within-spring warming speed over the Northern Hemisphere, Ecol. Indic., 50, 62–68, 2015.

Wang, D. and Hejazi, M.: Quantifying the relative contribution of the climate and direct human impacts on mean annual streamflow in the contiguous United States, Water Resour. Res., 47, W00J12, doi:10.1029/2010wr010283, 2011.

Wang, F., Notaro, M., Liu, Z., and Chen, G.: Observed local and remote influences of vegetation on the atmosphere across north America using a model-validated statistical technique that first excludes oceanic forcings, J. Climate, 27, 362–382, 2014.

Wi, S., Dominguez, F., Durcik, M., Valdes, J., Diaz, H. F., and Castro, C. L.: Climate change projection of snowfall in the Colorado River Basin using dynamical downscaling, Water Resour. Res., 48, W05504, doi:10.1029/2011wr010674, 2012.

Williams, I. N., Torn, M. S., Riley, W. J., and Wehner, M. F.: Impacts of climate extremes on gross primary production under global warming, Environ. Res. Lett., 9, 094011, doi:10.1088/1748-9326/9/9/094011, 2014.

Wood, A. W.: Long-range experimental hydrologic forecasting for the eastern United States, J. Geophys. Res., 107, 4429, doi:10.1029/2001jd000659, 2002.

Wood, A. W., Leung, L. R., Sridhar, V., and Lettenmaier, D. P.: Hydrologic implications of dynamical and statistical approaches to downscaling climate model outputs, Climatic Change, 62, 189–216, 2004.

Xu, X., Liu, W., Rafique, R., and Wang, K.: Revisiting continental US hydrologic change in the latter half of the 20th century, Water Resour. Manag., 27, 4337–4348, 2013.

Xue, Y., Janjic, Z., Dudhia, J., Vasic, R., and De Sales, F.: A review on regional dynamical downscaling in intraseasonal to seasonal simulation/prediction and major factors that affect downscaling ability, Atmos. Res., 147–148, 68–85, 2014.

Yao, S.: Effects of fire disturbance on forest hydrology, J. Forest. Res.-JPN, 14, 331–334, doi:10.1007/bf02857863, 2003.

Yu, M., Wang, G., Parr, D., and Ahmed, K. F.: Future changes of the terrestrial ecosystem based on a dynamic vegetation model driven with RCP8.5 climate projections from 19 GCMs, Climatic Change, 127, 257–271, 2014.

Zhang, F., Ju, W., Shen, S., Wang, S., Yu, G., and Han S.: How recent climate change influences water use efficiency in East Asia, Theor. Appl. Climatol., 116, 359–370, 2014.

Zhang, Y., Moran, M. S., Nearing, M. A., Ponce Campos, G. E., Huete, A. R., Buda, A. R., Bosch, D. D., Gunter, S. A., Kitchen, S. G., McNab, W. H., Morgan, J. A., McClaran, M. P., Montoya, D. S., Peters, D. P. C., and Starks, P. J.: Extreme precipitation patterns and reductions of terrestrial ecosystem production across biomes, J. Geophys. Res., 118, 148–157, doi:10.1029/2012JG002136, 2013.

Zhang, Y., Song, C., Zhang, K., Cheng, X., Band, L. E., and Zhang, Q.: Effects of land use/land cover and climate changes on terrestrial net primary productivity in the Yangtze River Basin, China, from 2001 to 2010, J. Geophys. Res., 119, 1092–1109, 2014.

11

Monitoring and modelling of soil–plant interactions: the joint use of ERT, sap flow and eddy covariance data to characterize the volume of an orange tree root zone

G. Cassiani[1], **J. Boaga**[1], **D. Vanella**[2], **M. T. Perri**[1], **and S. Consoli**[2]

[1]University of Padua, Department of Geosciences, Padua, Italy
[2]University of Catania, Department of Agriculture, Food and Environment, Catania, Italy

Correspondence to: G. Cassiani (giorgio.cassiani@unipd.it)

Abstract. Mass and energy exchanges between soil, plants and atmosphere control a number of key environmental processes involving hydrology, biota and climate. The understanding of these exchanges also play a critical role for practical purposes e.g. in precision agriculture. In this paper we present a methodology based on coupling innovative data collection and models in order to obtain quantitative estimates of the key parameters of such complex flow system. In particular we propose the use of hydro-geophysical monitoring via "time-lapse" electrical resistivity tomography (ERT) in conjunction with measurements of plant transpiration via sap flow and evapotranspiration (ET) from eddy covariance (EC). This abundance of data is fed to spatially distributed soil models in order to characterize the distribution of active roots. We conducted experiments in an orange orchard in eastern Sicily (Italy), characterized by the typical Mediterranean semi-arid climate. The subsoil dynamics, particularly influenced by irrigation and root uptake, were characterized mainly by the ERT set-up, consisting of 48 buried electrodes on 4 instrumented micro-boreholes (about 1.2 m deep) placed at the corners of a square (with about 1.3 m long sides) surrounding the orange tree, plus 24 mini-electrodes on the surface spaced 0.1 m on a square grid. During the monitoring, we collected repeated ERT and time domain reflectometry (TDR) soil moisture measurements, soil water sampling, sap flow measurements from the orange tree and EC data. We conducted a laboratory calibration of the soil electrical properties as a function of moisture content and porewater electrical conductivity. Irrigation, precipitation, sap flow and ET data are available allowing for knowledge of the system's long-term forcing conditions on the system. This information was used to calibrate a 1-D Richards' equation model representing the dynamics of the volume monitored via 3-D ERT. Information on the soil hydraulic properties was collected from laboratory and field experiments. The successful results of the calibrated modelling exercise allow for the quantification of the soil volume interested by root water uptake (RWU). This volume is much smaller (with a surface area less than 2 m^2, and about 40 cm thick) than expected and assumed in the design of classical drip irrigation schemes that prove to be losing at least half of the irrigated water which is not taken up by the plants.

1 Introduction

The system made of soil, vegetation and the adjacent atmosphere is characterized by complex patterns, structures and processes that act on a wide range of timescale and space scales. While the exchange of energy and water is continuous between compartments, the pertinent fluxes are strongly heterogeneous and variable in space and time and this makes their quantification particularly challenging. Plants are known to impact the terrestrial water cycle and underground water dynamics through evapotranspiration (ET) and root water uptake (RWU). The mechanisms of water flow in the root zone are controlled by soil physics, plant physiology and meteorological factors (S. R. Green et al., 2003). The translation of plant water use strategies into physically based models of RWU is a crucial issue in eco-hydrology and has fundamental consequence in the understanding and modelling of atmospheric as well as soil processes. Still, no consensus exists on the modelling of this process (Feddes et al., 2001; Raats, 2007). From a conceptual point of view, two main approaches exist today, which differ in the way of predicting the volumetric rate of RWU.

A first approach expresses water transport in plants as a chain process based on a resistance law. Coupled with a three-dimensional soil water flow model, this approach leads to fairly accurate RWU models at the plant scale (Doussan et al., 2006; Schneider et al., 2010), also under water stress conditions. The limitations of these models are the cost of characterizing parameters, such as root system architecture and conductance to water flow, and their computational demand. A second approach, mostly used in soil–vegetation–atmosphere transfer models, relies on "macroscopic parameters" and predicts RWU as a product of the potential transpiration rate, a spatially distributed root parameter (e.g. relative root length density), and a stress function, depending on soil water potential and a compensatory RWU function (Jarvis, 1989). The major drawback of this approach is the necessity to calibrate the macroscopic parameters, which introduces substantial uncertainties (Musters and Bouten, 2000). Note that the two approaches have indeed some formal links with each other (Couvreur et al., 2012; Javaux et al., 2008).

The complexity of RWU modelling is highly related to the uneven root distribution in the vertical and radial directions (Gong et al., 2006). This variability is partly induced by heterogeneities in the soil and localized soil compaction caused by both cultivation and irrigation patterns (Jones and Tardieu, 1998) that in turn cause heterogeneous water and nutrient distribution. Consequently, there is a clear need for the development of novel RWU modelling approaches (Couvreur et al., 2012; Feddes et al., 2001; Raats, 2007; Jarvis, 2011), as well as for accurate measurements techniques of soil water content and RWU dynamics.

In particular, soil moisture measurements are of paramount importance to calibrate RWU models. Traditionally, and especially beneath irrigated crops, soil moisture has been determined using methods such as neutron probes, time domain reflectometry (TDR) or capacitance systems. As these traditional techniques are point measurements, they do not provide sufficient information for reliable mass balance assessments; therefore, our understanding of RWU as a spatially distributed system remains fundamentally limited. In this respect the understanding of soil as a spatially heterogeneous system shares fundamental limitations with most of earth sciences. Therefore, much can be learnt looking at similar research fields.

Geophysical methods have long been established for the imaging of the soil subsurface at a variety of scales, from large-scale mining exploration (e.g. Parasnis, 1973) to the very small scale of soil mapping (e.g. Allred et al., 2008). The past 20 years, in particular, have seen the fast development of techniques that are useful in identifying structure and dynamics of the near surface, with particular reference to hydrological applications. This realm of research goes under the general name of hydro-geophysics (Binley et al., 2011; Rubin and Hubbard, 2005; Vereecken et al., 2006) and covers a wide range of applications from flow and transport in aquifers (e.g. Kemna et al., 2002; Perri et al., 2012) to the vadose zone (e.g. Daily et al., 1992), from catchment (e.g. Weill et al., 2013) and hillslope characterization (Cassiani et al., 2009a) to agriculture and eco-hydrological processes (Boaga et al., 2014; Ursino et al., 2014).

Possibly the most interesting results have been obtained when hydro-geophysical data have been coupled with distributed hydrological model predictions. The degree of integration of data and model range from trial and error calibration (e.g. Binley et al., 2002) to full data assimilation (e.g. Hinnell et al., 2010), but in all cases the availability of spatially extensive (and time intensive) data greatly improve the models' capability to identify within narrow ranges the relevant governing parameters, which in turn are of practical interest for hydrological predictions.

Relatively few hydro-geophysical applications, though, have been focussed on plant root system characterization (e.g. al Hagrey, 2007; al Hagrey and Petersen, 2011; Javaux et al., 2008; Jayawickreme et al., 2008; Werban et al., 2008), often limiting the analysis to a tentative identification of the main root location and extent. Electrical soil properties are a clear indication of soil moisture content distribution, and electrical and electromagnetic methods have been used to identify the effect of root activity (e.g. Cassiani et al., 2012; Shanahan et al., 2015). In particular, ERT has been used to characterize RWU and root systems (Garré et al., 2011; Michot et al., 2001, 2003; Srayeddin and Doussan, 2009). Amato et al. (2009, 2010) tested the ability of 3-D ERT for quantifying root biomass on herbaceous plants. Beff et al. (2013) used 3-D ERT for monitoring soil water content in a maize field during late growing seasons. Boaga et al. (2013) and Cassiani et al. (2015) demonstrated the reliability of the method in apple orchards.

In this paper we aim at applying hydro-geophysical techniques, with a combination of measurements and modelling, to a tree root system. This approach has, to the best of our knowledge, not been presented and analysed yet. In particular, we present the application of the time-lapse non-invasive 3-D electrical resistivity tomography (ERT) to monitor soil–plant interactions in the root zone of an orange tree located in the Mediterranean semi-arid Sicilian (southern Italy) context. The subsoil dynamics, particularly influenced by irrigation and RWU, have been characterized by the 3-D ERT measurements coupled with plant transpiration through sap flow measurements. The information contained in the ERT measurements in terms of vadose zone water dynamics was exploited by comparing the field results against a 1-D vadose zone model.

The specific goals of this paper are

1. to study the feasibility of a small-scale monitoring of root zone processes using time-lapse 3-D ERT;

2. to assess the value of the data above for a quantitative description of hydrological processes at the tens of centimetre scale;

3. to interpret these data with the aid of a physical hydrological model, in order to also derive information on the root zone physical structure and its dynamics.

2 Site description

The experiment was conducted in a 20 hectare orange orchard, planted with about 20 year-old trees (*Citrus sinensis*; *cv Tarocco Ippolito*) (Fig. 1). The field is located in Lentini (eastern Sicily; lat. 37°16′ N, long. 14°53′ E) in a Mediterranean semi-arid environment, characterized by an annual average precipitation of around 550 mm, very dry summers and average air temperature of 7 °C in winter and 28 °C in summer. The site presents conditions of crop homogeneity, flat slope, dominant wind speed direction for footprint analysis and quite a large fetch which are ideal for micro-meteorological measurements. The planting layout is 4.0 m × 5.5 m and the trees are drip irrigated with 4 in-line drippers per plant, spaced about 1 m apart, with $16\,\mathrm{L\,h^{-1}}$ of total discharge ($4\,\mathrm{L\,h^{-1}}$ per dripper); the crop is well watered by irrigation supplied every day from May to October, with an irrigation timing of $5\,\mathrm{h\,d^{-1}}$. The study area has a mean leaf area index (LAI) of about $4\,\mathrm{m^2\,m^{-2}}$, measured by a LAI-2000 digital analyser (LI-COR, Lincoln, Nebraska, USA). The LAI values are spatially averaged and refer to the ERT measurement period (October 2013). In the specific case of a mature orange orchard, LAI values result fairly constant in time in the region of interest.

The mean PAR (photosynthetic active radiation) light interception was 80 % within rows and 50 % between rows; the canopy height (h_c) is 3.7 m.

The soil characterization was performed via textural and hydraulic laboratory analyses, according to the USDA standards. The area, covered by mature orange orchards, was divided into regular grids, each having a $18 \times 32\,\mathrm{m^2}$ area, where undisturbed soil cores (0.05 m in height and 0.05 m in diameter) were collected at the 0–0.05 and 0.05–0.10 m depths for a total of 32 sampling points and 64 soil samples. The undisturbed soil cores were used to determine the soil bulk density, ρ_b ($\mathrm{Mg\,m^{-3}}$) and the initial water content, θ_i ($\mathrm{m^3\,m^{-3}}$), i.e. the θ value at the time of the field campaign. A total of 32 disturbed soil samples were also collected at the 0–0.05 m depth to determine its soil textural characteristics, using conventional methods following H_2O_2 pre-treatment to eliminate organic matter and clay deflocculation using sodium metaphosphate and mechanical agitation

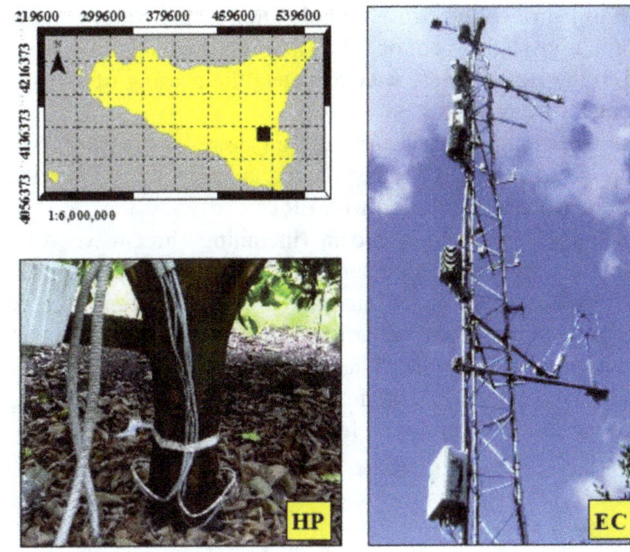

Figure 1. Bulgherano experimental site: the eddy covariance (EC) tower and a heat-pulse (HP) sap flow installation on an orange tree.

(Gee and Bauder, 1986). Three textural fractions according to the USDA standards, i.e. clay (0–2 μm), silt (2–50 μm) and sand (50–2000 μm), were used in the study to characterize the soil (Gee and Bauder, 1986). Most soil textures (i.e. 27 out of 32) were loamy sand and the remaining textures were sandy loam.

An undisturbed soil sample was collected from the surface soil layer (0–0.05 m depth) at each sampling location (sample size, $N = 32$), using stainless steel cylinders with an inner volume of $10^{-4}\,\mathrm{m^3}$ to determine the soil water retention curve. For each sample, the volumetric soil water content at 11 pressure heads, h, was determined by a sandbox ($h = 0.01, 0.025, 0.1, 0.32, 0.63, 1.0$ m) and a pressure plate apparatus ($h = 3$, 10, 30, 60, 150 m). For each sample, the parameters of the van Genuchten (1980, vG) model for the water retention curve with the Burdine (1953) condition were determined (Aiello et al., 2014).

Three soil water content profiles have been measured in the field using water content reflectometers (TDR) since 2009. Calibrated Campbell Scientific CS616 water content reflectometers (± 2.5 % of accuracy) were installed to monitor every 1 h the changes of volumetric soil water content ($\Delta\theta$). The TDR probe installation was designed to measure soil water content variations with time in the soil volume afferent to each plant. The location of the TDR probes is considered well suited with the specific characteristics of the micro-irrigation systems used in the area and the textural soil main features. For each location the TDR equipment consists of two sensors inserted vertically at 0.20 and 0.45 m depth and two sensors inserted horizontally at 0.35 m depth, with 0.20 m space in between. The water content reflectometer consists of two stainless steel rods connected to a printed circuit board. When the probe rods were inserted vertically into the soil surface they gave an indication of the water content in the upper 20–25 cm of soil. The probes installed horizontal to the surface were used to detect the passing of wetting fronts of water fluxes.

The data that are discussed here (see results section) correspond to the TDR probes located at about 1.5 m from the orange tree we monitored with ERT.

Hourly meteorological data (incoming short-wave solar radiation, air temperature, air humidity, wind speed and rainfall) are acquired by an automatic weather station located about 7 km from the orchard and managed by SIAS (Agrometeorological Service of the Sicilian Region). For the dominant wind directions, the fetch is larger than 550 m; for the other sectors the minimum fetch is 400 m (SE).

3 Methodology

3.1 Micrometeorological measurements

The experimental site is equipped with eddy covariance (EC) systems mounted on a micrometeorological fluxes tower (Fig. 1). Continuous energy balance measurements have been taken since 2009. In particular, net radiation (R_n, W m^{-2}) is measured with two CNR 1 Kipp & Zonen (Campbell Scientific Ltd) net radiometers at a height of 8 m. Soil heat flux density (G, W m^{-2}) is measured with three soil heat flux plates (HFP01, Campbell Scientific Ltd) placed horizontally 0.05 m below the soil surface. Three different measurements of G were selected: in the trunk row (shaded area), at one-third of the distance to the adjacent row, and at two-thirds of the distance to the adjacent row. The soil heat flux is measured as the mean output of three soil heat flux plates. Data from the soil heat flux plates are corrected for heat storage in the soil above the plates.

The air temperature and the three wind speed components are measured at two heights, 4 and 8 m, using fine wire thermocouples (76 μm diameter) and sonic anemometers (Windmaster Pro, Gill Instruments Ltd, at 4 m, and a CSAT, Campbell Sci., at 8 m). A gas analyzer (LI-7500, LI-COR) operating at 10 Hz was installed at 8 m. The raw data are recorded at a frequency of 10 Hz using two synchronized data loggers (CR3000, Campbell Sci.).

The EC measurement system and the data processing followed the guidelines of the standard EUROFLUX rules (Aubinet et al., 2000). A data quality check was applied during the post processing together with some routines to remove the common errors: running means for de-trending, three angle coordinate rotations and de-spiking. Stationarity and surface energy closure were also checked (Kaimal and Finningan, 1994).

Low-frequency measurements are taken for air temperature and humidity (HMP45C, Vaisala), wind speed and direction (05103 RM Young), and atmospheric pressure (CS106, Campbell Scientific Ltd) at 4, 8 and 10 m.

The freely distributed TK2 package (Mauder and Foken, 2004) is used to determine the first- and second-order statistical moments and fluxes on a half-hourly basis following the protocol used as a comparison reference described in Mauder et al. (2007).

3.2 Sap flow measurements

Heat-pulse techniques can be used to measure sap flow in plant stems with minimal disruption to the sap stream (Cohen et al., 1981; Green and Clothier, 1988; Swanson and Whitfield, 1981). The measurements are reliable, use inexpensive technology, provide a good time resolution of sap flow and they are well suited to automatic data collection and storage. Sequential or simultaneous measurements on numerous trees are possible, permitting the estimation of transpiration from whole stands of trees.

Measurements of water consumption at tree level (T_{SF}) have been taken using the HPV (heat pulse velocity) technique that is based on the measurement of temperature variations (ΔT), produced by a heat pulse of short duration (1–2 s), in two temperature probes installed asymmetrically on either side of a linear heater that is inserted into the trunk. For HPV measurements, two 4 cm sap flow probe with four thermocouples embedded (Tranzflo NZ Ltd., Palmerston North, New Zealand) were inserted in the trunks of the trees, belonging to the area of the micrometeorological EC tower footprint. The probes were positioned at the north and south sides of the trunk at 50 cm from the ground and wired to a data logger (CR1000, Campbell Sci., USA) for heat-pulse control and measurement; the sampling interval was 30 min. The temperature measurements are obtained by means of ultra-thin thermocouples that, once the probes are in place, are located at 5, 15, 25 and 45 mm within the trunk.

Data have been processed according to S. Green et al. (2003) to integrate sap flow velocity over sapwood area and calculate transpiration. In particular, the volume of sap flow (Q_{stem}) in the tree stem is estimated by multiplying the sap flow velocity by the cross sectional area of the conducting tissue. To this purpose, fractions of wood ($F_M = 0.48$) and water ($F_L = 0.33$) in the sapwood were determined on the trees where sap flow probes were installed. Wound-effect correction (Consoli and Papa, 2013; S. Green et al., 2003; Motisi et al., 2012) was done on a per-tree basis. Crop transpiration data have been available at the study site since 2009.

3.3 Electrical resistivity tomography

The key technique used to monitor the soil moisture content distribution in the volume surrounding the orange tree is ERT (e.g. Binley and Kemna, 2005). In particular, we installed a three-dimensional ERT system, consisting of 48 buried electrodes placed on 4 instrumented micro-boreholes, with 12 electrodes each (see Fig. 2). The electrodes are made of stainless steel wound around a 1 in. PVC pipe, and are spaced

Table 1. Times of acquisitions and irrigation schedule

Acquisition no.	Starting time (LT)	Ending time (LT)	Irrigation schedule	Date
0 (background)	10:40	11:00	11:30 to 16:30 $4\,\mathrm{l\,h^{-1}}$ from each dripper	2 October 2013
1	12:00	12:20		
2	13:00	13:20		
3	14:15	14:35		
4	15:00	15:20		
5	16:00	16:20		
6	17:00	17:20		
7	10:15	10:35	07:00 to 12:00 $4\,\mathrm{l\,h^{-1}}$ from each dripper	3 October 2013
8	11:05	11:25		
9	12:00	12:20		
10	13:00	13:20		
11	14:00	14:20		
12	15:00	15:20		
13	15:45	16:05		

10 cm along the pipe (see inset in Fig. 2); thus, the shallowest and the deepest are respectively at 0.1 and 1.2 m below the surface. Each electrode is made of a plate 3 cm wide. The boreholes are placed at the vertices of a square (with 1.3 m wide sides) that has the orange tree at its centre, and were inserted by percussion with the help of a pre-drilling with a smaller diameter in order to avoiding the disturbance of the electrical flow. The electrical contact is excellent for all 48 buried electrodes, as checked before each measurement. The four boreholes are water tight and in tight contact with the soil; therefore, they cannot act as pathways for preferential water infiltration. We focused our attention to an area slightly smaller than the square defined by the boreholes, in order to avoid the inevitable disturbance caused by borehole installation (slightly compacting the surrounding soil). The system is completed by 24 electrodes at the ground surface, placed along a square grid of about 0.21 m per side, covering the 1.3 m × 1.3 m square at the surface (Fig. 3): this set-up allows for a homogeneous coverage of the surface of the control volume. The chosen acquisition scheme was a skip-zero dipole–dipole configuration, i.e. a configuration where the current dipoles and potential dipoles are both of minimal size, i.e. they consist of neighbouring electrodes, e.g. along the boreholes. This set-up ensures maximal spatial resolution (as good as the electrode spacing, at least close to electrodes themselves) provided that the signal-to-noise ratio is sufficiently high. The data quality is assessed using a full acquisition of reciprocals to estimate the data error level (see e.g. Binley et al., 1995; Monego et al., 2010). Consistently, we used for the 3-D data inversion an Occam approach as implemented in the R3 software package (Binley, 2014) accounting for the error level estimated from the data themselves. The relevant three-dimensional computational mesh is shown in Fig. 3. At each time step, about 90–95 % of the dipoles survived the 10 % reciprocal error threshold. In order to build a time-consistent data set, only the dipoles surviving this error analysis for all time steps were subsequently used, reducing the number to slightly over 90 % of the total. The absolute inversions were run using the same 10 % error level. Time-lapse inversions were run at a lower error level equal to 2 % (consistently with the literature – e.g. Cassiani et al., 2006).

We conducted repeated ERT measurements using the above apparatus for about 2 days, starting on 2 October 2013 at 11:00 LT, and ending the next day at about 16:00 LT The schedule of the acquisitions and the irrigation times is reported in Table 1. Note that the background ERT survey was acquired on 2 October at 11:00 LT before the first irrigation period was started, so that all changes caused by irrigation and subsequent ET can be referred to that instant. Note that prior to 2 October 2013, irrigation had been suspended for at least 15 days. Note also that only one dripper – with a flow of about $4\,\mathrm{l\,h^{-1}}$ – is located at the surface of the control volume defined by the ERT set-up (Fig. 3).

4 Results and discussion

The paper presents results derived from both short-term (2 days) and long-term monitoring. The micrometeorological

Figure 2. 3-D ERT apparatus installed around one orange tree. The system is composed of four micro-boreholes carrying 12 electrodes each (see inset) and 24 surface electrodes – see text and Fig. 3 for geometry details.

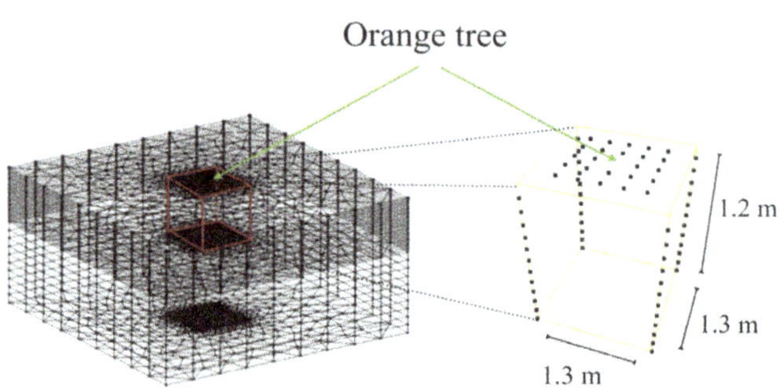

Figure 3. Electrode geometry around the orange tree and 3-D mesh used for ERT inversion.

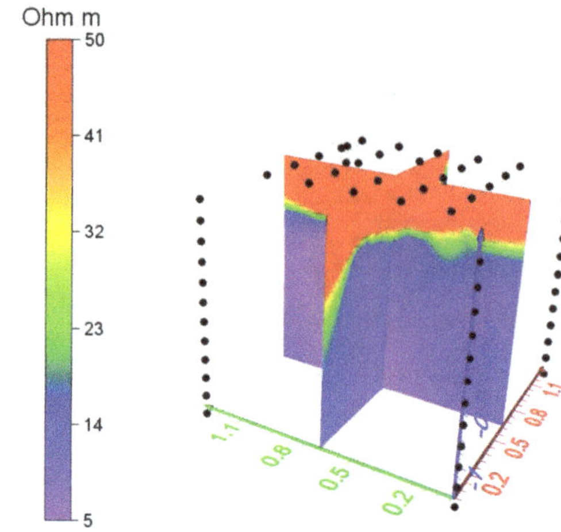

Figure 4. Cross sections of the ERT cube corresponding to the background acquisition of 2 October 2013, 11:00 LT. Note the very strong difference in electrical resistivity between the top 40 cm (above 50 Ohm) and the rest of the domain. The resistivity distribution is essentially one-dimensional with depth, with very limited horizontal variations.

data set (including the measurements of the energy balance components) and the sap flow data have been available since 2009. ERT measurements were carried out only during a 2-day period, but the state of the system at the time of the ERT measurements clearly depends on the past forcing acting on the system. In order to fully exploit the information content of this data set, we aimed at comparing data against simulations, as much as possible in a quantitative manner.

The ERT monitoring as described in Table 1 produced two clear results:

1. The initial conditions (11:00 LT of 2 October, before irrigation starts) around the tree show a very clear difference in electrical resistivity in the top 40 cm of soil with respect to the rest of the volume (Fig. 4). Specifically, the resistivity of the top layer ranges around 40–50 Ohm m, while the lower part of the profile is about 1 order of magnitude more conductive (about 5 Ohm m). As no apparent lithological difference is present at 40 cm depth (see also laboratory results below), we attributed this difference to a marked difference in soil moisture content. This was confirmed by all following evidence (see below).

2. The resistivity changes as a function of time, during the two irrigation periods, during the night interval, and afterwards, all show essentially the same pattern, with relatively small (but still clearly measurable) changes (Fig. 5). Two zone are identifiable: (a) a shallow zone (top 10–20 cm) where resistivity decreases with respect to the initial condition, and (b) a deeper zone (20–40 cm) where resistivity increases.

Qualitatively, both pieces of evidence can be easily explained in terms of water dynamics governed by precipitation, irrigation and RWU. Specifically, the shallower high resistivity zone in Fig. 4 can be correlated to a dry region where RWU manages to keep soil moisture content to minimal values, as an effect of the entire summer strong transpiration drive. The dynamics in Fig. 5, albeit small compared to the initial root uptake signal in Fig. 4, still confirm that the top 40 cm is home to strong root activity, to the point that irrigation cannot raise electrical conductivity of the shallow zone (10–20 cm) by no more than some 20 %, and the roots manage to make the soil even drier (with a resistivity increase by some 10 %) in the 20–40 cm depth layer (Fig. 5). Note that, in general, resistivity changes of the type observed here cannot be uniquely associated with soil moisture content changes, as porewater conductivity may play a key role (e.g. Boaga et al., 2013; Ursino et al., 2014). However, in the particular case at hand, care was taken to analyse the electrical conductivity of both the water used for irrigation and the porewater, purposely extracted at about 50 cm depth. Both waters showed an electrical conductivity value in the range of

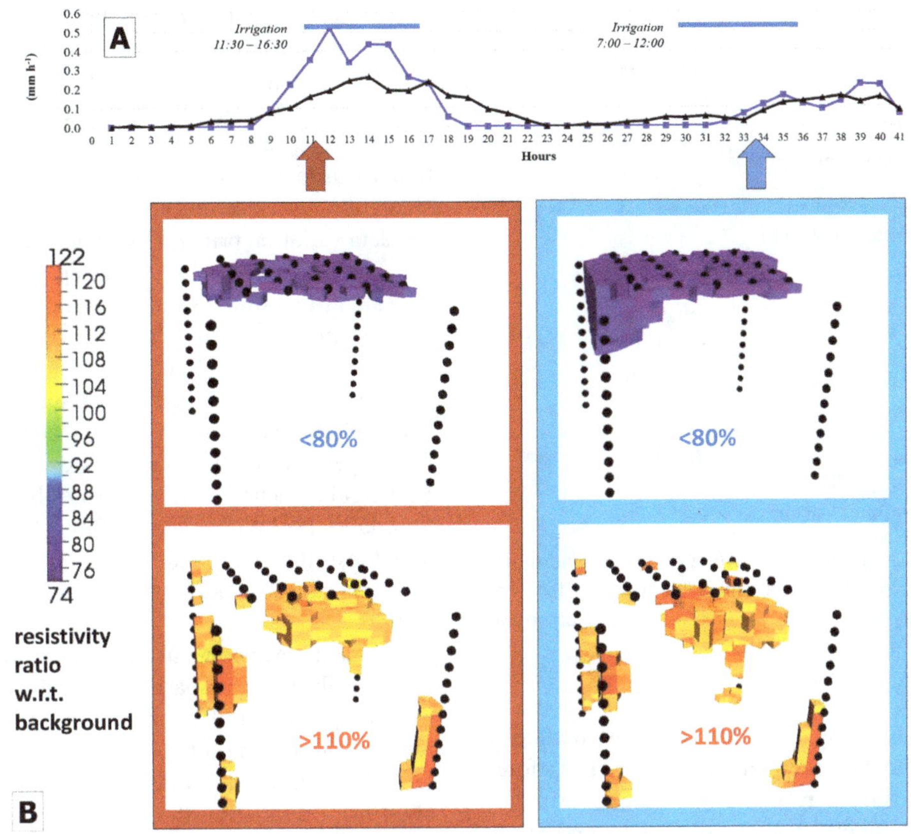

Figure 5. (a) Time series of sap flow (black line) and EC-derived total evapotranspiration (blue lines), both normalized in millimetres assuming an area of 20 m^2 pertaining to the orange tree monitored with ERT. Time is given in hours from midnight of 2 October. The two irrigation periods are shown by the blue bars. **(b)** 3-D ERT images of resistivity change with respect to background at two selected time instants shown by the arrows in **(a)**; the volumes corresponding to increase and decrease of resistivity above and below certain thresholds (80 and 110 %) are shown in separate panels, for clarity.

1300 $\mu S\,cm^{-1}$ (thus fairly high, fact that explains the overall small soil resistivity observed at the site). Therefore, in this particular case we can exclude porewater conductivity effects in the observed dynamics of the system. Once again it must be stressed that this is rather the exception than the rule.

A laboratory-based method was adopted for obtaining "unaltered" soil porewater through a column displacement technique (Knight et al., 1998). In particular, Rhizon soil moisture samplers (Cabrera, 1998) were used; they represent one of the latest developments in terms of tension samplers, where it is necessary to apply a suction to withdraw porewater with a vacuum tube (Tye et al., 2003).

The qualitative evidence above is, however, not very surprising and not particularly informative: the root activity dries the soil, this is not a discovery. Things become more interesting if we can translate the ERT data into quantitative estimates of soil moisture content, and if we can use these data to calibrate hydrological models of the root zone.

To this end, we tested Bulgherano soil samples in the laboratory to obtain a suitable constitutive relationship linking moisture content and resistivity, given the know porewater conductivity that was reproduced for the water used in the laboratory. All measurements were conducted using cylindrical Plexiglas cells equipped with a four-electrode configuration designed to allow for sample saturation and de-saturation with no sample disturbance, using an air injection apparatus at one end and a ceramic plate at the opposite end. The air entry pressure of the ceramic is 1 bar; thus, during all the experiments the plate remained under full water saturation, while allowing water outflow during de-saturation. At each de-saturation step, the electrical conductivity of the sample was measured under temperature-controlled conditions using a ZEL-SIP04 impedance metre (Zimmermann et al., 2008). A completed description of the set-up is given by Cassiani et al. (2009b).

Figure 6 shows two example experimental results on samples from two different depths. Note how in a wide range of soil moisture content (roughly from 5 % to saturation) the two curves in Fig. 6 lie practically on top of each other. The same applies for all tested samples. Note also that, even though some samples show the effect of the conductivity of the solid phase (through its clay fraction), at small saturation

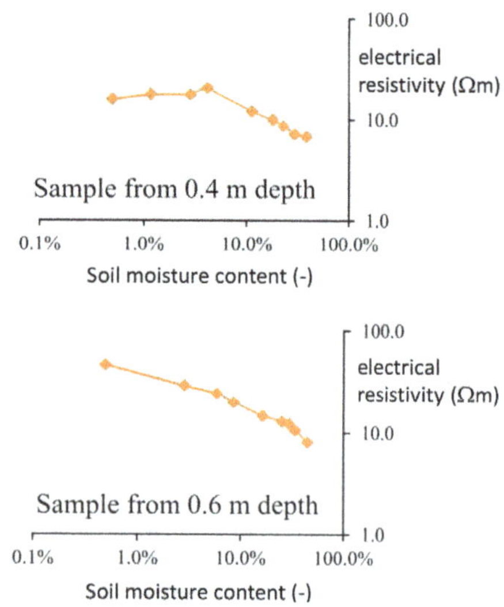

Figure 6. Experimental relationships between resistivity and moisture content determined in the lab on samples taken at two different depths at the Bulgherano site, using water having the same electrical conductivity measured in the porewater in situ.

(see sample from 0.4 m in Fig. 6) still the effect is small as it appears only at soil moisture smaller than 3–4 %. Therefore, we deemed it unnecessary to resort to constitutive laws that represent this solid phase effect, such as Waxman and Smits (1968) that has been used for similar purposes elsewhere (e.g. Cassiani et al., 2012), and we adopted the simpler formulation of Archie (1942). Consequently we translated resistivity into moisture content using the following relationship calibrated on the laboratory data, using water having the abovementioned electrical conductivity:

$$\theta = \frac{4.703}{\rho^{1.12}}, \tag{1}$$

where θ is volumetric soil moisture content (dimensionless) and ρ is electrical resistivity (in Ohm m). Equation (1) allows a direct translation of the 3-D resistivity distribution to a corresponding distribution of volumetric soil moisture content. However, it has long been established that inverted geophysical data may be accompanied by enough distortion of the true physical parameter field (Day-Lewis et al., 2005) as to induce violations of elementary physical principles, such as mass balance during tracer test monitoring experiments (e.g. Singha and Gorelick, 2005). This may cause substantial problems, particular when the use of data is expected to shift from a qualitative interpretation to a quantitative use in terms of data assimilation into hydrological models. For this reason, coupled versus uncoupled approaches have been proposed and discussed (Hinnell et al., 2010) even though their superiority seems to depend on the specific problem, as the information content of data even in a traditional, inverted approach may be sufficient (Camporese et al., 2011, 2014). Indeed, the geometry we are considering here is very effective to reconstruct the mass balance of irrigated water, as this comes as a quasi-one-dimensional infiltration front from the top, where, in addition, electrodes are located. The geometry is similar to the one used, e.g. Koestel et al. (2008), where mass balance was verified by comparison against very detailed TDR data collected in a lysimeter. In spite of these considerations, we decided to still limit ourselves to analysing the data variation principally as a function of depth, lumping the data horizontally by averaging estimated moisture content along two-dimensional horizontal planes. Note that the data set may lend itself to more complex analyses such as the one proposed by Manoli et al. (2014), especially if used in the context of a formal data assimilation, but we felt that one such an endeavour would exceed the scope of the current paper and deserves an ad hoc space. Note also that the ERT field evidence both in terms of background (Fig. 4) and time-lapse evolution (Fig. 5) of moisture content confirm the hypothesis that, within the control volume, the distribution of water in the soil is largely one-dimensional as a function of depth.

The data, once condensed in this manner, lend themselves more easily to a comparison with the results of infiltration modelling. We implemented a one-dimensional finite element model based on a Richards' equation solver ("Femwater" – details of this classical model are given by Lin et al., 1997), simulating the central square metre of the ERT-monitored control volume, down to a total depth of 2 m (much below the depth of the ERT boreholes), where we assumed that the water table is located (Dirichlet boundary condition). We applied at the top of the soil column a Neumann boundary condition consistent with the flux coming from irrigation that pertains the control volume (basically, the water coming from a single dripper). As Femwater is a 3-D simulator, the soil column is also bounded laterally by no-flow conditions, with the exception of the top 40 cm where we applied laterally a Neumann condition simulating the RWU (see below for details).

We considered only the central part of the ERT-controlled volume (1 m × 1 m), thus excluding the regions too close to the boreholes that, even though benefitting from the best ERT sensitivity, might have been altered from a hydraulic viewpoint by the drilling and installing operations. Correspondingly we horizontally averaged the ERT data only in this central region.

A very fine vertical discretization (0.01 m) and time stepping (0.01 h) ensures solution stability. The porous medium is homogeneous along the column and parameterized according to the Van Genuchten (1980) model. The relevant parameters have been derived independently from laboratory and field measurements, the latter particularly relevant for the definition of a reliable in situ saturated hydraulic conductivity estimate. The parameters used for the simulations are residual moisture content $\theta_r = 0.0$; porosity $\theta_s = 0.54$, $\alpha = 0.121\,m^{-1}$, $n = 1.6$; and saturated hydraulic conductivity $K_s = 0.002\,m\,h^{-1}$. We acknowledge that a more complete sensitivity analysis concerning the impact of the individual parameters would be beneficial, but this should be performed in a complete Monte Carlo manner in order to exclude identification trade-offs between the Van Genuchten parameters, the depth of the water table (known with some uncertainty) and the fluxes from irrigation, precipitation and ET. However, we feel that this endeavour shall be conducted also with regard to the effective 3-D spatial distribution of active roots, and is currently the subject of ongoing research.

The remaining elements of the predictive modelling exercise are initial and boundary conditions. As we focused primarily our attention on reproducing the state of the system at background conditions, we set the start of the simulation at the beginning of the year (1 January 2013), and we assumed for that time a condition drained to equilibrium. Given the van Genuchten parameters we used and the depth of the water table, this corresponds to a fairly wet initial condition. We verified a posteriori that moving the initial time back of 1 year or more did not alter the predicted results at the date of interest (3 October 2013). The dynamics during the year are sufficient to bring the system to the real, much drier condition in October. The forcing conditions on the system are all known: (a) irrigation is recorded, and only one dripper pertains to the considered square metre; (b) precipitation is measured; (c) sap flow is measured. Direct evaporation from the square metre of soil around the stem is neglected, considering the dense canopy cover and the consequent limited radiation received. Only 1 degree of freedom is left to be calibrated, i.e. the volume from which the roots uptake water. Thickness of the active root zone was estimated from the time-lapse observations (Fig. 5), and fixed to the top 0.4 cm after checking that limiting the root uptake to the 0.2 m to 0.4 m zone would produce results inconsistent with observations in the top 0.2 m. Therefore only the surface area of the root uptake zone remains to be estimated. We used the predictive model as a tool to identify the extent of this zone, that is of critical interest also for irrigation purposes.

Figure 7 shows the results of the calibration exercise. It is apparent that the total areal extent of the root uptake zone has a dramatic impact on the predicted moisture content profiles, as it scales the amount of water subtracted from the monitored square metre considered in the calibration. Even relatively small changes (±15 %) of the root uptake area produce very different soil moisture profiles. The value that allows for a good match of the observed profile is 1.75 m^2, while for areas equal to 1.5 m^2 and 2 m^2 the match is already unsatisfactory, leading respectively to underestimation and overestimation of the moisture content in the profile.

Another important fact that is apparent from Fig. 7 is that the estimated soil moisture in the shallow zone (roughly down to 0.4 m) is very small as an effect of RWU. However, this dry zone must have a limited areal extent (1.75 m^2, corresponding to a radius of about 0.75 m from the stem of the tree). Indeed this is indirectly confirmed by the soil moisture evolution measured by TDR. Figure 8 shows the TDR data

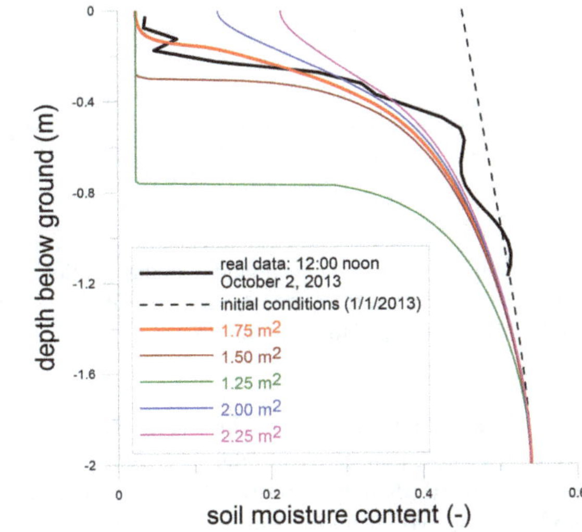

Figure 7. Results of 1-D Richards' equation simulations of the entire year 2013 up to 3 October 11:00 LT, i.e. in correspondence of the background ERT acquisition (the thick black line represents the resulting estimated moisture content profile obtained from averaging horizontally the central square metre of the ERT control volume). The different simulated curves correspond to different assumed areas of root water uptake (RWU), and show how 1.75 m^2 is the area that allows one to match the observed real profile with good accuracy. Note also the high sensitivity of the results to the estimated root uptake area.

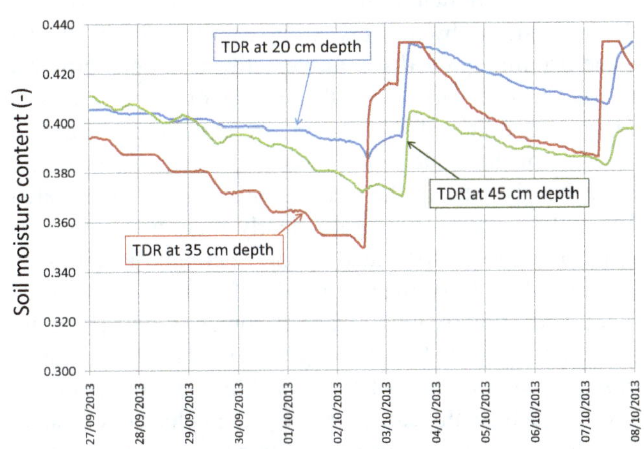

Figure 8. Moisture content time series from three TDR probes located about 1.5 m from the ERT-monitored tree. The signal coming from the irrigation experiment of 2 October 2013 is very clear. Before this experiment the system had been left without irrigation for about a fortnight.

from three probes located about 1.5 m from the monitored tree (thus outside our estimated root uptake zone). The signal coming from the irrigation experiment of 2 October 2013 is very apparent with an increase in moisture content of all three probes, located at different depths. Note that before this experiment the system had been left without irrigation for about a fortnight. The corresponding effect on the TDR data is ap-

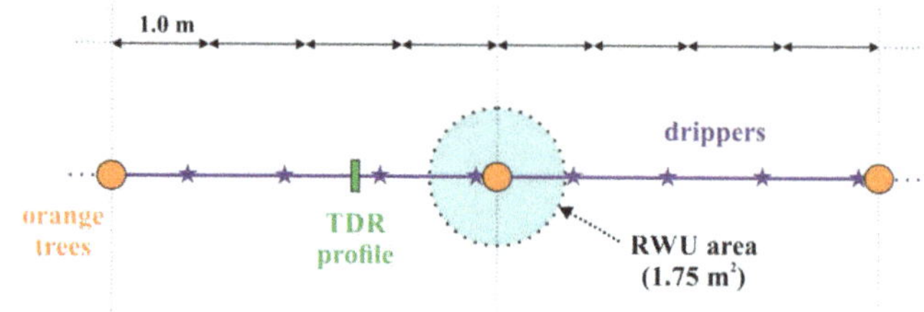

Figure 9. Scheme of the experimental field with the location of the main sensors. The radius of the root water uptake (RWU) zone, assumed to be circular, is equal to about 0.75 m.

parent: all three probes show a decline of moisture content during the day, with pauses overnight. The decline is more pronounced in the 0.35 m TDR probe, that lies at a depth we estimated to be nearly at the bottom of the RWU zone, and less pronounced above (0.2 m) and below (0.45 m). Note also that the TDR probes are close to another dripper, lying outside of the ERT-controlled volume (the drippers are spaced 1 m along the orange trees line, with the trees about 4 m from each other); thus, they reflect directly the infiltration from that dripper. However, at all three depths the moisture content is much higher than measured in the ERT-controlled block closer to the tree. This can be explained with the fact that in that region the root uptake is minimal or totally absent, while the decline of moisture content in time may well be an effect of water being drawn to the root zone by lateral movement, which is induced by the very strong capillary forces exerted by the dry fine grained soil in the active root zone closer to the tree. In order to clarify the impact of these results on our understanding of the system, we show the location of the trees, of the TDR probes and of the drippers in Fig. 9, where we also sketch the best estimate for the areal extent of the RWU zone. This figure clearly highlights how critical the information provided by ERT actually is. The scale at which RWU takes place is smaller (metre scale) than expected and often assumed when it comes to designing and implementing a field monitoring system. This has dramatic consequences in terms of how reliable conclusions can be drawn if such small-scale processes are neglected. Consider, e.g. what type of conclusions could be drawn on the basis of TDR data alone (Fig. 8) in light of the field situation as depicted in Fig. 9. The single, most important message that shall be conveyed by this paper is a warning to be particularly attentive to small-scale processes in soil–plant–atmosphere interactions, even in regular agricultural landscapes.

5 Conclusions

Near-surface geophysics is strongly affected by both static and dynamic soil/subsoil characteristics. This fact, if properly recognized, is potentially full of information on the soil/subsoil structure and behaviour. The information is maximized if geophysical data are collected in time-lapse mode. In the case of interactions with vegetation, its role should be properly modelled, and such models can be constrained by means (also) of geophysical data. This case study demonstrates that 3-D ERT is capable of characterizing the pathways of water distribution, and provides spatial information on root zone suction regions. The integration of modelling and data has proven, once again, a key component of this type of hydro-geophysical studies, allowing us to draw quantitative results of practical interest. In this case we had available a wealth of quantitative information about transpiration and soil moisture content that allowed the definition of the volume of soil affected by the RWU activity. This has obvious consequences for the possible improvement of irrigation strategies, as it is apparent how the monitored orange tree essentially drives water from one to two drippers out of the four in total that should pertain to its area in the plantation. This means that it is very likely that half of the irrigated water is indeed lost to deeper layers and brings no contribution to the plants. More advanced uses of this type of data are now considered, especially linking soil moisture distribution with plant physiological response and active root distribution in the soil. In the long-run studies of this type may give a fundamental contribution to our understanding of soil–plant–atmosphere interactions also in view of facing challenges coming from climatic changes.

Acknowledgements. We wish to acknowledge support from the EU FP7 project GLOBAQUA ("Managing the effects of multiple stressors on aquatic ecosystems under water scarcity") and the MIUR PRIN project 2010JHF437 "Innovative methods for water resources management under hydro-climatic uncertainty scenarios". We also wish to thank the Agro-meteorological Service of the Sicilian Region for supporting field campaigns.

Edited by: M. Vanclooster

References

Aiello, R., Bagarello, V., Barbagallo, S., Consoli, S., Di Prima, S., Giordano, G., and Iovino, M.: An assessment of the Beerkan method for determining the hydraulic properties of a sandy loam soil, Geoderma, 235–236, 300–307, 2014.

al Hagrey, S. A.: Geophysical imaging of root-zone, trunk, and moisture heterogeneity, J. Exp. Bot., 58, 839–854, 2007.

al Hagrey, S. A. and Petersen T.: Numerical and experimental mapping of small root zones using optimized surface and borehole resistivity tomography, Geophysics, 76, G25–G35, doi:10.1190/1.3545067.671, 2011.

Allred, B., Daniels, J. J., and Reza Ehsani, M.: Handbook of Agricultural Geophysics, CRC Press, USA, 432 pp., 2008.

Amato, M., Bitella, G., Rossi, R., Gomez, J. A., Lovelli, S., and Gomes, J. J. F.: Multi-electrode 3D resistivity imaging of alfalfa root zone, Eur. J. Agron., 31, 213–222, 2009.

Amato, M., Rossi, R., Bitella, G., and Lovelli, S.: Multielectrode Geoelectrical Tomography for the Quantification of Plant Roots, Ital. J. Agron./Riv. Agron., 3, 257–263, 2010.

Archie, G. E.: The electrical resistivity log as an aid in determining some reservoir characteristics, Trans. AIME 146, 54–67, 1942.

Aubinet, M., Grelle, A., Ibrom, A., Rannik, U., Moncrieff, J., Foken, T., Kowalski, P., Martin, P., Berbigier, P., Bernhofer, C., Clement, R., Elbers, J., Granier, A., Grunwald, T., Morgenster, K., Pilegaard, K., Rebmann, C., Snijders, W., Valentini, R., and Vesala, T.: Estimates of the annual net carbon and water exchange of Europeran forests: the EUROFLUX methodology, Adv. Ecol. Res., 30, 113–175, 2000.

Beff, L., Günther, T., Vandoorne, B., Couvreur, V., and Javaux, M.: Three-dimensional monitoring of soil water content in a maize field using Electrical Resistivity Tomography, Hydrol. Earth Syst. Sci., 17, 595–609, doi:10.5194/hess-17-595-2013, 2013.

Binley, A.: R3t code, http://www.es.lancs.ac.uk/people/amb/Freeware/R3t/R3t.htm (last access: August 2014), 2014.

Binley, A., Ramirez, A., and Daily, W.: Regularised image reconstruction of noisy electrical resistance tomography data, edited by: Beck, M. S., Hoyle, B. S., Morris, M. A., Waterfall, R. C., Williams, R. A., in: Process tomography, Proceedings of the 4th Workshop of the European Concerted Action on Process Tomography, Bergen, 6–8 April 1995, 401–410, 1995.

Binley, A. M. and Kemna, A.: DC resistivity and induced polarization methods, edited by: Rubin, Y. and Hubbard, S. S., in: Hydrogeophysics, Water Sci. Technol. Library, Ser. 50, Springer, New York, 129–156, 2005.

Binley, A. M., Cassiani, G., Middleton, R., and Winship, P.: Vadose zone flow model parameterisation using cross-borehole radar and resistivity imaging, J. Hydrol., 267, 147–159, 2002.

Binley, A. M., Cassiani, G., and Deiana, R.: Hydrogeophysics – Opportunities and Challenges, Bollettino di Geofisica Teorica ed Applicata, 51, 267–284, 2011.

Boaga, J., Rossi, M., and Cassiani, G.: Monitoring soil-plant interactions in an apple orchard using 3D electrical resistivity tomography, Conference on Four Decades of Progress in Monitoring and Modeling of Processes in the Soil-Plant-Atmosphere System: Applications and Challenges, Naples, 19–21 June 2013, Proc. Environ. Sci., 19, 394–402, 2013.

Boaga, J., D'Alpaos, A., Cassiani, G., Marani, M., and Putti M.: Plant-soil interactions in salt marsh environments: Experimental evidence from electrical resistivity tomography in the Venice Lagoon, Geophys. Res. Lett., 41, 6160–6166, doi:10.1002/2014GL060983, 2014.

Burdine, N. T.: Relative permeability calculation from pore size distribution data, Trans. Am. Inst. Min. Eng., 198, 71–78, 1953.

Cabrera, R. I.: Monitoring chemical properties of container growing media with small soil solution samplers, Sci. Horticult., 75, 113–119, 1998.

Camporese, M., Salandin, P., Cassiani G., and Deiana, R.: Impact of ERT data inversion uncertainty on the assessment of local hydraulic properties from tracer test experiments, Water Resour. Res., 47, W12508, doi:10.1029/2011WR010528, 2011.

Camporese, M., Cassiani, G., Deiana, R., Salandin, P., and Binley, A.: Coupled and uncoupled hydrogeophysical inversions using ensemble Kalman filter assimilation of ERT-monitored tracer test data, Water Resour. Res., accepted, doi:10.1002/2014WR016017, 2015.

Cassiani, G., Bruno, V., Villa, A., Fusi, N., and Binley, A. M.: A saline trace test monitored via time-lapse surface electrical resistivity tomography, J. Appl. Geophys., 59, 244–259, 2006.

Cassiani, G., Godio, A., Stocco, S., Villa, A., Deiana, R., Frattini, P., and Rossi, M.: Monitoring the hydrologic behaviour of steep slopes via time-lapse electrical resistivity tomography, Near Surface Geophysics, special issue on Hydrogeophysics – Methods and Processes, 475–486, doi:10.3997/1873-0604.2009013, 2009a. Cassiani, G., Kemna, A., Villa, A., and Zimmermann, E.: Spectral induced polarization for the characterization of free-phase hydrocarbon contamination in sediments with low clay content, Near Surface Geophysics, special issue on Hydrogeophysics – Methods and Processes, 547–562, doi:10.3997/1873-0604.2009028, 2009b.

Cassiani, G., Ursino, N., Deiana, R., Vignoli, G., Boaga, J., Rossi, M., Perri, M. T., Blaschek, M., Duttmann, R., Meyer, S., Ludwig, R., Soddu, A., Dietrich, P., and Werban, U.: Non-invasive monitoring of soil static characteristics and dynamic states: a case study highlighting vegetation effects, Vadose Zone Journal, Special Issue on SPAC – Soil-plant interactions from local to landscape scale, August 2012, V.11, vzj2011.0195, doi:10.2136/2011.0195, 2012.

Cassiani, G., Boaga, J., Rossi, M., Fadda, G., Putti, M., Majone, B., and Bellin, A.: Soil-plant interaction monitoring: small scale example of an apple orchard in Trentino, North-Eastern Italy, Sci. Total Environ., in press, 2015.

Cohen, Y., Fuchs, M., and Green, G. C.: Improvement of the heat-pulse method for determining sap flow in trees, Plant Cell Environ., 4, 391–397, 1981.

Consoli, S. and Papa, R.: Corrected surface energy balance to measure and model the evapotranspiration of irrigated orange orchards in semi-arid Mediterranean conditions, Irrigation Science September 2013, 31, 1159–1171, 2013.

Couvreur, V., Vanderborght, J., and Javaux, M.: A simple three-dimensional macroscopic root water uptake model based on the hydraulic architecture approach, Hydrol. Earth Syst. Sci., 16, 2957–2971, doi:10.5194/hess-16-2957-2012, 2012.

Daily, W., Ramirez, A., LaBrecque, D., and Nitao, J.: Electrical resistivity tomography of vadose zone movement, Water Resour. Res., 28, 1429–1442, 1992.

Day-Lewis, F. D., Singha, K., and Binley, A. M.: Applying petrophysical models to radar travel time and electrical resistivity tomograms: Resolution-dependent limitations, J. Geophys. Res.-Solid Earth, 110, B08206, doi:10.1029/2004JB003569, 2005.

Doussan, C., Pierret, A., Garrigues, E., and Pagès, L.: Water uptake by plant roots: II – Modelling of water transfer in thesoil root-system with explicit account of flow within the root system – Comparison with experiments, Plant Soil, 283, 99–117, doi:10.1007/s11104-004-7904-z, 2006.

Feddes, R. A., Hoff, H., Bruen, M., Dawson, T., de Rosnay, P., Dirmeyer, P., Jackson, R. B., Kabat, P., Kleidon, A., Lilly, A., and Pitman, A. J.: Modelling Root Water Uptake in Hydrological and Climate Models, B. Am. Meteor. Soc., 82, 2797–2809, 2001.

Garré, S., Javaux, M., Vanderborght, J., Pagès, L., and Vereecken, H.: Three-Dimensional Electrical Resistivity Tomography to

Monitor Root Zone Water Dynamics, Vadose Zone J., 10, 412–424, doi:10.2136/vzj2010.0079, 2011.

Gee, G. W. and Bauder, J. W.: Particle-size analysis, edited by: Klute, A., Methods of Soil Analysis, Part 1, Physical and Mineralogical Methods, Agronomy Monograph No. 9, 2 ed., 383–411, American Society of Agronomy/Soil Science Society of America, Madison, WI, 1986.

Gong, D., Shaozhong, K., Zhang, L., Taisheng, D., and Limin, Y.: A two-dimensional model of root water uptake for single apple trees and its verification with sap flow and soil water content measurements, Agr. Water Manage., 83, 119–129, 2006.

Green, S. R. and Clothier, B. E.: Water use of kiwifruit vines and apple trees by the heat-pulse technique, Exp. Bot., 39, 115–123, 1988.

Green, S. R., Vogeler, I., Clothier, B. E., Mills, T. M., and van den Dijssel, C.: Modelling water uptake by a mature apple tree, Austr. J. Soil Res., 41, 365–380, 2003.

Green, S., Clothier, B., and Jardine B.: Theory and Practical Application of Heat Pulse to Measure Sap Flow, Agronomy Journal; Nov/Dec 2003, 95, ProQuest Agricult. J., p. 1371, 2003.

Hinnell, A. C., Ferré, T. P. A., Vrugt, J. A., Huisman, J. A., Moysey, S., Rings, J., and Kowalsky, M. B.: Improved extraction of hydrologic information from geophysical data through coupled hydrogeophysical inversion, Water Resour. Res., 46, W00D40, doi:10.1029/2008WR007060, 2010.

Jarvis, N. J.: A simple empirical-model of root water-uptake, J. Hydrol., 107, 57–72, doi:10.1016/0022-1694(89)90050-4, 1989.

Jarvis, N. J.: Simple physics-based models of compensatory plant water uptake: concepts and ecohydrological consequences, Hydrol. Earth Syst. Sci., 15, 3431–3446, doi:10.5194/hess-15-3431-2011, 2011.

Javaux, M., Schroder, T., Vanderborght, J., and Vereecken, H.: Use of a Three- Dimensional Detailed Modeling Approach for Predicting Root Water Uptake, Vadose Zone J., 7, 1079–1088, 2008.

Jayawickreme, H., Van Dam, R., and Hyndman, D. W.: Subsurface imaging of vegetation, climate, and root-zone moisture interactions, Geophys. Res. Lett., 35, L18404, doi:10.1029/2008GL034690, 2008.

Jones, H. G. and Tardieu, F.: Modelling water relations of horticultural crops: a review, Sci. Hortic.-Amsterdam, 74, 21–46, 1998.

Kaimal, J. C. and Finnigan, J.: Atmospheric Boundary Layer Flows: Their Structure and Measurement, Oxford University Press, New York, 255–261, 1994.

Kemna, A., Vanderborght, J., Kulessa, B., and Vereecken, H.: Imaging and characterisation of subsurface solute transport using electrical resistivity tomography ERT and equivalent transport models, J. Hydrol., 267, 125–146, 2002.

Koestel J., Kemna, A., Javaux, M., Binley, A., and Vereecken, H.: Quantitative imaging of solute transport in an unsaturated and undisturbed soil monolith with 3-D ERT and TDR, Water Resour. Res., 44, W12411, doi:10.1029/2007WR006755, 2008.

Knight, B. P., Chaudri, A. M., McGrath, S. P., and Giller, K. E.: Determination of chemical availability of cadmium and zinc in soils using inert soil moisture samplers, Environ. Poll., 99, 293–298, 1998.

Lin, H. J., Richards, D. R., Talbot, C. A., Yeh, G.-T., Cheng, J., and Cheng, H.: FEMWATER: a three-dimensional finite element computer model for simulating density-dependent flow and transport in variably saturated media, US Army Corps of Engineers and Pennsylvania State University, Technical Report CHL-97-12, 1997.

Manoli, G., Bonetti, S., Domec, J. C., Putti, M., Katul, G., and Marani, M.: Tree root systems competing for soil moisture in a 3D soil-plant model, Adv. Water Res., 66, 32–42, doi:10.1016/j.advwatres.2014.01.006, 2014.

Mauder, M. and Foken, T.: Documentation and instruction manual of the eddy covariance software package TK2. Universität Bayreuth, Abt. Mikrometeorologie, Arbeitsergebnisse, http://www.geo.unibayreuth.de/mikrometeorologie/ARBERG (last access: August 2014), 26–44, 2004.

Mauder, M., Oncley, S. P., Vogt, R., Weidinger, T., Ribeiro, L., Bernhofer, C., Foken, T., Kosiek, W., De Bruin, H. A. R., and Liu, H.: The energy balance experiment EBEX-2000. Part II. Intercomparison of eddy-covariance sensors and post-field data processing methods, Bound.-Layer Meteorol., 123, 29–54, doi:10.1007/s10546-006-9139-4, 2007.

Michot, D., Dorigny, A., and Benderitter Y.: Determination of water flow direction and corn roots-induced drying in an irrigated Beauce CALCISOL, using electrical resistivity measurements, Comptes Rendus De L'Academie Des Sciences Serie Ii Fascicule a-Sciences De La Terre Et Des Planetes, 332, 29–36, 2001.

Michot, D., Benderitter, Y., Dorigny, A., Nicoullaud, B., King, D., and Tabbagh, A.: Spatial and temporal monitoring of soil water content with an irrigated corn crop cover using surface electrical resistivity tomography, Water Resour. Res., 39, p. 1138, 2003.

Musters, P. A. D. and Bouten, W.: A method for identifying optimum strategies of measuring soil water contents for calibrating a root water uptake model, J. Hydrol, 227, 273–286, 2000.

Monego, M., Cassiani, G., Deiana, R., Putti, M., Passadore, G., and Altissimo, L.: Tracer test in a shallow heterogeneous aquifer monitored via time-lapse surface ERT, Geophysics, 75, WA61–WA73, doi:10.1190/1.3474601, 2010.

Motisi, A., Consoli, S., Rossi, F., Minacapilli, M., Cammalleri, C., Papa, R., Rallo, G., and D'urso, G.: Eddy covariance and sap flow measurement of energy and mass exchange of woody crops in a Mediterranean environment, Acta Horticult., 951, 121–127, 2012.

Parasnis, D. S.: Mining geophysics, Elsevier Scientific Pub. Co., 395 pp., 1973.

Perri, M. T., Cassiani, G., Gervasio, I., Deiana, R., and Binley, A. M.: A saline tracer test monitored via both surface and cross-borehole electrical resistivity tomography: comparison of time-lapse results, J. Appl. Geophys., 79, 6–16, doi:10.1016/j.jappgeo.2011.12.011, 2012.

Raats, P. A. C.: Uptake of water from soils by plant roots, Transp. Porous. Med., 68, 5–28, 2007.

Rubin, Y. and Hubbard, S. S. (Eds.): Hydrogeophysics, Springer, Dordrecht, the Netherlands, 523 pp., 2005.

Schneider, C. L., Attinger, S., Delfs, J.-O., and Hildebrandt, A.: Implementing small scale processes at the soil-plant interface – the role of root architectures for calculating root water uptake profiles, Hydrol. Earth Syst. Sci., 14, 279–289, doi:10.5194/hess-14-279-2010, 2010.

Shanahan, P. W., Binley, A., Whalley, W. R., and Watts, C. W.: The use of electromagnetic induction to monitor changes in soil moisture profiles beneath different wheat genotypes, Soil Sci. Soc. Am. J., 79, 459–466, doi:10.2136/sssaj2014.09.0360, 2015.

Singha, K. and Gorelick, S. M.: Saline tracer visualized with three dimensional electrical resistivity tomography: Field-scale spatial moment analysis, Water Resour. Res., 41, W05023, doi:10.1029/2004WR003460, 2005.

Srayeddin, I. and Doussan, C.: Estimation of the spatial variability of root water uptake of maize and sorghum at the field scale by electrical resistivity tomography, Plant Soil, 319, 185–207, doi:10.1007/s11104-008-9860-5, 2009.

Swanson, R. H. and Whitfield, D. W.: A numerical analysis of heat pulse velocity theory and practice, J. Exp. Bot., 32, 221–239 1981.

Tye, A. M., Woung, S. D., Crout, N. M. J., Zhang, H., Preston, S., Barbosa-Jefferson, V. L., Davison, W., McGrath, S. P., Paton, G. I., Kilham, K., and Resende, L.: Predicting the activity of Cd^{2+} and Zn^{2+} in soil pore water from the radio-labile metal fraction, Geochim. Cosmochim. Acta, 67, 375–385, 2003.

Ursino, N., Cassiani, G., Deiana, R., Vignoli, G., and Boaga, J.: Measuring and modeling water-related soil-vegetation feedbacks in a fallow plot, Hydrol. Earth Syst. Sci., 18, 1105–1118, doi:10.5194/hess-18-1105-2014, 2014.

Van Genuchten, M. T.: A closed form equation for predicting the hydraulic conductivity of unsaturated soils, Soil Sci. Soc. Am. J., 44, 892–898, 1980.

Vereecken, H., Binley, A., Cassiani, G., Kharkhordin, I., Revil, A., and Titov, K.: Applied Hydrogeophysics, Springer-Verlag, Berlin, 1–8, 2006.

Waxman, M. H. and Smits L. J. M.: Electrical conductivities in oil-bearing shaly sands, Soc. Petr. Eng. J., 8, 107–122, 1968.

Weill, S., Altissimo, M., Cassiani, G., Deiana, R., Marani, M., and Putti, M.: Saturated area dynamics and streamflow generation from coupled surface–subsurface simulations and field observations, Adv. Water Resour., 59, 196–208, doi:10.1016/j.advwatres.2013.06.007, 2013.

Werban, U., al Hagrey, S. A., and Rabbel, W.: Monitoring of root-zone water content in the laboratory by 2D geoelectrical tomography, J. Plant Nutr. Soil Sci., 171, 927–935, doi:10.1002/jpln.200700145, 2008.

Zimmermann, E., Kemna, A., Berwix, J., Glaas, W., Münch, H. M., and Huisman, J. A.: A high-accuracy impedance spectrometer for measuring sediments with low polarizability, Meas. Sci. Technol., 19, 105603, doi:10.1088/0957-0233/19/10/105603, 2008.

12

Hydrogeochemical controls on brook trout spawning habitats in a coastal stream

Martin A. Briggs[1], **Judson W. Harvey**[2], **Stephen T. Hurley**[3], **Donald O. Rosenberry**[4], **Timothy McCobb**[5], **Dale Werkema**[6], **and John W. Lane Jr.**[1]

[1]U.S. Geological Survey, Hydrogeophysics Branch, 11 Sherman Place, Unit 5015, Storrs, CT 06269, USA
[2]U.S. Geological Survey, Water Cycle Branch, M.S. 430, Reston, VA 20192, USA
[3]Massachusetts Division of Fisheries and Wildlife, 195 Bournedale Road, Buzzards Bay, MA 02532, USA
[4]U.S. Geological Survey, National Research Program, M.S. 406, Bldg. 25, DFC, Lakewood, CO 80225, USA
[5]U.S. Geological Survey, 10 Bearfoot Road, Northborough, MA 01532, USA
[6]U.S. Environmental Protection Agency, Office of Research and Development, National Exposure Research Laboratory, Exposure Methods & Measurement Division, Environmental Chemistry Branch, Las Vegas, NV 89119 USA

Correspondence: Martin A. Briggs (mbriggs@usgs.gov)

Abstract. Brook trout (*Salvelinus fontinalis*) spawn in fall and overwintering egg development can benefit from stable, relatively warm temperatures in groundwater-seepage zones. However, eggs are also sensitive to dissolved oxygen concentration, which may be reduced in discharging groundwater (i.e., seepage). We investigated a 2 km reach of the coastal Quashnet River in Cape Cod, Massachusetts, USA, to relate preferred fish spawning habitats to geology, geomorphology, and discharging groundwater geochemistry. Thermal reconnaissance methods were used to locate zones of rapid groundwater discharge, which were predominantly found along the central channel of a wider stream valley section. Pore-water chemistry and temporal vertical groundwater flux were measured at a subset of these zones during field campaigns over several seasons. Seepage zones in open-valley sub-reaches generally showed suboxic conditions and higher dissolved solutes compared to the underlying glacial outwash aquifer. These discharge zones were cross-referenced with preferred brook trout redds and evaluated during 10 years of observation, all of which were associated with discrete alcove features in steep cutbanks, where stream meander bends intersect the glacial valley walls. Seepage in these repeat spawning zones was generally stronger and more variable than in open-valley sites, with higher dissolved oxygen and reduced solute concentrations. The combined evidence indicates that regional groundwater discharge along the broader valley bottom is predominantly suboxic due to the influence of near-stream organic deposits; trout show no obvious preference for these zones when spawning. However, the meander bends that cut into sandy deposits near the valley walls generate strong oxic seepage zones that are utilized routinely for redd construction and the overwintering of trout eggs. Stable water isotopic data support the conclusion that repeat spawning zones are located directly on preferential discharges of more localized groundwater. In similar coastal systems with extensive valley peat deposits, the specific use of groundwater-discharge points by brook trout may be limited to morphologies such as cutbanks, where groundwater flow paths do not encounter substantial buried organic material and remain oxygen-rich.

1 Introduction

The heat tracing of water can be used to map a distribution of spatially focused, or "preferential", groundwater-discharge zones throughout surface water systems at times of contrast between the surface and groundwater temperature. The measurement of the water temperature from the reach to watershed scale is now possible using thermal infrared and fiber-optic distributed temperature sensing (FO-DTS) methodology (Dugdale, 2016; Hare et al., 2015; Steel et al., 2017).

Remote infrared data collection throughout the river corridor has been enabled by handheld cameras, piloted aircraft, and the rapidly evolving capabilities of unmanned aerial systems. Researchers are capitalizing on the ongoing refinement of these technologies to identify zones of focused groundwater seepage to streams in order to map potential discrete preferential cold-water fish habitats such as summer thermal refugia (Dugdale et al., 2015). However, surface thermal surveys alone do not indicate groundwater flow path dynamics or the suitability of an interface aquatic habitat (Briggs et al., 2018a).

For example, dissolved oxygen (DO) concentration must be sufficiently high for cold groundwater seepage to provide support for fish life processes at the direct point of discharge to surface water (Ebersole et al., 2003), which is not apparent from thermal analysis alone. During warm summer periods in systems with suboxic groundwater, cold-water fish species such as salmonids can face a tradeoff between occupying discrete zones of preferred water temperatures with near-lethal DO levels and stream sections that are too warm for long-term survival (Matthews and Berg, 1997). The use of groundwater upwelling zones as thermal refugia is further complicated by competition with aggressive invasive species (to the northeastern USA) such as brown trout, which compete with native trout for resources (Hitt et al., 2017). Streams at higher elevations may support the persistence of reach-scale cold-water habitats where point-scale thermal refugia are not needed under current climatic conditions, serving as vital "climate refugia" against rising air temperatures (Isaak et al., 2015). In systems with reliably cold channel water in summer, which can also exist at low elevations when heavily influenced by discharging groundwater, salmonid fish may directly use groundwater-seepage zones for spawning rather than thermal refuge.

Brook trout (*Salvelinus fontinalis*) are a species of char that are native to eastern North America, from Georgia to Québec (MacCrimmon and Campbell, 1969). Populations have been stressed by warming temperatures and reduced water quality, particularly in low-elevation areas (Hudy et al., 2008). Stream network-scale tracking of fish has indicated that the brook trout directly utilize stream confluence mixing zones and preferential groundwater discharge to survive warm summer periods (Baird and Krueger, 2003; Petty et al., 2012; Snook et al., 2016). Additionally, brook trout spawn in the fall, and eggs deposited in redds develop over the winter before hatching in spring (Cunjak and Power, 1986). Oxygen use by the shallow buried embryos increases over the period of development (Crisp, 1981); therefore, DO concentration is a critical parameter of the pore water in which the eggs are bathed. Several studies have demonstrated the importance of hyporheic downwelling in increasing shallow oxygen concentrations, including for salmonid redds, where deeper stream-bed pore water is generally reduced in DO (e.g., Buffington and Tonina 2009; Cardenas et al., 2016; Harvey et al., 2013). Fine sediments can reduce the efficacy of hyporheic DO exchange in spawn zones (Obruca and Hauer, 2016) and are actively cleared by trout during the spawning process (Montgomery et al., 1996).

The importance of hyporheic exchange to salmonid spawning may be limited in the lowland streams that are expected to harbor native cold-water species in the 21st century, namely those with strong groundwater influence. Groundwater upwelling reduces the penetration of the hyporheic flow from surface water (Cardenas and Wilson, 2006) and may shut down hyporheic flushing in redds (Cardenas et al., 2016). While hyporheic exchange introduces oxygenated channel water into the shallow stream bed, the downward advection of heat associated with near-freezing surface water in winter will also cool streambed sediments (Geist et al., 2002), potentially impairing egg development. Coaster brook trout, a life-history variant of native brook trout exhibiting potadromous migrations within the Great Lakes, have been shown to specifically prefer groundwater-discharge zones for building redds (Van Grinsven et al., 2012). The development of trout in winter has been found to positively correlate with warmer stream water temperatures as influenced by groundwater seepage (French et al., 2017). Therefore, spatially discrete groundwater-discharge zones with adequate DO may form preferred brook trout spawning habitats (Curry et al., 1995).

Multiscale physical and biogeochemical factors influence temperature and DO concentrations along groundwater flow paths. In river valleys, discharge to the surface water of locally recharged groundwater is expected to emanate from more shallow, lateral flow paths controlled by the local topography (Modica, 1999; Winter et al., 1998). Shallow groundwater flow paths, particularly those within approximately 5 m of the land surface, will be more sensitive to annual air temperature patterns and long-term warming trends due to strong vertical conductive heat exchanges (Kurylyk et al., 2015b). The distance of seeps from upgradient groundwater recharge zones will also affect seepage temperature dynamics and associated aquatic ecosystems due to future changes in surface and recharge temperatures (Burns et al., 2017). Therefore, characterizing the hydrogeochemical attributes of discharging groundwater flow paths is critical in understanding the thermal stability of current and future point-scale preferential brook trout habitats (Briggs et al., 2018a). The complimentary methodology of geophysical remote sensing, geochemical sampling, and vertical bed temperature time series can indicate the physical and chemical properties of groundwater flow paths that source preferential discharge zones utilized routinely by fish for spawning.

Coarse-grained mineral-dominated aquifers with little fine particulate organic matter and low dissolved organic carbon supplies tend to result in generally oxic groundwater conditions (Back et al., 1993). The sandy surficial aquifer of Cape Cod, where our investigation took place, is a classic example of a mineral soil-dominated flow system (Frimpter and Gay, 1979). The flow of groundwater through near-stream organic deposits, however, can result in inverted redox gradients toward the upwelling interface, such that groundwater discharged to surface water is reduced in DO (Seitzinger et al., 2006). In sandy glacial terrain with superimposed peatland deposits, the specific flow patterns of groundwater to surface water in relation to buried peat will influence the groundwater-discharge biogeochemistry. Krause et al. (2013) found that stream-bed groundwater seepage was strongly reduced in DO in zones with peat deposits, likely due to an increase in both near-stream residence time and localized sources of dissolved organic carbon.

Interdisciplinary collaborations between physical and biological scientists are useful to better understand how cold-water species utilize the stream habitat influenced by groundwater discharge and the larger landscape-scale controls on discharge characteristics. While previous hydrogeological research in the coastal stream used for this study had focused on locating and quantifying discrete groundwater discharge (e.g., "cold anomalies", Hare et al., 2015; Rosenberry et al., 2016), here we endeavor to understand the hydraulic and biogeochemical controls on seepage zone distribution utilized directly by native brook trout. In this groundwater-dominated stream (e.g., likely climate refugia), brook trout do not need to occupy discrete inflows for summer thermal refugia but do favor certain upwelling zones for fall spawning. We compare over a decade of visual survey and electronic fish passive integrated transponder (PIT)-tag dropout data regarding repeat brook trout spawning locations to a comprehensive physical and chemical characterization of groundwater-seepage zones across 2 km of stream in order to do the following:

1. identify repeat brook trout spawning locations and determine if they are directly associated with the preferential discharge of groundwater through interface sediments, and
2. develop a hydrogeochemical characterization of trout-preferred groundwater-discharge zones that can aid in their identification in other less-studied systems and potential inclusion in stream habitat restoration efforts.

2 Site description and previous hydrogeologic characterization

Cape Cod is a peninsula in southeastern coastal Massachusetts, USA, composed primarily of highly permeable unconsolidated glacial moraine and outwash deposits. The largest of the Cape Cod sole-source aquifers occupies a western (landward) section of the peninsula (LeBlanc et al., 1986) and is incised by several linear valleys that drain groundwater south to the Atlantic Ocean via baseflow-dominated streams. Strong groundwater discharge to one such stream, the Quashnet River, supports a relatively stable flow regime that has averaged 0.49 ± 0.15 (SD) $m^3\,s^{-1}$ from 1986 to 2015 (Rosenberry et al., 2016). The lower Quashnet River emerges from a narrow sand and gravel valley into a broader area with well-defined lateral floodplains. Historical cranberry farming practices, abandoned in the 1950s, have modified the stream corridor (Barlow and Hess, 1993). Primary modifications included the straightening of the main channel (reducing natural sinuosity), installation of flood-control structures, incision of shallow groundwater drainage ditches in the lateral peatland floodplain, and widespread application of sand to the floodplain surface. The current bank-full width of the main channel averages approximately 4 m.

The Quashnet River has long been recognized as a critical habitat for a naturally reproducing population of native sea-run brook trout (Mullan, 1958) with a genetically distinct population (Annett et al., 2012). Efforts to restore trout habitats by the group Trout Unlimited and others have been ongoing for over 40 years (Barlow and Hess, 1993). These efforts include the removal of flood-control structures, the planting of trees along the main channel, and the addition of wood structures to stabilize banks and provide cover from airborne predators. Furthermore, the Commonwealth of Massachusetts purchased 12.5 ha in 1956 and an additional 146 ha along the lower Quashnet River in 1987 and 1988 to protect the area from development. The Massachusetts Division of Fisheries and Wildlife has been monitoring trout populations since 1988 and their movement since 2007.

The groundwater influence on stream temperature is pronounced, particularly over the 2 km reach above the U.S. Geological Survey gage no. 011058837, below which the stream stage is tidally affected. Ambient regional groundwater temperature is approximately 11 °C (Briggs et al., 2014), and strong conductive and advective exchange with the proximal aquifer maintains the surface water temperature well below the lethal threshold for brook trout (maximum weekly average temperature > 23.3 °C, Wehrly et al., 2007). Therefore, point-scale thermal refugia are not a current concern in this system, as the stream supports a system-scale cold-water habitat that is likely to persist into the future and serve as warming "climate refugia" (Briggs et al., 2018a). In winter, seepage zones can be located as relatively warm anomalies, increasing and buffering surface water temperatures from ambient atmospheric influence.

Previous work has measured relatively large net gains in streamflow over the lower Quashnet River (Barlow and Hess, 1993; Rosenberry et al., 2016), which are attributed to groundwater discharge through direct stream-bed seepage and the harvesting of groundwater from the floodplain platform via relic agricultural drainage ditches. Deployments of fiber-optic temperature sensing (FO-DTS) cables along the thalweg stream-bed interface indicate that the greatest density of focused seepage zones occurs along the broader valley area, approximately 1 km upstream of the U.S. Geological Survey gage (Fig. 1). This zone coincides with the largest gains in net streamflow (Hare et al., 2015). Based on the stream-bed interface temperature data presented by

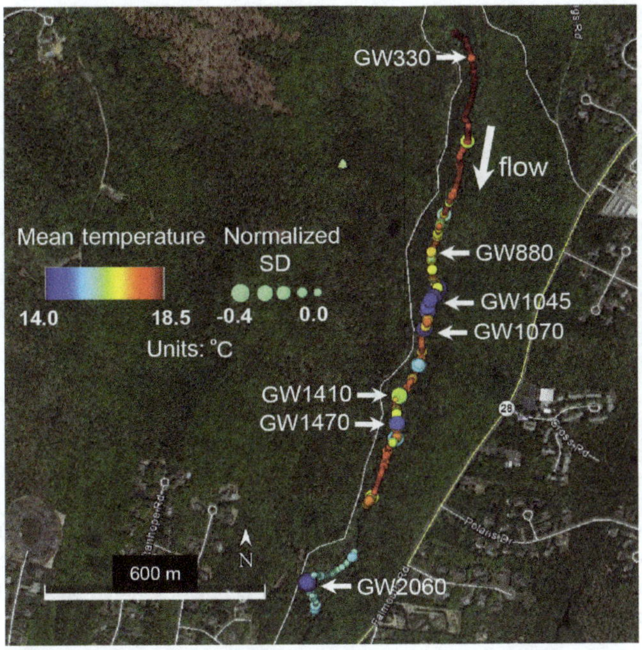

Figure 1. Fiber-optic distributed temperature data collected along the stream channel sediment–water interface over two days in July 2013 are summarized here using mean temperature (color) and temperature standard deviation normalized to known non-seepage locations (size). Locations of reduced mean temperature and the standard deviation of temperature can indicate zones of preferential groundwater upwelling. A subset of these apparent upwelling zones (labeled "GW" followed by the distance from upper reach boundary in meters) with varied thermal statistics was chosen for direct pore-water sampling and quantitative seepage measurements. This figure was modified from Rosenberry et al. (2016).

Rosenberry et al. (2016), Fig. 1 shows how temperature-sensitive fiber optic cables have been used to pinpoint possible groundwater-discharge zones based on an anomalously cold mean temperature and/or reduced thermal variance. A focused evaluation of FO-DTS anomalies with physical seepage meters and vertical temperature profilers confirmed localized, meter-scale seepage zonation along the streambed where discrete colder zones indicated through heat tracing showed approximately 5 times the groundwater-discharge rate of adjacent sandy bed locations only meters away (Rosenberry et al., 2016). The active heating of wrapped FO-DTS cables deployed vertically within an open-valley stream-bed seepage zone indicated the true vertical flow to at least 0.6 m into the bed sediments (Briggs et al., 2016), an expected characteristic of a more regional groundwater discharge (Winter et al., 1998), rather than that of a flow driven by the valley topography local to the river. Hyporheic exchange in the lower Quashnet River system is superimposed on the general upward hydraulic gradient to the stream, therefore being reduced to a thin, shallow hyporheic exchange zone (e.g., < 0.1 m depth) along the thalweg by these competing pressures (Briggs et al., 2014). Vertically compressed hyporheic zones such as these have been simulated for similar stream systems (e.g., Cardenas and Wilson, 2006).

3 Methods

A combination of fish tagging and visual spawning observations, heat tracing, geophysical surveys, and focused pore-water sampling was used to investigate the interplay between the locations of preferential brook trout spawning and the local hydrogeology. For consistency between varied methods and years of data collection, all sample locations are spatially referenced as downstream channel distances from the fish ladder river crossing at the upper end of the study reach (Fig. 2).

3.1 Observations regarding repeat spawning locations

Observations of discrete repeat brook trout spawning locations were made opportunistically as part of an ongoing PIT tagging study of the native reproducing population of the Quashnet River. Large-scale trout movements are continuously monitored in the lower Quashnet River at three stationary fish counting sites (Fig. 2a). However, the spatial resolution of these counting sites, separated by hundreds of meters, is not adequate in studying how brook trout utilize specific decimeter- to meter-scale zones of groundwater discharge. For this finer scale characterization, dropped fish tags have also been located through roving surveys using a handheld portable PIT antenna (Biomark, Inc.), which have been conducted in spring and fall since 2007. The dropout of PIT tags from the fish body is a process that is more likely to happen during spawning behavior in salmonids, so dropped tags were electronically and spatially mapped to reveal discrete zones of repeat spawning. Although these roving surveys do not yield the temporal continuity of the instream counting gates, the clustering of dropped tags can be mapped at the sub-meter scale, presumably directly at trout redds. In addition, spawning brook trout were located visually during annual fall data collection events by Massachusetts Fish and Wildlife Staff, with redd development behavior captured in one seepage feature by an underwater video in 2015 using a GoPro Hero camera (San Mateo, CA). We refer to the three most prominent sites of brook trout spawning within the study reach as Spawn 1 (113 m), Spawn 2 (146 m), and Spawn 3 (2062 m), from upstream to downstream, respectively (Fig. 2).

3.2 Spatial mapping of preferential groundwater discharge

To augment existing stream-bed interface thermal surveys for preferential groundwater discharge (e.g., Rosenberry et al., 2016; Fig. 1) and to investigate the bank dependence of the discharge location, ruggedized fiber-optic cables suit-

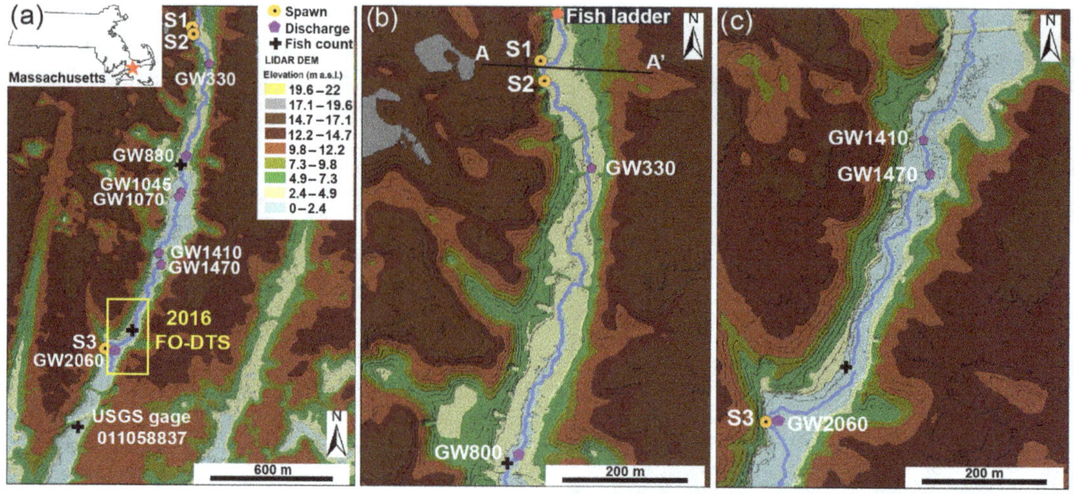

Figure 2. Lidar elevation data show the linear valley terrain of the Quashnet River study reach, as shown in panel **(a)** with Spawn (S1, S2, S3) locations and major open-valley seepage zones identified. The enlarged view of panel **(b)** shows the more narrow upper valley zone where Spawn 1 and 2 are located at the base of a steep cutbank and the topographic transecting point of Fig. 9 (A–A') is noted. Finally, panel **(c)** displays the lower open-valley reach where Spawn 3 is located along a major cutbank.

able for stream use were deployed in the river along the base of each bank from 1700 to 2160 m on 10 to 12 June 2016 (Fig. 2a). Two separate cables weighted with stainless steel armoring were installed directly along the foot of each bank on top of the stream-bed interface. Single-ended measurements made at the 1.01 m linear spatial sampling scale were integrated over 5 min intervals on each channel by an Oryx FO-DTS control unit (Sensornet Ltd.). During the same period, data were also collected along a high-resolution wrapped fiber-optic array for a dataset described in Kurylyk et al. (2017) but not shown here; this experimental setup resulted in measurements for each channel of four instrument channels, which were recorded at 20 min intervals. The calibration for dynamic instrument drift was performed automatically using approximately 30 m of cable for each channel, submerged in a continuously mixed ice bath and monitored with an independent Oryx T-100 thermistor.

3.3 Quantification of vertical groundwater discharge rates

Once preferential discharge locations are located along the stream bed with FO-DTS, actual vertical discharge rates can be assessed using a variety of methodologies (Kalbus et al., 2006). Temporal patterns in the groundwater-discharge flux rate can indicate source flow path hydrodynamics and can be derived from a bed-temperature time series using vertical temperature signal transport characteristics, as reviewed by Rau et al. (2013). Custom "1DTempProfilers" designed specifically for the quantification of groundwater discharge (Briggs et al., 2014) were used to monitor the stream-bed temperature over time along a shallow vertical profile. Profilers were deployed within a subset of the thermal anomalies previously identified with FO-DTS. The profiler deployment locations were chosen to represent a range of preferential groundwater-discharge rates and characteristics based on the on the observed FO-DTS temperature anomalies, e.g., anomalies of the varied mean temperature and buffering effect (Fig. 1) located at 330, 880, 1045, 1070, 1410, 1470, and 2060 m. These groundwater-discharge locations are referred to with the prefix "GW" followed by the meter mark for the remainder of the paper, such that the major stream-bed seep 330 m downstream of the fish ladder is referred to as "GW330". Data were collected at various locations from 11 June to 13 July 2014, 21 August to 13 September 2015, and 5 June to 9 July 2016. These deployments included the installation of 1DTempProfilers at the near-bank and channel sides of observed repeat spawning zones.

Individual thermal data loggers (iButton Thermochron DS1922L, Maxim Integrated) were waterproofed with silicone caulk and inserted horizontally into short slotted-steel pipes (0.025 m diameter). The shallow thermal profilers were driven vertically into the stream bed so that sensors were positioned at some combination of 0.01, 0.04, 0.07, and 0.11 m depths. Data were collected at temporal intervals of 0.5 h in 2014, 0.5 h in 2015, and 1 h in 2016. Rosenberry et al. (2016) found that when a subset of the 2014 stream-bed temperature data presented here were analyzed using the diurnal signal amplitude attenuation models employed by VFLUX2 (Irvine et al., 2015), a near 1 : 1 relation was found in comparison to physical seepage meter measurements of groundwater discharge ranging from 0.5 to 3 m d^{-1}. A similar diurnal signal-based stream-bed thermal parameter estimation is used here.

3.4 Stream-bed groundwater discharge and spawning zone pore-water characterization

Subsurface water samples were collected for chemical analysis at seven major open-valley seepage locations and three repeat spawn locations. Geochemical data collection occurred in 2014 and 2016 along with the 1DTempProfiler deployments, while stable water isotope data were collected in August 2017. For geochemical sampling, 0.0095 m (nominal) stainless steel drive points were inserted to depths of 0.3, 0.6, and/or 0.9 m and Masterflex Norprene tubing was attached to the drive point. A peristaltic pump was used to extract pore-water samples until they were free of obvious turbidity (typically requiring 3 min of pumping), after which the pumping rate was slowed and the groundwater samples were collected by pumping into 60 mL high-density polyethylene (HDPE) syringe barrels. First an unfiltered sample for specific conductivity was pushed from the syringe into a 30 mL HDPE Nalgene sample bottle. Second, a filtered sample for anion analysis was collected after attaching a 0.2 µm pore size (25 mm diameter) Pall polyethersulfone filter to the syringe. Lastly, the pumping rate was slowed again and an overflow cup was attached to the Norprene sample tubing and was held upright until it overflowed, at which point the DO was measured by a field colorimetric test using the manufacturer's evacuated reagent vials (Chemetrics V-2000). DO concentrations were read twice and the test was repeated using an alternative vial kit if results were near the concentration range limit or out of range. The collected samples were kept cool and out of the light and analyzed for Cl^- upon return to the laboratory using standard ion chromatographic techniques.

In addition to the drive point samples, pore-water samples were also collected in June 2016 from shallow depths 0.015, 0.04, 0.08 and 0.15 m below the stream-bed surface at locations GW1045 and Spawn 1, 2, 3 using MINIPOINT samplers (e.g., Harvey and Fuller, 1998). Water was pumped simultaneously from all depths using a multi-head pump that withdrew small-volume samples (15 mL) at low flow rates (1.5 mL min^{-1}) to minimize the disturbance of natural subsurface fluxes and chemical gradients. Pumped lines terminated at press-on luer fittings that were pushed onto 0.2 µm pore size (25 mm diameter) Pall polyethersulfone filters. Samples for specific conductivity were collected, whereas filtered samples were collected for anions in pre-labeled 20 mL LDPE plastic scintillation vials with Polyseal™ caps. Sample lines were then attached to overflow cups and dissolved oxygen concentrations were measured as described above.

During a follow-up field effort in August 2017, stream-bed pore-water samples were collected at the Spawn sites and at GW1045, GW1140 (approximately 70 m downstream of GW1070), and GW1470. Additionally, two large hillslope springs were identified along the edge of the riparian zone, upstream of Spawn 1, using a handheld thermal infrared camera (FLIR T640, FLIR Systems, Inc.). These exposed springs were sampled to identify a localized hillslope groundwater signature that would not be impacted by valley-floor peat deposits. Samples were drawn from push-point piezometers installed 0.2–0.44 m below the sediment interface, with deeper samples collected in the hillslope springs to avoid surface organic material. Pore water was evaluated for specific conductivity (SpC), DO, and stable water isotopes. Isotope samples were analyzed by the U.S. Geological Survey Stable Isotope Laboratory using dual-inlet isotope-ratio mass spectrometry. A substantial fraction of regional Cape Cod shallow groundwater exchanges with the numerous groundwater flow-through lakes as it discharges to the coast (Walter and Masterson, 2002). It is therefore assumed that the regional Cape Cod groundwater isotopic signature is likely to indicate evaporative processes (LeBlanc et al., 2008), offering a contrasting signal from locally recharged hillslope groundwater (no substantial evaporation). The local deuterium excess of contemporary water can indicate groundwater that has been influenced by evaporation in lakes and is therefore in disequilibrium with local meteoric water. Deuterium excess was determined here as $D_{xs} = \delta^2H - 8{\cdot}\delta^{18}O$ (Dansgaard, 1964).

As mentioned previously, historic cranberry farming practices extensively modified the Quashnet River valley, including the incision of drainage ditches into the floodplain. Some ditches extend from the valley wall to the main channel, whereas others are shorter or cut at angles. In addition to characterization of pore water, 34 major drainage ditches (observed flowing water) and a stream thalweg profile were spot-checked for specific conductivity on 16 June 2014 using the SmarTroll probe (YSI). At a subset of these ditch locations, filtered grab samples were collected and analyzed in the laboratory for Cl^- in a similar manner as the mini and drive point samples described above. In June 2016, the dataset was augmented for five ditch confluence locations upstream of Spawn 1.

3.5 Visualizing stream-bed sediment structure

Ground penetrating radars (GPR) have been successfully applied to several surface water and groundwater exchange studies to characterize underlying peat and sandy deposits (e.g., Lowry et al., 2009; Comas et al., 2011) due to strong expected differences in matrix porosity (water content), which can exceed 70 % in peat (Rezanezhad et al., 2016). An upstream to downstream GPR profile was collected on 7 July 2016 using a MALA HDR GX160 shielded antenna (MALA GPR, Sweden), towed down the stream center channel by hand with a small inflatable watercraft. The locations of major seep and spawning sites were specifically marked on the digital GPR record during data collection. The GPR data were processed using Reflexw software (Sandmeier, Germany) to convert reflection time to interface depth.

Figure 3. Several representative images of specific spawn zones and groundwater-discharge zones were collected in February 2016. The cutbank alcove at Spawn 1 is shown in **(a)**, while the open-valley seepage zone GW1045 is shown in **(b)**, and fresh cutbank slumping and visible seepage at Spawn 3 is shown in **(c)**. Underwater imagery collected at the Spawn 1 zone in fall 2015 is displayed in **(d)**, showing several fish clustered directly at the base of the cutbank where pore-water samples were obtained.

4 Results

The hydrogeochemical characterization of observed repeat trout spawning zones and other major stream-bed groundwater-discharge zones are contrasted below.

4.1 Observations regarding repeat spawning locations

Out of the dozens of preferential groundwater-discharge zones geolocated along the Quashnet River in this and previous work (e.g., Fig. 1), brook trout appear to consistently utilize only three discrete stream-bed locations for repeat spawning activity. These locations coincide with steep cutbanks where the river channel approaches the sand and gravel valley wall (Fig. 2b, c). Specifically, trout were found to occupy small "scalloped" alcove-bank features (Fig. 3a) that may be formed by groundwater sapping of fines and the subsequent slumping of sandy bank materials. In winter 2016, fresh slumping and direct seepage from the newly exposed sand wall was observed at Spawn 3 (Fig. 3c); a larger slump event had filled approximately one-third of the scalloped alcove at Spawn 2 by June 2016. Brook trout were observed clustered along the inner bank area at the Spawn 1 location in fall 2015 (Fig. 3d), and this spawning behavior was captured using an underwater video (Supplement).

Dropout PIT tags have been found repeatedly in each of the three preferential spawn zones. Seven dropout PIT tags were located in the Spawn 3 zone in March 2017, by far the most dropped tags found in any one location since the tracking program began in 2007. The only other obvious scalloped bank features along the 2 km study reach are located at GW1045 (Fig. 3b). Compared to the trout spawning zone alcoves along the valley-wall cutbanks (e.g., Fig. 3a), this open-valley seepage alcove was overgrown with watercress and thick (tens of centimeters), loose deposits of organic material.

4.2 Spatial mapping of preferential groundwater discharge

As shown in Fig. 1, previously collected FO-DTS data were used to guide data collection at a subset of representative preferential stream-bed groundwater discharges. Additionally, paired FO-DTS cables were deployed at the base of both stream banks through a lower reach section in 2016 (Fig. 2c), revealing differing thermal anomaly patterns (Fig. 4; Briggs et al., 2018b). The cable along the downstream-right bank captures a large, 8 m long cooler zone at Spawn 3 (Fig. 4b), and this seepage signature is spatially reduced but visible along the opposing bank (Fig. 4a). Other thermal anomalies observed along one bank show little or no signature along the other. Air temperature dropped noticeably over the final 1.5 days of deployment, and smaller cool anomalies that appeared on warm days were no longer captured by the stream-bed FO-DTS deployment, though the Spawn 3 signature is still visible along both cables.

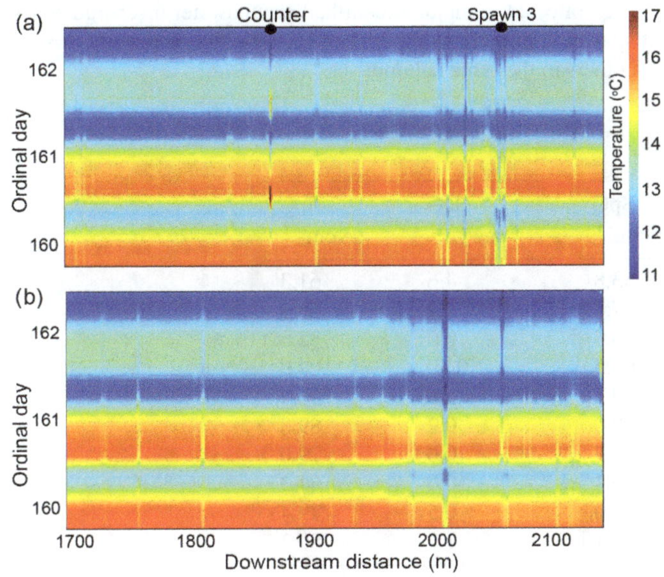

Figure 4. Fiber-optic-distributed temperature data collected from the approximate channel distance of 1700 to 2160 m along **(a)** the downstream right bank through the Spawn 3 meander bend area (see Fig. 2a for location), and **(b)** the downstream left bank along the same stream reach. The persistent vertical bands of relatively cool temperatures indicate discrete groundwater discharge. Some larger zones display a thermal signature on both bank cables, while smaller discharges may be specific to one bank.

4.3 Quantification of vertical groundwater-discharge rates

Ambient stream-bed temperature signal data can be used to measure stream-bed thermal conduction parameters (Luce et al., 2013), which is particularly important when applying heat-based methods to quantify upward vertical fluid flux (Rosenberry et al., 2016), compared to downward fluid-flux models that generally show less sensitivity to stream-bed thermal parameters. Diurnal signal-based thermal diffusivity measurements derived from a pair of 1DTempProfilers inserted in sandy channel sediments for a month in 2014 have the same geometric mean value of $0.11\,\mathrm{m^2\,d^{-1}}$, and this value is used here to model vertical groundwater discharge for all locations and data collection periods (Briggs et al., 2018b). Sub-daily groundwater-discharge fluxes evaluated over similar spring and early summer time periods in 2014 and 2016 show relatively stable patterns at open-valley seepage zones, generally $< 1\,\mathrm{m\,d^{-1}}$ (Fig. 6). At Spawn 1 and 3 seepage is stronger (2 to $3.5\,\mathrm{m\,d^{-1}}$) and more variable than at open-valley zones. The Darcy-based horizontal seepage estimate through the Spawn 3 bank, made using the bank piezometer, is $2.3\,\mathrm{m\,d^{-1}}$, which is similar to the temperature-based seepage rates at the Spawn 3 interface (Fig. 6), and indicates lateral discharge through the cutbank wall from a more localized groundwater flow path. The Spawn 2 zone shows a reduced and more stable discharge rate during summer 2016, and is likely impacted by a large bank slump into this zone that occurred during the winter of 2016, partially filling the alcove. Seepage patterns collected at Spawn 1 and 2 in late-summer 2015 show greater temporal stability, even though the stream stage at the downstream U.S. Geological Survey gage showed substantial variation. Discharge rates along the inner bank wall of the scalloped bank spawn zones were consistently higher than at bed areas located just a few meters away toward the channel.

4.4 Stream-bed groundwater discharge and spawning zone pore-water characterization

Based on previous characterization, the Cape Cod sand and gravel aquifer generally has high DO concentrations (9–11 $\mathrm{mg\,L^{-1}}$), relatively dilute specific conductance (SpC, 62 $\mathrm{\mu S\,cm^{-1}}$), and dilute chloride concentrations (Cl^-, 9.3 $\mathrm{mg\,L^{-1}}$) at depths ranging between 12 and 20 m (Savoie et al., 2012). The groundwater that discharges to the Quashnet River, however, is often strongly variable in all three of these parameters (Harvey et al., 2018). In June 2014, drive point data were primarily collected in open-valley seepage zones identified with FO-DTS (Fig. 1); these locations are suboxic to anoxic at 0.3 and 0.6 m stream-bed depths (Table 1). The highest stream-bed seepage DO is found at GW330 in the tighter upstream valley section (4.6 $\mathrm{mg\,L^{-1}}$ at both depths) and Spawn 3, where DO is 9.0 and 7.6 $\mathrm{mg\,L^{-1}}$ at 0.3 and 0.6 m depths, respectively (Table 1). SpC is also variable, but lowest and similar to the regional signal at GW330 and Spawn 3. Note that SpC and Cl^- are used here to indicate aquifer flow path hydrogeochemical properties and not unsuitable spawn habitats based on chemical concentration, as their range is well within general brook trout tolerances.

Drive point data collected at the 0.3 m depth in June 2016, primarily around spawn zones, generally show high DO and relatively low SpC at the interior of Spawn zones 1 and 3 near the cutbank (Table 1). Data collected a few meters toward the main channel from these near-bank spawn locations are reduced in DO with increased SpC. The Spawn 2 data were collected at the toe of the recent large sediment slump that had partially filled the alcove, and DO data are suboxic at 0.3 m (3.9 $\mathrm{mg\,L^{-1}}$) but more oxygen-rich at 0.9 m depth (7.2 $\mathrm{mg\,L^{-1}}$), indicating the potential for shallow stream-bed respiration that removes oxygen from discharging groundwater (assuming vertical flow) in the slumped material. In contrast to the spawn zones, the major open-valley seepage location GW1045 is nearly anoxic at all depths with SpC similar to the 2014 stream water profile grab samples ($n = 8$, $101.4 \pm 1.7\,\mathrm{\mu S\,cm^{-1}}$). Little difference was observed between near-bank and channel positions at GW1045 (both are suboxic) even though a large scalloped seepage bank feature was observed (Fig. 3b).

The drainage-ditch grab samples generally show Cl^- concentrations that are lower than the average 2014 channel grab samples ($n = 10$, $19 \pm 0.4\,\mathrm{mg\,L^{-1}}$), though the two most

Table 1. This table lists 2014 and 2016 drive point pore-water chemistry data collected in major stream-bed groundwater-discharge zones located with fiber-optic heat tracing and in zones of observed repeat trout spawning directly along the bank and farther toward the stream center channel. The italicized values indicate sample depths that differ from others in the same column.

Open valley groundwater discharges	0.3 m depth		0.6 m depth	
	DO $mg L^{-1}$	SpC $\mu S cm^{-1}$	DO $mg L^{-1}$	SpC $\mu S cm^{-1}$
GW330	4.6	53.8	4.6	61.3
GW880	1.4	97.7	3.4	65.1
GW1045	0.1	78.8	0.0	82.5
GW1045 (bank)	0.16	105.5	0.39	104.0
GW1045 (channel)	0.31	99.1	0.18	96.4
GW1070	0.2	100.0	0.2	89.8
GW1410	0.0	77.7	0.0	79.0
GW1470	0.1	69.1	0.0	64.3
GW2060	1.4	75.0	0.5	79.4
mean	0.9	84.1	1.0	80.2
Spawning locations (channel)	0.3 m depth		0.9 m depth	
Spawn 1 channel	4.41	143.9	5.68	143.2
Spawn 2 channel	5.25	139.3	n/a	n/a
Spawn 3 channel	1.76	82.1	2.68	79.9
mean	3.8	121.8	4.2	111.6
Spawning locations (bank)	0.3 m depth		0.9 m depth	
Spawn 1 bank	7.28	70.6	9.76	55.9
Spawn 2 bank	3.89	70.8	7.17	57.6
Spawn 3 bank (2016)	9.11	60.4	4.91	71.9
Spawn 3 bank (2014)	9.0	56.4	7.6 (*0.6 m*)	60.9 (*0.6 m*)
mean	7.3	64.6	7.4	61.6

n/a: not applicable.

upstream ditches are similar to stream water, and 2 open-valley ditches are appreciably higher in Cl^- (Fig. 7a). Spawn zones 1, 2, and 3 approximate the lowest Cl^- concentrations observed in drainage ditches, and Spawn 3 has a similar concentration to the adjacent 2016 stream-bank piezometer in both the 2014 and 2016 data. An analogous pattern is shown in the more widespread SpC data, with many drainage ditches and all spawn zones having concentrations around 60 $\mu S cm^{-1}$. However, several ditches cluster around the stream water average or higher, particularly in the open-valley area.

The shallow, shallow pore-water samples collected with the MINIPOINT system in discrete intervals show that stream-bed SpC is appreciably lower than stream water, even at the 0.02 m depth, at all near-bank spawn zones (Fig. 8a). Conversely, the shallow channel sediments at Spawn 1 and open-valley seepage at GW1045 approximate the stream water value for SpC. DO is high and stable along the shallow profiles (to 0.14 m) at the interior of Spawn zones 1 and 3 but suboxic at the Spawn 1 channel sample and Spawn 2 zones and essentially anoxic along the bank at GW1045. Center channel pore-water samples at GW1045 show moderate oxygen enrichment at 0.02 m (4.6 $mg L^{-1}$), which may result from hyporheic mixing, as deeper intervals along the same profile are nearly anoxic.

The underwater video collected here in the fall of 2015 indicates Quashnet River brook trout clustered tightly around an approximate 1 m^2 bed area in Spawn 1 (Fig. 3d, Supplement), directly at the base of the sandy cutbank. During the June 2016 collection of pore-water data, drive points were installed precisely in this area. A chemical analysis of 0.3 m deep pore water shows a strong gradient from the near-bank Spawn 1 zone to the outer alcove area, with specific conductance rising dramatically (70.6 to 143.9 $\mu S cm^{-1}$) and DO falling (7.28 to 4.41 $mg L^{-1}$) (Table 1). Spawn 3 shows a similar pattern from the near-bank zone toward the main channel (60.4 to 82.1 $\mu S cm^{-1}$ SpC; 9.11 to 1.76 $mg L^{-1}$ DO). Spawn 2, although complicated by the large slump during the previous winter, shows an increase in SpC from 70.6 to 139.3 $\mu S cm^{-1}$ from the inner to outer alcove. Conversely, pore water collected at 0.3, 0.6, and 0.9 m depths in the open-valley seepage alcove at GW1045 (pictured in Fig. 3b) are functionally anoxic with elevated SpC

Table 2. This table lists 2017 drive point pore-water chemistry and stable water isotope data collected in a subset of major stream-bed groundwater-seepage zones, zones of observed repeat trout spawning, and from springs located above the waterline along the same hillslope as the meander cutbanks of Spawn 1 and Spawn 2.

Location	Sample depth (m)	SpC ($\mu S\,cm^{-1}$)	DO ($mg\,L^{-1}$)	δ^2H (‰)	$\delta^{18}O$ (‰)	D_{xs} $\delta^2H - 8^{\bullet}\delta^{18}O$
Hillslope 1	40	74.82	5.004	−51.38	−8.2	14.22
Hillslope 2	44	60.59	9.318	−51.81	−8.73	18.03
Spawn 1	20	72.45	6.853	−48.9	−7.9	14.3
Spawn 2	20	51.75	5.419	−48.2	−7.95	15.4
Spawn 3	20	42.62	9.054	−44.32	−7.33	14.32
GW1045	20	109.8	0.043	−34.03	−4.93	5.41
GW1140	20	103.4	0.043	−32.56	−4.8	5.84
GW1470	20	97.68	0.04	−33.05	−4.72	4.71

compared to inner spawn zones and have little gradient from the bank to the channel.

Pore-water data collected in August 2017 indicate that all three Spawn sites are similar to emergent hillslope springs, characterized by relatively high DO and low SpC, compared to major open-valley stream-bed seepage zones that are anoxic with higher SpC (Table 2). Additionally, the stable isotopic signatures of the hillslope and Spawn zones are similar, but are contrasted by the lower deuterium excess metric determined for the open-valley seepages. This indicates that groundwater discharging through the stream bed away from the hillslope shows the evaporative signature of groundwater flow-through lakes and can therefore be considered regional discharge, compared to locally recharged hillslope groundwater apparently favored by trout for spawning.

4.5 Visualizing stream-bed sediment structure

Radar data were collected over most of the study reach length depicted in Fig. 2a, and although spatial reference data were not collected for each sample point due to integrated global positioning system failure, Spawn and groundwater-discharge zones of interest were precisely marked in the record (Fig. 5). The GPR data collected along the thalweg adjacent to Spawn 1 and 2 indicate that a contiguous thin layer of material underlies the sandy stream bed that may be peat deposited over deeper sands and gravels (Fig. 5a). The GPR profile through open-valley groundwater-discharge locations GW1045 and GW1070 shows the strongest radar signal reflectors of anywhere along the open-valley section (Fig. 5b). These discontinuous geologic structures are interpreted as layered sand and gravel, interspersed with thicker peat deposits. Otherwise, discontinuous reflections indicative of sediment-type interfaces of variable depths are observed near the downstream open-valley seepage zones and strongly attenuated GPR signals indicate thick lenses of buried peat with high water content (Fig. 5b,c).

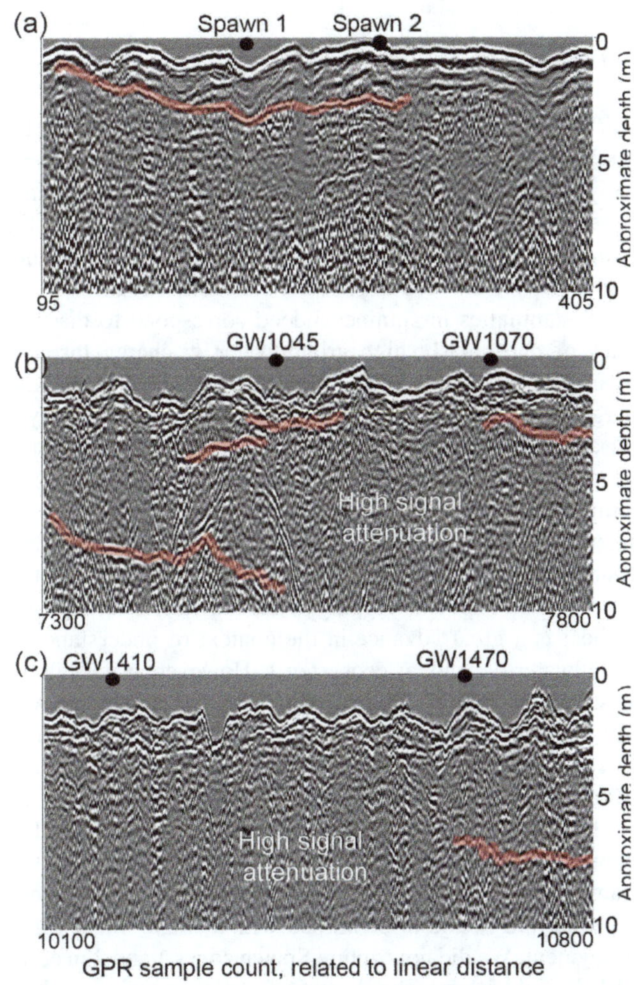

Figure 5. These images show ground-penetrating radar profiles collected down the center of the river channel to indicate peat, sand, and gravel layering in the stream bed. Stronger apparent radar reflectors are highlighted in red and likely indicate sediment layer boundaries (e.g., sand and gravel vs. peat). Spawn- and groundwater-discharge locations were directly marked in the radar data stream during collection and are shown for each sub-reach panel.

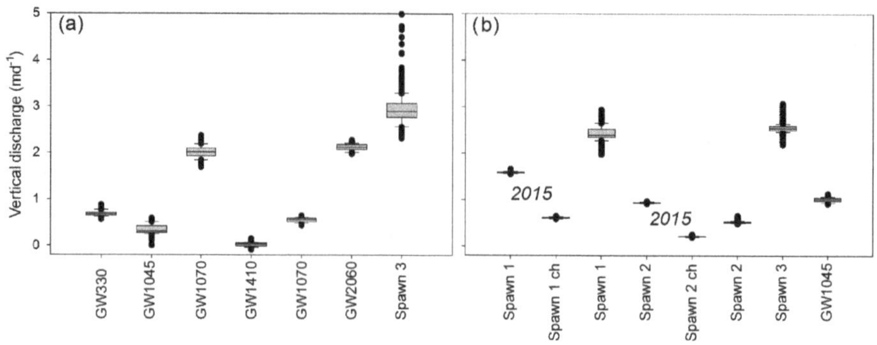

Figure 6. Summarizing box plots of sub-daily vertical groundwater-discharge rates modeled for the open-valley groundwater discharge and Spawn 3 bank locations for the 11 June to 13 July 2014 period are shown in panel **(a)**. Additionally, panel **(b)** displays discharge rates collected in Spawn and GW1045 locations directly against the cutbanks and farther out towards the channel (indicated by "ch") for the 21 August to 13 September 2015 and 5 June to 9 July 2016 periods.

5 Discussion

Heat tracing reconnaissance technologies, such as FO-DTS and thermal infrared, offer an efficient means to comprehensively characterize preferential groundwater-discharge points at the reach to watershed scale (Briggs and Hare, 2018). Using the groundwater-fed Quashnet River as an example, Rosenberry et al. (2016) showed that cold stream-bed interface anomalies in summer indeed correspond to discrete zones of particularly high groundwater discharge through stream-bed sediments. This spatial characterization of discharge points alone is not sufficient to understand the physical and chemical drivers of a niche habitat, but can efficiently guide additional data collection, as was done here. Compared to more randomly distributed stream-bed field parameter surveys and larger spatial scale evaluations of net groundwater discharge made with differential gaging, the comprehensive spatial mapping of groundwater discharge using heat is a great advance in the context of understanding groundwater-dependent ecosystems. However, in fast flowing streams, FO-DTS cable placement on the stream bed will likely impact which specific groundwater-discharge zones are captured with FO-DTS, as shown here by applying cables along opposite banks through the Spawn 3 area (Fig. 4). The largest seepage zones may have a spatial footprint that encompasses the stream-bed area from bank to bank (e.g., the Spawn 3 cold anomaly), but a subset of more discrete seepage zones are bound to be missed with a single linear cable deployment. We did not capture Spawn zones 1 and 2 in early FO-DTS field efforts (Fig. 1), but fish tracking indicated their importance in regards to trout spawning behavior. Therefore, in studies of niche stream habitats as influenced by preferential groundwater discharge, a combination of heat tracing and biological observation may be needed to both identify major discharge points and discern which points are directly used by the biota of interest (e.g., brook trout).

In a study of the regional Cape Cod aquifer condition, Frimpter and Gay (1979) state that groundwater is typically near DO saturation, except in the case of the downgradient of peat or river bottom sediments, where consumption of DO allows the mobilization of natural iron and manganese. Visible observations along the open-valley section, in addition to stream-bed sediment coring (Briggs et al., 2014), revealed the widespread coating of shallow stream-bed sediment grains with metal oxides, consistent with the conceptual model of organic material influence on near-surface groundwater (Fig. 9). Aquifer recharge passing through upgradient groundwater flow-through kettle lakes (e.g., Stoliker et al., 2016) may also serve to decrease the DO content of the regional flow paths that discharge vertically through the bed of the Quashnet River, although we hypothesize that localized peat deposits may be the primary control on both seepage zone distribution and chemistry.

Out of the dozens of preferential groundwater-discharge zones located along the lower Quashnet with heat tracing, most were suboxic to anoxic (Table 1). Brook trout consistently prefer three areas for fall spawning, all along meander bend cutbanks into the sand and gravel valley wall. Zones of locally enhanced seepage, likely controlled by subtle differences in sediment hydraulic conductivity, can lead to the groundwater sapping of fines, reduction in bank stability, and consequent slumping of bank material into the river; this process was observed in real time at the Spawn 3 meander in February 2016 (Fig. 3c). Slumping effectively forms *seepage-driven* alcoves outside of the main flow and are more suitable for redd placement, along with forming a more favorable coarse sand and gravel substrate (Bowerman et al., 2014; Hausle and Coble, 1976; Raleigh, 1982).

In other systems, trout have been observed to occupy microhabitat around and within groundwater-discharge zones, even being segregated by fish size and desirable temperature range (e.g., Fig. 2.4.1.2 in Torgersen et al., 2012). Here, real-time observation and visual imagery show trout clustering tightly against the bank in Spawn 3 (Fig. 3d, Supplement) where pore water was found to be more oxygen rich and lower in SpC. The month-long time series of vertical

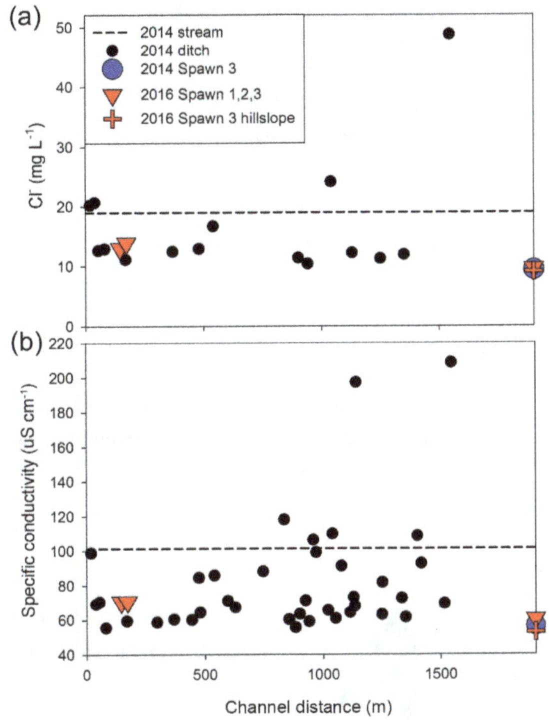

Figure 7. Drainage ditch chemistry throughout the lower Quashnet, showing **(a)** Cl^- and **(b)** specific conductance that was collected in June 2014, just above the confluence with the main channel. Data are plotted as the distance from the upper flood control structure in the narrow valley reach and are compared to groundwater-seepage data collected in preferential spawning locations and a hillslope piezometer.

groundwater-discharge rates are reduced considerably from the near-bank to the near-channel areas at all spawning zones (Fig. 6), indicating in part a reduction in stream-bed hydraulic conductivity as influenced by peat deposits under the main channel and as observed in GPR data (Fig. 5). The evidence of higher near-bank vertical groundwater flux rates and DO combined with lower SpC indicates limited interaction between the shallow groundwater flow paths and peat against the meander bend cutbanks. As observed in other systems, it appears that even short travel distances through organic deposits toward the center channel at Spawn 1 and 2 may be sufficient in increasing total dissolved solids, depleting DO (e.g., Levy et al., 2016), and rendering upwelling zones undesirable for redd construction. Therefore, near-surface channel sediments may need to be specifically characterized in preferential groundwater-discharge zones, as net chemical reactivity over the last ~ 1 m of transport may dominate net chemical change of the discharging groundwater.

The alcove seepage features utilized by trout in this study are apparently similar to the numerous cold-water alcove patches observed in another stream system by Ebersole et al. (2003). In that study of preferential salmonid habitats, alcoves were often located where streams converged on valley walls and were the most abundant type of discrete cold-water habitat type identified. Conversely, valley-wall alcoves were the least common type of seep morphology observed along the Quashnet River. It is likely that the artificial reduction in channel sinuosity along the Quashnet River by farming practices has reduced the number of natural higher-quality spawning locations.

Other bank and alcove features with strong groundwater discharge found along the open-valley section (Fig. 3b) were highly influenced by organic material deposition and did not apparently support spawning habitats. Our research indicates that in lowland systems with organic-rich floodplain sediments, valley-wall alcoves alone create a favorable brook trout spawning habitat via local mineral soil-dominated groundwater-discharge flow paths, as shown in conceptual Fig. 9. This finding might help inform future ecologically based stream restoration practices in using the natural landscape to predict desirable preferential groundwater-discharge points, as was recently done by Hare et al. (2017) to inform the engineering of a large-scale cranberry bog restoration.

The pore-water SpC, Cl^-, and DO data alone do not definitively show that seepage at the cutbank spawn sites is derived from more localized groundwater recharge, as opposed to regional groundwater that is unadulterated by buried peat lenses. However, the hydrodynamic data derived from long-term vertical temperature profiling in seepage zones does offer additional insight. In general, groundwater-discharge rates are more variable at cutbank spawn zones than in the open-valley stream-bed zones (Fig. 6), and this variability may be tied to shorter-term changes in local river stage and/or water table depth, impacting the local hydraulic gradient. The relatively stable patterns of open-valley groundwater discharge may be controlled by the regional gradient, where the flow path length term dominates the Darcy relation and is therefore relatively insensitive to local changes in river stage and water table fluctuations. Furthermore, the stable water isotope data display evaporative signatures at the open-valley stream-bed discharge sites, indicating regional groundwater that has passed through one or more upgradient flow-through lakes (Table 2). In contrast, the Spawn sites all show isotope signals that fall along the local meteoric waterline and therefore likely represent recharge to the hillslopes more local to the river. These localized groundwater flow systems would be expected to be less influenced by regional groundwater contamination, which is widespread in the regional Cape Cod aquifer (Walter and Masterson, 2002).

Groundwater drainage-ditch data collected along the river corridor indicate that low SpC/Cl^- conditions exist for the majority of ditches throughout the lower Quashnet River riparian areas (Fig. 7). The hillslope piezometer in sand and gravel at the down valley wall has a similar chemical signature along with high DO. This similarity further indicates that low-SpC groundwater discharges even to the lower por-

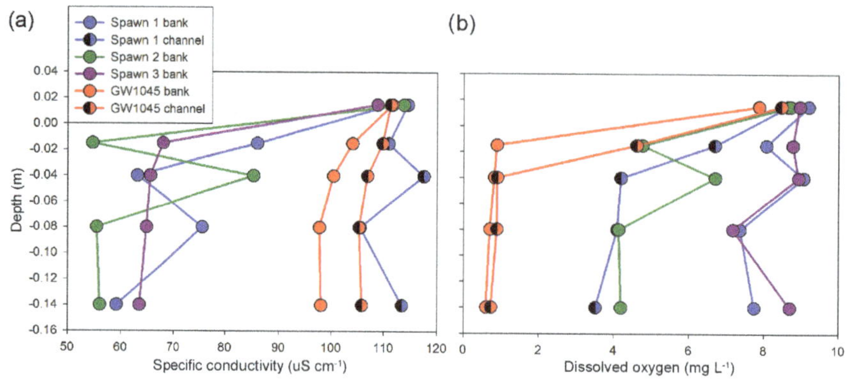

Figure 8. Minipoint pore-water chemistry data showing high spatial resolution profiles of **(a)** specific conductance and **(b)** dissolved oxygen, collected in June 2016 at the major seepage alcoves. Triangle symbols indicate data collected farther toward the thalweg from the respective alcove bank, and all profiles include a local stream water sample taken just above the stream-bed interface.

tion of the river corridor but is chemically modified by travel through near-stream organics. The relic drainage ditches allow the discharging groundwater to effectively short-circuit the valley floor peat deposits and remain high in DO, similar to the natural valley-wall springs and cutbank alcoves. Future restoration strategies that seek to actively enhance groundwater discharge (e.g., Kurylyk et al., 2015a) may consider capitalizing on this short circuit behavior, possibly by auguring through buried stream-bed peat or through the movement of the stream channel toward the valley wall to create more desirable brook trout aquatic habitat.

6 Conclusions

The three repeatedly utilized discrete spawning zone locations that have been identified for over a decade of observation have coupled strongly discharging groundwater with high DO concentration. A conceptual diagram of the hydrogeochemical setting of spawn zones vs. other non-favorable stream-bed locations of groundwater discharge is shown in Fig. 9. Spawn zones are located exclusively in side alcoves of the channel created by bank slumps along meanders, where the river cuts into steep hillslopes along the glacial sands and gravel valley wall. In the alcoves at the base of the cutbanks, hillslope groundwater with high DO concentrations is discharged through the stream bed without appreciable loss of oxygen. Just a few meters away toward the main channel, however, groundwater consistently discharges at lower rates, reduces in DO, and increases in SpC. The lowest oxygen concentrations in groundwater are associated with water emerging from the stream bed adjacent to the wide riparian areas that flank the Quashnet in the open-valley section of the study reach, even though groundwater-discharge rates were also relatively high. In the open valley, where the stream is not near the valley walls, proximity to the stream bank does not seem to control seepage chemistry, and GPR data indicated thick zones of discontinuous stream-bed peat. In this

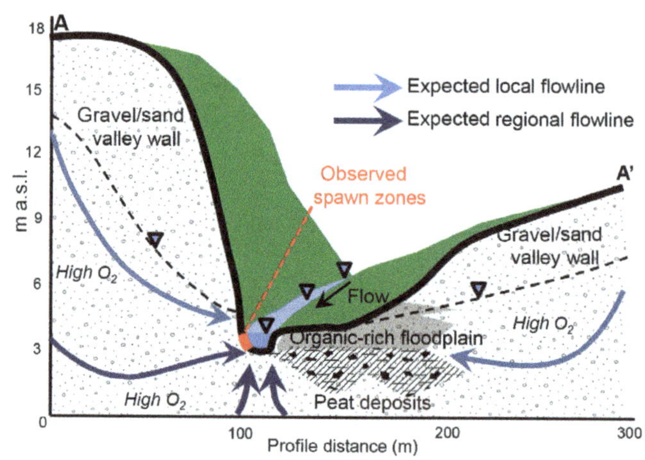

Figure 9. This conceptual model shows how valley-wall cutbank discharge zones are likely sourced by locally recharged hillslope groundwater that avoids substantial interaction with valley-floor organic material. The discharging groundwater remains oxygen-rich, therefore supporting trout spawning activity along discrete stream-bed sections at the meter scale. The topographic profile shown here (A–A') is derived from airborne lidar data and is oriented perpendicular to the stream at the Spawn 1 zone, as geolocated in Fig. 2b.

and other groundwater-dominated streams that are expected to serve as climate refugia for future native trout populations, hyporheic exchange will be limited by a strong upward hydraulic gradient. Therefore, preferential spawning habitat in such lowland valley systems may be primarily supported by discrete zones of oxic groundwater upwelling at the meter to sub-meter scale, as has been indicated by previous work (e.g., Curry et al., 1995).

In systems where all groundwater discharge is universally anoxic, preferential salmonid spawning zonation may be controlled by points of downwelling hyporheic water where shallow sediments remain high in DO (Buffington and Tonina, 2009; Cardenas et al., 2016). However, these hyporheic areas will deliver cold surface water to shallow sediments during winter, which may impair the overwintering of brook trout eggs (French et al., 2017). Here and in many other coastal systems, groundwater temperature is expected to range from approximately 10–12 °C, which is an ideal range for brook trout egg development (Raleigh, 1982). Points of oxic groundwater upwelling devoid of near-stream buried organics, combined with a recirculating side alcove and favorable sand and gravel sediments, may provide an ideal and unique and preferential spawning habitat for native trout.

Stream surface or stream-bed interface heat tracing of groundwater discharge offers an efficient means to locate discrete seepage zones but offers only limited insight into source groundwater flow path hydraulics and geochemistry. A combined toolkit that also includes spatially informed (using heat tracing) geochemical and isotope sampling and geophysical imaging can be used to trace groundwater flow paths back into the source aquifer, and develop a robust hydrogeochemical characterization. Additionally, as digital elevation models become more refined and combined with infrared data derived from unmanned aerial systems, the remote identification of relatively small features such as the seepage alcoves described here should be possible.

A comprehensive and process-based characterization of a niche stream habitat can be used to guide a stream ecological restoration design that directly incorporates the local preferential groundwater-discharge template.

Author contributions. All authors contributed to the analysis of field data and the development of this paper.

Competing interests. The authors declare that they have no conflict of interest.

Acknowledgements. Comments from anonymous reviewers and U.S. Geological Survey (USGS) reviews by Nathaniel Hitt and Paul Barlow are gratefully acknowledged. The U.S. Environmental Protection Agency (USEPA) through its Office of Research and Development partially funded and collaborated in the research described here under agreement number DW-14-92381701 to the USGS. The USGS authors were supported by the following USGS entities: the Office of Groundwater, Water Availability and Use Science Program, National Water Quality Program, and the Toxic Substances Hydrology Program. Field and laboratory assistance from Allison Swartz, Jay Choi, Jenny Lewis, Yao Du, Danielle Hare, Courtney Scruggs, Rayna Mitzman, David Rey, Geoff Delin, Eric White, MassWildlife Southeast District Staff, Jennifer Salas, and volunteers from Trout Unlimited is greatly appreciated. The paper has been subjected to Agency review and approved for publication. The views expressed in this article are those of the authors and do not necessarily represent the views or policies of the USEPA. Any use of trade, firm, or product names is for descriptive purposes only and does not imply endorsement by the U.S. government.

Edited by: Alberto Guadagnini

References

Annett, B., Gerlach, G., King, T. L., and Whiteley, A. R.: Conservation Genetics of Remnant Coastal Brook Trout Populations at the Southern Limit of Their Distribution: Population Structure and Effects of Stocking, T. Am. Fish. Soc., 141, 1399–1410, https://doi.org/10.1080/00028487.2012.694831, 2012.

Back, W., Baedecker, M. J., and Wood, W. W.: Scales in chemical hydrogeology: a historical perspective, in: Regional Ground-Water Quality, edited by: Alley, W. M., Van Nostrand Reinhold, New York, 111–128, 1993.

Baird, O. E. and Krueger, C. C.: Behavioral thermoregulation of brook and rainbow trout: comparison of summer habitat use in an Adirondack River, New York, T. Am. Fish. Soc., 132, 1194–1206, 2003.

Barlow, P. M. and Hess, K. M.: Simulated Hydrologic Responses of the Quashnet River Stream-Aquifer System to Proposed Ground-Water Withdrawals, Cape Cod, Massachusetts, USGS, Water-Resources Investigations Report, 93-4064, 51 pp., 1993.

Bowerman, T., Neilson, B. T., and Budy, P.: Effects of fine sediment, hyporheic flow, and spawning site characteristics on survival and development of bull trout embryos, Can. J. Fish. Aquat. Sci., 71, 1059–1071, 2014.

Briggs, M. A. and Hare, D. K.: Explicit consideration of preferential groundwater discharges as surface water ecosystem control points, Hydrol. Process., 2, 2435–2440, https://doi.org/10.1002/hyp.13178, 2018.

Briggs, M. A., Lautz, L. K., Buckley, S. F., and Lane, J. W.: Practical limitations on the use of diurnal temperature signals to quantify groundwater upwelling, J. Hydrol., 519, 1739–1751, https://doi.org/10.1016/j.jhydrol.2014.09.030, 2014.

Briggs, M. A., Buckley, S. F., Bagtzoglou, A. C., Werkema, D., and Lane, J. W.: Actively heated high-resolution fiber-optic distributed temperature sensing to quantify flow dynamics in zones of strong groundwater upwelling, Water Resour. Res., 52, 5179–5194, https://doi.org/10.1002/2015WR018219, 2016.

Briggs, M. A., Johnson, Z. C., Snyder, C. D., Hitt, N. P., Kurylyk, B. L., Lautz, L., Irvine, D. J., Hurley, S. T., and Lane, J. W.: Inferring watershed hydraulics and cold-water habitat persistence using multi-year air and stream temperature signals, Sci. Total Environ., 636, 1117–1127, https://doi.org/10.1016/j.scitotenv.2018.04.344, 2018a.

Briggs, M. A., Scruggs, C. R., Hurley, S. T., and White, E. A.: Temperature and geophysical data collected along the Quashnet

River, Mashpee/Falmouth MA, U.S. Geological Survey data release, https://doi.org/10.5066/F7PN93QF, 2018b.

Buffington, J. M. and Tonina, D.: A three-dimensional model for analyzing the effects of salmon redds on hyporheic exchange and egg pocket habitat A three-dimensional model for analyzing the effects of salmon redds on hyporheic exchange and egg pocket habitat, Can. J. Fish. Aquat. Sci., 66, 2157–2173, https://doi.org/10.1139/F09-146, 2009.

Burns, E. R., Zhu, Y., Zhan, H., Manga, M., Williams, C. F., Ingebritsen, S. E., and Dunham, J.: Thermal effect of climate change on groundwater-fed ecosystems, Water Resour. Res., 53, 3341–3351, https://doi.org/10.1002/2016WR020007, 2017.

Cardenas, M. B. and Wilson, J. L.: The influence of ambient groundwater discharge on exchange zones induced by current-bedform interactions, J. Hydrol., 331, 103–109, 2006.

Cardenas, M. B., Ford, A. E., Kaufman, M. H., Kessler, A. J., and Cook, P. L. M.: Hyporheic flow and dissolved oxygen distribution in fish nests: the effects of open channel velocity, permeability patterns, and groundwater upwelling, J. Geophys. Res.-Biogeosci., 121, 3113–3130, https://doi.org/10.1002/2016JG003381, 2016.

Comas, X., Slater, L., and Reeve, A. S.: Pool patterning in a northern peatland: Geophysical evidence for the role of postglacial landforms, J. Hydrol., 399, 173–184, https://doi.org/10.1016/j.jhydrol.2010.12.031, 2011.

Crisp, D. T.: A desk study of the relationship between temperature and hatching time for the eggs of five species of salmonid species, Freshwater Biol., 11, 361–368, https://doi.org/10.1111/j.1365-2427.1981.tb01267.x, 1981.

Cunjak, R. A. and Power, G.: Seasonal changes in the physiology of brook trout, *Salvelinus fontinalis* (Mitchill), in a sub-Arctic river system, J. Fish Biol., 29, 279–288, 1986.

Curry, R., Noakes, D. L. G., and Morgan, G. E.: Groundwater and the incubation and emergence of brook trout (*Salvelinus fontinalis*), Can. J. Fish. Aquat. Sci., 52, 1741–1749, 1995.

Dansgaard, W.: Stable isotopes in precipitation, Tellus, 16, 436–468, https://doi.org/10.3402/tellusa.v16i4.8993, 1964.

Dugdale, S. J.: A practitioner's guide to thermal infrared remote sensing of rivers and streams: recent advances, precautions and considerations, WIREs Water, 3, 251–268, https://doi.org/10.1002/wat2.1135, 2016.

Dugdale, S. J., Bergeron, N. E., and St-Hilaire, A.: Spatial distribution of thermal refuges analysed in relation to riverscape hydromorphology using airborne thermal infrared imagery, Remote Sens. Environ., 160, 43–55, https://doi.org/10.1016/j.rse.2014.12.021, 2015.

Ebersole, J. L., Liss, W. J., and Frissell, C. A.: Cold water patches in warm streams: physicochemical characteristics and the influence of shading, J. Am. Water Resour. As., 59860, 355–368, 2003.

French, W. E., Vondracek, B., Ferrington, L. C., Finlay, J. C., and Dieterman, D. J.: Brown trout (*Salmo trutta*) growth and condition along a winter thermal gradient in temperate streams, Can. J. Fish. Aquat. Sci., 74, 56–64, https://doi.org/10.1139/cjfas-2016-0005, 2017.

Frimpter, M. H. and Gay, F. B.: Chemical quality of ground water on Cape Cod, Massachusetts, U.S. Geological Survey, Water-Resources Investigations Report 79-65, 1979.

Geist, D. R., Hanrahan, T. P., Arntzen, E. V, Mcmichael, G. A., Murray, C. J., and Chien, Y.: Physicochemical Characteristics of the Hyporheic Zone Affect Redd Site Selection by Chum Salmon and Fall Chinook Salmon in the Columbia River, N. Am. J. Fish. Manage., 22, 1077–1085, 2002.

Hare, D. K., Briggs, M. A., Rosenberry, D. O., Boutt, D. F., and Lane, J. W.: A comparison of thermal infrared to fiber-optic distributed temperature sensing for evaluation of groundwater discharge to surface water, J. Hydrol., 530, 153–166, https://doi.org/10.1016/j.jhydrol.2015.09.059, 2015.

Hare, D. K., Boutt, D. F., Clement, W. P., Hatch, C. E., Davenport, G., and Hackman, A.: Hydrogeological controls on spatial patterns of groundwater discharge in peatlands, Hydrol. Earth Syst. Sci., 21, 6031–6048, https://doi.org/10.5194/hess-21-6031-2017, 2017.

Harvey, J. W. and Fuller, C. C.: Effect of enhanced manganese oxidation in the hyporheic zone on basin-scale geochemical mass balance, Water Resour. Res., 34, 623–636, 1998.

Harvey, J. W., Böhlke, J. K., Voytek, M. A., Scott, D., and Tobias, C. R.: Hyporheic zone denitrification: Controls on effective reaction depth and contribution to whole-stream mass balance, Water Resour. Res., 49, 6298–6316, https://doi.org/10.1002/wrcr.20492, 2013.

Harvey, J. W., Briggs, M. A., Buskirk, B., Swartz, A., Lewis, J., and Du, Y.: Surface water and groundwater water chemistry data collected along the Quashnet River, Mashpee/Falmouth, MA, U.S. Geological Survey data release, https://doi.org/10.5066/F7M044MF, 2018.

Hausle, D. A. and Coble, D. W.: Influence of sand in redds on survival and emergence of brook trout (*Salvelinus fontinalis*), T. Am. Fish. Soc., 105, 57–63, 1976.

Hitt, N. P., Snook, E. L., and Massie, D. L.: Brook trout use of thermal refugia and foraging habitat influenced by brown trout, Can. J. Fish. Aquat. Sci., 74, 406–418, https://doi.org/10.1139/cjfas-2016-0255, 2017.

Hudy, M., Thieling, T. M., Gillespie, N., and Smith, E. P.: Distribution, status, and land use characteristics of subwatersheds within the native range of brook trout in the Eastern United States, N. Am. J. Fish. Manage., 28, 1069–1085, 2008.

Irvine, D. J., Lautz, L. K., Briggs, M. A., Gordon, R. P., and Mckenzie, J. M.: Experimental evaluation of the applicability of phase, amplitude, and combined methods to determine water flux and thermal diffusivity from temperature time series using VFLUX 2, J. Hydrol., 531, 728–737, 2015.

Isaak, D. J., Young, M. K., Nagel, D. E., Horan, D. L., and Groce, M. C.: The cold-water climate shield: Delineating refugia for preserving salmonid fishes through the 21st century, Glob. Change Biol., 21, 2540–2553, https://doi.org/10.1111/gcb.12879, 2015.

Kalbus, E., Reinstorf, F., and Schirmer, M.: Measuring methods for groundwater – surface water interactions: a review, Hydrol. Earth Syst. Sci., 10, 873–887, https://doi.org/10.5194/hess-10-873-2006, 2006.

Krause, S., Tecklenburg, C., Munz, M., and Naden, E.: Streambed nitrogen cycling beyond the hyporheic zone: Flow controls on horizontal patterns and depth distribution of nitrate and dissolved oxygen in the upwelling groundwater of a lowland river, J. Geophys. Res.-Biogeosci., 118, 54–67, https://doi.org/10.1029/2012JG002122, 2013.

Kurylyk, B. L., Macquarrie, K. T. B., Linnansaari, T., Cunjak, R. A., and Curry, R. A.: Preserving, augmenting, and creating cold-water thermal refugia in rivers: concepts derived from research on the Miramichi River, New Brunswick (Canada), Ecohydrology, 8, 1095–1108, https://doi.org/10.1002/eco.1566, 2015a.

Kurylyk, B. L., MacQuarrie, K. T. B., Caissie, D., and McKenzie, J. M.: Shallow groundwater thermal sensitivity to climate change and land cover disturbances: derivation of analytical expressions and implications for stream temperature modeling, Hydrol. Earth Syst. Sci., 19, 2469–2489, https://doi.org/10.5194/hess-19-2469-2015, 2015b.

Kurylyk, B. L., Irvine, D. J., Carrey, S., Briggs, M. A., Werkema, D., and Bonham, M.: Heat as a hydrologic tracer in shallow and deep heterogeneous media: analytical solution, spreadsheet tool, and field applications, Hydrol. Process., 31, 2648–2661, https://doi.org/10.1002/hyp.11216, 2017.

LeBlanc, B. D. R., Massey, A. J., Cochrane, J. J., King, J. H., Smith, K. P., and Survey, U. S. G.: Distribution and Migration of Ordnance-Related Compounds and Oxygen and Hydrogen Stable Isotopes in Ground Water near Snake Pond, Sandwich, Massachusetts, 2001–2006, U.S. Geological Survey Scientific Investigations Report 2008-5052, 19 pp., 2008.

LeBlanc, D. R., Guswa, J. H., Frimpter, M. H., and Londquist, C. J.: Ground-water resources of Cape Cod, Massachusetts, U.S. Geological Survey, Hydrologic Atlas, 692, 4 sheets, 1986.

Levy, Z. F., Siegel, D. I., Glaser, P. H., Samson, S. D., and Dasgupta, S. S.: Peat porewaters have contrasting geochemical fingerprints for groundwater recharge and discharge due to matrix diffusion in a large, northern bog-fen complex, J. Hydrol., 541, 941–951, https://doi.org/10.1016/j.jhydrol.2016.08.001, 2016.

Lowry, C. S., Fratta, D., and Anderson, M. P.: Ground penetrating radar and spring formation in a groundwater dominated peat wetland, J. Hydrol., 373, 68–79, https://doi.org/10.1016/j.jhydrol.2009.04.023, 2009.

Luce, C. H., Tonina, D., Gariglio, F., and Applebee, R.: Solutions for the diurnally forced advection-diffusion equation to estimate bulk fluid velocity and diffusivity in streambeds from temperature time series, Water Resour. Res., 49, 488–506, https://doi.org/10.1029/2012WR012380, 2013.

MacCrimmon, H. R. and Campbell, S. C.: World Distribution of Brook Trout, *Salaelinus fontinalis*, J. Fish. Res. Board Can., 26, 1699–1725, 1969.

Matthews, K. R. and Berg, N. H.: Rainbow trout responses to water temperature and dissolved oxygen stress in two southern California stream pools, J. Fish Biol., 59, 50–67, 1997.

Modica, E.: Source and age of ground-water seepage to streams, U.S. Geological Survey, Fact Sheet Fact Sheet 063-99, 1999.

Montgomery, D. R., Buffington, J. M., Peterson, N. P., SchuettHames, D., and Quinn, T. P.: Stream-bed scour, egg burial depths, and the influence of salmonid spawning on bed surface mobility and embryo survival, Can. J. Fish. Aquat. Sci., 53, 1061–1070, https://doi.org/10.1139/cjfas-53-5-1061, 1996.

Mullan, J. W.: The sea run or "Salter" brook trout (*Salvelinus fontinalis*) fishery of the coastal streams of Cape Cod, Massachusetts, Massachusetts Division of Fisheries and Game, Bulletin No. 17, 1958.

Obruca, W. and Hauer, C.: Physical laboratory analyses of intergravel flow through brown trout redds (*Salmo trutta fario*) in response to coarse sand infiltration, Earth Surf. Proc. Land., 42, 670–680, https://doi.org/10.1002/esp.4009, 2016.

Petty, J. T., Hansbarger, J. L., Huntsman, B. M., and Mazik, P. M.: Transactions of the American Fisheries Society Brook Trout Movement in Response to Temperature, Flow, and Thermal Refugia within a Complex Appalachian Riverscape Brook Trout Movement in Response to Temperature, Flow, and Thermal Refugia within a Compl, T. Am. Fish. Soc., 141, 1060–1073, https://doi.org/10.1080/00028487.2012.681102, 2012.

Raleigh, R. F.: Habitat suitability index models: Brook trout, U.S. Fish and Wildlife Service, FWS/OBS, 82/10.24, 1982.

Rau, G. C., Andersen, M. S., McCallum, A. M., Roshan, H., and Acworth, R. I.: Heat as a tracer to quantify water flow in near-surface sediments, Earth-Sci. Rev., 129, 40–58, https://doi.org/10.1016/j.earscirev.2013.10.015, 2014.

Rezanezhad, F., Price, J. S., Quinton, W. L., Lennartz, B., Milojevic, T., and Cappellen, P. Van: Structure of peat soils and implications for water storage, flow and solute transport?: A review update for geochemists, Chem. Geol., 429, 75–84, https://doi.org/10.1016/j.chemgeo.2016.03.010, 2016.

Rosenberry, D. O., Briggs, M. A., Delin, G., and Hare, D. K.: Combined use of thermal methods and seepage meters to efficiently locate, quantify, and monitor focused groundwater discharge to a sand-bed stream, Water Resour. Res., 52, 4486–4503, https://doi.org/10.1002/2016WR018808, 2016.

Seitzinger, S., Harrison, J. A., Böhlke, J. K., Bouwman, A. F., Lowrance, R., Peterson, B., Tobias, C., and Van Drecht, G.: Denitrification across landscapes and waterscapes: a synthesis, Ecol. Appl., 16, 2064–2090, https://doi.org/10.1890/1051-0761(2006)016[2064:DALAWA]2.0.CO;2, 2006.

Snook, E. L., Letcher, B. H., Dubreuil, T. L., Zydlewski, J., Donnell, M. J. O., Whiteley, A. R., Hurley, S. T., and Danylchuk, A. J.: Movement patterns of Brook Trout in a restored coastal stream system in southern Massachusetts, Ecol. Freshw. Fish, 26, 360–375, https://doi.org/10.1111/eff.12216, 2016.

Steel, E. A., Beechie, T. J., Torgersen, C. E., and Fullerton, A. H.: Envisioning, Quantifying, and Managing Thermal Regimes on River Networks, BioScience, 67, 506–522, https://doi.org/10.1093/biosci/bix047, 2017.

Stoliker, D. L., Repert, D. A., Smith, R. L., Song, B., LeBlanc, D. R., Mccobb, T. D., Conaway, C. H., Hyun, S. P., Koh, D., Moon, H. S., and Kent, D. B.: Hydrologic Controls on Nitrogen Cycling Processes and Functional Gene Abundance in Sediments of a Groundwater Flow-Through Lake, Environ. Sci. Technol., 50, 3649–3657, https://doi.org/10.1021/acs.est.5b06155, 2016.

Van Grinsven, M., Mayer, A., and Huckins, C.: Estimation of Streambed Groundwater Fluxes Associated with Coaster Brook Trout Spawning Habitat, Groundwater, 50, 432–441, https://doi.org/10.1111/j.1745-6584.2011.00856.x, 2012.

Walter, B. D. A. and Masterson, J. P.: Simulated Pond-Aquifer Interactions under Natural and Stressed Conditions near Snake Pond, Cape Cod, U.S. Geological Survey, Water-Resources Investigations Report 99-4174, 34 pp., 2002.

Wehrly, K. E., Wang, L., and Mitro, M.: Field-based estimates of thermal tolerance limits for trout: incorporating exposure time and temperature fluctuation, T. Am. Fish. Soc., 136, 365–374, 2007.

Winter, T. C., Harvey, J. W., Franke, O. L., and Alley, W. M.: Ground water and surface water; a single resource, U.S. Geological Survey, Circular, 1139, 79 pp., 1998.

13

The socioecohydrology of rainwater harvesting in India: understanding water storage and release dynamics across spatial scales

Kimberly J. Van Meter[1], **Michael Steiff**[2], **Daniel L. McLaughlin**[3], **and Nandita B. Basu**[1,2]

[1]Department of Earth and Environmental Sciences, University of Waterloo, Waterloo, Ontario N2L 3G1, Canada
[2]Department of Civil and Environmental Engineering, University of Waterloo, Waterloo, Ontario N2L 3G1, Canada
[3]Department of Forest Resources and Environmental Conservation, Virginia Tech, Blacksburg, Virginia 24061, USA

Correspondence to: Nandita B. Basu (nandita.basu@uwaterloo.ca)

Abstract. Rainwater harvesting (RWH), the small-scale collection and storage of runoff for irrigated agriculture, is recognized as a sustainable strategy for ensuring food security, especially in monsoonal landscapes in the developing world. In south India, these strategies have been used for millennia to mitigate problems of water scarcity. However, in the past 100 years many traditional RWH systems have fallen into disrepair due to increasing dependence on groundwater. This dependence has contributed to accelerated decline in groundwater resources, which has in turn led to increased efforts at the state and national levels to revive older RWH systems. Critical to the success of such efforts is an improved understanding of how these ancient systems function in contemporary landscapes with extensive groundwater pumping and shifted climatic regimes. Knowledge is especially lacking regarding the water-exchange dynamics of these RWH tanks at tank and catchment scales, and how these exchanges regulate tank performance and catchment water balances. Here, we use fine-scale, water-level variation to quantify daily fluxes of groundwater, evapotranspiration (ET), and sluice outflows in four tanks over the 2013 northeast monsoon season in a tank cascade that covers a catchment area of 28 km^2. At the tank scale, our results indicate that groundwater recharge and irrigation outflows comprise the largest fractions of the tank water budget, with ET accounting for only 13–22 % of the outflows. At the scale of the cascade, we observe a distinct spatial pattern in groundwater-exchange dynamics, with the frequency and magnitude of groundwater inflows increasing down the cascade of tanks. The significant magnitude of return flows along the tank cascade leads to the most downgradient tank in the cascade having an outflow-to-capacity ratio greater than 2. At the catchment scale, the presence of tanks in the landscape dramatically alters the catchment water balance, with runoff decreasing by nearly 75 %, and recharge increasing by more than 40 %. Finally, while water from the tanks directly satisfies $\sim$40 % of the crop water requirement across the northeast monsoon season via surface water irrigation, a large fraction of the tank water is "wasted", and more efficient management of sluice outflows could lead to tanks meeting a higher fraction of crop water requirements.

1 Introduction

Issues of water stress are now estimated to impact more than one-third of the global population, and it is predicted that this fraction will nearly double as the world reaches peak population (Wada et al., 2014). Such increases in water stress are driven not only by a growing population, changing patterns of food consumption and climate-driven changes in water availability (Wiltshire et al., 2013) but also by spatial and temporal mismatches between water availability and water demand (Oki, 2006). From a spatial perspective, regional per capita water availability can vary drastically from more than 50 000 $m^3\ yr^{-1}$ to less than 500 $m^3\ yr^{-1}$ (Parish et al., 2012; Wada et al., 2014), with levels of water stress in one basin having little impact on that in another. Similarly, temporal mismatches, particularly in areas with high seasonal rainfall variability, can create high rates of runoff leading to flood events and high short-term availability during wet seasons, followed by severe water stress during dry periods (Haile, 2005). Such temporal mismatches, paired with a shortage of surface water storage, have been linked to both reduced incomes and a lack of food security (Gohar et al., 2013; Grey and Sadoff, 2007).

Both spatial and temporal mismatches in water stress and availability characterize the climatic regime of India. The monsoon-driven climate common to semi-arid areas of India results in remarkable temporal variation where it is common for half of the year's total rainfall to fall over a period of only 20 h (Keller et al., 2000). With such limited annual water availability and the extreme intra-annual rainfall variability, there have been ongoing efforts in India to increase storage capacity and additional water supplies for agricultural production and economic development (Grey and Sadoff, 2007). Over the last century, such efforts have focused primarily on large-scale projects designed to ensure higher levels of water storage and availability, such as the building of large dams and canal systems (Cullet and Gupta, 2009; Mehta, 2001). For millennia, however, India has met the demand for seasonal water storage and increased water availability at the local level via the building of village-scale rainwater harvesting (RWH) structures, often referred to as tanks (Van Meter et al., 2014).

It is estimated that more than 39 000 of these RWH tanks are present in the southern Indian state of Tamil Nadu, which is the focus of the present study (Van Meter et al., 2014). These RWH tanks, which commonly take the form of earthen impoundments, 20–40 ha in size (Gunnell and Krishnamurthy, 2003), are built up from natural depressions in the landscape and have historically been designed to meet the water needs of subsistence-level farmers for rice production via managed sluice channels for irrigation (Farmer, 1977). Tanks are often linked in a cascade with overflow from the upstream tanks spilling into surplus channels that lead to downstream tanks. The tank systems have fallen into decline in recent decades, primarily as a result of increasing reliance on groundwater pumping, and cheap access to electricity. This has led to declining groundwater levels, which, coupled with a growing demand for increased agricultural production, have led to renewed interest in these traditional systems (Kumar et al., 2008; Shah, 2004). Although the majority of existing RWH tanks still remain in a state of disrepair (Anbumozhi et al., 2001), it is estimated that reviving RWH systems at an all-India scale could potentially add as much as 125 $km^3 yr^{-1}$ to the country's current water supply, making them critical in meeting the projected water shortfall of 300 $km^3 yr^{-1}$ by 2050 (Gupta and Deshpande, 2004). Consequently, in India's Groundwater Recharge Master Plan (2005), the need for renovation or new construction of RWH structures was highlighted at a cost of approximately USD 6 billion, leading to high rates of revival of RWH structures across India (Agarwal and Narain, 1997; Shah et al., 2009)

With the renewed and large-scale interest in the use of RWH structures (Garg et al., 2013; Rockstrom et al., 2002; van der Zaag and Gupta, 2008), it is critically important to ask whether these ancient structures perform their intended purpose of significantly improving water availability in a basin (Batchelor et al., 2003; Bouma et al., 2011; Calder et al., 2008a, b; Garg et al., 2013). To do so requires quantifying the dominant tank inflows and outflows, specifically evapotranspiration (ET), groundwater recharge, and sluice outflows to irrigated fields. These water fluxes determine relative water allocation to aquifer supplies, irrigation needs, and atmospheric losses, and are influenced by a wide range of both natural and management controls, from climate and geology to the more direct anthropogenic controls (e.g., sluice outflow regulation). As such, a better understanding of tank fluxes and drivers of these fluxes is necessary when managing individual and cascades of tanks to meet both societal (irrigation demand) and environmental (increasing rates of groundwater recharge) needs (Glendenning et al., 2012; Neumann et al., 2004; Ngigi, 2003).

Unfortunately, there is a lack of empirical studies that quantify tank hydrologic fluxes, especially at the scale of watersheds comprising of multiple tanks (Glendenning et al., 2012; Mialhe et al., 2008; Pandey et al., 2011, 2013). One reason for the lack of information is that both groundwater recharge and ET are highly spatially variable, and thus difficult to accurately measure at the field scale (Glendenning et al., 2012). Most previous studies of RWH tanks estimate recharge as a residual term in the water-balance method (Glendenning et al., 2012); in arid environments, however, recharge magnitude is small compared to other fluxes (Bond, 1998), making estimates from water balance residuals vulnerable to errors in other measured components. Furthermore, water-balance methods used in RWH tanks estimate recharge using modeled values of tank evapotranspiration, another rarely measured but critically important water flux in these arid environments (Sharda et al., 2006). While there is consensus regarding the value of direct measurements of temporal variations in recharge and evapotranspiration fluxes from RWH structures, such data are difficult to obtain due to the inherent complexities in making these measurements, especially under resource constraints (Glendenning et al., 2012).

Here, we propose an innovative use of the White (1932) method as a cost-effective means of obtaining spatially integrated, direct measurements of both ET and groundwater exchange in flooded RWH tanks. The White method, which was originally developed to estimate the magnitude of groundwater consumption by phreatophytes (Loheide, 2008; Loheide et al., 2005), has since been used to estimate ET and groundwater exchange in small, surface water systems (Carlson Mazur et al., 2014; Hill and Durchholz, 2015; McLaughlin and Cohen, 2013). In these systems, diurnal variations

Table 1. Population and land-use data for the study cascade.

		Population				Land use					
				Farmers and agricultural laborers		Agriculture					
Tank no.	Village revenue district	Total population	Workforce	Total	% of workforce	Active	Fallow	Total	Forest	Settlements	Other
Tank 1	Pappinaickenpatti	3313	1986	1724	87 %	48 %	25 %	73 %	16 %	2 %	9 %
Tank 2	Kudipatti	2122	1300	1172	90 %	61 %	13 %	74 %	13 %	3 %	11 %
Tank 3						91 %	–	91 %	–	5 %	4 %
Tank 4	Ketuvarpatti	622	356	316	89 %	99 %	–	99 %	–	1 %	–
Cascade		6057	3642	3212	88 %	68 %	13 %	81 %	9 %	3 %	7 %

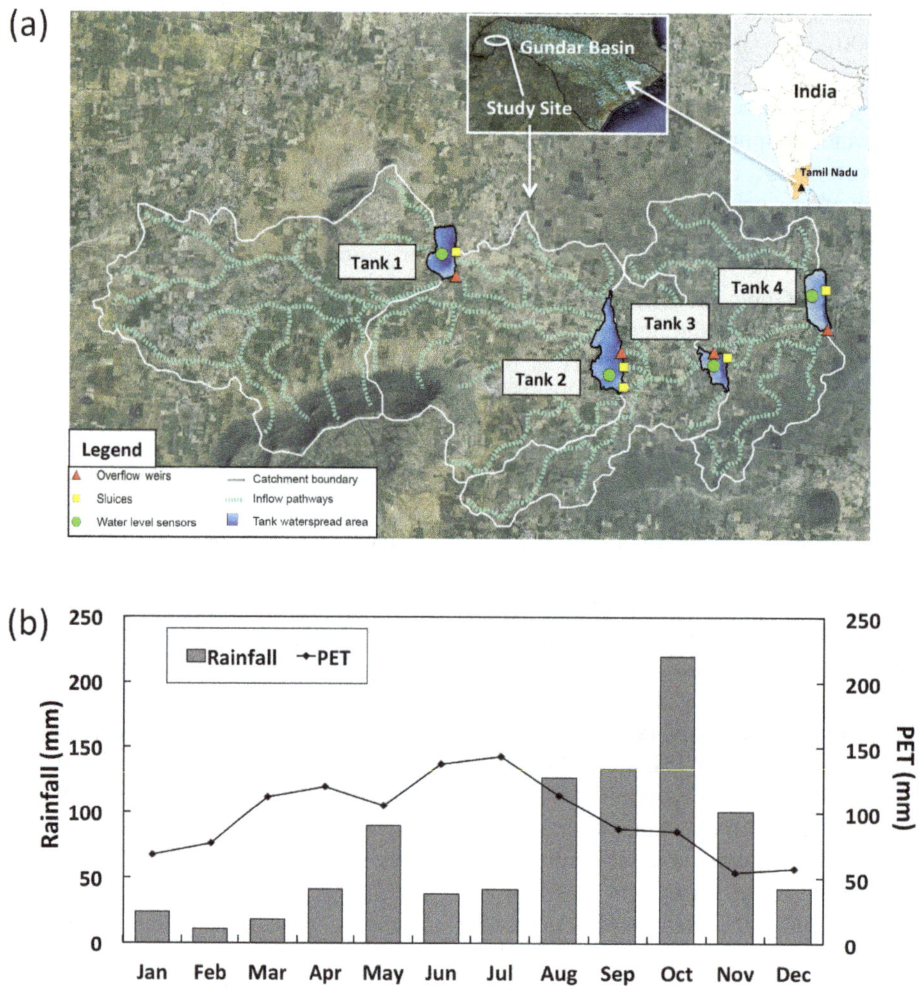

Figure 1. (a) Location of the Thirumal Samudram cascade within Tamil Nadu. The dotted lines indicate flowpaths calculated based on a digital elevation map (DEM) for the area; **(b)** average rainfall and potential evapotranspiration (PET) (1900–1970) measured at Peraiyur weather station, 10 km from the study cascade.

in high-resolution surface water level data are used to decouple ET dynamics from groundwater exchange. In this paper, we demonstrate an application of this method to RWH structures, which are more complex than the systems studied thus far in that they have additional outflows (overflow and sluice outflow), and are much larger in spatial extent (~ 1 ha vs. 20–60 ha). Furthermore, while most studies of RWH systems have focused on individual tanks, we explore how groundwater-exchange dynamics change along a tank cascade made up of four tanks, and scale up measured fluxes to estimate cumulative effects of tanks on catchment water balances. Our study has two linked objectives: (1) to quantify temporal patterns in groundwater exchange, ET, and sluice outflows over the northeast monsoon season; and (2) to describe spatial patterns of measured fluxes from upstream to downstream tanks in a cascade. Using these estimates, we attempt to answer the following questions:

- At the local scale, how do tanks partition water, and what is the spatial variability in this partitioning behavior along a tank cascade?
- At the catchment scale, how do tanks alter the water balance in a basin?
- What percentage of the irrigation requirements do tanks meet, and can they be managed more efficiently to increase this fraction?

2 Study area

2.1 Site description

The study site is located in the south Indian state of Tamil Nadu, in the foothills of the Western Ghats mountain range (Fig. 1a). The region surrounding the tank cascade is semi-arid, receiving a mean annual rainfall of 850 mm, with 50 % of that falling during the northeast monsoon season (October–December) (Government of Tamil Nadu, 2011; Vose et al., 1992). ET is greater than rainfall from January through July, while it is less than rainfall during the monsoon months (Fig. 1b). For the year in which the field study was done (2013), rainfall over the northeast monsoon season (October–December) was 355 mm, which is close to the 70-year average of 363 mm.

The focus of the study is the Thirumal Samudram (TS) tank cascade, a hydrologically connected group of four rainwater harvesting tanks that encompass an overall catchment area of 28 km^2, in the Madurai district of Tamil Nadu near the headwaters of the Gundar river basin (Fig. 1a). All four tanks in the cascade have undergone renovation through a joint effort of local stakeholders and the Development of Humane Action (DHAN) Foundation, an NGO group leading tank rehabilitation efforts across south India (DHAN, 2010), including regular desiltation, strengthening of tank bunds, and repair of surplus and sluice weirs. The four tanks provide irrigation water for three village revenue districts: Pappanaickenpatti (Tank 1), Kudipatti (Tanks 2 and 3), and Ketuvarpatti (Tank 4), from upstream to downstream. The population of the tank cascade area is 6057 (Government of India, 2011), and 88 % of the working population hold jobs either as farmers or agricultural laborers (Table 1).

The landscape surrounding the tank cascade has a gentle slope, ranging 0.5–1.0 %, and is characterized by Alfisol and Vertisol soils underlain by fractured rock of granitic origin (CGWB, 2012; ICRISAT, 1987; Palaniappan et al., 2009). Land use for the study area is primarily agricultural. Within the study cascade, 81 % of the land is devoted to agricultural use, with 42 % of this total being irrigated (Table 1) (DHAN, 2010). During the northeast monsoon season (October–January), paddy (rice) is the primary crop in the region, while during other periods of the year, a variety of other crops are cultivated, including cotton, groundnuts, and pulses (Government of Tamil Nadu, 2011).

2.2 Rainwater harvesting structures

Tanks in south India are created through the construction of an earthen dam (bund) across depressional areas in the landscape as a means of storing surface runoff (Van Meter et al., 2014) (Fig. 2). During elevated water levels, flooding extends beyond the main depressional area and into flatter, often farmed areas (i.e., tank water spread area). The bunds are constructed using locally available materials, usually a combination of amassed earth and stones, supported by the roots of trees and bushes growing along the bunds (Weiz, 2005). Sluices (typically sliding gates) are constructed within the tank bund and are used to control the release of water into irrigation channels, which then transport the stored water to agricultural fields in the downstream tank command area (i.e., fields irrigated by tank water. During heavy monsoon rains, water may spill over the tank's overflow weir into surplus channels leading to downstream tanks or to nearby waterways (Van Meter et al., 2014). Tanks are often linked through these surplus channels in chains or cascades that can range in size from several to more than a hundred tanks, forming a dense hydrological network across this intensively managed agricultural landscape.

Tank storage capacities vary across sites and time, with the latter due to siltation and desiltation cycles (Weiz, 2005). Historical data regarding maximum tank area and storage volumes for the four study tanks, obtained by the Public Works Department in India in approximately 1900, are summarized in Table 2 (DHAN, 2010). Information regarding the tank irrigated area, also known as the command area or "ayacut" (Weiz, 2005), is also provided. Although the maximum water depths of the four tanks are similar, ranging from 3–4 m at maximum fill, the historical data show that the tank areas vary significantly, ranging from 19.3 ha (Tank 3) to 58.7 ha (Tank 2). The ratio of command area to tank area historically ranged between 0.77 and 1.25 (Table 2), which is characteristic of tank systems found in this area (von Oppen and Subba Rao, 1987; Weiz, 2005). Table 2 also includes measurements made in the present study for comparison (discussed later).

3 Methods

3.1 Field methods: sensor installation and bathymetric survey

Tank water levels were continuously measured during and in the months immediately following the 2013 northeast monsoon season (October 2013–February 2014) using total pressure transducers (Solinst Levelogger Edge, accuracy

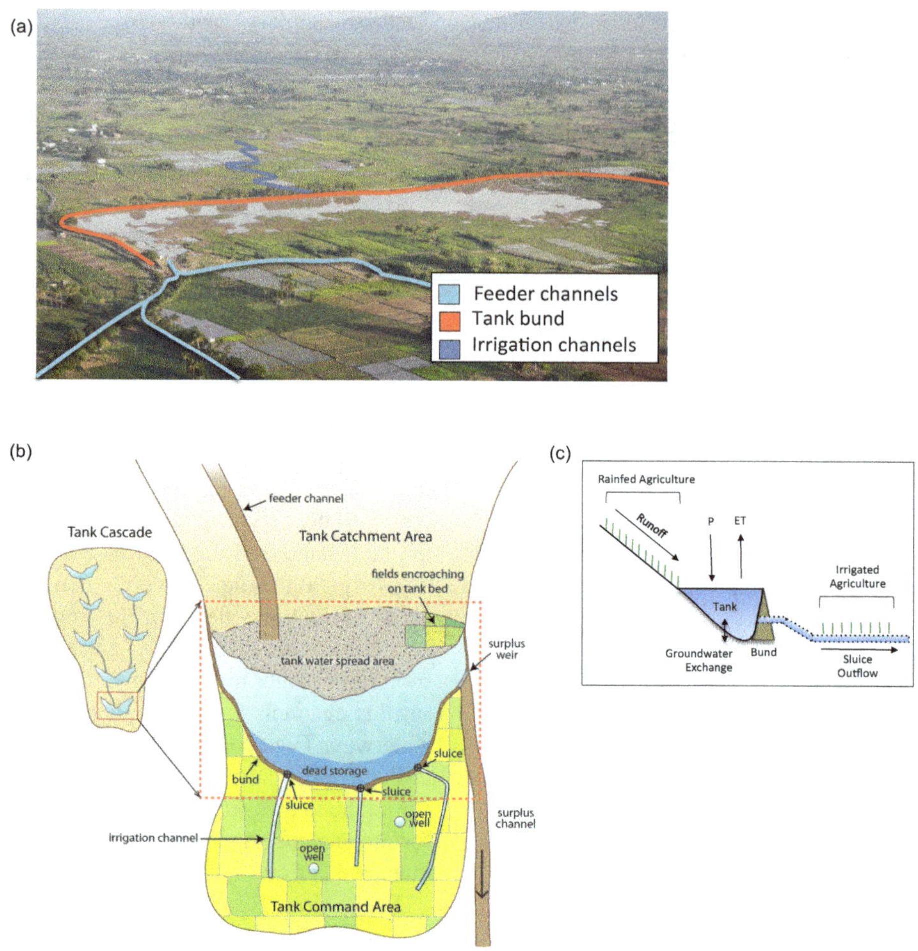

Figure 2. (a) Aerial view of Tank 4 in the Thirumal Samudram cascade; **(b)** plan view of a typical tank along with catchment and command area (Van Meter et al., 2014); **(c)** cross section showing tank water budget components.

Table 2. Summary of tank attributes based on historical tank data and the current study. Current tank capacity is based on our measurements, while historical tank capacity is based on Public Works Department data (DHAN, 2010).

Tank no.	Soil type	Maximum depth (m)	Maximum tank surface area (ha)	Tank command area (ha)	Command area/ surface area ratio	Tank capacity (m^3) Historical	Tank capacity (m^3) Current	Current capacity/ historical capacity
Tank 1	Alfisol	3.2	15	27	0.96	357 700	276 405	0.77
Tank 2	Vertisol	3.4	51	45	0.77	656 500	407 513	0.62
Tank 3	Vertisol	4.0	14	19	0.93	237 000	217 633	0.92
Tank 4	Vertisol	3.3	21	24	1.25	168 000	139 270	0.83

of ±0.3 cm, resolution of 0.01 cm; Solinst Canada, Georgetown, Ontario, Canada) installed in wells at the deepest point of each tank. Wells constructed of 10-gage PVC, were installed to a belowground depth of 70 cm and were screened above and below the ground surface. The pressure transducers measured total pressure (m H_2O) at 5 min intervals, and these measurements were corrected for variations in barometric pressure based on measurements collected at the same intervals with barometric pressure transducers (Solinst Barologger, accuracy of ±0.5 cm (±0.05 kPa), resolution of 0.001 cm (0.0001 kPa)). Slug tests were used to measure soil hydraulic conductivity (K_{sat}) in the installed wells using the Hvorslev (1951) method. Barometric pressure transducers were installed in dry wells open to atmospheric pressure but below ground to avoid changes in temperature and known temperature sensitivities (McLaughlin and Cohen, 2014). Corrected tank stage data were verified based on direct stage measurements at the study site. Pressure transducers were installed on 26 September before the start of the rainy season, and retrieved on 20 January for Tanks 1 and

2, and 7 March for Tanks 3 and 4, generally when wells became dry. Continuous precipitation was measured using Onset RG3-M automatic tipping bucket rain gages (Onset Computer Corporation, Bourne, MA) installed near each of the four tanks.

Bathymetric surveys were conducted using a combination of measured water depths in flooded areas (i.e., ground elevations relative to water surface) and a Trimble ProXRT2 GPS receiver paired with a Juno handheld computer for absolute ground elevations in exposed areas. Since Tank 4 had a large number of acacia trees that interfered with the accuracy of the Trimble, a Sokkia total station was used for ground elevation surveys. A total of 16–24 transects at a grid spacing of 40 m were taken in each tank, and all surveyed elevations were converted to ground elevations relative to the tank base (lowest point), which was defined as zero. The bathymetric data were used to create stage–volume and area–volume relationships for each tank, and estimate current tank capacities. The capacities estimated by this method led to reasonable values, with current capacities ranging between 62 and 92 % of the historical capacities (Table 2).

3.2 Sluice and overflow weir outflow estimates

There are six sluices in the study area, two in Tank 1, two in Tank 2, and one each in Tanks 3 and 4. Water release from the sluices is controlled by a sluice gate that can be opened to different degrees by a sluice rod. For our study tanks, the degree of sluice openness remained primarily unchanged during the period of study, and thus the major factor that controlled sluice discharge was the tank water level. To understand this relationship, sluice discharge was estimated at different tank water levels. Discharge was estimated by measuring the velocity and cross-sectional area over a chosen section of each outflow channel just downstream from the sluice outlet. This section was selected based on width uniformity and channel straightness. Approximately 20–40 measurements were made during each discharge measurement to obtain a reliable velocity estimate. Stage–discharge relationships developed for each sluice were used to estimate volumetric daily sluice outflow rates; these rates were then converted to area-normalized rates (S_o, cm d^{-1}) based on tank stage–area relationships (Sect. 3.1).

As described in Sect. 2.2, in addition to water loss via sluice outflow, water may also flow out of the tank by spillage through the overflow weir into surplus channels during large storm events. Overflow was observed during the study period only in the case of Tank 4 on 20 October, during the first major rains of the monsoon season. For this event, the surplus flow volume was estimated based on the observed drop in water levels between 20 and 21 October.

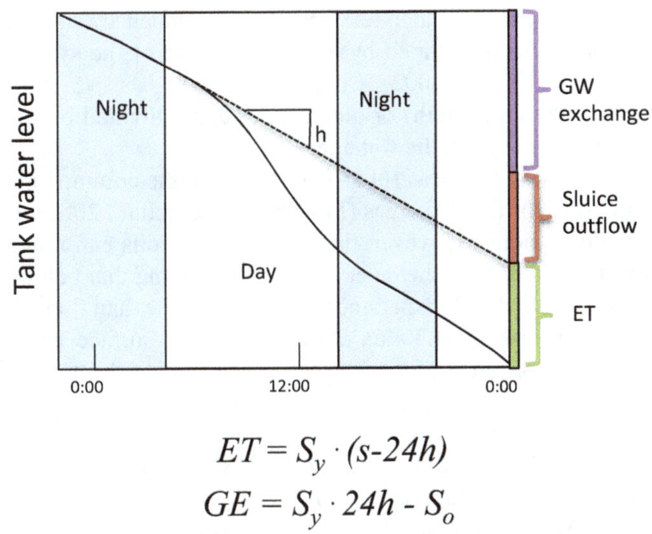

Figure 3. The White method for estimating ET and groundwater exchange using diurnal water level fluctuations. Grey bars denote nighttime.

3.3 Estimation of groundwater recharge and evapotranspiration

The White (1932) method was used to calculate daily ET and net groundwater exchange from high-resolution stage data on days with no rainfall (Fig. 3). The White method is based on two central assumptions: (1) ET (cm d^{-1}) fluxes are negligible at night, enabling groundwater flows to be estimated from nighttime stage changes, and (2) there is no diurnal variation in the groundwater exchange (GE; cm d^{-1}). Additionally, we have assumed that there is no surface inflow to the RWH tanks on days when it is not raining, which implies that overland flow occurs over very short time intervals. This is a reasonable assumption with the monsoonal rainfall dynamics that are characteristic of this region. Here, the White method was also modified to account for sluice outflow (S_o; cm d^{-1}), which occurred both night and day in our study.

ET and GE (cm d^{-1}; positive values indicate tank outflow or recharge) were estimated using the following equations:

$$\mathrm{ET} = S_y \times (s - 24\,h_n) \quad (1)$$

$$\mathrm{GE} = S_y \times 24 h_n - S_o, \quad (2)$$

where S_y is the specific yield (dimensionless), s (cm) is the 24 h stage change (positive values indicate net stage decline), and h_n (cm h^{-1}) is the slope of the nighttime decline in water level between 12 p.m. and 5 a.m. Since sluice outflow occurs throughout the day and night, the nighttime slope (h_n) includes both S_0 and GE, and thus ET can be estimated as the difference between the 24 h drop in water level (s) and h_n scaled to the daily rate (Eq. 1). Because only GE and S_0 occur at night, GE can be estimated by subtracting S_0 from the nighttime hourly slope h_n scaled to a daily rate, and after accounting for the specific yield. Specific yield (S_y) is defined as the volume of water released from or added to storage in porous media divided by the total volume of the system (Healy and Cook, 2002). On a per-unit-area basis, S_y represents the input (rain) or output (ET) depth divided by the observed change in the water level.

In our study, S_y was set to 1.0, following the common assumption for flooded areas (Mitsch and Gosselink, 2007). It should be noted, however, that S_y values in soils can range from 0.1 to 0.35 (Loheide et al., 2005), meaning that belowground water levels experience a greater decline than flooded areas for an equal ET flux. At the edge of a surface water body, this difference in water levels can lead to the formation of a hydraulic gradient, and thus to water subsidy from the flooded area to adjacent exposed areas. In soils allowing rapid equilibration of water levels, daytime declines from the flooded area would thus lead to subsidy to adjacent exposed areas (McLaughlin and Cohen, 2014). Under these circumstances, ET estimated via the White method using $S_y = 1.0$ would include both ET from standing water and any daytime flux to adjacent exposed areas to equilibrate greater ET-induced declines in belowground water levels. Such rapid equilibration between flooded areas and adjacent exposed areas was observed by McLaughlin and Cohen (2014) when applying the White method to estimate wetland ET in the sandy soils of Florida (hydraulic conductivity $K_{sat} = 1.13$–$6.42\,\mathrm{m\,d^{-1}}$). In our study area, however, soils are more clay dominated, and K_{sat} values for the tanks were measured to be 0.024–0.17 $\mathrm{m\,d^{-1}}$, 1–2 orders of magnitude lower than those for the Florida sites. These very low K_{sat} values suggest that any rapid equilibration (if any) would likely be limited to small edges, and thus a S_y value of 1.0 is a reasonable assumption. Moreover, measured losses in surface water are still valid and accurate components of the tank surface water budget regardless of the degree to which equilibration occurs. That is, if exposed areas are equilibrating with flooded areas, then the measured surface water decline will include both the direct flux (ET or GE) in the flooded area ($S_y = 1$) and the subsidy (indirect flux) to equilibrate those exposed areas where $S_y < 1$. In this case, the loss in surface water depth is still loss due to a particular flux (ET or GE), just over a greater footprint (i.e., direct fluxes in flooded areas + indirect losses to equilibrate flux-driven declines in adjacent areas). Therefore, when we convert ET and GE depth losses to surface water volume losses using stage-to-volume relationships, the estimates are accurate, and useful for discussing the proportions of stored surface water lost due to various water budget components.

3.4 Tank and catchment water balances

Volumetric water balance calculations were carried out at both the individual tank and the tank catchment scales across the northeast monsoon season to answer questions regarding the partitioning of rainfall into the various outflow components (e.g., S_o, ET, GE). For individual tank water balances, we utilized daily data for water levels, rainfall, S_o, ET, and GE. For nonrainfall days, ET and GE values were calculated using the White method. For rainfall days, ET and GE could not be calculated directly via the White method, as the method necessarily assumes a constant groundwater flow and therefore cannot account for rainfall-related inputs. This disruption in the continuity of the data set, without correction, would lead to gaps in the daily water balance and an underestimation of both ET and groundwater exchange across the monsoon season. To eliminate these gaps, we estimated ET values on rainfall days via interpolation between White method-estimated ET rates on days without rain. GE on rainfall days was estimated based on the residuals of the daily water balance, using the measured 24 h change in tank water levels, estimated ET rates, measured precipitation, and estimated runoff into the tank (McLaughlin and Cohen, 2013). Runoff was estimated using the Strange method (Shanmugham and Kanagavalli, 2013), an empirical method that was developed to predict runoff from catchments with irrigation tanks and small reservoirs and that is widely used throughout India by government departments dealing with irrigation (Latha et al., 2012). In this method, daily runoff is calculated as a percentage of daily rainfall, based on tabulated values in which runoff is expressed as a function of (a) rainfall on that day, (b) antecedent rainfall conditions, and (c) catchment characteristics (Shanmugham and Kanagavalli, 2005). The Strange method has been shown to provide results comparable to those obtained with the more commonly used SCS curve number method (Latha et al., 2012), but is more representative of the south Indian conditions that are the focus of our study. Stage-to-area relationships (Sect. 3.1) were used to convert daily stage change and estimated fluxes (ET, GE, and S_o) into volumes, which were calculated for each tank. Note that the water balances for all tanks are calculated for the period of 17 October 2013–13 January 2014, a period that spans the entire monsoon season and for which water-level data were available for all four tanks.

Water balances were also calculated at the catchment scale using a nested catchment design for four catchments (Fig. 4): (1) Catchment 1 (C1): Tank 1 (T1) and its contributing catchment; (2) Catchment 2 (C2): Tank 2 (T2) and its contributing catchment, which includes Tank 1 and its catchment area and command area; (3) Catchment 3 (C3): Tank 3 (T3) and its contributing catchment, which includes Tanks 1 and 2, and their catchment and command areas; and (4) Catchment 4 (C4): Tank 4 (T4) and its contributing catchment, which includes Tanks 1, 2, and 3 and their catchment and command areas. This nested catchment design enabled us to explore the effect of varying catchment sizes and tank-to-catchment ratios on the water partitioning.

Further, in order to understand the impact of the tanks at the catchment scale, we explored two scenarios for each of the four catchments scales (i.e., C1–C4): (1) a with-tank (WT) scenario to represent current conditions within the catchment (i.e., four existing tanks); and (2) a no-tank (NT) scenario, with all other conditions (e.g., rainfall, ET on the catchment area) being the same. For the NT case, catchment-scale runoff was calculated using the Strange method (Shanmugham and Kanagavalli, 2013) and daily rainfall over the monsoon season. Remaining rainfall was assumed to exit the system through ET and groundwater recharge. For the WT case, we assumed the sluice outflow from the most downstream tank in the catchment (T1 for C1, T2 for C2, T3 for C3, and T4 for C4) to represent the Q value for each catchment. For T4, a surplus overflow event occurred at the start of the season, the volume of which was estimated based on stage–volume relationships; this volume was added to the sluice outflow to estimate the Q for C4. The Q values for the NT and WT scenarios were compared for all four catchments to understand the effect of tanks on the catchment runoff.

To understand the effect of tanks on catchment-scale groundwater recharge, we assumed the mean recharge to be 17 % of the mean annual rainfall for the NT case following Anurag et al. (2006). For the WT case, the landscape was assumed to include three different domains, with separate recharge fractions being assumed for each domain: (1) tank bed area: GE (Sect. 3.2) was used; (2) tank command area: 50 % of the sum of rainfall and sluice outflow (based on typical values for paddy fields (Hundertmark and Facon, 2003); and (3) the rest of the watershed: 17 % of rainfall (Anurag et al., 2006). The command area and the tank bed area estimates for the four tanks are provided in Table 2.

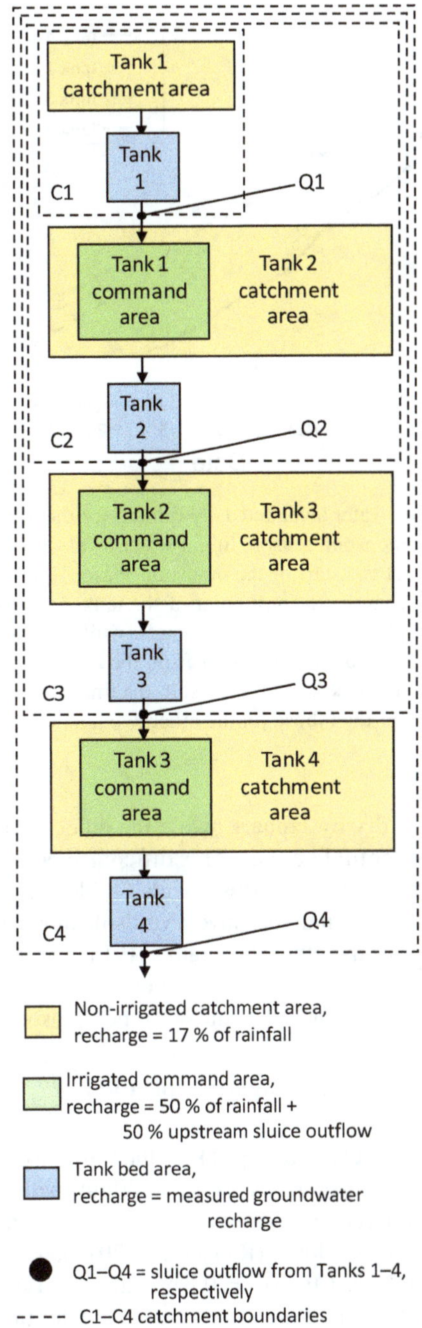

Figure 4. Schematic for the catchment-scale water balance calculations. The dotted lines represent the boundaries, and points Q1–Q4 represent the outlet points for the four nested catchments C1–C4.

4 Results and discussion

The current section is divided into two broad subsections. In the first, we report measurements of tank water levels and fluxes (ET and GE), and use these data as a basis for discussing tank water level dynamics across the monsoon season. In the second, we provide analysis of these and complementary data to answer questions regarding controls on the tank and catchment water balances and the ability of tank rainwater harvesting systems to meet irrigation water demand.

4.1 Tank water-exchange dynamics

4.1.1 Tank water levels over the northeast monsoon season

Water levels in the tanks rose sharply in mid-October following the monsoon rains, and then dropped over the next 3 months as water left the tanks through ET, sluice outflow, and groundwater recharge (Fig. 5). Note that although the northeast monsoon rains began in early September, the tanks started filling only in mid-October. This time lag is likely due to a threshold effect, where runoff to the tanks occurs after cumulative rain volumes begin to exceed catchment infiltration capacity. Two distinct fill events can be observed, one on 16 October and the second on 17 November for all tanks except Tank 1, for which the second fill event is not as apparent. Between 16 October and 17 November, the trajectories of Tanks 1 and 3 parallel each other, while those of Tanks 2 and 4 are similar to each other. Towards the latter part of the season, the water-level trajectories of the four tanks approximately parallel each other. Tank 1 loses its water the earliest

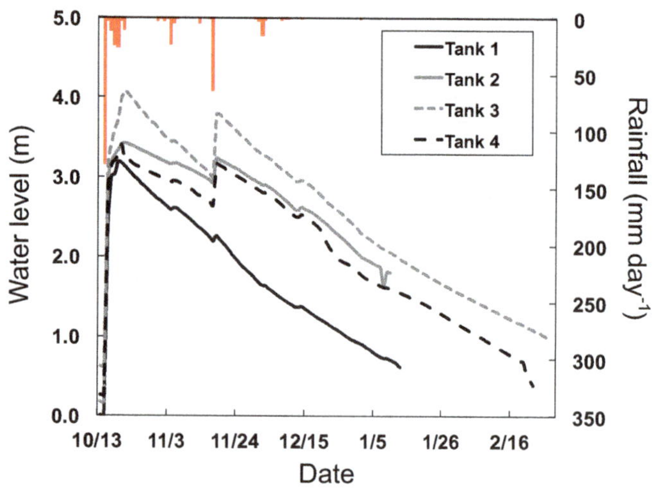

Figure 5. Tank water level and daily rainfall for the four tanks over the northeast monsoon season. Initial water levels are zero, reflecting dry tanks at the start of the monsoon season. Tank water level is measured from the deepest point of the tank. In Tanks 3 and 4, sensors were placed at the deepest points, while in Tanks 1 and 2, sensor wells were offset somewhat from the deepest points due to vegetation in the tank beds. As a result, the time series for Tanks 1 and 2 end earlier than those for the other two tanks.

and is mostly dry by January, while the other three tanks retain some water till February. This difference possibly occurs because Tank 1 overlies Alfisol soils (Table 2), which have higher K_{sat} values than the clayey Vertisol soils that make up the tank beds of the other three tanks (Pathak et al., 2013). In the following sections, we explore how the outflow fluxes in the four tanks vary over the course of the monsoon season.

4.1.2 Estimation of evapotranspiration

ET fluxes estimated with Eq. (1) for the four tanks are shown in Fig. 6. ET rates derived with the White method are reasonable for the region and season potential ET (PET) ca. 3–12 mm d^{-1} for Madurai (Rao et al., 2012), ranging from 5.5 ± 1.0 for Tank 1 to 10.1 ± 0.8 mm d^{-1} for Tank 3 during periods when the tank inundated area is greater than 25 % of maximum area. Below this 25 % threshold (shown in Fig. 6 with dashed line), ET estimates for the tanks exceed PET rates by factors of 2–3. These very high late-season ET values are likely the result of the tanks, at this stage, existing as small areas of flooding surrounded by comparatively extensive areas of exposed soil. Such conditions, particularly in arid regions, can create an oasis effect (Drexler et al., 2004; Paraskevas et al., 2013), in which advection of dry air from exposed areas can increase ET rates in flooded areas beyond typical values (and PET). The magnitude of the oasis effect is known to become greater when the soil is dry and surrounding vegetation is at higher moisture stress (Holmes and Robertson, 1958), thus explaining the significant increase in measured ET values as the landscape dries out late in the monsoon season.

It may also be that an overestimation of the S_y term in Eqs. (1) and (2) could be a contributor to the high measured ET values during late season and low water level conditions. White method calculations by McLaughlin and Cohen (2014) showed ET rates exceeding PET by a factor of 5 or more when flooded areas were small, compared to $\mathrm{ET}/\mathrm{PET} \approx 1.0$ at moderate to maximum flooded areas. As discussed in Sect. 3.3, however, the very low K_{sat} values in our study area, compared to the much higher values in the McLaughlin and Cohen (2014) study, suggest that any equilibration occurring at the tank edges would be very small in magnitude. Accordingly, the oasis effect, described above, appears to be the most likely explanation of our high ET values.

4.1.3 Estimation of groundwater exchange

The temporal pattern of net GE, estimated using Eq. (2), is presented in Fig. 7 together with trends in tank water levels and daily precipitation. GE rates across the monsoon season appear to be driven by a combination of both tank water levels and the occurrence and magnitude of rainfall events. Tank 2, for example, has relatively lower recharge rates (positive values in Fig. 7) in the earlier part of the season, with values decreasing with the occurrence of each major rainfall event, and then increasing incrementally over time until the next rainfall. The last period of significant rainfall occurs in mid-December, and shortly after this time, recharge magnitudes for Tank 2 reach a peak, and then slowly decrease with decreasing tank water levels. A similar pattern can be seen for Tank 4, where the peak recharge value occurs during the mid-December period, followed by a steady decline in recharge magnitudes as tank water levels decrease. In contrast, Tanks 1 and 3 appear to be less impacted by rainfall events; for these tanks, recharge magnitudes begin to decrease with decreases in tank water levels much earlier in the season, after the last major rainfall (64 mm) on 17 November. In the last few weeks of the monsoon season, Tanks 2–4 all switch over to a groundwater inflow regime (negative GE values). Lower recharge rates as well as these switches to groundwater inflow towards the end of the season may be due to tank water levels consistently having greater declines compared to the surrounding aquifer, resulting in decreases and potential reversals of hydraulic head gradients. This period is also, however, punctuated by some distinct, very high groundwater outflow events that may correspond to observed groundwater pumping in the vicinity, highlighting a potential direct human influence to tank recharge rates.

To better characterize the dominant drivers for the magnitude and direction of GE, with the overall goal of generalizing these observations to larger scales, we plotted GE as a function of days since last rainfall for all four tanks (Fig. 8a). For Tanks 2 and 4, there is a threshold value of days since

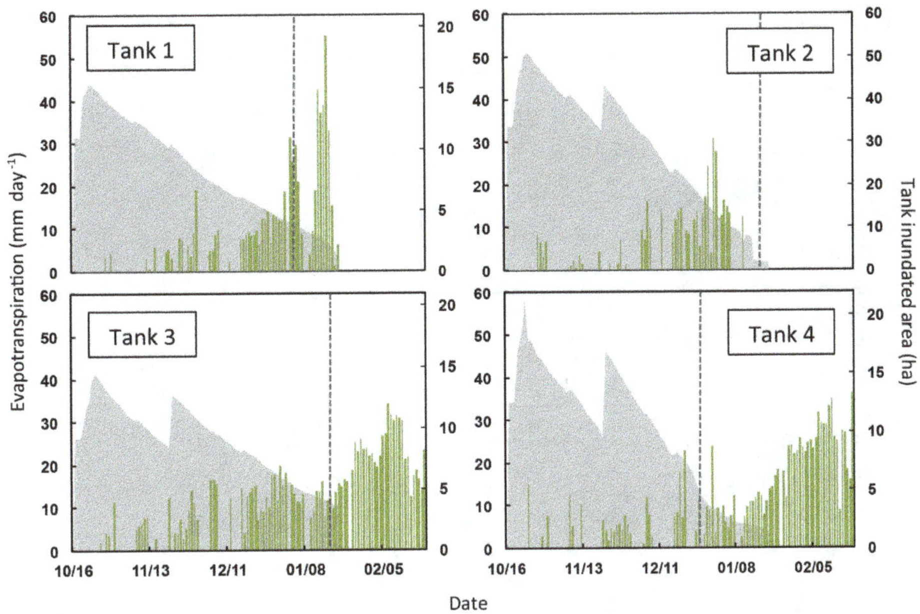

Figure 6. The temporal variation in daily ET over the monsoon season, shown as green bars. There are data gaps in the figure since estimates were made using the White method only on nonrainfall days. ET increases towards the later part of the season, coincident with decreases in tank surface area (shown as the grey shaded area). ET rates are reasonable for the region and season when the inundated area is greater than 25 % of maximum area, as indicated by the dashed line.

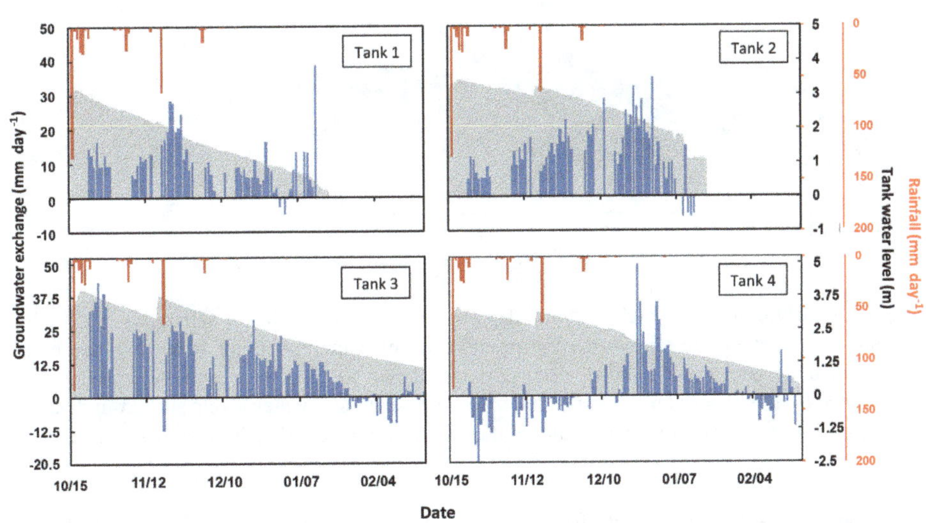

Figure 7. Daily groundwater exchange (mm d^{-1}) over the course of the northeast monsoon season (blue bars). Positive values indicate groundwater outflow (recharge) from the tank, while negative values indicate inflow (discharge) into the tank. Groundwater exchange magnitudes generally decrease and even switch from outflow to inflow towards the latter part of the season, when tank water levels (shown in grey and plotted on the secondary y axes) are low. There are in some cases some very high groundwater outflow events near the end of the season corresponding to pumping in the vicinity. Rainfall is shown as red bars.

rain (14 days for Tank 2, and 16 days for Tank 4) that separates rainfall–GE relationships. That is, there is significant scatter in the rainfall–GE relationship at values less than this threshold, but strong negative relationships emerge between the two variables at higher values of day since rain (Fig. 8a). In contrast, Tanks 1 and 3 have much lower threshold values of only 1 and 3 days, respectively. This pattern of decreasing recharge with days since last rainfall is reasonable, as water levels in the tank steadily decrease over time, leading to decreased hydraulic head and thus lower rates of recharge. In contrast, immediately following a rain event, the system becomes more dynamic, and recharge is a function of not only tank water levels but also the short-term response of the local surrounding aquifer. When plotted for all tanks, GE was

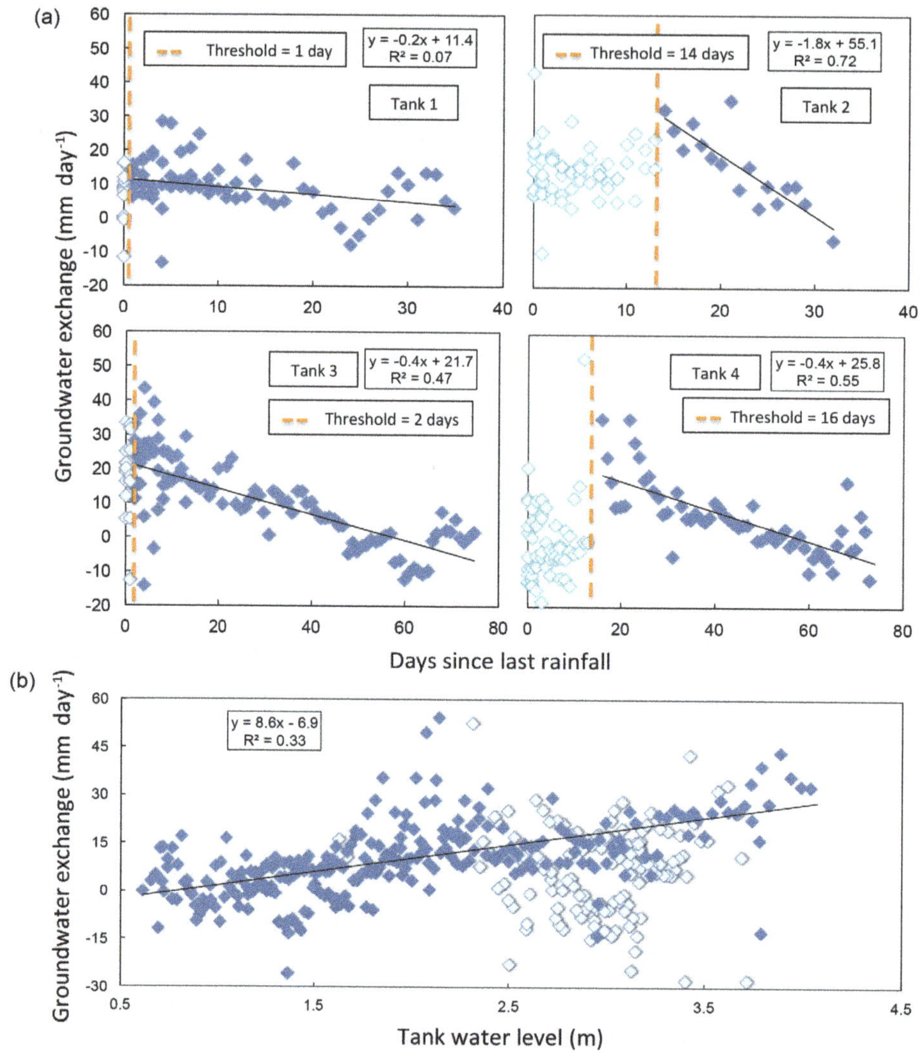

Figure 8. (a) Relationship between groundwater exchange and days since last rainfall, shown separately for the four tanks. The threshold line (dashed orange) separates the more erratic rainfall-driven groundwater exchange behavior following rain events (shown as light blue diamonds) from the more predictable behavior typical of drier periods (shown as dark blue diamonds), when GE is driven primarily by hydraulic head values determined by tank water levels. **(b)** Relationship between tank water levels and groundwater exchange shown for all four tanks combined. Lighter blue diamonds correspond to the rainfall values below the threshold shown above in Fig. 7a.

also found to respond linearly to tank water levels for most days throughout the monsoon season, except in the hydrologically dynamic periods after rain events, when the behavior was more erratic (Fig. 8b).

In addition to these patterns of groundwater exchange across the monsoon season, differences can also be seen along the tank cascade, from top (Tank 1) to bottom (Tank 4). First, while recharge dominates the exchange dynamics of Tanks 1–3, Tank 4 is more discharge driven. As shown in Fig. 9a, close to 90 % of all days throughout the monsoon show net recharge behavior for Tanks 1–3, while Tank 4 is split almost equally between net recharge and net discharge days. From a volume perspective, the discharge-to-recharge ratio for the tanks shows a general trend from smaller (0.3 in Tank 1) to larger (1.2 in Tank 4) across the tank cascade (Fig. 9b), with Tank 4 demonstrating net discharge behavior. Tank 4 is the most downgradient tank, suggesting the possibility that aquifer levels adjacent to Tank 4 are higher (possibly due to upstream tanks' recharge) for a longer period of time than the other three tanks, leading to more frequent groundwater inflow.

Our finding of a distinct spatial pattern in groundwater exchange and sluice outflow dynamics across the tank cascade is a novel contribution of the present study. Most studies that have explored the recharge–discharge functions of tanks (Glendenning et al., 2012) have focused on individual tanks, with no consideration of the position of the tank in a cascade as an important control on its functioning. Our results

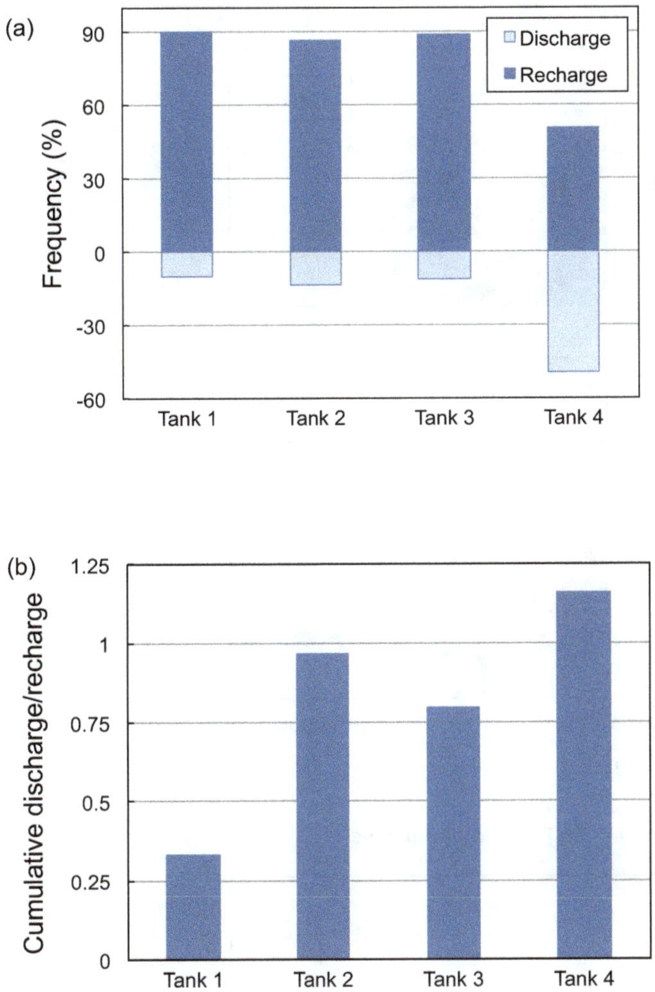

Figure 9. (a) The frequency of daily recharge (outflow) and discharge (inflow) events over the northeast monsoon season, and **(b)** the ratios of cumulative discharge to cumulative recharge magnitudes. The results for the four tanks indicate that all tanks function as both recharge and discharge systems, but that Tank 4 is much more dominated by discharge behavior based on both frequency and overall magnitudes.

indicate that in order to upscale tank-scale information to understand catchment and regional-scale impact of tanks, more studies should focus on exploring the spatial arrangement of tanks in the landscape.

4.2 Exploring biophysical vs. management controls on tank water balance at the tank and catchment scales

Three questions were posed in the introduction regarding the partitioning of water within a tank cascade, the ways in which tanks alter the catchment water balance, and the ability of tanks to meet irrigation requirements in the semi-arid landscapes of south India. Below, we use our measured data to provide answers to these questions in the context of a discussion of physical vs. management controls on tank functionality.

4.2.1 Water balance at the tank scale

The first question we asked was how tanks partition incoming surface water (direct rainfall on tank and surface runoff from tank catchment) into various outflow components, namely ET, groundwater outflow/inflow, and sluice outflow to the fields in the tank command area. The flow volumes corresponding to these components for each tank over the duration of the northeast monsoon season are plotted by week in Fig. 10a and are summarized in Table 3. Notably, recharge to groundwater is a significant component of tank outflows. Although the primary function of tanks in south India has historically been to provide surface water for irrigation, and despite the high clay content of soils in the area, groundwater recharge is the primary outflow mechanism in Tanks 1–3 (46–59 % of total outflows). For Tank 4, however, which is dominated by discharge behavior, the primary outflow mechanism is sluice outflow, which directly provides irrigation water to the tank command area. As seen in Fig. 10a, sluice outflows and recharge are the greatest early in the season, when tank levels are at their highest, and then decrease over time, ceasing entirely by mid-December for all four tanks.

Although the surface water volume lost to ET is substantial (0.48–1.64 million m^3 over the 83-day study period), it is a relatively small fraction of the overall water budget. On a cumulative scale (Table 3), ET values range from 13 % of total outflows for Tank 1 to 22 % for Tanks 2 and 3. These smaller percentages of ET compared to recharge contradict the established view of tanks losing a significant fraction of their water through ET (Kumar et al., 2006). In addition, although the tanks have been constructed in soils with a high clay content, all but Tank 4, which has a high discharge–recharge ratio, have high relatives rates of groundwater recharge. For Tanks 2 and 3, recharge is the largest outflow component (57–59 %) and is more than double the values for sluice outflow and evapotranspiration. For Tank 1, recharge is also the largest outflow component (47 %), although it is similar in magnitude to sluice outflows (41 %). The differences in flow partitioning between the four tanks can be attributed to differences in both natural (e.g., topographical position of the tank along the cascade) and human (e.g., sluice management) factors.

Interestingly, a trend can be seen in the relationship between total tank outflows over the monsoon season and the maximum tank capacity (Fig. 10b). As we move down the cascade of tanks, the outflow-to-capacity ratio increases, from 1.06 for Tank 1 to as high as 2.25 for Tank 4. The outflow-to-capacity ratio is an indication of how many times a tank fills up during the season, and the increase in values along the cascade of tanks is a function of increasing return flows from upstream command areas entering the downstream tanks. For Tank 4 in particular, groundwater discharge

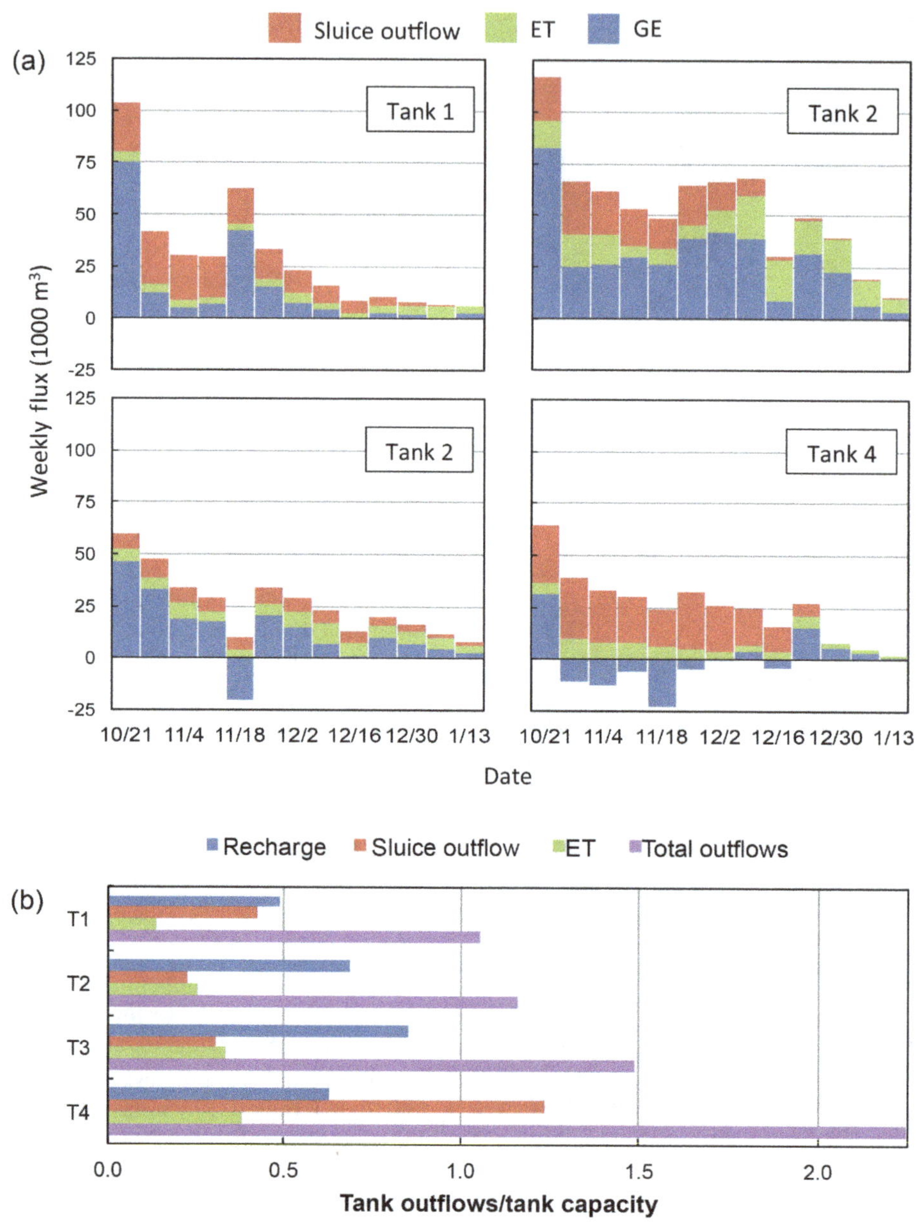

Figure 10. **(a)** Tank outflow dynamics (ET in green, sluice outflow in red, and GE in blue) shown as weekly integrated volumes for all four tanks. **(b)** Tank water outflows as a fraction of the tank capacity, with total outflows calculated as the sum of ET, S_0, and groundwater recharge. The outflow-to-capacity ratios increase down the cascade, such that total outflows for Tank 4 over the study period are more than double the total tank capacity.

provides a significant input of water into the tank (Fig. 9). Accordingly, Tank 4 has relatively greater amounts of water available for surface water irrigation throughout the season, with sluice outflow alone accounting for 1.2 times the total tank capacity. This increase in the outflow-to-capacity ratio along the cascade of tanks is an important feature of the tank cascade system, and highlights the need to study the tanks not in isolation, but in relation to their position along the cascade. Biophysical controls (e.g., weeds or sediments in tank beds of upgradient tanks) or management choices (e.g., planting crops with lower or high water requirements in upgradient tanks) can completely alter the water availability in a downstream tank. Thus, rehabilitation efforts and tank management should focus on maximizing benefits at the cascade scale instead of only at the individual tank scale.

4.2.2 Water balance at the catchment scale

The second question we asked was how tanks alter the partitioning of rainfall into runoff at the catchment outlet (Q)

Table 3. Partitioning of tank outflows across the northeast monsoon season.

	Tank 1	Tank 2	Tank 3	Tank 4
Total outflows (m^3)	376 794	762 483	352 934	377 257*
Evapotranspiration				
total (m^3)	48 291	164 423	78 745	64 358
Percent of total outflows	13 %	22 %	22 %	17 %
Sluice outflow				
total (m^3)	153 038	146 612	72 279	207 636
Percent of total outflows	41 %	19 %	20 %	55 %
Recharge				
Total (m^3)	175 465	451 448	201 910	105 263
Percent of total outflows	47 %	59 %	57 %	28 %

* Note that the total outflow volume given here for Tank 4 does not include the 20 October overflow event at the start of the monsoon season. As water exiting the tank via the overflow weir passes directly out of the tank catchment, bypassing the tank command area and thus not remaining as a source for irrigation or groundwater exchange within the tank cascade, we considered it separately from other flows.

and recharge within the catchment. Water balance calculations were done at the tank and catchment scales for the four nested catchment scenarios described in Sect. 3.4. Further, we simulated scenarios both with and without tanks to understand the contribution of tanks towards altering catchment scale water partitioning.

Our results show a dramatic difference between the with-tank and no-tank scenarios, and a distinct spatial pattern of response in the four nested catchments. We found a significant decrease in Q at the four nested scales, from 22 % of rainfall in the no-tank scenario to 5–9 % of rainfall with tanks (Table 4). At the largest catchment scale (C4), the runoff decreased from approximately 2.29 million m^3 in the NT scenario to only 0.69 million m^3 in the presence of tanks (Table 4). This approximately 70 % decrease is consistent with other work showing large decreases in runoff due to the presence of tanks (Kumar et al., 2008). Conversely, catchment-scale net recharge was observed to increase from 17 % of rainfall without tanks to 24–27 % with tanks (Table 4), which corresponds to an overall increase in net groundwater recharge of 40 %, highlighting the potential beneficial role tanks may play in augmenting groundwater resources.

Despite this strong link between the presence of tanks and groundwater recharge, tank maintenance has declined across south India as farmers have become increasingly reliant on groundwater irrigation sources (Balasubramanian and Selvaraj, 2003). With tank-irrigated area across Tamil Nadu having decreased from 940 000 ha in 1960 to approximately 503 000 ha in 2010, some suggest that current tanks are operating at only 30 % of their potential capacity (Amarasinghe et al., 2009; Government of Tamil Nadu, 2011; Palanisami and Meinzen-Dick, 2001). This degradation of tank functionality is eliminating or significantly degrading the primary mechanism for aquifer recharge in an area where, without rainwater harvesting, the majority of monsoon rainfall will leave a catchment as runoff within hours of falling. Our water balance calculations show that tanks, with adequate maintenance, provide a mean groundwater recharge benefit of 5600 m^3 per hectare of tank waterspread area. At the scale of the Gundar basin, with its 2276 village-scale RWH tanks, each covering an area of approximately 40 ha (DHAN, 2010), these results suggest that fully functional tanks could provide a groundwater recharge benefit of 522 million m^3. With a population of approximately 3 million this difference translates to a difference in water availability throughout the Gundar basin of 174 m^3 per capita. It is currently estimated that all of India is experiencing some degree of water stress, with per capita availability ranging from 1000–1700 $m^3\,yr^{-1}$ (Amarasinghe et al., 2005). Accordingly, maintaining tanks at full functionality has the potential to increase per capita water availability in the Gundar by approximately 10–15 %.

It should be noted that the recharge benefit suggested by the results in our tank cascade is significantly larger than that reported for a watershed in Gujarat, a state in western India, where it was shown that the construction of new rainwater harvesting structures would lead to a 60 % decrease in catchment runoff, but only a 5 % increase in recharge (Sharma and Thakur, 2007). In the Gujarat catchment, however, annual rainfall is approximately half that in our south India catchment, and ET rates are estimated at more than 50 $mm\,d^{-1}$, suggesting that variations in climate can strongly impact the contribution of rainwater harvesting structures to groundwater recharge.

4.2.3 Management controls on irrigation efficiency

While the first two questions focused on the physical controls on tank water dynamics, our third question focused on understanding how tank water management affects water balances and, in doing so, contributes to meeting the irrigation require-

Table 4. Water balance summary at the tank catchment scale.

	Catchment 1	Catchment 2	Catchment 3	Catchment 4
Area (km^2)	5.0	16.2	22.5	28.4
Precipitation P (million m^3)	1.8	5.8	8.1	10.2
Runoff Q (million m^3)				
with tanks	0.15	0.30	0.37	0.69
without tanks	0.40	1.31	1.81	2.29
Recharge R (million m^3)				
with tanks	0.48	1.44	1.97	2.42
without tanks	0.31	0.99	1.37	1.73
Q / P				
with tanks	0.09	0.05	0.05	0.07
without tanks	0.22	0.22	0.22	0.22
R / P				
with tanks	0.27	0.25	0.24	0.24
without tanks	0.17	0.17	0.17	0.17

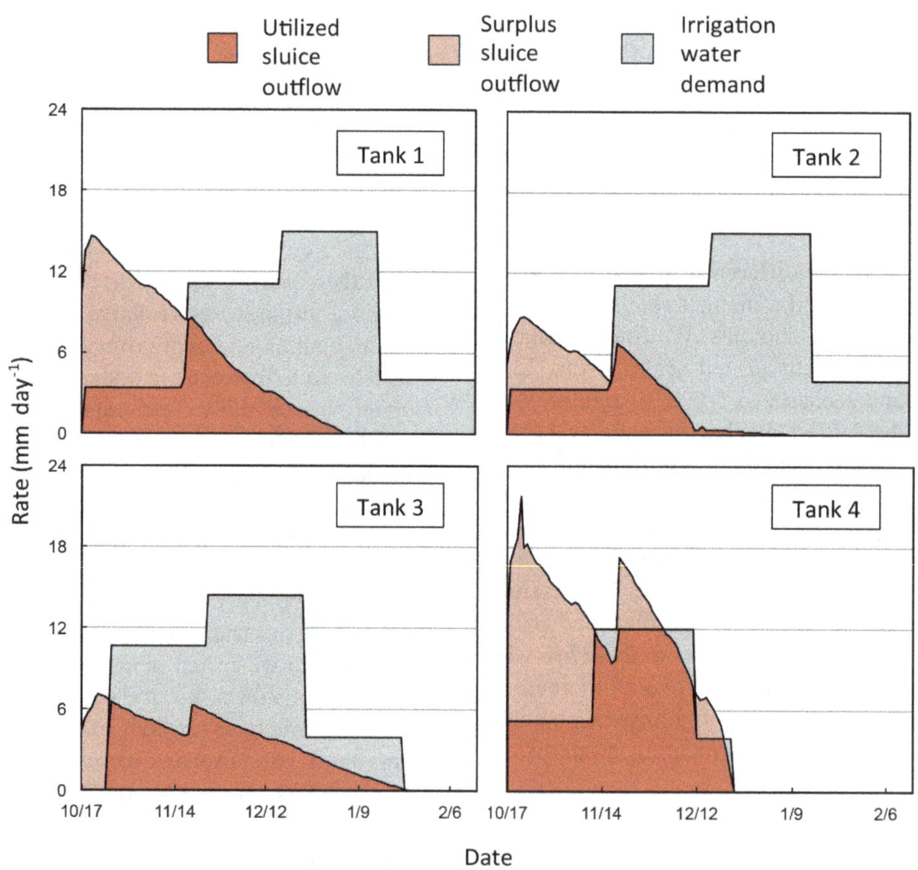

Figure 11. Water supply-and-demand portraits in our tank cascade. The grey area represents the irrigation water demand (IWD), calculated as the difference between crop water requirements and rainfall (Brouwer et al., 1989). Planting dates were 17 October, 17 October, 25 September, and 13 September, for Tanks 1, 2, 3, and 4, respectively. The darker red area corresponds to the portion of sluice outflow that is utilized to meet the irrigation water demand, while the light red area corresponds to the portion of sluice outflow that is not directly utilized by crops in the tank's command area.

ments of the tank command areas. To answer this question we have plotted supply-and-demand curves over the growing season (Fig. 11). The supply curves are the sluice outflow volumes from the four tanks. The demand curve in this

Table 5. Sluice outflows and irrigation water demand (IWD).

	Tank 1	Tank 2	Tank 3	Tank 4
Planting date	17/10	17/10	25/09	9/13
Sluice water				
Total (mm)	570	326	391	861
Utilized (mm)	283	210	333	516
Surplus (mm)	287	116	58	345
Percent surplus	50 %	36 %	15 %	40 %
Irrigation water demand				
Total (mm)	996	996	872	752
Unmet demand (mm)	713	786	540	235
Percent unmet	72 %	79 %	62 %	31 %

case is the crop water requirement in mm d^{-1}, which is adjusted by the available rainfall to get the irrigation water demand (IWD is equivalent to the rainfall subtracted from the crop water requirement). The crop water requirement data in mm d^{-1} were obtained from Brouwer et al. (1989) for the four growing stages of paddy. Paddy planting dates, which differed dramatically between the four tanks (17 October, 17 October, 25 September, and 13 September for Tanks 1, 2, 3, and 4), are based on field observations. The earlier planting dates in the command areas of Tanks 3 and 4 were most likely due to the availability of borewell water for those areas. As can be seen in Fig. 11, the difference in planting dates leads to different demand curves for the four tanks.

The supply-and-demand curves assess the ability of the tanks to meet paddy water demand by comparing IWDs to sluice outflows. The darker red areas in Fig. 11 denote sluice water used to meet the IWD, while the lighter red areas represent sluice water that is flowing out at a time when crops are not requiring that water. The grey areas in the figure represent the IWD unmet by sluice outflow. Notably, large quantities of surplus sluice water leave the tank soon after it fills. These surplus sluice outflows are not needed by the crops at the time they leave the tank and will either leave the catchment as evaporation or downstream runoff, or will recharge groundwater over the course of the outflow channel and become available to downstream users. Because the sluices are for the most part not actively managed or appropriately maintained in our study area, the sluices remaining perpetually open and outflows are purely a function of water levels in the tank rather than a timed need for irrigation water. As reported in Table 5, it was found that anywhere from 31–79 % of IWD within the study cascade remains unmet, while approximately 15–50 % of available sluice outflows leave the tank unutilized by crops in the tank command areas. This remaining irrigation water demand in many cases must be met by farmers using groundwater pumping to supplement tank water, and in other cases remains unmet, leading to reduced yields or crop failure. In the case of groundwater pumping, it should be noted that a significant portion of the tank water does leave the tanks as groundwater outflow, and is subsequently extracted by groundwater wells for irrigation, thus helping to meet the crop water requirements by an indirect route. The magnitude of this contribution of tank outflows to the crop water budget, however, is difficult to ascertain, and thus has not been included herein.

The timing of planting also has a significant impact on the ability of the tanks to meet crop water requirements (Fig. 11), with the later planting dates in Tanks 1 and 2 leading to more than 70 % of the IWD being unmet by sluice outflows (Table 5). Conversely, Tank 4, with its much earlier planting time (13 September), more effectively meets crop water requirements with sluice outflow. First, the early planting time leads to the lowest total IWD of all the tanks (752 mm), as more of the crop water requirements can be met by rainfall. In addition, there is a better temporal match for Tank 4 between the unregulated sluice outflows at high tank water levels (Fig. 11) and the crop water needs of the plants. Accordingly, more than 500 mm of the IWD is met by sluice outflows, and only 31 % of the overall demand remains unmet. These results suggest that, to optimize tank operations and to maximize the water-provisioning capabilities of the tanks, earlier planting times could be adopted by farmers, with supplemental irrigation from groundwater being utilized until the tanks fill. Such a change in management, however, would be dependent on both groundwater availability and the economics of groundwater pumping. Indeed, interactions with the villagers revealed that the earlier planting dates in the downgradient tanks could be attributed to the greater availability of groundwater in that region, which enables the farmers to plant before the monsoons have arrived.

5 Conclusion

In recent decades, there has been growing interest in the revival and expanded use of rainwater harvesting tanks across the agricultural landscapes of India and other semi-arid regions to address issues of water scarcity and aquifer depletion. While it is well established that these tanks can increase local water availability, leading to higher crop yields and direct socioeconomic benefits (Palanisami et al., 2010), the impact of widespread use of small, distributed storage reservoirs on the catchment-scale partitioning of water resources is still an open question. Furthermore, while significant resources are being used to rehabilitate tanks, there is a lack of understanding regarding how these ancient structures function in a modern landscape, under current socioeconomic and environmental pressures. The hydrology of these tanks is so intricately tied with the social system in which they are embedded that only a system approach, accounting for interactions between natural and human systems, can allow us to fully understand and manage these systems. Accordingly, any full analysis of tank water dynamics must be carried out within the domain of the emerging science of sociohydrology (Sivapalan et al., 2012).

In this paper we have used high-resolution monitoring of tank water levels to help quantify daily fluxes of evapotranspiration, groundwater recharge, and sluice outflows from the tanks, and have coupled this information with village-level data on planting dates and irrigated areas, to further our understanding of natural and human controls on water partitioning at both tank and catchment scales. At the tank scale, groundwater recharge and sluice outflow were observed to be the largest components of the tank water budget, with ET accounting for only 13–22 % of the outflows, including open water evaporation and ET of plants transpiring in the tank bed. At the catchment scale, our results demonstrate that the presence of tanks within the catchment decreases runoff by approximately 70 %, increases recharge by 40 %, and directly satisfies approximately 40 % of crop water requirements across the northeast monsoon season via surface water irrigation. These findings suggest that village-scale rainwater harvesting tanks can dramatically increase water availability at a local or village scale, but also may have negative impacts on downstream users due to large decreases in catchment runoff. Our results also highlight that, despite ongoing the efforts toward tank rehabilitation in our study cascade, a lack of sluice maintenance leads to a large fraction of tank water not being available for use in the tank command area. Thus, a more efficient management of sluice outflows, and better maintenance of the sluices themselves, could lead to the tanks meeting a higher fraction of crop water requirements.

An interesting and novel attribute of our study is the exploration of biophysical and social controls on tank water dynamics as a function of the location of the tank along a cascade, in a four-tank cascade system. We observe a distinct spatial pattern in groundwater-exchange dynamics with the most downgradient tank being mostly driven by groundwater inflow, while the other tanks are more outflow driven. Consequently the most downgradient tank has a much greater outflow-to-capacity ratio, and is able to provide a much larger volume of sluice outflow compared to its capacity. The ability of the most downgradient tank to provide more irrigation water is a function of the return flow from the command areas of the upstream tanks and highlights the need to study tanks, not in isolation, but as a part of a cascade. There is also a distinct pattern in the crop planting dates in the four tanks, with the more downgradient tanks having earlier planting dates, due to greater availability of groundwater, thus leading to a more efficient use of tank water. This dynamic highlights the feedbacks between the natural and human systems, where a greater availability of water at the catchment outlet leads to farmers deciding on earlier planting dates, which in turn leads to a more efficient use of the available water.

In conclusion, our results demonstrate the significant role that tanks can play in addressing challenges of limited water availability, by both increasing groundwater recharge as well as the water available for irrigation. However, they also draw attention to the potentially negative environmental impacts of tanks with respect to reducing downstream flows. These findings highlight the need to understand the spatiotemporal patterns in tank water dynamics at the basin scale, especially within the framework of a coupled natural and human systems approach that allows a more complete understanding of how tanks alter the sociohydrological dynamics of water-stressed landscapes. Thus, ongoing rehabilitation efforts of tanks must be complemented by more studies that quantify the functioning of these rehabilitated tanks and their impacts in altering basin-scale water dynamics, with the overall goal of appropriately managing tradeoffs between socioeconomic benefits and environmental costs.

Author contributions. The field study was carried out by Michael Steiff under the guidance of Daniel L. McLaughlin. Data analysis was carried out by Kimberly J. Van Meter and Michael Steiff. Drafting of the manuscript was led by Kimberly J. Van Meter and Nandita B. Basu, with contributions by Daniel L. McLaughlin. Nandita B. Basu conceived of the project and was instrumental to the basic experimental design.

Acknowledgements. This research is financially supported by the US National Science Foundation (1211968), Dynamics of Coupled Natural-Human Systems and by the University of Waterloo start-up funds. We thank the DHAN Foundation for generously sharing their experiences, data, and hospitality during the two field visits that allowed us to carry out this project. We would also like to thank K. Palanisami for his help with establishing connections for the research team within India and with his guidance regarding field site selection.

Edited by: A. Mejia

References

Agarwal, A. and Narain, S.: Dying Wisdom: Rise, Fall and Potential of India's Traditional Water Harvesting Systems, (Centre for Science and Environ, 1997.

Amarasinghe, U., Sharma, B. R., Aloysius, N., Scott, C., Smakhtin, V., and Fraiture, C. de: Spatial Variation in Water Supply and Demand Across River Basins of India, IWMI, 2005.

Amarasinghe, U. A., Singh, O. P., Sakthivadivel, R., and Palanisami, K.: State of irrigation in Tamil Nadu: trends and turning points, Strateg. Anal. Natl. River Link. Proj. NRLP India Ser. 5, 275–304, 2009.

Anbumozhi, V., Matsumoto, K., and Yamaji, E.: Towards improved performance of irrigation tanks in semi-arid regions of India: modernization opportunities and challenges, Irrig. Drain. Syst., 15, 293–309, 2001.

Anuraga, T. S., Ruiz, L., Kumar, M. S. M., Sekhar, M., and Leijnse, A.: Estimating groundwater recharge using land use and soil data: A case study in South India, Agr. Water Manage., 84, 65–76, 2006.

Balasubramanian, R. and Selvaraj, K. N.: Poverty, private property, and common pool resource management: the case of irrigation tanks in South India, South Asian Network for Development and Environmental Economics, Kathmandu, 1–61, 2003.

Batchelor, C. H., Rama Mohan Rao, M. S., Manohar Rao, S., and others: Watershed development: A solution to water shortages in semi-arid India or part of the problem, Land Use Water Resour. Res., 3, 1–10, 2003.

Bouma, J. A., Biggs, T. W., and Bouwer, L. M.: The downstream externalities of harvesting rainwater in semi-arid watersheds: An Indian case study, Agric. Water Manag., 98, 1162–1170, doi:10.1016/j.agwat.2011.02.010, 2011.

Brouwer, C., Prins, K., and Heibloem, M.: Irrigation Water Management: Irrigation Scheduling, FAO, available at: http://www.fao.org/docrep/t7202e/t7202e00.HTM (last access: 17 October 2015), 1989.

Calder, I., Gosain, A., Rao, M. S. R. M., Batchelor, C., Snehalatha, M., and Bishop, E.: Watershed development in India. 1. Biophysical and societal impacts, Environ. Dev. Sustain., 10, 537–557, doi:10.1007/s10668-006-9079-7, 2008a.

Calder, I., Gosain, A., Rao, M. S. R. M., Batchelor, C., Garratt, J., and Bishop, E.: Watershed development in India. 2. New approaches for managing externalities and meeting sustainability requirements, Environ. Dev. Sustain., 10, 427–440, doi:10.1007/s10668-006-9073-0, 2008b.

Carlson Mazur, M. L., Wiley, M. J., and Wilcox, D. A.: Estimating evapotranspiration and groundwater flow from water-table fluctuations for a general wetland scenario, Ecohydrology, 7, 378–390, doi:10.1002/eco.1356, 2014.

CGWB (Central Ground Water Board): Aquifer Systems of Tamil Nadu and Puducherry, Central Ground Water Board, Chennai, 1–78, 2012.

Cullet, P. and Gupta, J.: India: evolution of water law and policy, in: The Evolution of the Law and Politics of Water, 157–173, Springer, available at: http://link.springer.com/chapter/10.1007/978-1-4020-9867-3_10 (last access: 7 June 2015), 2009.

DHAN: Development of Humane Action (DHAN), available at: http://www.dhan.org/themes/vtadp.php (last access: 21 July 2015), 2010.

Drexler, J. Z., Snyder, R. L., Spano, D., and Paw, U. K. T.: A review of models and micrometeorological methods used to estimate wetland evapotranspiration, Hydrol. Process., 18, 207–2101, 2004.

Farmer, B. H.: Green revolution? Technology and change in rice – growing areas of Tamil Nadu and Sri Lanka, Westview Press, Boulder, CO, 1–429, 1977.

Garg, K. K., Wani, S. P., Barron, J., Karlberg, L., and Rockstrom, J.: Up-scaling potential impacts on water flows from agricultural water interventions: opportunities and trade-offs in the Osman Sagar catchment, Musi sub-basin, India, Hydrol. Process., 27, 3905–3921, doi:10.1002/hyp.9516, 2013.

Glendenning, C. J., van Ogtrop, F. F., Mishra, A. K., and Vervoort, R. W.: Balancing watershed and local scale impacts of rain water harvesting in India – A review, Agric. Water Manag., 107, 1–13, doi:10.1016/j.agwat.2012.01.011, 2012.

Gohar, A. A., Ward, F. A., and Amer, S. A.: Economic performance of water storage capacity expansion for food security, J. Hydrol., 484, 16–25, doi:10.1016/j.jhydrol.2013.01.005, 2013.

Government of India: Census of India: Census Data Online, [online] Available at: http://censusindia.gov.in/2011-common/censusdataonline.html (last access: 6 August 2015), 2011.

Government of Tamil Nadu: Season and Crop Report Tamil Nadu 2009–10, 352, 2011.

Grey, D. and Sadoff, C. W.: Sink or Swim? Water security for growth and development, Water Policy, 9, 545–571, doi:10.2166/wp.2007.021, 2007.

Gunnell, Y. and Krishnamurthy, A.: Past and present status of runoff harvesting systems in dryland peninsular India: A critical review, AMBIO J. Hum. Environ., 32, 320–324, 2003.

Gupta, S. K. and Deshpande, R. D.: Water for India in 2050: first-order assessment of available options, Curr. Sci., 86, 1216–1224, 2004.

Haile, M.: Weather patterns, food security and humanitarian response in sub-Saharan Africa, Philos. Trans. R. Soc. B Biol. Sci., 360, 2169–2182, doi:10.1098/rstb.2005.1746, 2005.

Healy, R. W. and Cook, P. G.: Using groundwater levels to estimate recharge, Hydrogeol. J., 10, 91–109, 2002.

Hill, A. J. and Durchholz, B.: Specific Yield Functions for Estimating Evapotranspiration from Diurnal Surface Water Cycles, J. Am. Water Resour. Assoc., 51, 123–132, doi:10.1111/jawr.12237, 2015.

Holmes, R. M. and Robertson, G. W.: Conversion of latent evaporation to potential evapotranspiration, Can. J. Plant Sci., 38, 164–172, 1958.

Hundertmark, W. and Facon, T.: Options for effective rice water management, in Proceedings of the 20th session of the International Rice Commission, Ganfkok, Thailand, 23–26 July 2002, Sustainable rice production for food security, 145–170, Food and Agriculture Organization of the United Nations, Bangkok, Thailand, available at: http://www.fao.org/docrep/006/y4751e/y4751e0j.htm (last access: 7 October 2015), 2003.

Hvorslev, M. J.: Time lag and soil permeability in ground-water observations, US Corps of Engineers, Vicksburg, Mississippi, available at: http://trid.trb.org/view.aspx?id=124794 (last access: 22 March 2016), 1951.

Keller, A. A., Sakthivadivel, R., and Seckler, D. W.: Water Scarcity and the Role of Storage in Development, IWMI, 2000.

Kumar, M. D., Ghosh, S., Patel, A., Singh, O. P., and Ravindranath, R.: Rainwater harvesting in India: some critical issues for basin planning and research, Land Use Water Resour. Res., 6, 1–17, 2006.

Kumar, M. D., Patel, A., Ravindranath, R., and Singh, O. P.: Chasing a Mirage: Water Harvesting and Artificial Recharge in Naturally Water-Scarce Regions, Econ. Polit. Wkly., 43, 61–71, doi:10.2307/40278725, 2008.

Latha, M., Rajendran, M., and Murugappan, A.: Comparison of GIS based SCS-CN and Strange table Method of Rainfall-Runoff Models for Veeranam Tank, Tamil Nadu, India., Int. J. Sci. Eng. Res., 3, 1–5, 2012.

Loheide, S. P.: A method for estimating subdaily evapotranspiration of shallow groundwater using diurnal water table fluctuations, Ecohydrology, 1, 59–66, doi:10.1002/eco.7, 2008.

Loheide, S. P., Butler, J. J., and Gorelick, S.: Estimation of groundwater consumption by phreatophytes using diurnal water table fluctuations: A saturated-unsaturated flow assessment, Water Resour. Res., 41, doi:10.1029/2005WR003942, 2005.

McLaughlin, D. L. and Cohen, M. J.: Realizing ecosystem services: wetland hydrologic function along a gradient of ecosystem condition, Ecol. Appl., 23, 1619–1631, 2013.

McLaughlin, D. L. and Cohen, M. J.: Ecosystem specific yield for estimating evapotranspiration and groundwater exchange from diel surface water variation, Hydrol. Process., 28, 1495–1506, doi:10.1002/hyp.9672, 2014.

McLaughlin, D. L., Kaplan, D. A., and Cohen, M. J.: A significant nexus: Geographically isolated wetlands influence landscape hydrology, Water Resour. Res., 50, 7153–7156, doi:10.1002/2013WR015002, 2014.

Mehta, L.: The manufacture of popular perceptions of scarcity: dams and water-related narratives in Gujarat, India, World Dev., 29, 2025–2041, 2001.

Mialhe, F., Gunnell, Y., and Mering, C.: Synoptic assessment of water resource variability in reservoirs by remote sensing: General approach and application to the runoff harvesting systems of south India, Water Resour. Res., 44, W05411, doi:10.1029/2007WR006065, 2008.

Mitsch, W. J. and Gosselink, J. G.: Wetlands, Hoboken, John Wiley & Sons, Inc., 2007.

Neumann, I., MacDonald, D., and Gale, I.: Numerical approaches for approximating technical effectiveness of artificial recharge structures, Commissioned Report, British Geological Survey, Keyworth, Nottingham, available at: http://nora.nerc.ac.uk/3230/1/RR05001.pdf (last access: 2 February 2014), 2004.

Ngigi, S. N.: What is the limit of up-scaling rainwater harvesting in a river basin?, Phys. Chem. Earth, 28, 943–956, doi:10.1016/j.pce.2003.08.015, 2003.

Oki, T.: Global Hydrological Cycles and World Water Resources, Science, 313, 1068–1072, doi:10.1126/science.1128845, 2006.

Palaniappan, S. P., Balasubramanian, R., Ramesh, T., Chandrasekaran, A., Mani, K. G., Velayutham, M., and Lal, R.: Sustainable Management of Dryland Alfisols (Red Soils) in South India, J. Crop Improv., 23, 275–299, doi:10.1080/15427520902809888, 2009.

Palanisami, K. and Meinzen-Dick, R.: Tank performance and multiple uses in Tamil Nadu, South India, Irrig. Drain. Syst., 15, 173–195, 2001.

Palanisami, K., Meinzen-Dick, R., and Giordano, M.: Climate change and water supplies: options for sustaining tank irrigation potential in India, Rev. Agric. Econ. Polit. Wkly., 45, 183–190, 2010.

Pandey, P. K., Soupir, M. L., Singh, V. P., Panda, S. N., and Pandey, V.: Modeling Rainwater Storage in Distributed Reservoir Systems in Humid Subtropical and Tropical Savannah Regions, Water Resour. Manag., 25, 3091–3111, doi:10.1007/s11269-011-9847-5, 2011.

Pandey, P. K., van der Zaag, P., Soupir, M. L., and Singh, V. P.: A new model for simulating supplemental irrigation and the hydro-economic potential of a rainwater harvesting system in humid subtropical climates, Water Resour. Manag., 27, 3145–3164, 2013.

Paraskevas, C., Georgiou, P., Ilias, A., Panoras, A. and Babajimopoulos, C.: Evapotranspiration and simulation of soil water movement in small area vegetation, Int. Agrophys., 27, 445–453, 2013.

Parish, E. S., Kodra, E., Steinhaeuser, K., and Ganguly, A. R.: Estimating future global per capita water availability based on changes in climate and population, Comput. Geosci., 42, 79–86, doi:10.1016/j.cageo.2012.01.019, 2012.

Pathak, P., El-Swaify, S. A., Singh, S., Raghavan, N., and Beckerman, S. R.: Alfisols in the Semi-Arid Tropics. in Proceedings of the Consultants' Workshop on the State of the Art and Management Alternatives for Optimizing the Productivity of SAT Alfisols and Related Soils 1–44, International Crops Research Institute for the Semi-Arid Tropics., 1987.

Rockstrom, J., Barron, J., and Fox, P.: Rainwater management for increased productivity among small-holder farmers in drought prone environments, Phys. Chem. Earth, 27, 949–959, 2002.

Shah, T.: Water and Welfare: Critical Issues in India's Water Future, Econ. Polit. Wkly., 39, 1211–1213, doi:10.2307/4414796, 2004.

Shanmugham, C. R. and Kanagavalli, J.: Technology of Tanks: The Traditional Water Bodies of Rural India, Reflection Publications Trust, Madurai, India, 2013.

Sharda, V. N., Kurothe, R. S., Sena, D. R., Pande, V. C., and Tiwari, S. P.: Estimation of groundwater recharge from water storage structures in a semi-arid climate of India, J. Hydrol., 329, 224–243, doi:10.1016/j.jhydrol.2006.02.015, 2006.

Sharma, A. K. and Thakur, P. K.: Quantitative assessment of sustainability of proposed watershed development plans for kharod watershed, western India, J. Indian Soc. Remote Sens., 35, 231–241, 2007.

Sivapalan, M., Savenije, H. H. G., and Blöschl, G.: Socio-hydrology: A new science of people and water, Hydrol. Process., 26, 1270–1276, doi:10.1002/hyp.8426, 2012.

van der Zaag, P. and Gupta, J.: Scale issues in the governance of water storage projects, Water Resour. Res., 44, W10417, doi:10.1029/2007WR006364, 2008.

Van Meter, K. J. V., Basu, N. B., Tate, E., and Wyckoff, J.: Monsoon Harvests: The Living Legacies of Rainwater Harvesting Systems in South India, Environ. Sci. Technol., 48, 4217–4225, doi:10.1021/es4040182, 2014.

von Oppen, M. and Subba Rao, K. V.: Tank Irrigation in Semi-Arid Tropical India, Research Bulletin, International Crops Research Institute for the Semi-Arid Tropics, Patancheru, Andrha Pradesh, 1987.

Vose, R. S., Schmoyer, R. L., Steurer, P. M., Peterson, T. C., Heim, R., Karl, T. R. and Eischeid, J.: The Global Historical Climatology Network: Long-term monthly temperature, precipitation, sea level pressure, and station pressure data, Oak Ridge National Laboratory Oak Ridge, Tennessee. [online] available at: ttp://www.nativefishlab.net/library/textpdf/12377.pdf (Accessed 7 October 2015), 1992.

Wada, Y., Gleeson, T., and Esnault, L.: Wedge approach to water stress, Nat. Geosci., 7, 615–617, 2014.

Weiz, B.: Water Reservoirs in South India: An Anthropological Approach, Diss., 30, 1–299, 2005.

Wiltshire, A., Gornall, J., Booth, B., Dennis, E., Falloon, P., Kay, G., McNeall, D., McSweeney, C., and Betts, R.: The importance of population, climate change and CO_2 plant physiological forcing in determining future global water stress, Global Environ. Change, 23, 1083–1097, 2013.

14

Quantifying energy and water fluxes in dry dune ecosystems of the Netherlands

B. R. Voortman[1], R. P. Bartholomeus[1], S. E. A. T. M. van der Zee[2], M. F. P. Bierkens[3,4], and J. P. M. Witte[1,5]

[1]KWR Watercycle Research Institute, P.O. Box 1072, 3430 BB Nieuwegein, the Netherlands
[2]Soil Physics and Land Management, Environmental Sciences Group, Wageningen University, P.O. Box 47, 6700 AA Wageningen, the Netherlands
[3]Department of Physical Geography, Faculty of Geosciences, Utrecht University, P.O. Box 80115, 3508 TC Utrecht, the Netherlands
[4]Deltares, P.O. Box 85467, 3508 AL Utrecht, the Netherlands
[5]VU University, Institute of Ecological Science, Department of Systems Ecology, de Boelelaan 1085, 1081 HV Amsterdam, the Netherlands

Correspondence to: B. R. Voortman (bernard.voortman@kwrwater.nl)

Abstract. Coastal and inland dunes provide various ecosystem services that are related to groundwater, such as drinking water production and biodiversity. To manage groundwater in a sustainable manner, knowledge of actual evapotranspiration (ET_a) for the various land covers in dunes is essential. Aiming at improving the parameterization of dune vegetation in hydrometeorological models, this study explores the magnitude of energy and water fluxes in an inland dune ecosystem in the Netherlands. Hydrometeorological measurements were used to parameterize the Penman–Monteith evapotranspiration model for four different surfaces: bare sand, moss, grass and heather. We found that the net longwave radiation (R_{nl}) was the largest energy flux for most surfaces during daytime. However, modeling this flux by a calibrated FAO-56 R_{nl} model for each surface and for hourly time steps was unsuccessful. Our R_{nl} model, with a novel submodel using solar elevation angle and air temperature to describe the diurnal pattern in radiative surface temperature, improved R_{nl} simulations considerably. Model simulations of evaporation from moss surfaces showed that the modulating effect of mosses on the water balance is species-dependent. We demonstrate that dense moss carpets (*Campylopus introflexus*) evaporate more (5 %, +14 mm) than bare sand (total of 258 mm in 2013), while more open-structured mosses (*Hypnum cupressiforme*) evaporate less (−30 %, −76 mm) than bare sand. Additionally, we found that a drought event in the summer of 2013 showed a pronounced delayed signal on lysimeter measurements of ET_a for the grass and heather surfaces, respectively. Due to the desiccation of leaves after the drought event, and their feedback on the surface resistance, the potential evapotranspiration in the year 2013 dropped by 9 % (−37 mm) and 10 % (−61 mm) for the grass and heather surfaces, respectively, which subsequently led to lowered ET_a of 8 % (−29 mm) and 7 % (−29 mm). These feedbacks are of importance for water resources, especially during a changing climate with an increasing number of drought days. Therefore, such feedbacks need to be integrated into a coupled plant physiological and hydrometeorological model to accurately simulate ET_a. In addition, our study showed that groundwater recharge in dunes can be increased considerably by promoting moss vegetation, especially of open-structured moss species.

1 Introduction

Coastal and inland sand dunes are major drinking water production sites in the Netherlands. Approximately 23 % of Dutch drinking water originates from aquifers in these dunes, which are replenished by both natural groundwater recharge and artificial infiltration of surface waters. Another ecosystem service of groundwater in dune systems is that shallow groundwater tables sustain nature targets with a very high conservation value. Such targets, like wet dune slacks and oligotrophic pools, are often legally enforced, e.g., by the European Habitat Directive and by the Water Framework Directive. Furthermore, a deep layer of fresh groundwater in coastal dunes protects the hinterland from the inflow of saline groundwater.

Under a warming climate, summers are expected to become dryer and the water quality of surface waters may degrade (Delpla et al., 2009), especially during dry periods with low river discharge rates (Zwolsman and van Bokhoven, 2007; van Vliet and Zwolsman, 2008). To maintain current drinking water quality and production costs, water production in the future may have to rely more on natural groundwater recharge. This implies that drinking water companies need to search for new water production sites or intensify current groundwater extractions, while protecting groundwater-dependent nature targets.

For sustainable management of renewable groundwater resources, groundwater extractions should be balanced with the amount of precipitation that percolates to the saturated zone, the groundwater recharge. Knowledge of actual evapotranspiration (ET_a, here defined as the sum of plant transpiration, soil evaporation and evaporation from canopy interception) for the various land covers is essential to quantify the amount of recharge. Inland dune systems are predominantly covered with deciduous and pine forest. Well-developed hydrometeorological models are available to simulate ET_a for these forest ecosystems (Dolman, 1987; Moors, 2012). Other ecosystems, such as heathland and bare sand colonized by algae, mosses, tussock-forming grasses or lichens, received less attention. However, heathland and drift sand ecosystems have a higher conservation value than forest plantations, in particular those of coniferous trees. Nature managers are therefore often obligated to protect and develop certain heathland and drift sand ecosystems at the expense of forest ecosystems (The European Natura 2000 policy). A better parameterization of heathland and drift sand ecosystems in hydrometeorological models would aid in the sustainable management of important groundwater resources and would allow quantifying the cost and benefit of nature conservation in terms of groundwater recharge.

To this end, this study explores diurnal patterns in energy and water fluxes in a dry dune ecosystem on an elevated sandy soil in the Netherlands. Our study aims to improve the parameterization of dune vegetation in hydrometeorological models based on field measurements, focusing on four different surfaces: bare sand, moss (*Campylopus introflexus*), grass (*Agrostis vinealis*) and heather (*Calluna vulgaris*). A second objective is to quantify the effect of moss species on the water balance. Mosses and lichens are present in most successional stages in dry dune ecosystems, either as pioneer species or as understory vegetation. Voortman et al. (2014) hypothesized that moss-covered soils could evaporate less than a bare soil, since the unsaturated hydraulic properties of moss layers reduce evaporation under relatively moist conditions. Such hydraulic behavior could have large implications on the ecological interactions between vascular and nonvascular plants in water-limited ecosystems, as the presence of a moss cover could facilitate the water availability for rooting plants. Such interactions are of importance to groundwater resources, as the resilience of plant communities to drought determines the succession rate and biomass, which subsequently feedback on evapotranspiration.

A third objective is to gain insight into the delayed effect of dry spells on potential and actual evapotranspiration for heathlands and grasslands. To quantify the evapotranspiration loss term, many hydrological modeling frameworks use the concept of potential evapotranspiration ET_p (Federer et al., 1996; Kay et al., 2013; Zhou et al., 2006), defined as the maximum rate of evapotranspiration from a surface where water is not a limiting factor (Shuttleworth, 2007). ET_p is input to modeling frameworks and reduces to ET_a in cases of water stress. However, if dry spells result in a vegetation dieback, the simulated ET_p should be adjusted to account for the smaller transpiring leaf area after the dry spell. The model simulations presented in this paper give some guidance on the magnitude of errors in simulated ET_a if feedbacks of dry spells on ET_p are neglected.

The knowledge presented in this paper will help to improve and interpret the simulations of water recharge in sand dunes by hydrological models, and will sustain rainwater harvesting in dunes by vegetation management.

2 Measurements and methods

2.1 General setup

A field campaign started in August 2012 to measure energy and water fluxes in the drinking water supply area "Soestduinen", situated on an elevated sandy soil (an ice-pushed ridge) in the center of the Netherlands (52.14° latitude, 5.31° longitude). Due to deep groundwater levels, the vegetation in this region is groundwater-independent, i.e., relying solely on rainwater (on average 822 mm rain per year, 40 % falling in the first 6 months of the year and 60 % falling in the last 6 months of the year). The reference evapotranspiration according to Makkink (1957) is on average 561 mm per year. The field data were used to parameterize the Penman–Monteith equation, to calculate ET_p and to perform hydrological model simulations of ET_a, based on the actual availability of soil moisture. The Penman–Monteith equation is given by

$$ET_p = \frac{\Delta (R_n - G) + \rho_a c_p (e_s - e_a)/r_a}{\left(\Delta + \gamma\left(1 + \frac{r_s}{r_a}\right)\right)\lambda \rho_w}, \quad (1)$$

where ET_p is the potential evapotranspiration ($mm\,s^{-1}$), Δ is the slope of the saturation vapor pressure vs. temperature

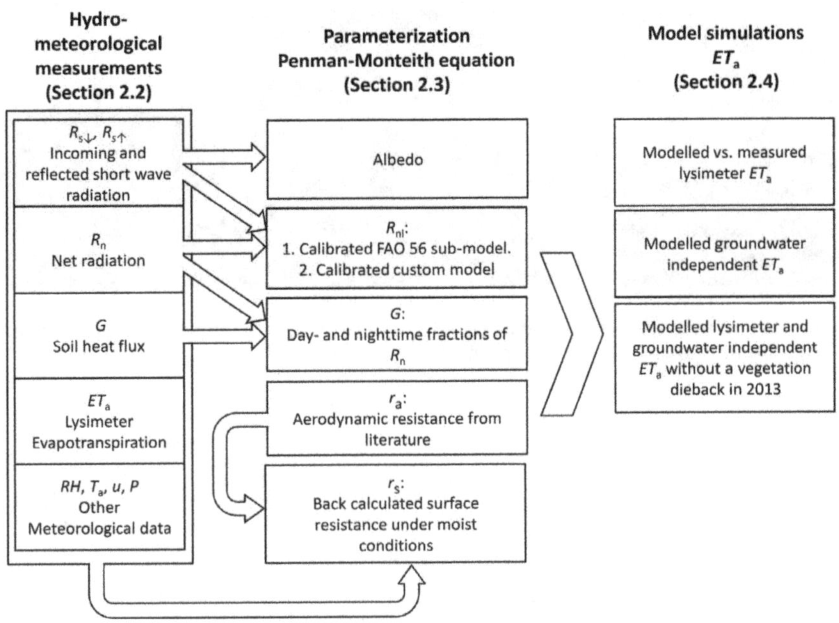

Figure 1. Organization of the research from measurements to model simulations.

curve (kPa °C^{-1}), R_n is the net radiation (J m^{-2}), G is the soil heat flux (J m^{-2}), ρ_a is the air density (kg m^{-3}), c_p is specific heat of moist air (J kg^{-1} °C^{-1}), e_s is the saturation vapor pressure of the air (kPa), e_a is the actual vapor pressure of the air (kPa), r_a is aerodynamic resistance to turbulent heat and vapor transfer (s m^{-1}), γ is the psychrometric constant (kPa °C^{-1}), λ is the latent heat of vaporization (J kg^{-1}) an d ρ_w is the density of liquid water (kg m^{-3}). Results of Irmak et al. (2005) suggest that estimates of ET_p on hourly time steps are more accurate than estimates on a daily timescale. Furthermore, Liu et al. (2005) showed that the use of daily input values leads to a systematic overestimation of ET_a, especially for sandy soils.

Hence, energy fluxes in the Penman–Monteith equation are preferably simulated at subdiurnal timescales. Furthermore, understanding and sim-ulation of plant physiological processes requires knowledge of the diurnal variation of environmental variables (Nozue and Maloof, 2006). Therefore, field data were aggregated to hourly time steps to maintain the diurnal pattern and to analyze our field results at the same time interval as commonly available climate data.

In this paper evapotranspiration is defined as the sum of transpiration, soil evaporation and evaporation from canopy interception, expressed in mm per time unit. Radiative and soil heat fluxes are expressed in W m^{-2}. Figure 1 shows the procedures followed to translate field data (Sect. 2.1) to submodels of the Penman–Monteith equation (Sect. 2.2) and to subsequently calculate ET_p and simulate ET_a (Sect. 2.3).

2.2 Hydrometeorological measurements

Four homogeneous sites of bare sand, moss (*Campylopus introflexus*), grass (*Agrostis vinealis*) and heather (*Calluna vulgaris*) (Fig. 2) were selected to measure actual evapotranspiration (ET_a), the net radiation (R_n), the soil heat flux (G) and the albedo. Other meteorological variables such as wind speed (u, at 2 m above the surface), relative humidity (RH, 1.5 m above the surface), air temperature (T_a, 1.5 m above the surface) and rain (P) were measured at a weather station, installed in-between the measurement plots at a maximum distance of 40 m from each plot. Measurements were collected with data loggers (CR1000, Campbell Scientific Inc.) at a 10 s interval and aggregated to minutely values. Field measurements of bare sand, moss and grass were collected between August 2012 and November 2013. The field measurements in the heather vegetation were collected between June 2013 and November 2013.

The net radiation was measured with net radiometers (NR-Lite2 Kip & Zonen B.V.). The net radiometers were installed at a relatively low height of 32, 40, 40 and 50 cm above the bare sand, moss, grass and heather surfaces, respectively (relative to the average vegetation height), to limit the field of view to a homogenous surface. The incoming solar radiation ($R_{s\downarrow}$) and reflected solar radiation ($R_{s\uparrow}$) were measured with an albedo meter (CMA6, Kip & Zonen B.V.) that was rotated between the four surfaces. It was installed next to each R_n sensor. Due to a snow cover (winter months) or sensor maintenance (October 2012, May 2013), some periods were omitted (Fig. 3).

Eight self-calibrating heat flux plates (HFP01SC, Hukseflux B.V.) (two for each site) were installed 8 cm below the

Figure 2. The vegetation types studied in this paper, **(a)** the moss surface with an approximately 2 cm thick layer of *Campylopus introflexus* (inset), **(b)** the grass surface, primarily *Agrostis vinealis* and **(c)** the heather surface, *Calluna vulgaris*.

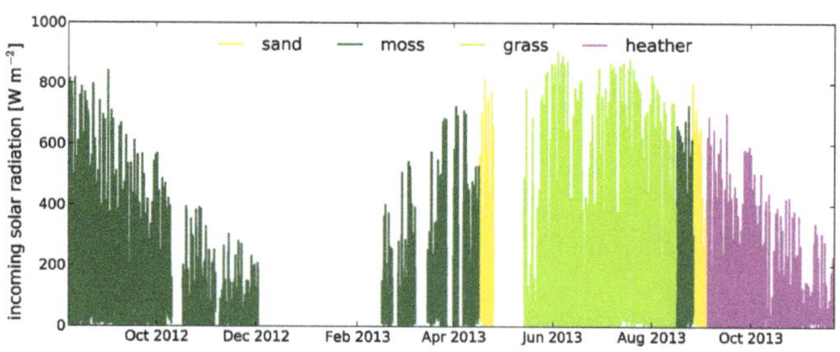

Figure 3. Measured incoming solar radiation $R_{S\downarrow}$ at the four different surfaces. Periods with snow cover or sensor maintenance were omitted.

soil surface near the net radiometers. These heat flux plates were programmed to calibrate themselves for 15 min at 6 h time intervals, based on a known heat flux supplied by an integrated heater. Besides each soil heat flux plate, an averaging thermocouple (TCAV, Campbell Scientific, Inc.) was installed at 2 and 6 cm depth and a soil moisture probe (CS616, Campbell Scientific, Inc) was installed at 4 cm depth to estimate the change in heat storage (S) above the heat flux plates. The sum of the measured soil heat flux at 8 cm depth and S represents the heat flux at the soil surface. Sensor installation and procedures to calculate S were followed according to the Campbell Scientific Inc. (2014) HFP01SC instruction manual.

Within each surface, one weighing lysimeter was installed. The lysimeters (Fig. 4) had a 47.5 cm inner diameter and were 50 cm deep. Intact soil monoliths were sampled by hammering the PVC tube into the soil, alternated with excavating the surrounding soil to offset soil pressures. The lysimeters were turned upside down, to level the soil underneath and to close this surface with a PVC end cap. To allow water to drain out of the lysimeter bottom plate, a 2.5 cm diameter hole was made in the base plate. A 15 cm long fiberglass wick (Pepperell 2 × 1/2 inch) was installed in the PVC end cap to guide drainage water through the hole into a tipping bucket (Davis 7852) below the lysimeter. The wick, together with two sheets of filter cloth (140–150 μm, Eijkelkamp Agrisearch Equipment), placed at the bottom of the lysimeter tank, prevented soil particles from flushing out

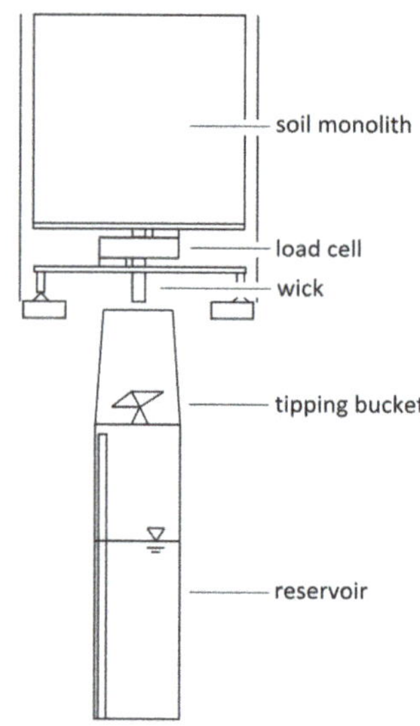

Figure 4. Lysimeter design.

of the lysimeter. The tipping bucket below the lysimeter had a resolution of 0.2 mm for the intercepting area of the tipping bucket, which was equal to 0.024 mm for the cross-sectional area of the lysimeter. Drainage water was collected in a reservoir installed below the lysimeter.

The lysimeters were weighted with temperature compensated single point load cells (Utilcell 190i, max 200 kg). These load cells were initially connected to the full bridge data ports of the data loggers. However, the measurement resolution of the data loggers was too coarse to fully compensate for temperature effects on weight measurements. Fluctuations of 0.333 μV due to temperature effects were within the data logger measurement resolution, which equals 36 g in weight change, i.e., 0.2 mm of evaporation. To increase the lysimeter precision, digitizers (Flintec LDU 68.1) were installed in May 2013 to process and digitize the load cell signals without interference of the data logger. In this setup, a measurement resolution of 10 g was achieved, i.e., 0.06 mm equivalent water depth, which is adequate for measuring ET_a for daily time periods (subtracting two values would lead to a maximum error of 0.06 mm caused by the measurement resolution). Analysis of measured ET_a was therefore limited to the period after installation of the digitizers.

After a rain event on 7 September 2013, the tipping buckets below the grass and heather lysimeters became partly clogged with beetles nesting underneath the lysimeters. This led to a continuous drainage signal which was out of phase with the weight measurements. Without accurate drainage measurements, lysimeter weight signals cannot be transferred to evapotranspiration. Therefore, ET_a data on days with a poor drainage signal after 7 September 2013 were disregarded in the analyses for the grass and heather lysimeters.

2.3 Parameterization of the Penman–Monteith equation

2.3.1 Net radiation (R_n)

The net radiation (R_n) is defined as

$$R_n = R_{ns} + R_{nl} = (1 - \text{albedo})\, R_{s\downarrow} + \left(\varepsilon_s R_{l\downarrow} - R_{l\uparrow}\right), \tag{2}$$

where R_{ns} is the net shortwave radiation, R_{nl} is the net longwave radiation, $R_{s\downarrow}$ is the incoming solar radiation, $R_{l\downarrow}$ is the downwelling longwave radiation from the atmosphere to the surface, $R_{l\uparrow}$ is the emitted longwave radiation by the surface into the atmosphere and ε_s is the surface emissivity representing the reflected downwelling longwave radiation. The albedo in Eq. (2) was determined by linear regression between measured $R_{s\downarrow}$ and $R_{s\uparrow}$. Based on the albedo obtained this way, R_{nl} follows from measurements of R_n by subtracting calculated R_{ns} from measured R_n. Throughout this paper, this back-calculated R_{nl} is referred to as the measured R_{nl}.

In hydrometeorological models, R_{nl} is commonly estimated under clear sky conditions and multiplied by a factor to correct for clouds (Irmak et al., 2010; Gubler et al., 2012; Blonquist Jr. et al., 2010; Temesgen et al., 2007). A similar approach was followed in this study in which the Stefan–Boltzmann law is substituted into Eq. (2) for $R_{l\downarrow}$ and $R_{l\uparrow}$ under clear sky conditions (Saito and Šimůnek, 2009; Van Bavel and Hillel, 1976) and multiplied by a cloudiness function to obtain R_{nl}:

$$R_{nl} = \left(\varepsilon_s \varepsilon_a \sigma T_a^4 - \varepsilon_s \sigma T_s^4\right) f_{cd}, \tag{3}$$

where ε_a is the clear sky emissivity of the atmosphere (–), ε_s is the surface emissivity (–), σ is the Stefan–Boltzmann constant (5.67×10^{-8} Wm^{-2}K^{-1}), T_a is the air temperature (K), T_s is the surface temperature (K) and f_{cd} is a cloudiness function (–; described later). For vegetated surfaces, $\varepsilon_s = 0.95$ was used (based on Jones, 2004), and $\varepsilon_s = 0.925$ was used for bare sand (based on Fuchs and Tanner, 1968). Estimating ε_a has a long history and numerous parameterizations are available. In this study, the empirical relationship found by Brunt (1932) was used:

$$\varepsilon_a = 0.52 + 0.065\sqrt{e_a}, \tag{4}$$

where e_a is the water vapor pressure measured at screen level (hPa). The cloudiness function f_{cd} in Eq. (3) is limited to $0.05 \leq f_{cd} \leq 1$ and equal to

$$f_{cd} = \frac{R_{s\downarrow}}{R_{s0}}, \tag{5}$$

where R_{s0} is the estimated clear sky solar radiation. We estimated R_{s0} following the FAO irrigation and drainage paper No. 56 (Allen et al., 1998). Since f_{cd} is undefined during the night, an interpolation of f_{cd} between sunset and sunrise is required. According to Gubler et al. (2012) f_{cd} can be best linearly interpolated between the 4 to 6 h average before sunset and after sunrise. We adopted this approach, applying a 5 h average.

An estimate of T_s is required to fully parameterize Eq. (3). We developed a new approach to simulate the diurnal pattern in T_s. Using Eq. (3), we back-calculated $T_s - T_a$ based on measured R_{nl} for clear hours ($f_{cd} > 0.9$). Generally, $T_s - T_a$ will be negative during nighttime (when solar elevation β (radians) < 0), and will gradually increase to positive values during daytime ($\beta > 0$ 0). We describe this pattern by (Fig. 5):

$$T_s - T_a = f_{cum}\left(\beta, \mu_\beta, \sigma_\beta\right)\left(T_{s,\,amp} + \beta T_{s,\,slope}\right) + T_{s,\,offset}, \tag{6}$$

where f_{cum} is a cumulative normal distribution function with mean μ_β and standard deviation σ_β, describing the moment at which the surface becomes warmer than the air temperature (μ_β) and the speed at which the surface warms up or cools down (σ_β) as a function of solar elevation angle (β). $T_{s,amp}$ is the amplitude of T_s (K), $T_{s,slope}$ is the slope between β and $T_s - T_a$ during daytime (K radians^{-1}) and $T_{s,offset}$ is the average value of $T_s - T_a$ during nighttime (K). The parameters of Eq. (6), except $T_{s,offset}$, were fitted to the data by minimizing the root mean square error (RMSE) by generalized reduced gradient nonlinear optimization. The $T_{s,offset}$ was determined as the average nighttime $T_s - T_a$ to limit the number of parameters during the optimization. Equation (6) was substituted for T_s in Eq. (3) to estimate R_{nl}. This novel approach to derive R_{nl} was compared to the R_{nl} model of the FAO-56 approach (Allen et al., 1998), originally derived to obtain daily estimates of R_{nl} (using minimum and maximum daily T_a divided by 2 instead of T_a in Eq. 7) but commonly applied at hourly timescales (ASCE-EWRI, 2005; Perera et al., 2015; Gavilán et al., 2008; López-Urrea et al., 2006):

$$R_{nl} = -\sigma T_a^4\left(a - b\sqrt{e_a}\right)\left(1.35\frac{R_s}{R_{s0}} - 0.35\right), \tag{7}$$

where the first term between brackets represents the net emittance, which should compensate for the fact that T_s is not measured. The empirical parameters a and b can be calibrated for a specific climate and/or vegetation. The second term between brackets is a cloudiness function. The default parameter values for a and b are 0.34 and 0.14, respectively (Allen et al., 1998). We calibrated these parameters for every site by linear least squares regression for clear days ($R_s/R_{s0} > 0.9$) and compared the performance of both R_{nl} models (Eqs. 3, 7).

2.3.2 Soil heat flux (G)

The soil heat flux is commonly expressed as a fraction of R_n, particularly on large scales using remote sensing (Su, 2002; Bastiaanssen et al., 1998; Kustas et al., 1998; Kustas and Daughtry, 1990; Friedl, 1996). We adopted the same approach, making a distinction between daytime (F_{day}) and nighttime (F_{night}) fractions, determined by linear least squares regression between R_n and the average of the two sets of soil heat flux measurements.

2.3.3 Aerodynamic resistance (r_a)

The aerodynamic resistance under neutral stability conditions can be estimated by (Monteith and Unsworth, 1990)

$$r_a = \frac{\ln\left[\frac{z_m - d}{z_{om}}\right]\ln\left[\frac{z_h - d}{z_{oh}}\right]}{k^2 u_z}, \quad (8)$$

where z_m is the height of wind speed measurements (m), d is the zero plane displacement height (m), z_{om} is the roughness length governing momentum transfer (m), z_h is the height of the humidity measurements (m), z_{oh} is the roughness length governing transfer of heat and vapor (m), k is the von Karman's constant (0.41 (–)) and u_z is the wind speed at height z_m ($\mathrm{m\,s^{-1}}$). For grass, empirical equations are developed (FAO-56 approach) to estimate d, z_{om} and z_{oh}:

$$d = 0.66V \quad (9)$$

$$z_{om} = 0.123V \quad (10)$$

$$z_{oh} = 0.1z_{om}, \quad (11)$$

where V is the vegetation height. Wallace et al. (1984) found comparable coefficients for heather: $d = 0.63V$ and $z_{om} = 0.13V$ and therefore Eqs. (9)–(11) were applied for both surfaces using a constant vegetation height of 7 and 31 cm for the grass and heather surfaces, respectively. For the moss surface, we used a vegetation height of 2 cm, which is equal to the thickness of the moss mat. For the bare sand surface we assumed $d = 0$ m, and used typical surface roughness values published by Oke (1978): $z_{oh} = 0.001$ m and $z_{om} = z_{oh}$.

2.3.4 Surface resistance (r_s) and canopy interception

Canopy interception was simulated as a water storage which needs to be filled before rainwater reaches the soil surface. A maximum storage capacity of 0.50 mm was defined for heather following the study of Ladekarl et al. (2005). To our knowledge no literature value of the interception capacity of the specific grass species (*Agrostis vinealis*) is published. Considering the relatively low vegetation height, we assumed a maximum interception capacity of 0.25 mm.

We distinguished wet (r_{swet}) and dry canopy surface resistance (r_s), since interception water evaporates without the interference of leaf stomata. During canopy interception (i.e., if the interception store is fully or partly filled), we used a surface resistance of $0\,\mathrm{s\,m^{-1}}$, reducing Eq. (1) to the Penman equation (Penman, 1948; Monteith and Unsworth, 1990). After the canopy storage is emptied, the surface resistance switches to r_s. The r_s was back-calculated for daytime periods for the heather and grass lysimeters by substituting measured R_n, G, ET_a, e_s and e_a and simulated r_a into Eq. (1) under nonstressed conditions (i.e., $ET_p = ET_a$). Nighttime evaporation was assumed to be equal to 0 mm. To make sure that the back-calculated r_s was based on days at which evapotranspiration occurred at a potential rate, it was back-calculated for every two consecutive days after precipitation events and after emptying of the (calculated) interception store. The surface resistance (r_s) of bare sand and moss was assumed to be equal to $10\,\mathrm{s\,m^{-1}}$, i.e., similar to the surface resistance under well-watered conditions of bare soil found by Van de Griend and Owe (1994).

During the summer of 2013, a dry spell (from 4 until 25 July 2013) resulted in a vegetation dieback of grass and heather. Surface resistances were back-calculated for periods before and after the drought event. The drought event had 22 consecutive dry days with a cumulative reference evapotranspiration of 85 mm according to Makkink (1957). Drought events of similar magnitude have been recorded 12 times during the past 57 years (from 1958 until 2014) at climate station "de Bilt" located in the center of the Netherlands (52.1° latitude, 5.18° longitude), 10 km from the measurement site. The measurements in the heather vegetation started a week before the drought event. During this week, there were 2 days (30 June and 1 July 2013) for which r_s could be back-calculated. The estimated r_s for these days was 35 and $107\,\mathrm{s\,m^{-1}}$ respectively. We selected the r_s value of the second day to use in our model simulations ($107\,\mathrm{s\,m^{-1}}$) because it was in close agreement with the median surface resistance found by Miranda et al. (1984) of $110\,\mathrm{s\,m^{-1}}$ in a comparable heather vegetation. After the drought event, r_s increased to $331\,\mathrm{s\,m^{-1}}$ ($N = 14$, standard error $= 102\,\mathrm{s\,m^{-1}}$). For the grass vegetation, the surface resistance before the drought event was $181\,\mathrm{s\,m^{-1}}$ ($N = 9$, standard error $= 68\,\mathrm{s\,m^{-1}}$). After the drought event, the surface resistance increased to $351\,\mathrm{s\,m^{-1}}$ ($N = 4$, standard error $= 47\,\mathrm{s\,m^{-1}}$). Since mosses of these habitats are desiccation-tolerant and quickly rehydrate after drought (Proctor et al., 2007), we did not assess the effect of the dry spell on the surface resistance of the moss surface.

The parameters thus obtained were used to parameterize the Penman–Monteith equation and to calculate hourly ET_p values for each surface.

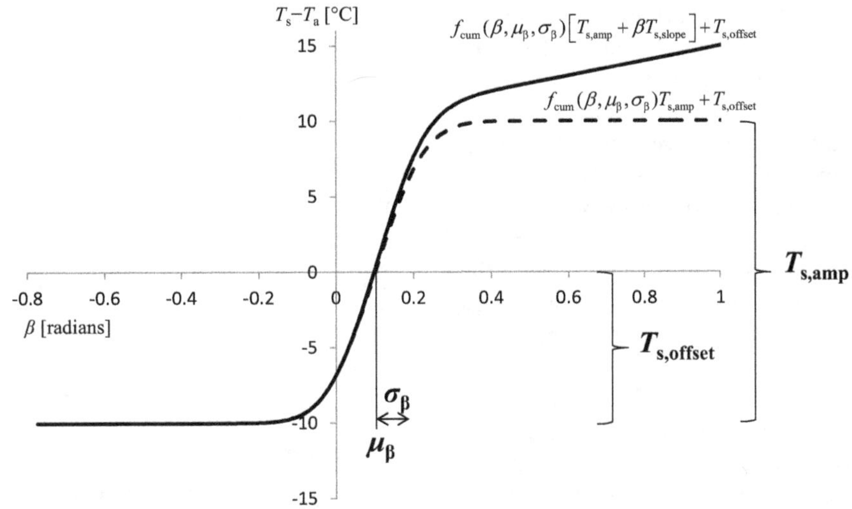

Figure 5. Equation (6) and associated parameters to describe the surface–air temperature difference, substituted for T_s in R_{nl} (Eq. 3).

2.4 Model simulations of ET_a

Using hourly ET_p for the year 2013 (876 mm precipitation), we used Hydrus 1D (Šimůnek et al., 2008) to simulate ET_a. If meteorological data of the local weather station were missing due to snow cover or sensor maintenance, the meteorological data of weather station "de Bilt" were used for the calculation of ET_p.

First, we simulated ET_a for the lysimeter surfaces and compared our results with the lysimeter measurements of ET_a. The lower boundary condition in the model was a seepage face with hydraulic pressure equal to 0 at a depth of 65 cm below the surface (50 cm soil and 15 cm wick). This boundary condition assumes that the boundary flux will remain zero as long as the pressure head is negative. When the lower end of the soil profile becomes saturated, a zero pressure head is imposed at the lower boundary and outflow calculated accordingly. Second, we simulated ET_a for the groundwater-independent surroundings. We expected that the availability of soil moisture in the lysimeter tanks to be larger than in the groundwater-independent surroundings, because the lowest sections of the lysimeters need to be saturated before drainage occurs. To estimate the yearly ET_a of dune vegetation in environments with deep groundwater levels, we used a free drainage boundary condition (i.e., a pressure head gradient of 0 and an elevation head of 1) located 2.5 m below the surface. Third, we investigated the magnitude of the vegetation dieback in the summer of 2013 on both ET_p and ET_a, by using two different surface resistances: one derived from the period before, and one for the period after the vegetation dieback.

Soil hydraulic properties in the hydrological model were described by the Van Genuchten relationships (Van Genuchten, 1980). Soil samples ($100\,cm^3$) collected next to each lysimeter at 5 and 15 cm depth were used to derive the drying retention function. The average drying retention parameters (of the two samples collected next to each lysimeter) were used in the hydrological model, taking hysteresis into account by assuming the wetting retention curve parameter (α_{wet}) to be twice as large as the drying retention curve parameter (α_{dry}) (Šimůnek et al., 1999). The unsaturated hydraulic properties (parameters l and K_0) were estimated using the Rosetta database and pedotransfer functions, providing the fitted drying retention curve parameters as input (Schaap et al., 2001). The hydraulic properties of the 15 cm long wick, guiding drainage water below the lysimeter into the tipping bucket, were taken from Knutson and Selker (1994).

Since mosses have neither leaf stomata nor roots, ET_a from the moss surface is limited by the capacity of the moss material to conduct water to the surface. This passive evaporation process is similar to the process of soil evaporation, i.e., evaporation becomes limited if the surface becomes too dry to deliver the potential rate. The unsaturated hydraulic properties of the dense *Campylopus introflexes* moss mat covering the lysimeter soil were based on the hydraulic properties derived by Voortman et al. (2014) and used in the first 2 cm of the model domain. Macro pores in the moss mat were neglected by Voortman et al. (2014), which implies that direct implementation of these hydraulic properties would result in large amounts of surface runoff generation or ponding, since the unsaturated hydraulic conductivity (K_0) of the moss mat is lower than $0.28\,cm\,d^{-1}$. Therefore, the dual porosity model of Durner (1994) was used to add $1000\,cm\,d^{-1}$ to the hydraulic conductivity curve of Voortman et al. (2014) between -1 and 0 cm pressure head (Appendix A). This permits the infiltration of rainwater at high intensity rain showers without affecting the unsaturated hydraulic behavior at negative pressure heads. Because of the complex shape of the retention function of the moss mat, hysteresis in the soil hydraulic functions in the underlying soil was neglected for the simulation of evaporation from moss surfaces. The sensitivity of this simplification on the model outcomes was investigated by adjusting the soil hydraulic function of the soil from the drying to the wetting curve. This had a negligible effect (< 1 mm) on the simulated yearly ET_a (data not shown). Besides simulations of moss evaporation with a cover of *Campylopus introflexus*, soil physical characteristics of *Hypnum cupressiforme* were used in the first 2 cm of the model domain to analyze the effect of different moss species on the water balance. Soil parameters used in the model are explained in more detail in Appendix A.

Since the grass and heather lysimeters fully covered the soil, soil evaporation was neglected for these surfaces. The root profile for the grass and heather lysimeters was 30 cm deep, with the highest concentration of roots in the upper layer decreasing linearly with depth. A water stress reduction function (Feddes et al., 1978) was used to simulate the closure of leaf stomata during water-stressed periods. Vegetation parameters are explained in more detail in Appendix B. Modeled actual evapotranspiration ($ET_{a,mod}$) was aggregated to daily values and compared to field measurements of ET_a during moist ($ET_{a,mod} = ET_p$) and dry conditions ($ET_{a,mod} \neq ET_p$).

2.5 Model performance assessment

Model performance of R_{ns}, R_{nl}, G and $ET_{a,mod}$ simulations were tested with the Nash–Sutcliffe model efficiency coefficient (NSE):

$$\mathrm{NSE} = 1 - \frac{\sum_{t=1}^{N}\left(x_{o,t} - x_{m,t}\right)^2}{\sum_{t=1}^{N}\left(x_{o,t} - \overline{x}_o\right)^2}. \tag{12}$$

where N is the total number of observations, $x_{m,t}$ is the model-simulated value at time step t, $x_{o,t}$ is the observed value at time step t, and $\bar{x}$ is the mean of the observations. NSE = 1 corresponds to a perfect match of modeled to observed data. If NSE < 0, the observed mean is a better predictor than the model. To assess the magnitude of error of model simulations, the root mean square error (RMSE), the mean difference (MD) and the mean percentage difference (M%D) were used.

3 Results and discussion

3.1 Parameterization of the Penman–Monteith equation

3.1.1 Net shortwave radiation

The measured incoming and reflected solar radiation were used to compute the albedo of the four surfaces by linear regression (Fig. 6; Table 5). This single value for the albedo slightly overestimates the reflected solar radiation at large

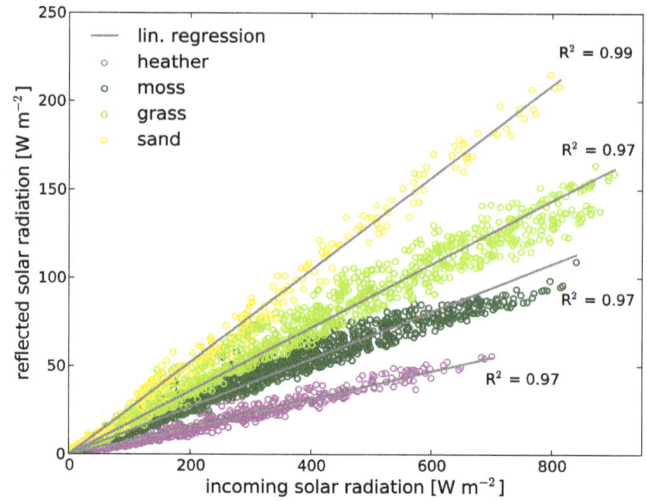

Figure 6. Linear regression between incoming and reflected solar radiation.

incoming solar radiation (Fig. 7) because of a dependency of the albedo on solar elevation angle β (Yang et al., 2008; Zhang et al., 2013). Nonetheless the use of a single value for the albedo hardly affects the error in modeled R_{ns}; the mean difference (MD) between measured and modeled R_{ns} lies between −0.23 and 1.63 $\mathrm{W m^{-2}}$ (Table 1), which is equal to the energy required to evaporate 0.008 to 0.057 $\mathrm{mm\,d^{-1}}$. The NSE for estimating R_{ns} is close to 1 (Table 1), showing almost a perfect match of modeled to observed data.

The dense moss mat *Campylopus introflexes* entirely covers the underlying mineral soil, which results in a low albedo (0.135) due to the dark green surface. The albedo of bare sand (0.261) is comparable to values found in literature for bare dry coarse soils (Qiu et al., 1998; Van Bavel and Hillel, 1976; Linacre, 1969; Liakatas et al., 1986) and the albedo for grass (0.179) is consistent with values reported in other studies during summer time (Hollinger et al., 2010) or for dried grass (Van Wijk and Scholte Ubing, 1963). Heather has a somewhat lower albedo (0.078) than was found in the literature: Miranda et al. (1984) report an albedo of 0.13 (*Calluna*, LAI ca. 4); Wouters et al. (1980) report an albedo of 0.102 (*Calluna*). The heather vegetation in our study was in a later successional stage with aging shrubs having a relatively large fraction of twigs and a smaller LAI (3.47) than found by Miranda et al. (1984). Furthermore, the albedo data of heather vegetation were collected primarily past the growing season from September till November. The darker surface after the growing season and the lower LAI explains the small albedo compared to other studies.

3.1.2 Net longwave radiation

The fitted function of Eq. (6) describes the dynamics of the surface temperature relative to air temperature (Fig. 8, Table 5). All surfaces have a similar average nighttime surface

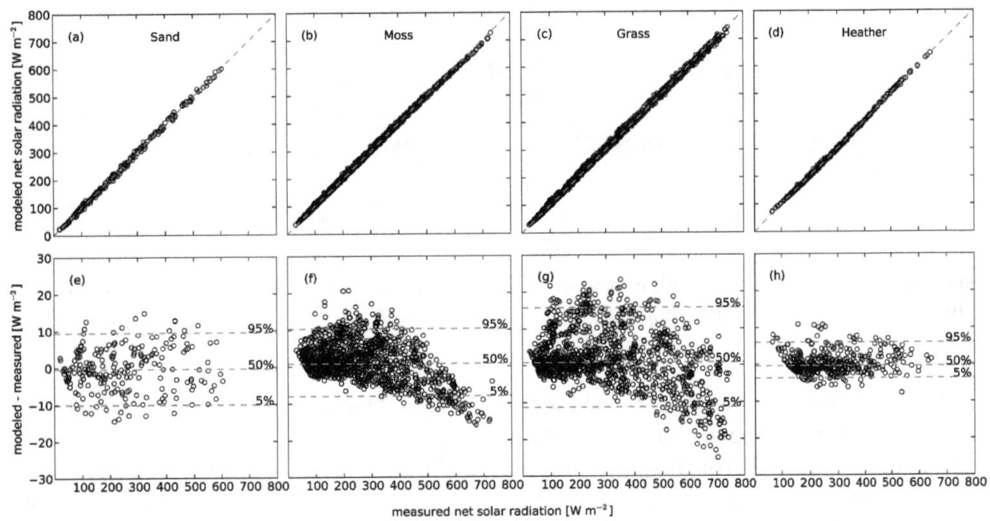

Figure 7. Modeled compared to measured net solar radiation (panels **a–d**, dashed lines are 1 : 1 lines) and deviations from the 1 : 1 line (panels **e–h**; dashed lines indicate 5, 50 and 95 percentiles).

Table 1. Model performance of R_{ns} simulations.

Surface	N	NSE	RMSE ($W m^{-2}$)	MD ($W m^{-2}$)	M%D (%)
Sand	218	0.998	5.99	−0.23	−0.10
Moss	1317	0.999	5.46	1.18	0.46
Grass	1203	0.998	7.78	1.63	0.55
Heather	407	0.999	3.00	0.24	0.09

Table 2. Calibrated net emissivity parameters of the FAO-56 R_{nl} submodel (Eq. 7).

	a	b
Sand	0.31	−0.00
Moss	0.33	0.02
Grass	0.36	−0.06
Heather	0.24	0.02

temperature ($T_{s,offset}$) relative to T_a, ranging between −7.47 and −10.21 °C. The solar elevation angle at which the surfaces become warmer than the air temperature (μ_β), as well as the speed at which the surface warms up or cools down (σ_β), are comparable between the surfaces. The main difference between the surfaces is observed at high solar elevation angles. Sand and moss show a clear increasing slope during the day, while grass and heather are able to attenuate the increase in surface temperature, possibly due to a larger latent heat flux (Fig. 8). The moss surface shows the largest increase in surface temperature during the day. Although organic layers, e.g., dry peat, have a larger specific heat ($1600\,J kg^{-1} K^{-1}$) than dry sand ($693\,J kg^{-1} K^{-1}$) (Gavriliev, 2004), the energy required to heat up the moss material is much smaller than for sand, because of the small dry bulk density of ca. $26.8\,g L^{-1}$ (derived for *Campylopus introflexus* from Voortman et al., 2014). Therefore, the surface temperature and the emitted longwave radiation are largest for the moss surface.

Our R_{nl} model (Eqs. 3 and 6) simulates R_{nl} much better than the calibrated (Table 2) FAO-56 R_{nl} submodel (Table 3). For the natural grass surface, the NSE even becomes negative using the calibrated FAO-56 approach. Several studies

Table 3. Model performance of R_{nl} simulations for hourly time steps.

Surface	N	NSE	RMSE ($W m^{-2}$)	MD ($W m^{-2}$)	M%D (%)
			Using Eq. (3)		
Sand	5891	0.65	27.37	0.92	1.52
Moss	5997	0.74	28.57	3.73	5.19
Grass	6113	0.71	25.66	1.41	2.36
Heather	2424	0.63	27.63	−0.21	−0.40
			Using FAO-56 Eq. (7)		
Sand	5891	0.41	35.39	4.34	7.14
Moss	5997	0.31	46.67	14.84	20.65
Grass	6113	−0.07	49.41	−18.23	−30.38
Heather	2424	0.29	38.24	10.50	19.54

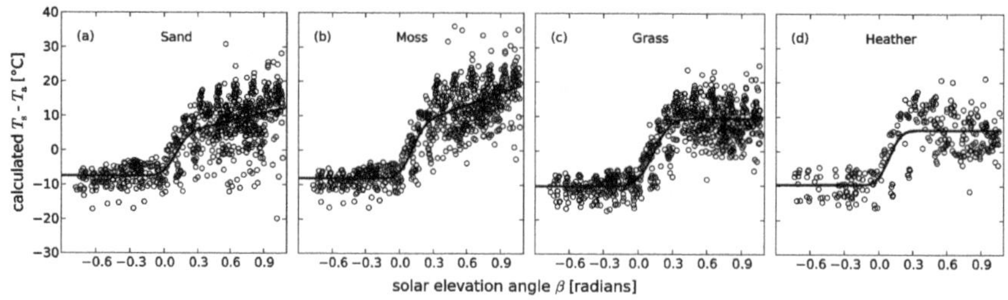

Figure 8. Measured surface temperature relative to air temperature (T_s–T_a) for clear hours ($f_{cd} > 0.9$) as a function of solar elevation angle β. Relationships (red lines) were fitted to the data using Eq. (6).

showed that the FAO-56 R_{nl} submodel underestimates the magnitude of R_{nl} for reference grass vegetation and poorly describes the diurnal pattern (Matsui, 2010; Blonquist Jr. et al., 2010; Yin et al., 2008; Temesgen et al., 2007). As mentioned, the FAO-56 R_{nl} submodel was originally developed for reference grass vegetation under well-watered conditions for daily time steps, but is commonly applied at hourly timescales (ASCE-EWRI, 2005; Perera et al., 2015; Gavilán et al., 2008; López-Urrea et al., 2006; Irmak et al., 2005). At daily time steps, T_s is close to T_a, since the warmer daytime T_s is compensated by the cooler nighttime T_s. For hourly time steps, the assumption that T_s follows T_a is not valid, which explains the poor performance of the FAO-56 R_{nl} model for hourly time steps. This poor performance cannot be compensated by calibrating the net emissivity parameters, since the diurnal pattern remains unaffected.

In this analysis a typical pattern in T_s relative to T_a is used to estimate T_s (Eq. 6), and subsequently R_{nl} (Eq. 3). This relationship (Fig. 8) is sensitive to local weather conditions, which implies that the parameters of Eq. (6) (Table 5) are not directly transferable to other locations or climates. The applicability of the presented approach to simulate R_{nl} should be tested before it is used for other surfaces or climates. It should be noted that the number of parameters that are required to simulate R_{nl} is relatively large. However, μ_β as well as σ_β, are comparable between the surfaces. These parameters might be assumed similar for every surface, reducing the species-specific model parameters to three (one more than the FAO-56 approach). More data of different vegetation types are required to generalize these results and to assess the number of parameters that are required to accurately simulate R_{nl}.

3.1.3 Soil heat flux

The soil heat flux G as a fraction of R_n (F_{day} and F_{night}) decreases with vegetation cover (Table 5). The nighttime fractions are larger than the daytime fractions, as R_n becomes smaller in magnitude during the night, which simultaneously corresponds to a change in direction of R_n and G, from downward (positive) to upward (negative). Relatively small systematic errors are made using daytime and nighttime fractions of R_n to simulate G (MD between 1.92 and 0.69 $\mathrm{W\,m^{-2}}$) (Table 4). In remote sensing algorithms G is often simulated as fraction of R_n, depending on the LAI or the fractional vegetation cover. In e.g., the SEBS algorithm, the soil heat flux fraction (F) is interpolated between 0.35 for bare soil and 0.05 for a full vegetation canopy (Su, 2002). These limits are close to the bare sand (0.270) and heather (0.066) F_{day} fractions (Table 5). The heather F_{day} (0.066) was close to the value found by Miranda et al. (1984) of 0.04.

The analysis of the relationship between R_n and G was based on the average of two sets of soil heat flux plates per surface. These sets of measurements showed on average a good agreement: a MD below 1.07 $\mathrm{W\,m^{-2}}$, with a RMSE ranging between 5.02 and 9.40 $\mathrm{W\,m^{-2}}$.

Table 4. Model performance of G simulations.

Surface	N	NSE	RMSE ($\mathrm{W\,m^{-2}}$)	MD ($\mathrm{W\,m^{-2}}$)	M%D (%)
Sand	6080	0.820	20.06	1.92	22.16
Moss	5335	0.901	12.02	1.65	24.29
Grass	6046	0.868	8.97	1.60	43.42
Heather	2028	0.641	11.39	0.69	40.27

Table 5. Parameters of the four different surfaces used for the calculation of ET_p for hourly time steps.

Parameter	Sand	Moss	Grass	Heather
Albedo (–)	0.261	0.135	0.179	0.078
μ_β (radians)	0.10	0.10	0.13	0.09
σ_β (radians)	0.09	0.09	0.11	0.08
$T_{s,amp}$ (°C)	11.26	14.21	19.70	15.89
$T_{s,offset}$ (°C)	−7.47	−8.14	−10.21	−9.67
$T_{s,slope}$ (°C radians^{-1})	7.83	11.82	0.00	0.00
F_{day} (–)	0.270	0.211	0.129	0.066
F_{night} (–)	0.761	0.647	0.527	0.462
r_{swet} ($\mathrm{s\,m^{-1}}$)	–	–	0	0
r_s ($\mathrm{s\,m^{-1}}$) before drought	10	10	181	107
r_s ($\mathrm{s\,m^{-1}}$) after drought	10	10	351	331

3.1.4 Energy balance

All the terms in the energy balance can be defined using daily lysimeter measurements of LE (latent heat flux) and an estimate of the sensible heat flux (H) as a residual term of the energy balance. For daytime measurements (between sunrise and sunset), the LE, H, G, $R_{s\uparrow}$ and R_{nl} can be expressed as fraction of the $R_{s\downarrow}$. Table 6 summarizes the average fraction of $R_{s\downarrow}$ attributed to these five different energy fluxes during the measurement campaign. The net longwave radiation is for most surfaces the largest energy flux during daytime (Table 6).

The LE of most surfaces is the second largest flux during daytime, of which fraction increases with vegetation cover. Despite the large difference in albedo between bare sand and moss, the moss surface has only a slightly larger LE fraction than bare sand (Table 6). This is primarily caused by the larger R_{nl} flux of moss, which compensates the smaller amount of reflected solar radiation.

Table 6. Average fractionation of the incoming shortwave radiation ($R_{s\downarrow}$) between different energy fluxes during daytime.

Surface	LE	H	G	$R_{s\uparrow}$	R_{nl}
Sand	0.22	0.13	0.10	0.26	0.28
Moss	0.24	0.17	0.09	0.14	0.36
Grass	0.27	0.21	0.06	0.18	0.29
Heather	0.35	0.20	0.05	0.08	0.32

3.2 Potential and actual evapotranspiration

The modeled ET_a is in agreement with the measured ET_a, with some exceptions at the onset of dry out events (Fig. 9). In general, the reduction of ET_p to ET_a is modeled a few days later than it emerges from measurements. The cumulative $ET_{a,mod}$ over the measurement period (May–October 2013) deviates 21 mm (13 %), −13 mm (−7 %), 5 mm (2 %) and −3 mm (−2 %) from the measured ET_a of the sand, moss, grass and heather lysimeters, respectively. The results of modeled vs. measured ET_a for non-water-stressed ($ET_a = ET_p$) and water-stressed conditions ($ET_{a,mod} < ET_p$) are summarized in Table 7.

We did not calibrate our model, e.g., by adjusting soil hydraulic properties, because several processes outlined by Allen et al. (1991) and wall flow (Cameron et al., 1992; Corwin, 2000; Till and McCabe, 1976; Saffigna et al., 1977) affect lysimeter measurements of ET_a and drainage. We suspect that wall flow caused the slightly earlier reduction of ET_p to ET_a at the onset of dry out events than was simulated by the model. Wall flow leads to a quicker exfiltration of rainwater and a subsequent lower moisture content in the lysimeter, and therefore a slightly earlier timing of drought compared to the model. Since wall flow does not occur in the undisturbed vegetation outside the lysimeters, calibrating e.g., soil hydraulic properties using measured surface and drainage fluxes in the objective function could lead to biased characterizations of the soil hydraulic properties and erroneous simulations of soil water flow and ET_a.

In our simulations, we neglected vapor flow within the soil and moss layer. Due to temperature and potential gradients, vapor fluxes may occur through the soil and moss layer in upward and downward direction by diffusion. Vapor flow may occur by advection as well, e.g., through macropores. Water and vapor flows act together and are hard to distinguish.

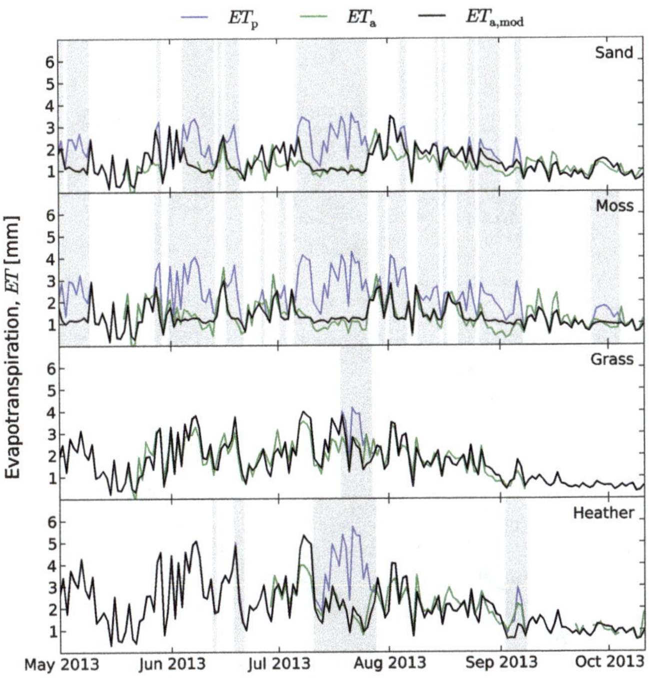

Figure 9. Measured and modeled daily ET for the four lysimeters. Gray bars indicate time periods where $ET_{a,mod}$ is smaller than ET_p, i.e., when evapotranspiration was water-limited.

Modeling and lab experiments show a minor cumulative effect of vapor flow on evaporation for moist and temperate climates. Soil evaporation in a temperate climate for loamy sand in Denmark was only slightly smaller (1.5 %) than a simulation excluding vapor flow (Schelde et al., 1998). Experiments of Price et al. (2009) show that only 1 % of the total water flux was caused by vapor flow in columns of *Sphagnum* moss. Nevertheless, for a dry and warm Mediterranean climate – different from ours – Boulet et al. (1997) found a dominant vapor flux down to a depth of 25 cm in a bare soil during 11 days in a dry and warm Mediterranean climate. Because large temperature and potential gradients occur when $ET_a \neq ET_p$, vapor flow could especially become dominant in the water-limited phase of evaporation. We compared the model performance between dry ($ET_{a,mod} \neq ET_p$) and wet ($ET_{a,mod} = ET_p$) days in Fig. 10. The model performance in both moisture conditions is comparable (RMSE of sand when dry was 0.40, when wet 0.46; RMSE of moss when dry was

Table 7. Modeled ET_p and ET_a for different surfaces in a lysimeter (lys.) and for a situation with deep groundwater levels (gw. ind.) for the year 2013.

	ET_p (mm)	ET_a lys. (mm)	ET_a gw. ind. (mm)
Bare sand	400	295	258
Moss (*Campylopus int.*)	468	312	272
Moss (*Hypnum cup.*)	468	–	182
Grass	392	350	333
Grass, no dieback	429 (+9 %)	382 (+9 %)	362 (+9 %)
Heather	549	460	391
Heather, no dieback	610 (+11 %)	499 (+8 %)	420 (+7 %)

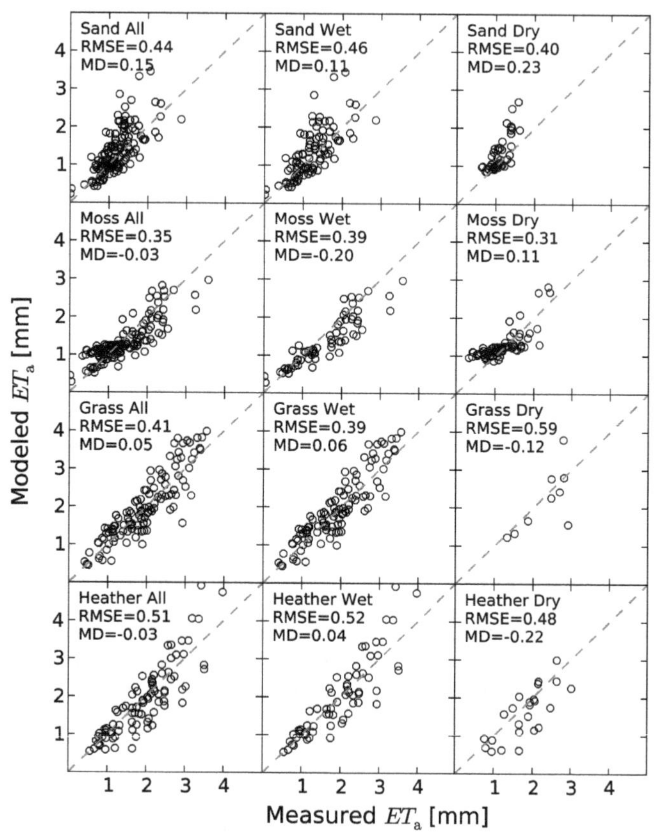

Figure 10. Measured vs. modeled ET_a of the lysimeters for all, wet ($ET_{a,mod} = ET_p$) and dry ($ET_{a,mod} \neq ET_p$) days. Dotted lines represent the 1 : 1 lines.

0.30, when wet 0.39), suggesting that our simplified model could describe the dominant processes and the simulation of vapor flow was not required for the temperate climate of our study area.

One would expect oasis effects to occur in the vicinity of the lysimeters, because freely draining lysimeters must saturate at the bottom of the lysimeter tank before water drains out. This enlarges the water availability inside the lysimeters compared to its groundwater-independent surroundings and occasionally leads to a situation in which the vegetation inside the lysimeters is still transpiring, while the vegetation outside the lysimeters becomes water-stressed and heats up. In such a situation, advection of sensible heat generated in the vicinity of the lysimeters could contribute to the available energy for lysimeter evapotranspiration. However, calculated ET_p was seldom smaller than measured lysimeter ET_a, indicating that oasis effects were absent. Furthermore, if oasis effects were prominent, systematic underestimation of modeled lysimeter ET_a would occur, since we ignored the possible contribution of heat advection. Note that it is very unlikely that oasis effects affected the back-calculated surface resistances (Table 5), since these were based on days after rain events for which we may assume ET_a to be equal to ET_p for both the lysimeters and their surroundings.

Neglecting feedbacks of drought on the transpiring leaf area and thereby the surface resistance (i.e., using a fixed r_s) of heather and dry grassland vegetation leads to an overestimation of cumulative ET_a of 7–9 % for years with relatively severe drought (Table 7). The delayed drought response of these vegetation types is therefore of importance to water balance studies, especially when, according to the expectations, summers become dryer as a result of a changing climate. Longer recordings of ET_a in heathland and grassland are required to understand and parameterize the drought response of these vegetation types in coupled plant physiological and hydrometeorological models.

To our knowledge, this paper describes for the first time the evaporation characteristics of a moss surface in a dune ecosystem in a temperate climate. The evaporation rate of the dense moss mat *Campylopus introflexus* is 5 % larger than the evaporation rate of bare sand. *Campylopus introflexus* forms dense moss mats and of the moss species investigated by Voortman et al. (2014), it has the largest water-holding capacity. Voortman et al. (2014) hypothesized that moss-covered soils could be more economical with water than bare soils, since the unsaturated hydraulic properties of moss layers reduce the magnitude of evaporation under relatively moist conditions. Our simulations of evaporation from the more open-structured *Hypnum cupressiforme* moss species (common in coastal dunes), which primarily differs in moisture content near saturation compared to *Campylopus introflexus* (0.20 instead of 0.61), confirms this hypothesis. The simulated evaporation rate for this species was 29 % lower than the evaporation rate of bare soil. From both our measurements and model simulations, xerophytic (drought-tolerant) mosses appear to be very economical with water; their evaporation rate is comparable with that of bare sand, or lower.

Campylopus introflexus is considered an invasive species in the Northern Hemisphere and was first discovered in Europe in 1941 (Klinck, 2010). Considering the large difference in yearly evaporation between *Hypnum cupressiforme* and *Campylopus introflexus* species (90 mm), the invasion of the *Campylopus introflexus* could have had negative impacts on water resources in specific areas which were previously dominated by more open-structured moss species with poorer water retention characteristics. For sustainable management of groundwater resources in coastal and inland sand dunes, an accurate estimate of the groundwater recharge is required. For consultancy about the availability of water, moss species cannot be categorized in a singular plant functional type, since the modulating effect of the moss cover is species-specific. However, in terms of water retention characteristics, the species investigated by Voortman et al. (2014) are distinguished from each other by the water-holding capacity near saturation (θ_0, Appendix A), which is easily measured in a laboratory. Moss species could be categorized by this characteristic.

Mosses and lichens are common in early successional stages after colonizing and stabilizing drift sand or as understory vegetation in heathlands or grasslands. Vascular plants might benefit from the presence of certain moss species as more water may be conserved in the root zone. On the other hand, field observations show that moss- and lichen-rich vegetation can persist for many decades (Daniëls et al., 2008). Detailed measurements of understory evaporation in heathlands and grasslands are required to unravel the ecological interactions between mosses and vascular plants.

4 Conclusions

In this study, the net longwave radiation (R_{nl}) appeared to be one of the largest energy fluxes in dune vegetation. The poor performance of the calibrated FAO-56 approach for simulating R_{nl} for hourly time steps illustrates that this energy flux has attracted insufficient attention in evapotranspiration research. The novel approach presented in this study to simulate R_{nl} outperformed the calibrated FAO-56 approach and forms an accurate alternative for estimating R_{nl}.

A relatively simple hydrological model could be used to simulate evapotranspiration of dry dune vegetation with satisfactory results. Improvements in terms of climate robustness would be especially achieved if plant physiological processes were integrated in the hydrometeorological model. Without considering the effects of dry spells on the surface resistance (r_{s}) of grassland and heathland vegetation, $\mathrm{ET_a}$ would be overestimated with 9 and 7 % for years with relatively severe drought (drought events with a reoccurrence of once per 5 years).

Moss species are very economical with water. The evaporation of moss surfaces is comparable or even lower than bare sand. By promoting moss-dominated ecosystems in coastal and inland dunes, the evapotranspiration could be reduced considerably, to the benefit of the groundwater system. Differences in evaporation between moss species are large and should be considered in water balance studies.

Long-term measurements of $\mathrm{ET_a}$ in heathland and grassland are required to study feedbacks between climate and plant physiological processes in order to integrate the drought response of natural vegetation in coupled plant physiological and hydrometeorological models. To understand the ecological interaction between mosses and vascular plants, detailed measurements of understory evaporation in heathlands and grasslands are required.

Appendix A: Soil hydraulic properties for the simulation of unsaturated flow with Hydrus-1D

Unsaturated flow in Hydrus 1D is described by a modified form of Richards' equation:

$$\frac{\partial \theta}{\partial t} = \frac{\partial}{\partial z}\left(K\frac{\partial h}{\partial z} + K\right), \tag{A1}$$

where K is the unsaturated conductivity ($\mathrm{L\,T^{-1}}$), z is the vertical coordinate (L) and t is the time (T). The soil hydraulic properties were assumed to be described by the Mualum van Genuchten functions:

$$\theta(h) = \theta_{\mathrm{r}} + \frac{\theta_0 - \theta_{\mathrm{r}}}{[1+|\alpha h|^n]^m} \tag{A2}$$

$$K(\theta) = K_0 S_{\mathrm{e}}^l\left[1-\left(1-S_{\mathrm{e}}^{1/m}\right)^m\right]^2 \tag{A3}$$

with

$$S_{\mathrm{e}}(h) = \frac{\theta(h)-\theta_{\mathrm{r}}}{\theta_0-\theta_{\mathrm{r}}}, \tag{A4}$$

where θ is the volumetric water content ($\mathrm{L^3\,L^{-3}}$), h is the soil water pressure head (L), θ_0 is an empirical parameter matching measured and modeled θ ($\mathrm{L^3\,L^{-3}}$), θ_{r} is the residual water content ($\mathrm{L^3\,L^{-3}}$) and α ($\mathrm{L^{-1}}$) and n (–) are empirical shape parameters of the retention function. K_0 is an empirical parameter, matching measured and modeled K ($\mathrm{L\,T^{-1}}$), S_{e} is the effective saturation (–), l is the pore-connectivity parameter (–) and m ($=1-1/n$) (–) is an empirical parameter. Drying retention data of two soil samples collected next to each lysimeter at 5 and 15 cm depth were used to fit a retention function with the RETC code (Van Genuchten et al., 1991). Hysteresis in the retention function was accounted for by assuming the retention curve parameter α for the wetting curve (α_{wet}) to be twice as large as α of the drying retention curve (α_{dry}) (Šimůnek et al., 1999). The unsaturated hydraulic conductivity parameters l and K_0 were estimated using the Rosetta database and pedotransfer functions, providing the fitted drying retention curve parameters as input (Schaap et al., 2001). Average parameter values per lysimeter are summarized in Table A1.

The hydraulic properties of the 15 cm long wick, guiding drainage water below the lysimeter into the tipping bucket, were taken from Knutson and Selker (1994) who analyzed the same brand and type of wick, i.e., Peperell 1/2 inch. The K_0 of the wick was adjusted to correct for the smaller cross-sectional area of the wick compared to the cross-sectional area of the lysimeter in the 1-D model simulation (Table A1).

The heterogeneous pore structure of the moss material was described by the functions of Durner (1994):

$$S_\mathrm{e} = w_1\left(1+[\alpha_1 h]^{n_1}\right)^{-m_1} + w_2\left(1+[\alpha_2 h]^{n_2}\right)^{-m_2} \tag{A5}$$

$$K\left(S_\mathrm{e}\right) = K_\mathrm{s}\frac{\left(w_1 S_{\mathrm{e}_1} + w_2 S_{\mathrm{e}_2}\right)^l \left(w_1\alpha_1\left[1-\left(1-S_{\mathrm{e}_1}^{1/m_1}\right)^{m_1}\right] + w_2\alpha_2\left[1-\left(1-S_{\mathrm{e}_2}^{1/m_2}\right)^{m_2}\right]\right)^2}{\left(w_1\alpha_1 + w_2\alpha_2\right)^2}, \tag{A6}$$

where w_1 and w_2 are weighting factors for two distinct pore systems of the moss layer, a capillary pore system (subscript 1) and a macropore system active near saturation ($h > -1$ cm, subscript 2), and K_s is the hydraulic conductivity at saturation. Average hydraulic parameters of the capillary pore system and the volumetric portion of the macropore system of the moss species *Campylopus introflexus* and *Hypnum cupressiforme* were taken from Voortman et al. (2014) (illustrated with dotted lines in Figs. A1 and A2). The α_2 parameter was fitted to the functions of Voortman et al. (2014) using $K_\mathrm{s} = 1000\,\mathrm{cm\,d^{-1}}$ and $n_2 = 2$ by minimizing the RMSE by generalized reduced gradient nonlinear optimization. Hydraulic parameter values are listed in Table A2.

Table A1. Hydraulic parameter values of lysimeter soils.

	θ_r (–)	θ_0 (–)	α_dry ($\mathrm{cm^{-1}}$)	α_wet ($\mathrm{cm^{-1}}$)	n (–)	K_0 ($\mathrm{cm\,h^{-1}}$)	L (–)
Bare sand	0.01	0.367	0.023	0.046	2.945	1.042	−0.401
Moss	0.01	0.397	0.019	–	2.335	0.734	−0.173
Grass	0.01	0.401	0.025	0.050	2.071	1.119	−0.278
Heather	0.01	0.392	0.018	0.036	2.581	0.679	−0.186
Wick	0.00	0.630	0.098	0.196	3.610	2.180	0.500

Table A2. Hydraulic parameter values of the two moss species.

	θ_r (–)	θ_s (–)	α_1 ($\mathrm{cm^{-1}}$)	n (–)	K_s ($\mathrm{cm\,h^{-1}}$)	l (–)	w_2 (–)	α_2 ($\mathrm{cm^{-1}}$)	n_2 (–)
Campylopus int.	0.060	0.936	0.080	2.25	41.67	−2.69	0.371	45.89	2.00
Hypnum cup.	0.010	0.971	0.013	2.17	41.67	−2.37	0.800	16.61	2.00

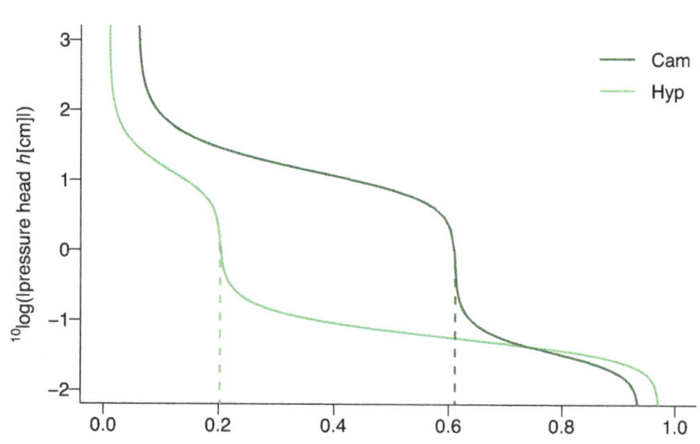

Figure A1. Water retention functions of two moss species: *Campylopus introflexus* and *Hypnum cupressiforme*. The dotted lines indicate the contribution of the capillary pore system, characterized by Voortman et al. (2014).

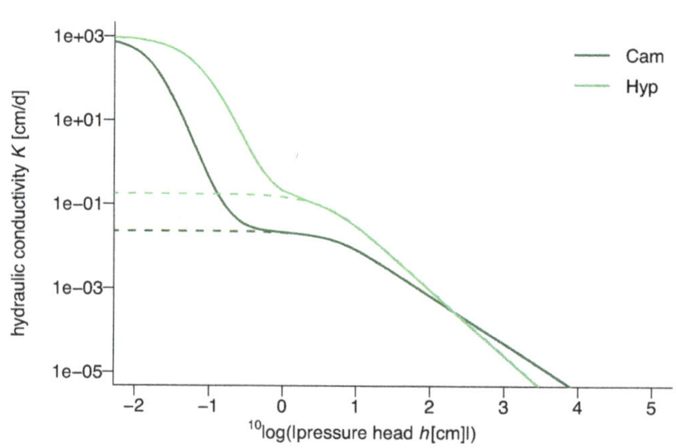

Figure A2. Hydraulic conductivity functions for two moss species: *Campylopus introflexus* and *Hypnum cupressiforme*. The dotted lines indicate the contribution of the capillary pore system, characterized by Voortman et al. (2014).

Appendix B: Feddes function used in the Hydrus 1-D model to simulate the closure of leaf stomata during water-stressed periods

The Feddes function (Feddes et al., 1978) describes the relative transpiration rate in relation to the soil water pressure head (Fig. B1) (being 0 if transpiration ceases and 1 if it equals potential rate). Near-positive pressure heads, root water uptake ceases due to oxygen stress (P0). At the dry end of the function, root water uptake ceases (P3). The moment at which transpiration becomes limited due to moisture stress is dependent on the potential transpiration rate. At a high potential transpiration rate (5 mm d^{-1} in the model simulation), leaf stomata start to close earlier (P2H) than under a low potential transpiration rate (P2L, 1 mm d^{-1} in the model simulation). Values for the parameters of Fig. B1 are listed in Table B1.

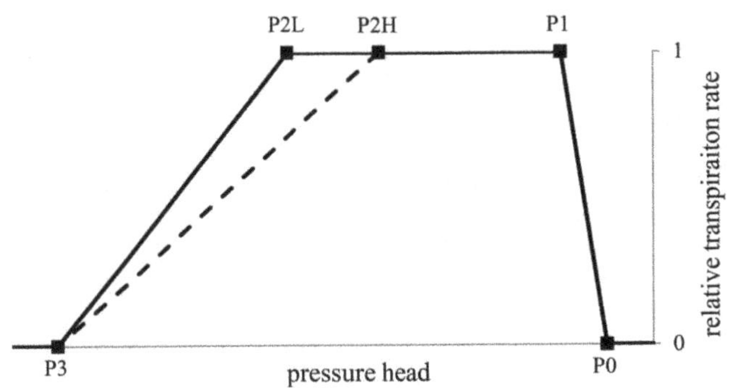

Figure B1. The relative transpiration rate as a function of soil water pressure head according to Feddes et al. (1978).

Table B1. Parameters of the water stress reduction function used in the Hydrus 1-D model.

P0 (cm)	P1 (cm)	P2H (cm)	P2L (cm)	P3 (cm)	r2H ($mm h^{-1}$)	r2L ($mm h^{-1}$)
−10	−25	−300	−1000	−8000	5	1

Acknowledgements. This study was carried out within the joint research programme of the Dutch Water Utility sector (http://www.kwrwater.nl/BTO), and the project Climate Adaptation for Rural Areas (CARE), which was funded by the Knowledge for Climate Program (http://knowledgeforclimate.climateresearchnetherlands.nl/climateadaptationforruralareas).
We thank the staff of Vitens for their permission to perform hydrometeorological measurements in one of their drinking water extraction sites.

Edited by: I. Neuweiler

References

Allen, R. G., Pruitt, W. O., and Jensen M. E.: Environmental requirements of lysimeters, in: Lysimeters for evapotranspiration and environmental measurements, edited by: Allen, R. G., Howell, T. A., Pruitt, W. O., Walter, I. A., and Jensen, M. E., American Society of Civil Engineers, New York, 170–181, 1991.

Allen, R. G., Pereira, L. S., Raes, D., and Smith, M.: Crop Evapotranspiration-Guidelines for Computing Crop Water Requirements, FAO irrigation and drainage paper 56, United Nations Food and Agriculture Orginization, Rome, 1998.

ASCE-EWRI: The ASCE Standardized Reference Evapotranspiration Equation, Environmental and Water Resources Institute of the American Society of Civil Engineers, Reston, Verginia, USA, 2005.

Bastiaanssen, W. G. M., Menenti, M., Feddes, R. A., and Holtslag, A. A. M.: A remote sensing Surface Energy Balance Algorithm for Land (SEBAL). 1. Formulation, J. Hydrol., 212–213, 198–212, doi:10.1016/S0022-1694(98)00253-4, 1998.

Blonquist Jr., J. M., Allen, R. G., and Bugbee, B.: An evaluation of the net radiation sub-model in the ASCE standardized reference evapotranspiration equation: implications for evapotranspiration prediction, Agr. Water Manage., 97, 1026–1038, doi:10.1016/j.agwat.2010.02.008, 2010.

Boulet, G., Braud, I., and Vauclin, M.: Study of the mechanisms of evaporation under arid conditions using a detailed model of the soil-atmosphere continuum. Application to the EFEDA I experiment, J. Hydrol., 193, 114–141, doi:10.1016/S0022-1694(96)03148-4, 1997.

Brunt, D.: Notes on radiation in the atmosphere. I, Q. J. Roy. Meteor. Soc., 58, 389–420, doi:10.1002/qj.49705824704, 1932.

Cameron, K. C., Smith, N. P., McLay, C. D. A., Fraser, P. M., McPherson, R. J., Harrison, D. F., and Harbottle, P.: Lysimeters without edge flow: an improved design and sampling procedure, Soil Sci. Soc. Am. J., 56, 1625–1628, doi:10.2136/sssaj1992.03615995005600050048x, 1992.

Campbell Scientific Inc.: Model HFP01SC Self-Calibrating Soil Heat Flux Plate, Logan, Utah, 24 pp., 2014.

Corwin, D. L.: Evaluation of a simple lysimeter-design modification to minimize sidewall flow, J. Contam. Hydrol., 42, 35–49, doi:10.1016/S0169-7722(99)00088-1, 2000.

Daniëls, F. J. A., Minarski, A., and Lepping, O.: Dominance pattern changes of a lichen-rich corynephorus grassland in the inland of the Netherlands, Annali di Botanica, VIII, 9–19, 2008.

Delpla, I., Jung, A. V., Baures, E., Clement, M., and Thomas, O.: Impacts of climate change on surface water quality in relation to drinking water production, Environ. Int., 35, 1225–1233, doi:10.1016/j.envint.2009.07.001, 2009.

Dolman, A. J.: Predicting evaporation from an oak forest, PhD thesis, University of Groningen, Netherlands, 91 pp., 1987.

Durner, W.: Hydraulic conductivity estimation for soils with heterogeneous pore structure, Water Resour. Res., 30, 211–223, doi:10.1029/93wr02676, 1994.

Feddes, R. A., Kowalik, P. J., and Zaradny, H.: Simulation of Field Water Use and Crop Yield, Pudoc, Wageningen, Netherlands, 1978.

Federer, C. A., Vörösmarty, C., and Fekete, B.: Intercomparison of methods for calculating potential evaporation in regional and global water balance models, Water Resour. Res., 32, 2315–2321, doi:10.1029/96wr00801, 1996.

Friedl, M. A.: Relationships among remotely sensed data, surface energy balance, and area-averaged fluxes over partially vegetated land surfaces, J. Appl. Meteorol., 35, 2091–2103, doi:10.1175/1520-0450(1996)035<2091:rarsds>2.0.co;2, 1996.

Fuchs, M. and Tanner, C. B.: Surface temperature measurements of bare soils, J. Appl. Meteorol., 7, 303–305, doi:10.1175/1520-0450(1968)007<0303:stmobs>2.0.co;2, 1968.

Gavilán, P., Estévez, J., and Berengena, J.: Comparison of standardized reference evapotranspiration equations in Southern Spain, J. Irrig. Drain. E.-ASCE, 134, 1–12, 2008.

Gavriliev, R. I.: Therma properties of soils and surface covers, in: Thermal Analysis, Construction, and Monitoring Methods for Frozen Ground, edited by: Esch, D. C., American Society of Civil Engineers, Restion, Verginia, USA, 277–295, 2004.

Gubler, S., Gruber, S., and Purves, R. S.: Uncertainties of parameterized surface downward clear-sky shortwave and all-sky longwave radiation., Atmos. Chem. Phys., 12, 5077–5098, doi:10.5194/acp-12-5077-2012, 2012.

Hollinger, D. Y., Ollingerw, S. V., Richardsonw, A. D., Meyersz, T. P., Dail, D. B., Martinw, M. E., Scott, N. A., Arkebauerk, T. J., Baldocchi, D. D., Clark, K. L., Curtis, P. S., Davis, K. J., Desai, A. R., Dragonikk, D., Goulden, M. L., Gu, L., Katulzzz, G. G., Pallardy, S. G., Pawu, K. T., Schmid, H. P., Stoy, P. C., Suyker, A. E., and Verma, S. B.: Albedo estimates for land surface models and support for a new paradigm based on foliage nitrogen concentration, Glob. Change Biol., 16, 696–710, 2010.

Irmak, S., Howell, T. A., Allen, R. G., Payero, J. O., and Martin, D. L.: Standardized ASCE Penman–Monteith: impact of sum-of-hourly vs. 24 h timestep computations at reference weather station sites, T. ASABE, 48, 1063–1077, 2005.

Irmak, S., Mutiibwa, D., and Payero, J. O.: Net radiation dynamics: performance of 20 daily net radiation models as related to model structure and intricacy in two climates, T. ASABE, 53, 1059–1076, 2010.

Jones, H. G.: Application of Thermal Imaging and Infrared Sensing in Plant Physiology and Ecophysiology, Adv. Bot. Res., 41, 107–163, doi:10.1016/S0065-2296(04)41003-9 2004.

Kay, A. L., Bell, V. A., Blyth, E. M., Crooks, S. M., Davies, H. N., and Reynard, N. S.: A hydrological perspective on evaporation: historical trends and future projections in Britain, Journal of Water and Climate Change, 4, 193–208, doi:10.2166/wcc.2013.014, 2013.

Klinck, J.: NOBANIS Invasive Alien Species Fact Sheet Campylopus introflexus, from: Online Database of the North European and Baltic Network on Invasive Alien Species – NOBANIS, available at: www.nobanis.org, last access: 9 February 2015, 2010.

Knutson, J. H. and Selker, J. S.: Unsaturated hydraulic conductivities of fiberglass wicks and designing capillary wick pore-water samplers, Soil Sci. Soc. Am. J., 58, 721–729, 1994.

Kustas, W. P. and Daughtry, C. S. T.: Estimation of the soil heat flux/net radiation ratio from spectral data, Agr. Forest Meteorol., 49, 205–223, doi:10.1016/0168-1923(90)90033-3, 1990.

Kustas, W. P., Zhan, X., and Schmugge, T. J.: Combining optical and microwave remote sensing for mapping energy fluxes in a semiarid watershed, Remote Sens. Environ., 64, 116–131, doi:10.1016/S0034-4257(97)00176-4, 1998.

Ladekarl, U. L., Rasmussen, K. R., Christensen, S., Jensen, K. H., and Hansen, B.: Groundwater recharge and evapotranspiration for two natural ecosystems covered with oak and heather, J. Hydrol., 300, 76–99, doi:10.1016/j.jhydrol.2004.05.003, 2005.

Liakatas, A., Clark, J. A., and Monteith, J. L.: Measurements of the heat balance under plastic mulches. Part I. Radiation balance and soil heat flux, Agr. Forest Meteorol., 36, 227–239, doi:10.1016/0168-1923(86)90037-7, 1986.

Linacre, E. T.: Net radiation to various surfaces, J. Appl. Ecol., 6, 61–75, doi:10.2307/2401301, 1969.

Liu, S., Graham, W. D., and Jacobs, J. M.: Daily potential evapotranspiration and diurnal climate forcings: influence on the numerical modelling of soil water dynamics and evapotranspiration, J. Hydrol., 309, 39–52, 2005.

López-Urrea, R., Olalla, F. M. D. S., Fabeiro, C., and Moratalla, A.: An evaluation of two hourly reference evapotranspiration equations for semiarid conditions, Agr. Water Manage., 86, 277–282, 2006.

Makkink, G. G.: Testing the Penman formula by means of lysimeters, J. Inst. Wat. Engrs., 11, 277–288, 1957.

Matsui, H.: Comparison of net longwave radiation equation in Penman-type evapotranspiration equation, Transactions of The Japanese Society of Irrigation, Drain. Rural Eng., 78, 531–536, doi:10.11408/jsidre.78.531, 2010.

Miranda, A. C., Jarvis, P. G., and Grace, J.: Transpiration and evaporation from heather Moorland, Bound.-Lay. Meteorol., 28, 227–243, doi:10.1007/bf00121306, 1984.

Monteith, J. L. and Unsworth, M. H.: Principles of Envoronmental Physics, Edward Arnold, London, 1990.

Moors, E. J.: Water Use of Ferest in the Netherlands, PhD thesis, Vrije Universiteit Amsterdam, Amsterdam, the Netherlands, 2012.

Nozue, K. and Maloof, J. N.: Diurnal regulation of plant growth, Plant Cell Environ., 29, 396–408, doi:10.1111/j.1365-3040.2005.01489.x, 2006.

Oke, T. R.: Boundary Layer Climates, Methuen & Co Ltd, London, 372 pp., 1978.

Penman, H. L.: Natural evaporation from open water, bare soil and grass, P. Roy. Soc. Lond. A-Mat., 193, 120–145, doi:10.1098/rspa.1948.0037, 1948.

Perera, K. C., Western, A. W., Nawarathna, B., and George, B.: Comparison of hourly and daily reference crop evapotranspiration equations across seasons and climate zones in Australia, Agr. Water Manage., 148, 84–96, doi:10.1016/j.agwat.2014.09.016, 2015.

Price, J. S., Edwards, T. W. D., Yi, Y., and Whittington, P. N.: Physical and isotopic characterization of evaporation from Sphagnum moss, J. Hydrol., 369, 175–182, 2009.

Proctor, M. C. F., Oliver, M. J., Wood, A. J., Alpert, P., Stark, L. R., Cleavitt, N. L., and Mishler, B. D.: Desiccation-tolerance in bryophytes: A review, Bryologist, 110, 595–621, doi:10.1639/0007-2745(2007)110[595:DIBAR]2.0.CO;2, 2007.

Qiu, G. Y., Yano, T., and Momii, K.: An improved methodology to measure evaporation from bare soil based on comparison of surface temperature with a dry soil surface, J. Hydrol., 210, 93–105, 1998.

Saffigna, P. G., Kenney, D. R., and Tanner, C. B.: Lysimeter and field measurements of chloride and bromide leaching in an uncultivated loamy sand, Soil Sci. Soc. Am. J., 41, 478–482, 1977.

Saito, H. and Šimůnek, J.: Effects of meteorological models on the solution of the surface energy balance and soil temperature variations in bare soils, J. Hydrol., 373, 545–561, doi:10.1016/j.jhydrol.2009.05.019, 2009.

Schaap, M. G., Leij, F. J., and Van Genuchten, M. T.: Rosetta: a computer program for estimating soil hydraulic parameters with hierarchical pedotransfer functions, J. Hydrol., 251, 163–176, doi:10.1016/S0022-1694(01)00466-8, 2001.

Schelde, K., Thomsen, A., Heidmann, T., Schjønning, P., and Jansson, P. E.: Diurnal fluctuations of water and heat flows in a bare soil, Water Resour. Res., 34, 2919–2929, doi:10.1029/98wr02225, 1998.

Shuttleworth, W. J.: Putting the "vap" into evaporation, Hydrol. Earth Syst. Sci., 11, 210–244, doi:10.5194/hess-11-210-2007, 2007.

Šimůnek, J., Kodešová, R., Gribb, M. M., and Van Genuchten, M. T.: Estimating hysteresis in the soil water retention function from cone permeameter experiments, Water Resour. Res., 35, 1329–1345, doi:10.1029/1998wr900110, 1999.

Šimůnek, J., Šejna, M., Saito, H., Sakai, M., and Van Genuchten, M. T.: The Hydrus-1D Software Package for Simulating the movement of water, heat, and multiple solutes in variably saturated media, version 4.0, HYDRUS Softwater Series 3, Department of Environmental Sciences, University of California Riverside, Riverside, 2008.

Su, Z.: The Surface Energy Balance System (SEBS) for estimation of turbulent heat fluxes, Hydrol. Earth Syst. Sci., 6, 85–100, doi:10.5194/hess-6-85-2002, 2002.

Temesgen, B., Eching, S., and Frame, K.: Comparing net radiation estimation methods: CIMIS versus Penman–Monteith, J. Irrig. Drain. E.-ASCE, 133, 265–271, doi:10.1061/(ASCE)0733-9437(2007)133:3(265), 2007.

Till, A. R. and McCabe, T. P.: Sulfur leaching and lysimeter characterization, Soil Sci., 121, 44–47, 1976.

Van Bavel, C. H. M. and Hillel, D. I.: Calculating potential and actual evaporation from a bare soil surface by simulation of concurrent flow of water and heat, Agr. Meteorol., 17, 453–476, doi:10.1016/0002-1571(76)90022-4, 1976.

Van de Griend, A. A. and Owe, M.: Bare soil surface resistance to evaporation by vapor diffusion under semiarid conditions, Water Resour. Res., 30, 181–188, doi:10.1029/93wr02747, 1994.

Van Genuchten, M. T.: A closed-form equation for predicting the hydraulic conductivity of unsaturated soils, Soil Sci. Soc. Am. J., 44, 892–898, doi:10.2136/sssaj1980.03615995004400050002x, 1980.

Van Genuchten, M. T., Leij, F. J., and Yates, S. R.: The RETC Code for Quantifying the Hydraulic Functions of Unsaturated Soils, Version 1.0, Salinity Laboratory, USDA, ARS, Riverside, California, 1991.

van Vliet, M. T. H. and Zwolsman, J. J. G.: Impact of summer droughts on the water quality of the Meuse river, J. Hydrol., 353, 1–17, doi:10.1016/j.jhydrol.2008.01.001, 2008.

Van Wijk, W. R. and Scholte Ubing, D. W.: Radiation, in: Physics of Plant Environment, edited by: Van Wijk, W. R., North-Holland Publishing, Amsterdam, the Netherlands, 62–101, 1963.

Voortman, B. R., Bartholomeus, R. P., Van Bodegom, P. M., Gooren, H., Van Der Zee, S. E. A. T. M., and Witte, J. P. M.: Unsaturated hydraulic properties of xerophilous mosses: towards implementation of moss covered soils in hydrological models, Hydrol. Process., 28, 6251–6264, doi:10.1002/hyp.10111, 2014.

Wallace, J. S., Lloyd, C. R., Roberts, J., and Shuttleworth, W. J.: A comparison of methods for estimating aerodynamic resistance of heather (*calluna vulgaris* (L.) hull) in the field, Agr. Forest Meteorol., 32, 289–305, doi:10.1016/0168-1923(84)90055-8, 1984.

Wouters, D. S., Keppens, H., and Impens, I.: Factors determining the longwave radiation exchange over natural surfaces, Arch. Meteor. Geophy. B., 28, 63–71, doi:10.1007/bf02243835, 1980.

Yang, F., Mitchell, K., Hou, Y.-T., Dai, Y., Zeng, X., Wang, Z., and Liang, X.-Z.: Dependence of land surface albedo on solar zenith angle: observations and model parameterization, J. Appl. Meteorol. Clim., 47, 2963–2982, doi:10.1175/2008jamc1843.1, 2008.

Yin, Y., Wu, S., Zheng, D., and Yang, Q.: Radiation calibration of FAO56 Penman–Monteith model to estimate reference crop evapotranspiration in China, Agr. Water Manage., 95, 77–84, doi:10.1016/j.agwat.2007.09.002, 2008.

Zhang, Y.-F., Wang, X.-P., Hu, R., Pan, Y.-X., and Zhang, H.: Variation of albedo to soil moisture for sand dunes and biological soil crusts in arid desert ecosystems, Environ. Earth Sci., 71, 1281–1288, doi:10.1007/s12665-013-2532-7, 2013.

Zhou, M. C., Ishidaira, H., Hapuarachchi, H. P., Magome, J., Kiem, A. S., and Takeuchi, K.: Estimating potential evapotranspiration using Shuttleworth–Wallace model and NOAA-AVHRR NDVI data to feed a distributed hydrological model over the Mekong River basin, J. Hydrol., 327, 151–173, 2006.

Zwolsman, J. J. G. and van Bokhoven, A. J.: Impact of summer droughts on water quality of the Rhine River – a preview of climate change?, Water Sci. Technol., 56, 45–55, 2007.

15

The spatial distribution and temporal variation of desert riparian forests and their influencing factors in the downstream Heihe River basin, China

Jingyi Ding[1], Wenwu Zhao[1,2], Stefani Daryanto[2], Lixin Wang[2], Hao Fan[1], Qiang Feng[1,3], and Yaping Wang[1]

[1]State Key Laboratory of Earth Surface Processes and Resource Ecology, Faculty of Geographical Science, Beijing Normal University, Beijing 100875, P. R. China

[2]Department of Earth Sciences, Indiana University – Purdue University, Indianapolis (IUPUI), Indianapolis, Indiana 46202, USA

[3]College of Forestry, Shanxi Agricultural University, Taigu, Shanxi 030801, P. R. China

Correspondence to: Wenwu Zhao (zhaoww@bnu.edu.cn) and Lixin Wang (lxwang@iupui.edu)

Abstract. Desert riparian forests are the main restored vegetation community in Heihe River basin. They provide critical habitats and a variety of ecosystem services in this arid environment. Since desert riparian forests are also sensitive to disturbance, examining the spatial distribution and temporal variation of these forests and their influencing factors is important to determine the limiting factors of vegetation recovery after long-term restoration. In this study, field experiment and remote sensing data were used to determine the spatial distribution and temporal variation of desert riparian forests and their relationship with the environmental factors. We classified five types of vegetation communities at different distances from the river channel. Community coverage and diversity formed a bimodal pattern, peaking at the distances of 1000 and 3000 m from the river channel. In general, the temporal normalized difference vegetation index (NDVI) trend from 2000 to 2014 was positive at different distances from the river channel, except for the region closest to the river bank (i.e. within 500 m from the river channel), which had been undergoing degradation since 2011. The spatial distribution of desert riparian forests was mainly influenced by the spatial heterogeneity of soil properties (e.g. soil moisture, bulk density and soil particle composition). Meanwhile, while the temporal variation of vegetation was affected by both the spatial heterogeneity of soil properties (e.g. soil moisture and soil particle composition) and to a lesser extent, the temporal variation of water availability (e.g. annual average and variability of groundwater, soil moisture and runoff). Since surface (0–30 cm) and deep (100–200 cm) soil moisture, bulk density and the annual average of soil moisture at 100 cm obtained from the remote sensing data were regarded as major determining factors of community distribution and temporal variation, conservation measures that protect the soil structure and prevent soil moisture depletion (e.g. artificial soil cover and water conveyance channels) were suggested to better protect desert riparian forests under climate change and intensive human disturbance.

1 Introduction

Riparian zone is the linkage between terrestrial and aquatic ecosystems (Naiman and Décamps, 1997), which is usually defined as the stream channel between the low- and the high-water marks, in addition to the terrestrial landscape above the high-water mark. Consequently, vegetation in the riparian zone is likely to be influenced by elevated water tables or extreme flooding and by the ability of soils to hold water (Nilsson and Berggren, 2000). Riparian zone plays an important role in numerous ecological processes and provides a variety of ecosystem services, such as sand stabilization and carbon sequestration (Décamps et al., 2004). Desert riparian forests, also known as "Tugai forests", are mainly located in the floodplains of major central Asian rivers. They are considered the core of riparian zone in hyperarid areas (Gärtner et al., 2014) by providing critical habitats for various species and functioning as an "ecological shelter" against desertification (Ding and Zhao, 2016). However, due to their low biodiversity and weak resilience, desert riparian forests are sensitive to disturbance and likely to be threatened by climate change and human disturbance (Li et al., 2013).

Desert riparian forests are the main communities in Heihe River basin, the second largest inland river in China (Feng, 2015). During the past century, increasing human population and over-exploitation of the mid-stream water resources have led to a significant degradation in the downstream desert riparian forests (Wang et al., 2014). Since 2000, the ecological water conveyance project (EWCP), a restoration project aiming to deliver water downstream, has been implemented to restore the ecosystems of Heihe River basin (Wang et al., 2011). Every year, approximately 300 billion m^3 of water is delivered using concrete channels, which were built perpendicular to the river to expand the range of the river impact and to deliver water for irrigation. The influence of ecological water conveyance may reach as far as 2000 m from the river (Si et al., 2005; Zeng et al., 2016). While most of vegetation in the downstream Heihe River has been restored (Lü et al., 2015), nearly 20 % of its oasis area covered by desert riparian forests remains degraded despite better downstream water conditions and the rising of groundwater table (Y. Zhang et al., 2011). To understand what may cause such variations in this desert riparian forest, we need to examine the spatial distribution of community characteristics along an ecological gradient, which can reflect how the spatial distribution of vegetation has been shaped by multiple environmental factors (Goebel et al., 2012; Li et al., 2013).

In hyperarid zone, groundwater is regarded the major driving factor of vegetation distributions and groundwater table should be between 2 and 4 m deep to support vegetation growth (Zheng et al., 2005). Study in Tarim River, for example, shows that species diversity peaks where groundwater depth is around 2–4 m. Once groundwater tables fall beyond 4.5 m, a deficiency in soil moisture occurs, followed by degradation of vegetation communities (Li et al., 2013). While groundwater drops rapidly away from the river bank to approximately 6 m deep at a distance of 1000 m from Tarim River channel (Aishan et al., 2013), the groundwater table in downstream Heihe River, where most of the desert riparian forest are distributed, remains above 4 m even at the distance of 3800 m from the river channel (Wang et al., 2011; Fu et al., 2014). Yet some sites have not been completely restored in the Heihe riparian zones, and its downstream vegetation community is still dominated by shrubs instead of multiple layers of tree (He and Zhao, 2006; Y. Zhang et al., 2011). A study by Zhu et al. (2013) shows that Patrick's richness index and the Shannon–Wiener diversity index for the downstream vegetation form a bimodal pattern instead of a unimodal pattern with groundwater depth in the Heihe River basin, indicating that there may be other factors affecting the spatial distribution of desert riparian forests.

Apart from groundwater, soil properties, such as soil moisture, soil physical and chemical properties also shape the community characteristics and species vitality (Stirzaker et al., 1996; Salter and Williams, 1965). Soil moisture, influenced by precipitation and groundwater, is the direct water source for desert riparian forests (Wang et al., 2012). As different depths of soil moisture affect species diversity in each community layer (D'Odorico et al., 2007; Fang et al., 2016), a decline in soil moisture may reduce the diversity of drought-sensitive species (e.g. herbs), resulting in a community shift towards drought-tolerant vegetation types with increasing distance from the river channel (Zhu et al., 2014). Study also finds that variation in soil properties explains the evolutionary trajectory of dominant species in arid areas and the changes in soil nutrients contribute greatly to species diversity (Yang et al., 2008).

The temporal variation in vegetation can reflect how communities respond to the changing environment during ecological restoration (Bakker et al., 1996; Scott et al., 2002). Due to the lack of long-term field-based observational data, long-term series of remote sensing data (e.g. MODIS-NDVI, SPOT-NDVI) have been widely used to explore vegetation changes and to evaluate the effectiveness of ecological restoration (Wang et al., 2014; Geng et al., 2014). Since the implementation of ecological restoration, the normalized difference vegetation index (NDVI) in the downstream Heihe River basin has significantly increased and the temporal variation in environmental factors, especially water availability (e.g. runoff and groundwater) has been reported as the major driving factor of vegetation recovery (Jia et al., 2011; Zeng et al., 2016). While large-scale remote sensing data (e.g. MODIS-NDVI, SPOT-NDVI) can capture the general trend of the whole study area, they fail to accurately delineate the temporal variation of desert riparian forest vegetation at the fine scale (i.e. < 100 m). However, community variation is a result of long-term interactive effects between vegetation and the environment, influenced by both spatial heterogeneity and temporal variation factors during ecological restoration (Zhu et al., 2013; Xi et al., 2016). Until recently, very few studies have incorporated both spatial and temporal variation of desert riparian forests into a single study due to the inconsistency in scale between fine field sampling and coarse remote sensing analysis. As desert riparian forest is the main community that maintains ecosystem functions under hyperarid conditions, comprehensive research that simultaneously examines the spatial and temporal variation of vegetation is crucial for the restoration of degraded riparian zones.

In this study, we aim to (i) explore both the spatial distribution and temporal variation of Heihe desert riparian forest at different distances from the river channel and (ii) disentangle the impacts of spatial heterogeneity of soil properties and temporal variation of water availability on the vegetation community. We studied 3000 m transects running perpendicular to the river channel to include different distances from the river channel and to depict the spatial distribution of vegetation (e.g. changes in floristic composition, community structure and diversity) along the distance from the river channel. At the same time, we used NDVI variations at different distances from the river channel, derived from high-resolution images (e.g. 30 m resolution) from 2000 to 2014, to depict the temporal variation of the vegetation. Spatial heterogeneity of soil properties (e.g. soil moisture, soil physical properties and soil nutrition) and temporal variation of water availability (e.g. annual average and annual variability of groundwater, soil moisture and runoff) were used to explain the vegetation community variance.

2 Data and methods

2.1 Study area

The study was conducted in the downstream Heihe River (40°20′–42°30′ N, 99°30′–102°00′ E) in the Ejina Oasis, Inner Mongolia, northwest China. The oasis covers an area of 3×10^4 km^2, with declining surface elevation (i.e. 1127 to 820 m above sea level) from the south-west to the north-east. This region has a typical continental arid climate with a mean annual temperature of 8.77 °C. Its maximum and minimum temperatures usually occur in July (41 °C) and January (−36 °C) (Wen et al., 2005). The mean annual precipitation is <39 mm, 84% of which occurs during the growing season (May to September), while the mean annual potential evaporation is > 3390 mm (Chen et al., 2014).

The Heihe River originates from rainfall and snowmelt in the Qilian Mountains. It branches into the Donghe River and the Xihe River at Langxinshan Mountain and ultimately flows into East Juyan Lake and West Juyan Lake in Ejina. The population of the Ejina Oasis is 32 410 (Ejina statistical office, 2012). The local economy mainly depends on cantaloupe farming and animal husbandry (e.g. sheep, cattle and camels). The Ejina Oasis is one of China's most important tourist attractions with respect to desert riparian forests, attracting almost 200 000 visitors per year during September to October (Hochmuth et al., 2015). One primary road is built parallel to the river channel, and another runs across the south of the oasis.

The desert riparian forests are the main components of the Ejina Oasis. They mainly grow along the river banks and spread across the fluvial plain, with the dominant vegetation including *Populus euphratica*, *Tamarix ramosissima*, *Lycium ruthenicum*, *Sophora alopecuroides*, *Karelinia caspica*, and *Peganum harmala*. Sparse and drought-tolerant desert species, such as *Reaumuria soongarica*, *Zygophyllum xanthoxylon* and *Calligonum mongolicum* are mainly distributed in the Gobi Desert. The main soil types in the area are Gypsi Sali-Orthic Aridosols and Calcaric Aridi-Orthic Primosols. Para-alkalic Aqui-Orthic Halosols and Calcaric Ochri-Aquic Cambosols also exist in the lake basins and lowlands (Chen et al., 2014; Gong, 2014).

2.2 Spatial field survey and experimental design

In the downstream Heihe River basin, desert riparian forests make up the core of the desert oasis, mainly composed of tree, shrub and grass communities. However, the spatial extent of the riparian zone is difficult to precisely delineate due to the heterogeneity of the landforms (Décamps et al., 2004). Although previous studies indicate that the forests are distributed between 0 and 2000 m from the river channel, corresponding to the influence range of ecological water conveyance (Si et al., 2005), our study extended beyond that range (i.e. up to 3000 m from the river channel) to fully cover the distribution pattern of the desert riparian forests in downstream Heihe River.

Our field survey was conducted in July 2015 after the ecological water conveyance was delivered. Ecological water conveyance is implemented according to the water dispatching scheme and is conducted in April, July, August, September and November with a scheduled discharge (Feng, 2015). Due to the regulated water discharge, the flooding only affects the sites near the river bank (within a 100 m radius) (Liu et al., 2008). We conducted vegetation and soil sampling perpendicular to the river channel and sampled at distances of 100, 500, 1000, 1500, 2000, 2500 and 3000 m from the river channel. Five replicates were sampled at each distance from the river channel and a total of 35 sampling sites were established (Table S1 in the Supplement). These sites were far from farmlands, irrigated channels and reservoirs to minimize the impact of human disturbance and other water resources. Although there is a main road extending across the oasis and almost parallel to the river channel (Fig. 1), the vegetation community growing near the road is considered to be undisturbed by the road, as the road is separated from the surroundings by iron wire.

Three tree quadrats (30 m × 30 m) and shrub quadrats (10 m × 10 m) were established randomly at each site. The number of each species (tree and shrub), plant height, coverage and the diameter at breast height of the trees (≥ 2 m) were recorded individually. Four herb quadrats (2 m × 2 m) were established at each corner of the tree or shrub quadrat to collect data on the number of herb species, vegetation cover and height.

At each site, soil samples for soil moisture measurement were randomly collected in three replicates using an auger (5 cm in diameter) at 5, 10, 15, 20, 30, 40, 50, 60, 70, 80, 100, 120, 140, 160, 180 and 200 cm deep. The soil samples were sealed in a freezer and the gravimetric soil water content (SWC) was determined via oven drying at 105 °C to a constant weight. Bulk density (BD) was measured by collecting undisturbed soil cores from the surface layer using a stainless-steel cutting ring (100 cm^3 in volume) with three replicates at each site, which were then oven dried at 105 °C to a constant weight. Soil samples were collected every 20 cm at a depth of 100 cm in each site to determine the soil composition and chemical properties. Sur-

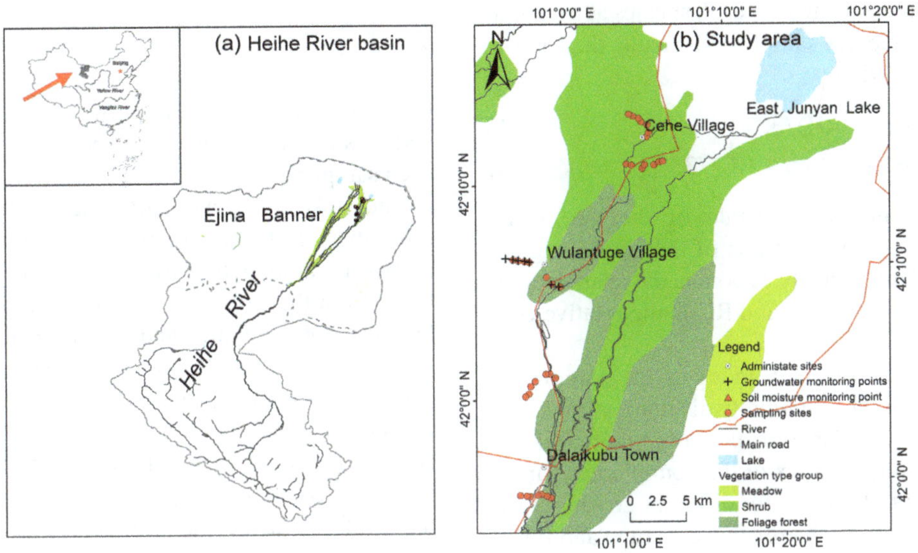

Figure 1. The Heihe River basin in China **(a)** and the location of sampling points in the study area **(b)**.

face soil samples (from a depth of 0–20 cm) were subsequently analysed in the laboratory to determine their clay (< 0.002 mm), silt (0.002–0.05 mm), sand (0.05–2 mm) and gravel (> 2 mm) content using a Malvern Mastersizer 2000. Soil organic matter (SOM) was measured using the $K_2Cr_2O_7$ method (Liu, 1996). Total nitrogen (TN) was determined using the Kjeldahl procedure (ISSCAS, 1978). Total phosphorus (TP) was determined using a UT–1810PC spectrophotometer (PERSEE, Beijing, China), after H_2SO_4–$HClO_4$ digestion (ISSCAS, 1978). Total salt content (TS) was determined by oven method (Liu, 1996).

2.3 Temporal data collection and processing

To understand the long-term vegetation variation since the implementation of ecological water conveyance, we analysed NDVI data from 2000 to 2014. As the NDVI measures the vegetation status, including coverage and vigour, we used the maximum NDVI during growing season as the indicator of vegetation community characteristics (Wang et al., 2014). NDVI data during 2000–2014 were calculated using ENVI (5.0) based on the Landsat TM/ETM image (30 m) acquired from Geospatial Data Cloud (http://www.gscloud.cn/). We calculated the NDVI at each distance from the river channel based on the NDVI derived from the sampling sites. The variable environmental factors, such as 2 cm soil moisture, 100 cm soil moisture and groundwater at each site during the research period, were extracted from remotely sensed data (1000 m resolution) (doi:10.3972/heihe.0034.2016.db), generated by the land model CLM4.5 using high-resolution ASTER DEM dataset, the multi-source integrated Chinese land cover map (MICLCover), the Heihe watershed allied telemetry experimental research land cover map (HiWATER Land Cover Map) and the China soil characteristics dataset (Zeng et al., 2016). The validation of the retrieved remote sensing data is provided in the Supplement S4. Land use change information of 2000 and 2014 was extracted from land use data with a scale of 1 : 100 000 (doi:10.3972/heihe.020.2013.db; doi:10.3972/hiwater.155.2014.db) (Zhong et al., 2015). The spatial variation of groundwater table was obtained from groundwater monitoring data recorded by seven wells (i.e. 7.62–9.66 m deep) in the Ejina Oasis, located at 50, 300, 2200, 2700, 3200 and 3700 m from the river centre (Fig. 1). These monitoring data of the groundwater table were recorded as 18 h averages with a three decimal places accuracy using a HOBO automatic groundwater table gauge from October 2010 to December 2014 (Fu et al., 2014). The diurnal and annual variations of soil moisture were obtained from the monitoring data of soil moisture, recorded at 0.5 Hz, as a 10 min average from 2013 to 2015 using a suite of micrometeorological sensors (CR800, CR23X, CR23XTD, Campbell Scientific Inc.) installed at a site that is located within Heihe riparian forest, about 1500 m from the Heihe river channel (Fig. 1) (doi:10.3972/hiwater.241.2015.db; doi:10.3972/hiwater.318.2016.db) (Liu et al., 2011). The retrieved remote sensing data, monitoring data, land use data, groundwater monitoring data and runoff data at Donghe station (i.e. a hydrological station in the downstream Heihe) (doi:10.3972/heihe.1009.2013.db) were acquired from the Environmental & Ecological Science Data Center for West China, National Natural Science Foundation of China (http://westdc.westgis.ac.cn).

2.4 Statistical analysis

The importance value (P), an index that characterizes the relative importance of plant species in the community (Zhang and Dong, 2010), was calculated for each species at each tree shrub and herb layer in every sampling site using the following formulas (Zhang, 2011):

$$P_{\text{Tree}} = (\text{RDen} + \text{RDom} + \text{RH})/3, \tag{1}$$

$$P_{\text{Shrub or Grass}} = (\text{RDen} + \text{RDom} + \text{RC})/3, \tag{2}$$

where P_{Tree} is the importance value of the tree layer. $P_{\text{Shrub or Grass}}$ is the importance value of the shrub or grass layer. RDen is the relative density, RDom is the relative dominance, RH is the relative height and RC is the relative coverage.

In our study, the total diversity index of the community was employed to depict the community diversity at each site. According to the characteristics of the community vertical structure, the total diversity index of the community is measured using the weight of indices for different growth types. The weight is the average of the relative coverage and thickness of the leaf layer (Fan et al., 2006). We applied the following formula (Gao et al., 1997):

$$W_i = (C_i/C + h_i/h)/2, \tag{3}$$

where C is the total coverage of the community ($C = \sum C_i$); $i = 1$ for the tree layer, 2 for the shrub layer and 3 for the herb layer; h is the thickness of the leaf layer of various growth types ($h = \sum h_i$); W_i is the weighted parameter of the diversity index for the ith growth type; C_i is the coverage of the ith growth type; and h_i is the average thickness of the leaf layer of the ith growth type. Among the different growth types, the thickness of the tree leaf layer is calculated at 33.3 % the height of the tree layer, the shrub layer at 50 % and the herb layer at 100 %.

The total diversity index of the community was calculated according to the following formula:

$$A = \sum W_i A_i, \tag{4}$$

where A is the total diversity index of the community; $i = 1$ for the tree layer, 2 for the shrub layer and 3 for the herb layer; W_i is the weighted parameters of the diversity index for the ith growth type; A_i is the diversity index of the ith growth type.

Species diversity indices were calculated (Liu et al., 1997), including the Shannon–Wiener diversity index (H)

$$H = -\sum_{i=1}^{S} (P_i \ln P_i). \tag{5}$$

Simpson dominance index (D) was calculated as

$$D = 1 - \sum_{i=1}^{S} P_i^2 \tag{6}$$

and Pielou evenness index (J_{sw}) was calculated as

$$J_{\text{sw}} = H/(\ln(S)). \tag{7}$$

Finally, Patrick richness index (R) was calculated as

$$R = S, \tag{8}$$

where P_i is the relative importance value of the ith species, and S is the total number of species in each growth type at each sampling site.

The least significant difference test was used to determine the significance of the variability in vegetation characteristics among five transects (Table S2). For each distance from the river channel, vegetation community characteristics, soil moisture and soil properties of the five sites were calculated as the mean ± standard error of the mean. The monitoring data of soil moisture in desert riparian forest showed that soil moisture under 20 cm was relatively stable and could represent water condition at the sampling site (Fig. S1 in the Supplement). Thus, to depict the water condition at the sampling sites and to illustrate the vertical structure of soil moisture, soil water content was divided into three layers: 0–30 cm soil moisture (SWC1), 30–100 cm soil moisture (SWC2), and 100–200 cm soil moisture (SWC3) in accordance with the distribution of fine roots herbs, trees and shrubs in this area (Fu et al., 2014). We averaged the soil moisture at each corresponding finer increment to obtain the value of SWC1, SWC2 and SWC3. The soil chemical properties, however, were analysed using the mean values of 0–100 cm due to the low vertical variation. The NDVI change rate was calculated based on the percentage change of NDVI from 2000 to 2014. The annual average value and annual variability were used to depict the temporal variation of community characteristics and environmental factors. The annual averages of NDVI (NDVI_a), groundwater (GWT_a), 2 cm soil moisture (SWC2cm_a) and 100 cm soil moisture (SWC100cm_a) were calculated using the mean values from 2000 to 2014. The annual variability of NDVI (NDVI_c), groundwater (GWT_c), 2 cm soil moisture (SWC2cm_c) and 100 cm soil moisture (SWC100cm_c) were calculated using the mean values of change rate for each year.

Regression analysis was used to examine the variation pattern. Exponential and polynomial regressions were fit to the data to explain the statistical relationship between community characteristics and distance from the river channel. Pearson correlation was used to determine the strengths of possible relationships between community characteristics and environmental factors. Significant differences were evaluated at the 0.05 and 0.01 levels. Statistical analysis was performed using SPSS (version 18.0).

Two-way indicator species analysis (TWINSPAN, in WinTWINSPAN, version 2.3), a method of community hierarchical classification (Hill, 1979), was used to classify the possible desert riparian forest community types. The importance value data for all plant species, obtained from the vegetation survey, were used in this analysis and the cutoff levels of the importance value for each class were set as 0, 0.1, 0.2, 0.4, 0.6 and 0.9. To further separate the key influencing factors of the 18 environmental variables, the marginal and conditional effects of the various variables were calculated through Monte Carlo forward selection in RDA (redundancy analysis), which directly showed the significance and the percentage of vegetation variance explained by each factor. Marginal effects reflect the effects of the environmental variables on the community characteristics, while conditional effects reflect the effects of the environmental variables on the community characteristics after the anterior variable is eliminated by the forward selection method. Since the redundant variables are eliminated and a group of key environmental factors is identified through forward selection, this method allows key variables to be determined through the strength of their effects and significance. The variation of community characteristics explained by the different group of environmental factors was analysed using variation partitioning analysis. The significance of the resulting ordination was evaluated by 499 Monte Carlo permutations (Zhang and Dong, 2010). The Monte Carlo test and variation partitioning analysis were performed using the software program CANOCO (ver. 5.0) (Microcomputer Power, USA) (Ter Braak and Simlauer, 2012).

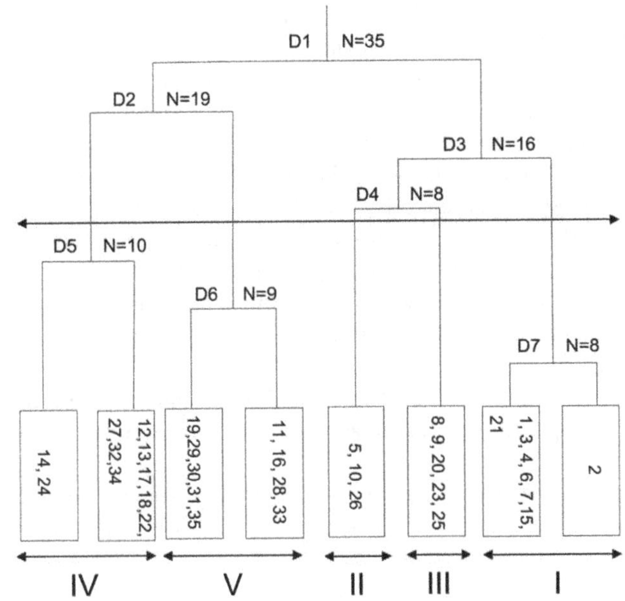

Figure 2. The dendrogram of the sampling sites based on the TWINSPAN classification. Note: numbers 1–35 represent the site numbers of the sampling sites, D indicates the classification levels and N indicates the number of sampling sites for the classification. I to V represent communities I to V. Arrows indicates that all the sites were divided into five major groups after the fourth classification.

3 Results

3.1 Vegetation community types and temporal changes in vegetation composition

The species composition at each site in the downstream Heihe River basin is shown in Table S1, and the following five plant community types distributed across the 3000 m transect from the river channel were obtained based on the TWINSPAN classification (Fig. 2):

i. Community I was an association of (Ass.) *Populus euphratica – Tamarix ramosissima* + herbs. Characterized by multiple layers of tree–shrub–herb, the community coverage was low (38.05 %) and dominated by the tree species *Populus euphratica* with sparse understory vegetation. This community was mainly distributed near the river bank, mostly within 500 m from the river channel.

ii. Community II was Ass. *Tamarix ramosissima – Lycium ruthenicum* + herbs. This community was composed of shrub and herb layers with high community coverage (81.43 %). *Tamarix ramosissima* was the dominant shrub species of the shrub layer, while the herbs were dominated by both hygrophyte and xerophyte species, such as *Kochia scoparia* and *Peganum harmala*. This community was mainly distributed near the river bank (approximately 1000 m from the river channel).

iii. Community III included *Tamarix ramosissima*. This community was mainly composed of shrub layers and dominated by *Tamarix ramosissima* with average community coverage of 75.93 %. It was mainly distributed at the distance between 1000 and 2000 m from the river channel.

iv. Community IV was Ass. *Lycium ruthenicum – Tamarix ramosissima* + xerophyte herbs. This community was mainly composed of shrub and herb layers with average community coverage of 68.86 %. *Lycium ruthenicum* was the dominant shrub species (importance value = 0.42–0.77), while the dominant xerophytic herb species were *Sophora alopecuroides* and *Suaeda salsa*. It was mainly distributed between 1500 and 2500 m from the river channel.

v. Community V was Ass. *Tamarix ramosissima – Lycium ruthenicum – Reaumuria soongarica*. This community was the transition community from desert riparian shrub forests to desert shrubs, indicated by the presence of *Reaumuria soongarica*, a typical desert shrub. *Tamarix ramosissima* was the dominant species of the shrub layer, mainly existing in the form of shrub dunes. This community was mainly distributed approximately 2500–3000 m from the river channel, with low community coverage (54.40 %).

Between 2000 and 2014, changes in the vegetation composition of each community type (I to V) were obtained from changes in the land use map (Fig. 3). Among the five community types, community V underwent the most changes, with

22.22 % of the sites changing from sparse forest to grassland, 22.22 % from grassland to shrubland and 22.22 % from bare land to grassland. The majority (> 60 %) of the vegetation composition remained unchanged in communities I to IV, with the following exceptions: (i) 37.5 % of the sites in community I changed from shrubland to sparse forest and from bare land to grassland, (ii) 33 and 20 % of the sites in communities II and III changed from bare land to grassland and from sparse forest to grassland, respectively, and (iii) 20 % of the sites in community IV changed from sparse forest to grassland and another 20 % from grassland to shrubland (Fig. 3).

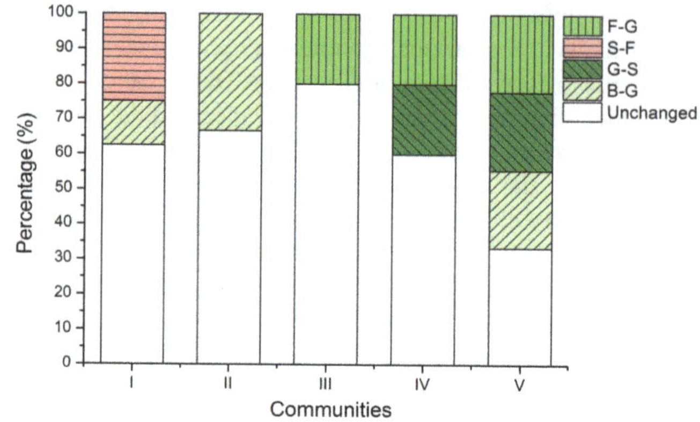

Figure 3. The percent changes of vegetation composition in each community from 2000 to 2014. Note: F–G change from sparse forest to grassland, S–F change from shrubland to sparse forest, G–S change from grassland to shrubland, and B–G change from bare land to grassland.

3.2 The spatial distribution and temporal variation of community characteristics in desert riparian forest

Community characteristics varied with distance from the river channel (Fig. 4). Vegetation community height and density dropped rapidly after 500 m (Fig. 4a and b), while community coverage formed a bimodal pattern, peaking at the distances of 500–1000 and 3000 m (Fig. 4c). The variation of vertical structure was depicted by the following hierarchical coverage (Fig. 4d): (i) the tree layer mainly existed within 1000 m, (ii) the shrub layer peaked at approximately 1500–2000 m and (iii) the herb layer fluctuated with distance from the river channel, peaking at 500 and 2500–3000 m from the river channel. All diversity indices showed a bimodal pattern with distance from the river channel. The Shannon–Wiener diversity index, Pielou evenness index and Patrick richness index peaked at 1000 and 3000 m (Fig. 4e–g). Simpson dominance index, however, formed an opposite trend to the other three diversity indices, by peaking at 500 and 2000 m, where the other indices were at their low level (Fig. 4h).

The temporal variation in community characteristics was depicted by the variation in NDVI (Fig. 5). At different distances from the river channel, the NDVI showed an increasing trend throughout the research period except for a small decrease during the initial years (2000–2002). The NDVI decreased with distance from the river channel, peaked at 100 and 500 m from the river channel, and reached the lowest value at the furthest distance from the river channel (3000 m) (Fig. 5a). The annual variability in the NDVI, however, showed a contrasting trend, increasing with distance from the river channel (Fig. 5b).

3.3 Pearson correlation between community characteristics and environmental factors

The result of the Pearson correlation analysis between community characteristics and environmental factors is shown in Table 1. The community density showed significant positive correlations with SWC2, SWC3, SWC2cm_a and SWC100cm_a, but negative correlations with BD and GWT_a. The community coverage was positively correlated with all the three soil moisture layers ($P < 0.0.1$) but negatively correlated with BD. The tree and shrub layers were influenced by GWT_a and BD, respectively, while the herb layer was positively correlated with SWC1 and SCW3. Among the diversity indices, the Patrick richness index was significantly correlated with SOM and gravel, while Simpson dominance index was significantly correlated with sand and silt. For the temporal variation of community characteristics, NDVI_a was mainly influenced by soil moisture (SWC1, SWC2 and SWC3), soil particle composition (clay, gravel) and BD, while NDVI_c was significantly correlated with SWC3, gravel and TS.

The correlation coefficient between NDVI and runoff was measured to examine the relationship between runoff and NDVI (Fig. 6). The relationship between runoff and the NDVI of the following year was significantly stronger than the relationship between runoff and the NDVI of the same year, indicated by the higher correlation coefficient and significance in the 1-year lag NDVI–runoff correlation compared to the NDVI–runoff correlation.

3.4 Key environmental factors that influenced community characteristics

To further examine the key environmental factors that controlled the variation in vegetation indices, redundant variables were eliminated by a forward selection method. Table 2 shows the key influencing factors based on the marginal and conditional effects of 18 variables under the Monte Carlo test in the process of forward selection. All the environmental factors explained 74 % of the total variance. In the Monte Carlo test of forward selection ($P < 0.05$), SWC1, SWC3, BD and SWC100cm_a were regarded the key environmental factors that influenced the variation of community characteristics. Spatial heterogeneity factors explained 43.5 % of the variation of community characteristics and accounted for 98.4 % of the variance explanation explained by all the inves-

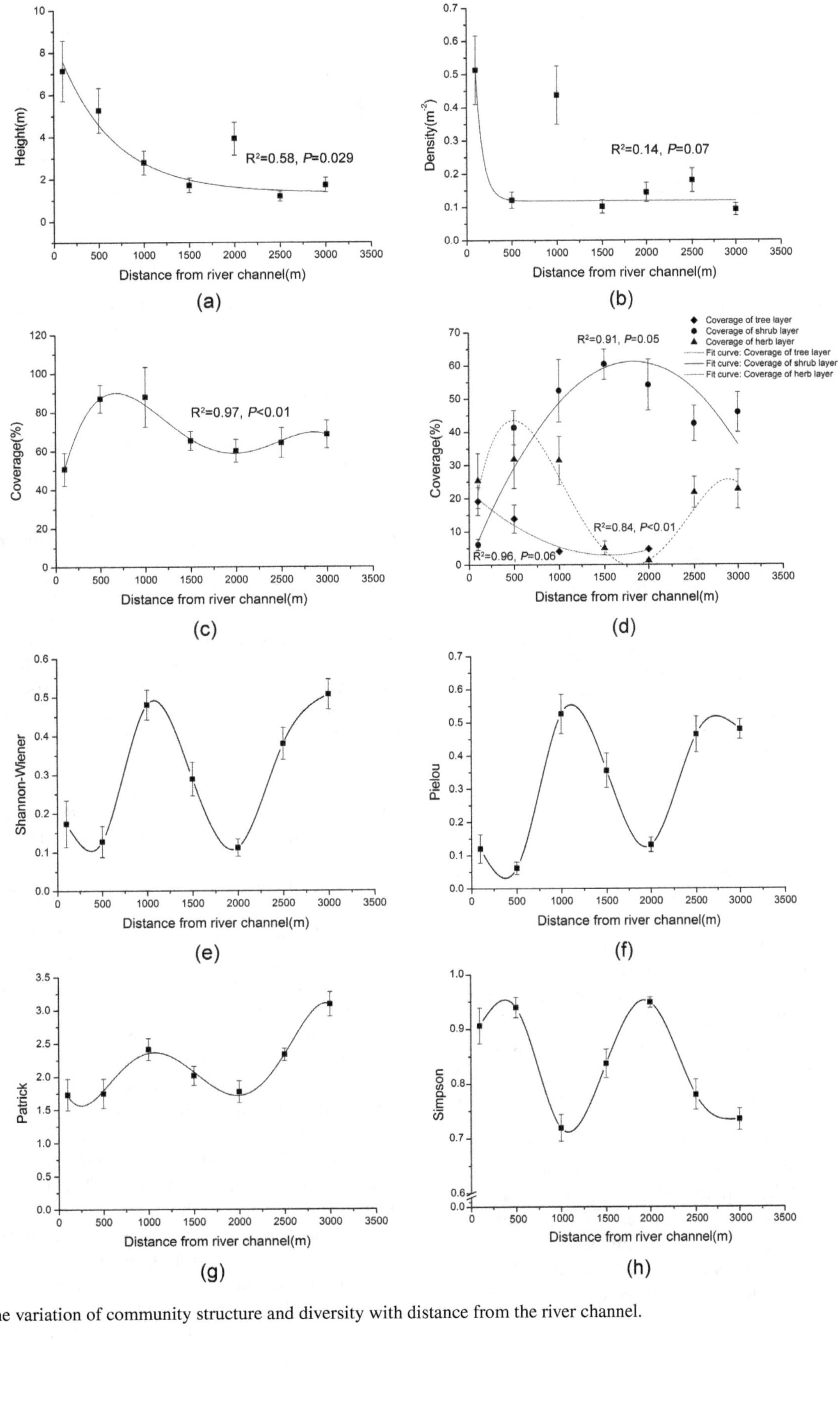

Figure 4. The variation of community structure and diversity with distance from the river channel.

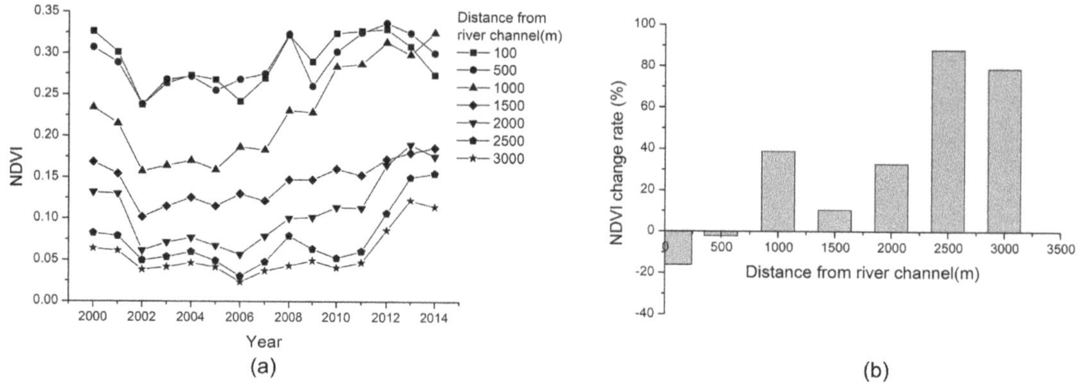

Figure 5. The temporal variation of NDVI **(a)** and annual variability of NDVI **(b)** from 2000 to 2014 at different distances from the river channel.

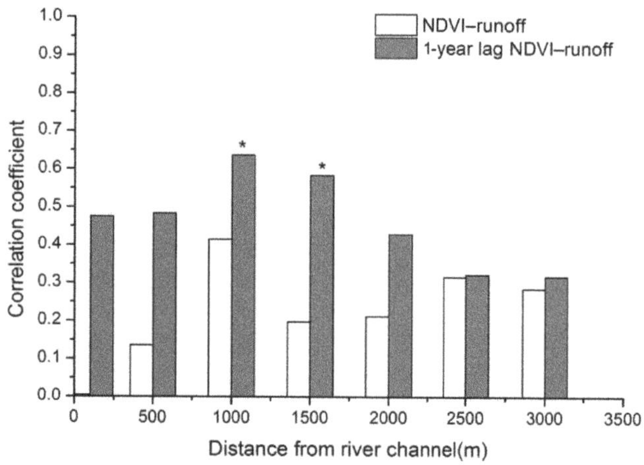

Figure 6. Pearson correlation coefficients of NDVI–runoff and 1-year lag NDVI–runoff at different distances from the river channel. Note: * above the bar indicates significant correlations ($P < 0.05$).

tigated environmental factors (Table 3). This result indicated that spatial heterogeneity of environmental factors was the major driving force of the spatio-temporal variation in this desert riparian forest. In contrast, temporal variation factors only explained 15.9 % of the variation of community characteristics, accounting for 35.9 % of the total explanation that was explained by all the investigated environmental factors (Table 3). This result suggested that temporal variation of the environmental factors exerted less impact on community characteristics compared to spatial heterogeneity of environmental factors in this desert riparian forest. While each factor group affected community characteristics separately, the spatial heterogeneity factors and temporal variation factors also jointly shaped the variation of community characteristics in downstream Heihe riparian forests. When combined, these factors explained 15.2 % of the community characteristics variation, accounting for 34.3 % of the total explanation that was explained by all the investigated environmental factors (Table 3, Supplement S6).

4 Discussion

4.1 The spatial distribution in Heihe desert riparian forest and its influencing factors

In the downstream Heihe River basin, community height and density declined significantly as the dominant species changed from trees to riparian and desert shrubs with increasing distance from the river channel. Community coverage reached two local maxima at 1000 and 3000 m from the river, with diverse shrub and herb layers. The spatial distribution of community diversity can illustrate how vegetation responds and interacts with numerous environmental factors along different ecological gradients (Oksanen and Minchin, 2002). Different from studies in the relatively humid regions (e.g. coastal regions or boreal forest), which suggest that riparian forest species diversity either decreases or forms a unimodal pattern with increasing distance from the stream (Pabst and Spies, 2011; Macdonald et al., 2014), our findings in the downstream Heihe (i.e. hyperarid condition) showed that community diversity formed a bimodal pattern along the distance from the river channel. We attributed fine-textured soils found in the upper soil layer to the high species diversity at sites located about 1000 m from the river channel. A previous study shows that fine-textured soils increase the soil water-holding capacity and improve the gravimetric moisture content in the upper soil layer, which provide suitable habitat for the growth of diverse herb species with shallow rooting systems (Liu et al., 2008). At the same time, other studies find that the presence of the deep-rooted tree, *Populus euphratica*, can redistribute deep soil water to the shallow layer as a strategy of mutualism, benefiting the growth of shallow-rooted herbaceous species (Hao et al., 2013). A similar mechanism also occurs at a further distance (i.e. 3000 m) from the river. Although situated in the transition region (from riparian forest to desert shrubs), soils at the distance of 3000 m from the river channel were still rich in fine particles (clay and silt; 35.6 %) (Table S3), brought by the interaction between wind erosion and shrubs (Ravi et al., 2009). The presence of

Table 1. Pearson correlations between community characteristics and environmental factors.

	H	R	C	J_{SW}	Height	Density	Cover-a	Cover-t	Cover-s	Cover-h	NDVI_a	NDVI_c
SWC1	0.255	0.167	−0.286	0.182	−0.088	0.251	**0.545**	−0.017	0.168	**0.514**	**0.430**	0.188
SWC2	0.046	−0.072	−0.098	0.067	−0.114	**0.382**	**0.439**	0.007	0.280	0.263	**0.469**	0.254
SWC3	0.142	0.157	−0.147	0.111	−0.242	**0.362**	**0.448**	−0.142	0.175	**0.382**	**0.445**	**0.506**
Clay	0.112	0.005	−0.128	0.045	0.048	0.290	0.204	0.037	−0.093	0.272	**0.398**	0.125
Silt	0.308	0.117	**−0.344**	0.311	−0.121	0.111	0.321	−0.071	0.247	0.168	0.185	−0.115
Sand	−0.327	−0.148	**0.354**	−0.306	0.130	−0.165	−0.307	0.076	−0.166	−0.217	−0.212	0.125
Gravel	0.226	**0.350**	−0.155	0.179	−0.284	−0.081	−0.185	−0.173	−0.179	0.011	**−0.413**	**−0.396**
BD	0.174	0.282	−0.127	0.123	−0.041	**−0.353**	**−0.350**	0.049	**−0.465**	0.063	**−0.354**	−0.050
SOM	−0.256	**−0.398**	0.187	−0.102	0.193	0.058	−0.192	0.116	−0.121	−0.296	−0.025	−0.009
TN	−0.191	−0.333	0.138	−0.060	0.101	0.032	−0.278	0.112	−0.296	−0.223	−0.006	0.108
TP	−0.238	−0.303	0.198	−0.098	0.116	0.022	−0.181	0.084	−0.090	−0.288	−0.018	0.194
TS	−0.139	−0.125	0.111	−0.099	−0.184	0.271	0.011	−0.086	0.034	−0.131	−0.140	**0.382**
GWT_c	0.094	−0.028	−0.133	0.228	−0.074	0.001	−0.137	0.102	−0.060	−0.189	−0.286	0.040
SWC2cm_c	0.113	0.085	−0.117	0.084	−0.161	0.098	−0.027	−0.093	−0.029	−0.024	−0.177	0.119
SWC100cm_c	0.171	0.185	−0.165	0.109	−0.116	−0.080	0.073	−0.096	0.107	0.038	−0.198	0.141
GWT_a	−0.022	−0.226	−0.050	0.127	0.300	**−0.343**	−0.092	**0.352**	0.017	−0.131	0.042	0.004
SWC2cm_a	−0.169	−0.270	0.129	−0.096	0.013	**0.405**	−0.184	0.103	−0.224	−0.183	0.160	0.144
SWC100cm_a	−0.085	−0.194	0.047	−0.014	−0.094	**0.403**	−0.137	−0.046	−0.206	−0.150	0.090	0.140

Note: significant correlations with $P < 0.05$ and $P < 0.01$ are shown in bold and in bold with underline, respectively. H, Shannon–Wiener diversity index; R, Patrick richness index; C, Simpson dominance index; J_{SW}, Pielou evenness index; a, total plant community; t, tree layer; s, shrub layer; h, herb layer; NDVI_a, annual average of NDVI;NDVI_c, annual variability of NDVI; SWC1, 0–30 cm soil moisture; SWC2, 30–100 cm soil moisture; SWC3, 100–200 cm soil moisture; BD, bulk density; SOM, soil organic matter; TN, total nitrogen; TP, total phosphorus; TS, total salt content. GWT_a, annual average of groundwater table; SWC2cm_a, annual average of 2 cm soil moisture; SWC100cm_a, annual average of 100 cm soil moisture; GWT_c, annual variability of groundwater table; SWC2cm_c, annual variability of 2 cm soil moisture; SWC100cm_c, annual variability of 100 cm soil moisture.

Table 2. The selection of the key influencing factors based on the marginal and conditional effects obtained from the forward selection of the Monte Carlo test.

Marginal effects		Conditional effects			
Environmental factors	Percentage of variance explained (%)	Environmental factors	Percentage of variance explained (%)	*P* value	*R* value (%)
SWC1	20.2	SWC1	20	0.002	27.03
SWC3	18.8	SWC3	14	0.004	18.92
SWC2	12.3	BD	10	0.006	13.51
BD	11.4	SWC100cm_a	9	0.018	12.16
TN	7.1	GWT_a	4	0.078	–
Silt	7.0	GWT_c	3	0.096	–
Sand	6.1	TP	2	0.250	–
SOM	4.1	Clay	2	0.282	–
Clay	3.8	TN	2	0.296	–
SWC2cm_a	3.7	SWC2cm_a	2	0.308	–
TP	3.6	SWC100cm_c	1	0.444	–
Gravel	2.6	SWC2cm_c	3	0.112	–
SWC100cm_a	2.5	SWC2	1	0.620	–
GWT_c	1.8	Silt	1	0.636	–
GWT_a	1.4	TS	< 0.1	0.788	–
SWC100cm_c	0.6	SOM	< 0.1	0.932	–
TS	0.5	Sand	< 0.1	0.992	–
SWC2cm_c	0.1	Gravel	< 0.1	0.960	–
		Total	74	0.036	–

Note: *R* value represents the relative proportion of individual explanation to the total variance explanation. SWC1, 0–30 cm soil moisture; SWC2, 30–100 cm soil moisture; SWC3, 100–200 cm soil moisture; BD, bulk density; SOM, soil organic matter; TN, total nitrogen; TP, total phosphorus; TS, total salt content. GWT_a, annual average of groundwater table; SWC2cm_a, annual average of 2 cm soil moisture; SWC100cm_a, annual average of 100 cm soil moisture; GWT_c, annual variability of groundwater table; SWC2cm_c, annual variability of 2 cm soil moisture; SWC100cm_c, annual variability of 100 cm soil moisture.

Table 3. The percentage of community characteristic variation explained by different groups of environmental factors.

Fraction	Variation	% of all	% of explained	*F*	*P*
a	0.435	43.5	98.4	5.9	0.008
b	0.159	15.9	35.9	4.0	0.088
c	0.152	15.2	34.3	2.2	0.016
Total explained	0.442	44.2	100	–	–

Note: *a*: spatial distribution factors, including 0–30 cm soil moisture (SWC1), 30–100 cm soil moisture (SWC2), 100–200 cm soil moisture (SWC3), bulk density (BD), clay, silt, sand, gravel, soil organic matter (SOM), total nitrogen (TN), total phosphorus (TP) and total salt content (TS); *b*: temporal factors, including annual average of groundwater table (GWT_a), annual average of 2 cm soil moisture (SWC2cm_a), annual average of 100 cm soil moisture (SWC100cm_a), annual variability of groundwater table (GWT_c), annual variability of 2 cm soil moisture (SWC2cm_c) and annual variability of 100 cm soil moisture (SWC100cm_c); *c*: variation that jointly explained by group *a* and *b*. Variation: the variance explained by different fraction when the total variance is 1; % of all: the proportion of variation explained by different fraction; % of explained: the relative proportion of individual explanation to the total explanation.

fine soil particles can increase the soil water-holding capacity and soil nutrients around the shrub patches ("fertile islands") (Ravi et al., 2010), allowing for the growth of some xerophytic herbs and increasing the community diversity in arid region (Stavi et al., 2008). By contrast, Simpson dominance index peaked at the distances of 500 and 2000 m (Fig. 4h), likely due to inter-species competition for water and nutrient resources reported by previous studies (Maestre et al., 2005). At these sites, dominant species with high importance value (i.e. trees and shrubs at 500 and 2000 m, respectively) often have high competition for resources, thus hindering the growth of other species (i.e. herbs) and increasing the dominance index of the community (Koerselman and Meuleman, 1996).

Among different environmental factors, changes in water availability associated with soil properties are considered to be the most important selective forces, which shape ecosystem stability in hyperarid zones (Rosenthal et al., 2005; Ravi et al., 2010; Feng, 2015). Our study showed that spatial heterogeneity of soil properties was the major driving force for the spatial distribution of vegetation, with SWC1, SWC3 and BD contributing 59.46 % of the total explanation of vegetation variance (Table 2). As the fine roots of most herb species are mainly distributed within the top 30 cm of the surface soils (Fu et al., 2014), surface soil moisture (0–30 cm soil moisture; SWC1) likely became the main water source for the herb layers and contributed to a high coverage of dominant herb species, such as *S. alopecuroides* and *K. caspica*. Meanwhile, deep soil moisture (SWC2 and SWC3) mainly influenced the community density and coverage. SWC2 (30–100 cm soil moisture) is the main water resource for shrubs such as *T. ramosissima*. SWC3 (100–200 cm soil moisture), recharged by the flood-raised groundwater table (Liu et al., 2015), is the main water source for phreatophytes such as *P. euphratica* or desert shrubs (Yi et al., 2012). As trees and shrubs contributed greatly to community composition, the increase in SWC2 and SWC3 could significantly promote vegetation growth, community density and coverage.

Apart from soil moisture, soil physical properties were also important in determining the spatial distribution of vegetation community in our study. Bulk density and soil composition (clay, silt, gravel), which are critical for water and nutrient-holding capacity (Stirzaker et al., 1996; Meskinivishkaee et al., 2014), mainly influenced community density, community coverage, shrub coverage and diversity indices (Table 1). Previous study shows that soil with low bulk density usually has high water and nutrient-holding capacity associated with fine soil particles as opposed to soil with high bulk density (Ravi et al., 2010). Low water-holding capacity associated with coarse soil particles can induce water stress and limit vegetation growth, especially for herb species, which contribute greatly to community coverage, density and diversity (Stirzaker et al., 1996). Interestingly, we found that soil nutrients explained no more than 0.1 % of the vegetation variance and that SOM was negatively correlated with the species richness (Tables 1 and 2). Our findings were different from the positive relationship commonly found between SOM and species richness in semiarid zones (e.g. the Loess Plateau) (Jiao et al., 2011; Yang et al., 2014). Although SOM determines soil nutrient storage and the supply of available nutrients, our sites in the hyperarid zone were characterized by barren soil with less than 1 % SOM (Fig. S5d, Table S3). Such a low amount of SOM might not be able to boost the growth of various species in desert riparian forests. At the same time, the dominant species (i.e. *P. euphratica* and *T. ramosissima*), despite producing a high amount of litter, also has high competition for resources, hindering the diversity and growth of other species (Su and Zhao, 2003).

4.2 The temporal variation in Heihe desert riparian forest and its influencing factors

Our results showed that NDVI generally has increased since the implementation of the ecological water conveyance project in 2000, except for an initial decrease between 2000 and 2002, likely due to the lagging effect and the relatively low amount of runoff during these years (Jin et al., 2008). With better water availability (e.g. increased surface soil moisture and groundwater table), the conversion of sparsely forested land or bare land to shrubland and grassland likely contributed to the increase of NDVI, especially at the furthest distance from the river (2000–3000 m) (Fig. 3). By contrast, NDVI near the river bank underwent a slight decrease during some recent years (2012–2014) (Fig. 5), likely due to the conversion of shrubland to sparse forest land (Fig. 3). Other factors, such as increasing grazing pressure and tourism pressures as reported by Hochmuth et al. (2015), may also contribute to the decrease in NDVI around the periphery of the river.

Our results indicated that the NDVI at different distances from the river channel was affected by the spatial heterogeneity of soil properties, particularly soil moisture and bulk density, explaining 45.95 and 13.51 % of the vegetation variance, respectively (Table 2). All three soil moisture layers (SWC1, SWC2 and SWC3) positively affected the annual average of NDVI (NDVI_a) by supplying water to both shallow- and deep-rooted riparian vegetation (Fu et al., 2014). Deep soil moisture (SWC3), recharged by the increasing groundwater table, is particularly important for the shrub and tree layers (e.g. *P. euphratica* and *T. ramosissima*) (Yi et al., 2012), which contributed largely to the increase rate of NDVI in Heihe desert riparian forest. Bulk density accounted for 13.51 % of the total explanation of the vegetation variance and was negatively correlated with the annual average of NDVI (NDVI_a) and the annual variability of NDVI (NDVI_c). Study shows that, in the hyperarid zone, soils with high bulk density are often characterized by a high proportion of coarse soil particles but low clay content (Ravi et al., 2009). In general, low water-holding capacity in the upper soil layer may constrain the growth of shallow-rooted vegetation, lower the average NDVI and inhibit the NDVI increase.

Apart from soil, the temporal variation in runoff driven by the ecological water conveyance played an important role in influencing the annual variability of the NDVI in this desert riparian forest. The correlation between runoff and NDVI was stronger with that of the following year than that of the same year (Fig. 6), consistent with previous findings (Jin et al., 2008), which indicate the lagging effects of runoff. The lagging effects were unsurprising, considering that the increase in runoff needs to undergo a series of hydrological (e.g. seepage, interflow, groundwater movement) and ecological processes (e.g. vegetation water uptake and water utilization) to increase groundwater, soil moisture and vegetation growth (Williams et al., 2006; Liu et al., 2012). Compared to the significant relationship between the spatial heterogeneity of soil moisture (e.g. SWC3) and the annual variability of NDVI (i.e. NDVI_c) (Table 1), there was weaker relationship between the annual variability in water availability (i.e. SWC2cm_c, SWC100cm_c, GWT_c) and the annual variability of NDVI (i.e. NDVI_c). We suggested that a combination of hydrological and ecological processes in the Heihe riparian zone may result in a more stable increase of groundwater and soil moisture during 2000–2014 (Fig. S6a–c) compared to the amount of runoff that showed significant annual fluctuations (Fig. S6d), thus contributing less to the NDVI variability. In addition, temporal variation of soil moisture and groundwater was derived from retrieved remote sensing dataset. The deviations between simulations and observations in temporal variation of soil moisture and groundwater could also partly account for its weak correlations with the community characteristics.

Compared to the spatial heterogeneity of the soil properties, the temporal variation factors only accounted for 35.9 % of the total explanation (vs. 98.4 % for the spatial heterogeneity factors) (Table 3). Except for the possible influence of the accuracy of the retrieved remote sensing dataset, lack of data regarding the temporal variation of soil properties could also partly accounted for the low explanatory power of temporal influencing factors. Although studies show that most of the physical and chemical properties of the soils remain unchanged during the 15-year ecological restoration (Cao et al., 2010; Chen et al., 2014), some of those soil properties, such as soil organic matter, might change due to vegetation recovery during ecological restoration. Future studies that include the temporal variation in soil properties might therefore be required to comprehensively address how temporal variation in environmental factors impacts on vegetation variation.

4.3 Community resilience of desert riparian forests and implications for ecological restoration

Although community diversity was generally low at most sites in the downstream Heihe River basin, our findings suggested that there were some sites with significantly higher community diversity, for example those at distances of 1000 and 3000 m from the river channel (Fig. 4). Consistent with other finding (Hao et al., 2013), high resistance to water stress was observed at these distances from the river channel, where trees and shrubs can lifting water up from the deeper to shallower layers as a strategy for mutualism. Different species contribute differently to the ecosystem functions (e.g. trees and shrubs with large crowns mainly contribute to sand fixation, while diverse herbs contribute to biodiversity), maintaining a stable habitat after drought stress and/or human disturbance (Krieger, 2001). Indeed, species-rich communities can maintain ecosystem functions during stress-based perturbations due to the complementary of functional traits and ecological redundancy (Luck et al., 2003; Isbell et al., 2015).

In contrast, communities at the other distances from the river channel could easily undergo degradations, such as that at the distance of 500 m from the river channel, indicated by the decreasing NDVI in recent years (Fig. 5a). Exposure to human disturbance (e.g. tourism, grazing) may exacerbate the effects of projected rise in temperature and drought frequency in regions that are far from the river channel and experience the limited influence of ecological water conveyance (Si et al., 2005; Zhang et al., 2015) by changing soil physical properties and reducing ecosystem services such as water and soil conservation (Zhao et al., 2012; Daryanto et al., 2013). Our study showed that water availability and spatial heterogeneity of soil properties were the main driving forces for the spatial distribution and temporal variation of restored desert riparian forest at Heihe River basin. To halt degradation in this critical zone, we suggested to build natural channels perpendicular to the river to fully extend the influencing scope of the ecological water conveyance and benefit the regions far from the river bank (Y. C. Zhang et al., 2011). At the same time, multiple conservation measures, such as establishing critical fenced areas for ecological protection and constructing artificial shields or establishing straw checker boards on the bare land to prevent erosion, were recommended around the periphery of the river.

5 Conclusions

Through extensive field observations at multiple desert riparian forest locations and analyses of long-term remote sensing images, we found that community characteristics differed spatially and temporally with distance from the river channel. In locations with high diversity indices, high community resilience could be maintained by the multiple interactions between vegetation and soil properties. In contrast, regions with low diversity might face greater challenges under climate change and intensive human disturbance. Since the influence of ecological water conveyance is currently limited to a distance of 1000 or 2000 m from the river (Si et al., 2005; Zeng et al., 2016), extending the distance of ecological water conveyance was recommended to recharge the surface soil moisture and benefit the growth of ground cover (i.e. herb species), which contribute greatly to species diversity. In addition, multiple conservation measures that protect the soil structure (e.g. artificial soil cover and livestock grazing exclusion) were recommended for regions close to the river to reduce the adverse effects of grazing on soil properties.

Data availability. The spatial field experiment data in this study are available on request. The temporal data used in this study were obtained from reliable public data repositories. Landsat TM/ETM image (30 m) were acquired from Geospatial Data Cloud (http://www.gscloud.cn/). The retrieved remote sensing dataset, land use data, soil moisture monitoring data, groundwater monitoring data and runoff data were obtained from the Environmental & Ecological Science Data Center for West China, National Natural Science Foundation of China (http://westdc.westgis.ac.cn).

Competing interests. The authors declare that they have no conflict of interest.

Acknowledgements. This work was supported by the National Natural Science Foundation of China (no. 91425301), the Program for Changjiang Scholars and Innovative Research Team in University (no.IRT_15R06) and the State key Laboratory of Earth Surface Processes and Resource Ecology. Wenwu Zhao acknowledges the support from China Scholarship Council (no. 201406045031). We are grateful for reviews from the editor, Kevin Bishop, and four anonymous reviewers, who significantly improved the quality of this article. We are thankful to Environmental & Ecological Science Data Center for West China, National Natural Science Foundation of China for providing us the retrieved remote sensing data, monitoring data, land use data, groundwater monitoring data and runoff data. We are also thankful to the Alashan Research Station of Cold and Arid Region Environment and Engineering Research Institution, Chinese Academy of Sciences for their support and contributions to the fieldwork.

Edited by: K. Bishop

References

Aishan, T., Halik, Ü., Cyffka, B., Kuba, M., Abliz, A., and Baidourela, A.: Monitoring the hydrological and ecological response to water diversion in the lower reaches of the Tarim River, Northwest China, Quatern. Int., 311, 155–162, 2013.

Bakker, J. P., Willems, J. H., and Zobel, M.: Long-term vegetation dynamics: Introduction, J. Veg. Sci., 7, 146, 1996.

Cao, S. K., Feng, Q., Si J. H., Zhang, X. F., Liu, W., and Chang, Z. Q.: Research on vegetation and environmental factors in Ejina Oasis: A review, J. Desert Res., 30, 1416–1423, 2010.

Chen, X., Duan, Z., and Luo, T.: Changes in soil quality in the critical area of desertification surrounding the Ejina Oasis, Northern China, Environ. Earth Sci., 72, 2643–2654, 2014.

Daryanto, S., Eldridge, D. J., and Wang, L.: Spatial patterns of infiltration vary with disturbance in a shrub-encroached woodland, Geomorphology, 194, 57–64, 2013.

Décamps, H., Pinay, G., Naiman, R. J., Petts, G. E., Mcclain, M. E., Hillbricht-Ilkowska, A., Hanley, T. A., Holmes, R. M., Quinn, J., Gibert, J., Planty-Tabacchi, A. M., Schiemer, F., Tabacchi, E., and Zalewski, M.: Riparian zones: Where biogeochemistry meets biodiversity in management practice, Polish J. Ecol., 52, 3–18, 2004.

Ding, J, Y. and Zhao, W. W.: Comparing Chinese and international studies of riparian forests: A bibliometric survey (1981–2014), Acta Ecol. Sin., 36, 377–385, 2016.

D'Odorico, P., Caylor, K., Okin, G. S., and Scanlon, T. M.: On soil moisture–vegetation feedbacks and their possible effects on the dynamics of dryland ecosystems, J. Geophys. Res.-Biogeo., 112, 231–247, 2007.

Ejina statistical office: The main data bulletin on sixth demographic census of Ejina oasis in 2010, http://www.ejnqtj.gov.cn/List.asp?ID=583 (last access: 22 September 2016), 2012.

Fan, W. Y., Wang, X. A., Wang, C., Guo, H., and Zhao, X. J.: Niches of major plant species in Malan forest standing in the Loess Plateau, Acta Bot. Boreali-Occident. Sin., 26, 157–164, 2006.

Fang, X., Zhao, W., Wang, L., Feng, Q., Ding, J., Liu, Y., and Zhang, X.: Variations of deep soil moisture under different vegetation types and influencing factors in a watershed of the Loess Plateau, China, Hydrol. Earth Syst. Sci., 20, 3309–3323, doi:10.5194/hess-20-3309-2016, 2016.

Feng, Q.: Ecological water requirements and ecological water regulation in the lower reaches of Heihe, Science Press, Beijing, 2015.

Fu, A. H., Chen, Y. N., and Li, W. H.: Water use strategies of the desert riparian forest plant community in the lower reaches of

Heihe River Basin, China, Sci. China Earth Sci., 44, 693–705, 2014.

Gao, X. M., Huang, J. H., and Wan, S. Q.: Ecological studies on the plant community succession on the abandoned cropland in Taibaishan, Qinling Mountains. I. The community a diversity feature of the successional series, Acta Ecol. Sin., 17, 619–625, 1997.

Gärtner, P., Förster, M., Kurban, A., and Kleinschmit, B.: Object based change detection of Central Asian Tugai vegetation with very high spatial resolution satellite imagery, Int. J. Appl. Earth Obs. Geoinf., 31, 110–121, 2014.

Geng, L., Ma, M., Wang, X., Yu, W., Jia, S., and Wang, H.: Comparison of eight techniques for reconstructing multi-satellite sensor time-series NDVI Data Sets in the Heihe River Basin, China, Remote Sensing, 6, 2024–2049, 2014.

Goebel, P. C., Palik, B. J., and Pregitzer, K. S.: Structure and composition of riparian forests in an old-growth northern hardwood–hemlock watershed, Forest Ecol. Manage., 280, 52–61, 2012.

Gong, Z. T.: Chinese soil geography, Science Press, Beijing, 523–524, 2014.

Hao, X. M., Li, W. H., Guo, B., and Ma, J. X.: Simulation of the effect of root distribution on hydraulic redistribution in a desert riparian forest, Ecol. Res., 28, 653–662, 2013.

He, Z. and Zhao, W.: Characterizing the spatial structures of riparian plant communities in the lower reaches of the Heihe River in China using geostatistical techniques, Ecol. Res., 21, 551–559, 2006.

Hill, M. O.: A fortran program for arranging multivariate data in an ordered two–way table by classification of the individuals and attributes, Section of Ecology and Systematics, Cornell University, Ithaca, New York, 1979.

Hochmuth, H., Thevs, N., and He, P.: Water allocation and water consumption of irrigation agriculture and natural vegetation in the Heihe River watershed, NW China, Environ. Earth Sci., 73, 5269–5279, 2015.

Isbell, F., Craven, D., Connolly, J., Loreau, M., Schmid, B., Beierkuhnlein, C., Bezemer, T. M., Bonin, C., Bruelheide, H., Luca, E. D., Ebeling, A., Griffin, J. N., Guo, Q. F., Hautier, Y., Hector, A., Jentsch, A., Kreyling, J., Lanta, V., Manning, P., Meyer, S. T., Mori, A. S., Naeem, S., Niklaus, P. A., Polley, H. W., Reich, P. B., Roscher, C., Seabloom, E. W., Smith, M. D., Thakur, M. P., Tilman, D., Tracy, B. F., Putten, W. H. V. D., Ruijven, J. V., Weigelt, A., Weisser, W. W., Wilsey, B., and Eisenhauer, N.: Biodiversity increases the resistance of ecosystem productivity to climate extremes, Nature, 526, 574–577, 2015.

ISSCAS – Institute of Soil Sciences, Chinese Academy of Sciences: Physical and chemical analysis methods of soils, Shanghai Science Technology Press, Shanghai, 1978.

Jia, L., Shang, H., Hu, G., and Menenti, M.: Phenological response of vegetation to upstream river flow in the Heihe Rive basin by time series analysis of MODIS data, Hydrol. Earth Syst. Sci., 15, 1047–1064, doi:10.5194/hess-15-1047-2011, 2011.

Jiao, F., Wen, Z. M., and An, S. S.: Changes in soil properties across a chronosequence of vegetation restoration on the Loess Plateau of China, Catena, 86, 110–116, 2011.

Jin, X. M., Hu, G. C., and Li, W. M.: Hysteresis effect of runoff of the Heihe River on vegetation cover in the Ejina oasis in northwestern China, Earth Sci. Front., 15, 198–203, 2008.

Koerselman, W. and Meuleman, A.: The vegetation N:P ratio : a new model tool to detect the nature of nutrient limitation, J. Appl. Ecol., 33, 1441–1450, 1996.

Krieger, D. J.: Economic value of forest ecosystem services: A review, The Wilderness Society, Washington, D.C., 2001.

Li, W., Zhou, H., Fu, A., and Chen, Y.: Ecological response and hydrological mechanism of desert riparian forest in inland river, northwest of China, Ecohydrology, 6, 949–955, 2013.

Liu, C. R., Ma, K. P., Yu, S. L., and Wang, W.: Plant community diversity in Dongling mountain, Beijing, China: Effect of sample size on diversity measures, Acta Ecol. Sin., 17, 584–592, 1997.

Liu, D., Tian, F., Hu, H., Lin, M., and Cong, Z.: Ecohydrological evolution model on riparian vegetation in hyperarid regions and its validation in the lower reach of Tarim River, Hydrol. Process., 26, 2049–2060, 2012.

Liu, G. S.: Soil physical and chemical analysis and description of soil profiles, Chinese Standards Press, Beijing, 1996.

Liu, H., Zhao, W., He, Z., and Liu, J.: Soil moisture dynamics across landscape types in an arid inland river basin of Northwest China, Hydrol. Process., 29, 3328–3341, 2015.

Liu, S. M., Xu, Z. W., Wang, W. Z., Jia, Z. Z., Zhu, M. J., Bai, J., and Wang, J. M.: A comparison of eddy-covariance and large aperture scintillometer measurements with respect to the energy balance closure problem, Hydrol. Earth Syst. Sci., 15, 1291–1306, doi:10.5194/hess-15-1291-2011, 2011.

Liu, W., Wang, Z. J., and Xi, H. Y.: Variations of physical and chemical properties of water and soil and their significance to ecosystem in the Lower reaches of Heihe River, J. Glaciol. Geocryol., 30, 688–696, 2008.

Lü, Y., Zhang, L., Feng, X., Zeng, Y., Fu, B., Yao, X., Li, J., and Wu, B.: Recent ecological transitions in China: greening, browning, and influential factors, Scient. Rep., 5, 8732, 2015.

Luck, G. W., Daily, G. C., and Ehrlich, P. R.: Population diversity and ecosystem services, Trends Ecol. Evol., 18, 331–336, 2003.

Macdonald, R. L., Chen, H. Y. H., Palik, B. P., and Prepas, E. E.: Influence of harvesting on understory vegetation along a boreal riparian–upland gradient, Forest Ecol. Manage., 312, 138–147, 2014.

Maestre, F. T., Valladares, F., and Reynolds, J. F.: The stress-gradient hypothesis does not fit all relationships between plant-plant interactions and abiotic stress: further insights from arid environments, J. Ecol., 94, 17–22, 2005.

Meskinivishkaee, F., Mohammadi, M. H., and Vanclooster, M.: Predicting the soil moisture retention curve, from soil particle size distribution and bulk density data using a packing density scaling factor, Hydrology and Earth System Sciences, 18, 4053–4063, doi:10.5194/hess-18-4053-2014, 2014.

Naiman, R. J. and Décamps, H.: The ecology of interfaces: riparian zones, Ann. Rev. Ecol. Syst., 28, 621–658, 1997.

Nilsson, C. and Berggren, K.: Alterations of riparian ecosystems caused by river regulation, BioScience, 50, 783–792, 2000.

Oksanen, J. and Minchin, P. R.: Continuum theory revisited: what shape are species responses along ecological gradients?, Ecol. Model., 157, 119–129, 2002.

Pabst, R. J. and Spies, T. A.: Distribution of herbs and shrubs in relation to landform and canopy cover in riparian forests of coastal Oregon, Can. J. Bot., 76, 298–315, 2011.

Ravi, S., D'Odorico, P., Wang, L., White, C. S., Okin, G. S., Macko, S. A., and Collins, S. L.: Post-fire resource redistribution in desert grasslands: A possible negative feedback on land degradation, Ecosystems, 12, 434–444, 2009.

Ravi, S., Breshears, D. D., Huxman, T. E., and D'Odorico, P.: Land degradation in drylands: Interactions among hydrologic–aeolian erosion and vegetation dynamics, Geomorphology, 116, 236–245, 2010.

Rosenthal, D. M., Ludwig, F., and Donovan, L. A.: Plant responses to an edaphic gradient across an active sand dune/desert boundary in the great basin desert, Int. J. Plant Sci., 166, 247–255, 2005.

Salter, P. J. and Williams, J. B.: Influence of texture on moisture characteristics of soils, Eur. J. Soil Sci., 16, 310–317, 1965.

Scott, W. A., Adamson, J. K., Rollinson, J., and Parr, T. W.: Monitoring of aquatic macrophytes for detection of long-term change in river systems, Environ. Monit. Assess., 73, 131–153, 2002.

Si, J. H., Feng, Q., Zhang, X. Y., Su, Y. H., and Zhang, Y. W.: Vegetation changes in the lower reaches of the Heihe river after its water import, Acta Bot. Bor.-Occident. Sin., 25, 631–640, 2005.

Stavi, I., Ungar, E. D., Lavee, H., and Sarah, P.: Surface microtopography and soil penetration resistance associated with shrub patches in a semiarid rangeland, Geomorphology, 94, 69–78, 2008.

Stirzaker, R. J., Passioura, J. B., and Wilms, Y.: Soil structure and plant growth: Impact of bulk density and biopores, Plant Soil, 185, 151–162, 1996.

Su, Y. Z. and Zhao, H. L.: Soil properties and plant species in an age sequence of Caragana microphylla plantations in the Horqin Sandy Land, north China, Ecol. Eng., 20, 223–235, 2003.

Ter Braak, C. J. F. and Smilauer: Canoco reference manual and user's guide: software for ordination, version 5.0, Microcomputer Power, Ithaca, USA, 2012.

Wang, L., D'Odorico, P., Evans, J. P., Eldridge, D. J., McCabe, M. F., Caylor, K. K., and King, E. G.: Dryland ecohydrology and climate change: critical issues and technical advances, Hydrol. Earth Syst. Sci., 16, 2585–2603, doi:10.5194/hess-16-2585-2012, 2012.

Wang, P., Zhang, Y., Yu, J., Fu, G., and Fei, A.: Vegetation dynamics induced by groundwater fluctuations in the lower Heihe River Basin, northwestern China, J. Plant Ecol., 4, 77–90, 2011.

Wang, Y., Roderick, M. L., Shen, Y., and Sun, F.: Attribution of satellite-observed vegetation trends in a hyper-arid region of the Heihe River basin, Western China, Hydrol. Earth Syst. Sci., 18, 3499–3509, doi:10.5194/hess-18-3499-2014, 2014.

Wen, X., Wu, Y., Su, J., Zhang, Y., and Liu, F.: Hydrochemical characteristics and salinity of groundwater in the Ejina Basin, Northwestern China, Environ. Geol., 48, 665–675, 2005.

Williams, D. G., Scott, R. L., Huxman, T. E., Goodrich, D. C., and Lin, G.: Sensitivity of riparian ecosystems in and semiarid environments to moisture pulses, Hydrol. Process., 20, 3191–3205, 2006.

Xi, H., Feng, Q., Zhang, L., Si, J., Chang, Z., Yu, T., and Guo, R.: Effects of water and salinity on plant species composition and community succession in Ejina Desert Oasis, northwest China, Environ. Earth Sci., 75, 138, 2016.

Yang, L. X., Chen, S. F., An, J. J., Zhao, F. Z., Han, X. H., Feng, Y. Z., Yang, G. H., and Ren, G. X.: Relationship among community diversity and soil organic matter, total nitrogen under different vegetation types in the gully region of Loess region, Acta Agrest. Sin., 22, 291–298, 2014.

Yang, Y. H., Chen, Y. N., and Li, W. H.: Soil properties and their impacts on changes in species diversity in the lower reaches of Tarim River, Xinjiang, China, Acta Ecol. Sin., 28, 602–611, 2008.

Yi, L., Zhao, L., Ruan, Y., Xiao, H., Cheng, G. D., Zhou, M., Wang, F., and Li, C.: Study of the replenishment sources of typical ecosystems water and dominant plant water in the lower reaches of the Heihe, China, J. Glaciol. Geocryol., 34, 1478–1486, 2012.

Zeng, Y., Xie, Z., Yu, Y., Liu, S., Wang, L., Jia, B., Qin, P., and Chen, Y.: Ecohydrological effects of stream–aquifer water interaction: a case study of the Heihe River basin, northwestern China, Hydrol. Earth Syst. Sci., 20, 2333–2352, doi:10.5194/hess-20-2333-2016, 2016.

Zhang, A., Zheng, C., Wang, S., and Yao, Y.: Analysis of streamflow variations in the Heihe River Basin, northwest China: Trends, abrupt changes, driving factors and ecological influences, J. Hydrol. Reg. Stud., 3, 106–124, 2015.

Zhang, J. T.: Quantitative ecology, Science Press, Beijing, 20–21, 2011.

Zhang, J. T. and Dong, Y.: Factors affecting species diversity of plant communities and the restoration process in the loess area of China, Ecol. Eng., 36, 345–350, 2010.

Zhang, Y., Yu, J., Wang, P., and Fu, G.: Vegetation responses to integrated water management in the Ejina basin, northwest China, Hydrological Processes, 25, 3448–3461, 2011.

Zhang, Y. C., Yu, J. J., Qiao, M. Y., and Yang, H. W.: Effects of eco-water transfer on changes of vegetation in the lower Heihe River basin, Journal of Hydraulic Engineering, 42, 757–765, 2011.

Zhao, W. W., Fu, B. J., and Chen, L. D.: A comparison between soil loss evaluation index and the C-factor of RUSLE: a case study in the Loess Plateau of China, Hydrol. Earth Syst. Sci., 16, 2739–2748, doi:10.5194/hess-16-2739-2012, 2012.

Zheng, D., Li, W. H., Chen Y. N., and Liu, J. Z.: Relations between groundwater and natural vegetation in the arid zone, Resour. Sci., 27, 160–167, 2005.

Zhong, B., Yang, A., Nie, A., Yao, Y., Zhang, H., Wu, S., and Liu, Q.: Finer resolution land-cover mapping using multiple classifiers and multisource remotely sensed data in the Heihe river basin, IEEE J. Select. Top. Appl. Earth Obs. Rem. Sens., 8, 4973–4992, 2015.

Zhu, J. T., Yu, J. J., Wang, P., Yu, Q., and Eamus, D.: Distribution patterns of groundwater-dependent vegetation species diversity and their relationship to groundwater attributes in northwestern China, Ecohydrology, 6, 191–200, 2013.

Zhu, J. T., Yu, J. J., Wang, P., Yu, Q., and Eamus, D.: Variability in groundwater depth and composition and their impacts on vegetation succession in the lower Heihe River Basin, north-western China, Mar. Freshwater Res., 65, 206–217, 2014.

Permissions

All chapters in this book were first published in HESS, by Copernicus Publications; hereby published with permission under the Creative Commons Attribution License or equivalent. Every chapter published in this book has been scrutinized by our experts. Their significance has been extensively debated. The topics covered herein carry significant findings which will fuel the growth of the discipline. They may even be implemented as practical applications or may be referred to as a beginning point for another development.

The contributors of this book come from diverse backgrounds, making this book a truly international effort. This book will bring forth new frontiers with its revolutionizing research information and detailed analysis of the nascent developments around the world.

We would like to thank all the contributing authors for lending their expertise to make the book truly unique. They have played a crucial role in the development of this book. Without their invaluable contributions this book wouldn't have been possible. They have made vital efforts to compile up to date information on the varied aspects of this subject to make this book a valuable addition to the collection of many professionals and students.

This book was conceptualized with the vision of imparting up-to-date information and advanced data in this field. To ensure the same, a matchless editorial board was set up. Every individual on the board went through rigorous rounds of assessment to prove their worth. After which they invested a large part of their time researching and compiling the most relevant data for our readers.

The editorial board has been involved in producing this book since its inception. They have spent rigorous hours researching and exploring the diverse topics which have resulted in the successful publishing of this book. They have passed on their knowledge of decades through this book. To expedite this challenging task, the publisher supported the team at every step. A small team of assistant editors was also appointed to further simplify the editing procedure and attain best results for the readers.

Apart from the editorial board, the designing team has also invested a significant amount of their time in understanding the subject and creating the most relevant covers. They scrutinized every image to scout for the most suitable representation of the subject and create an appropriate cover for the book.

The publishing team has been an ardent support to the editorial, designing and production team. Their endless efforts to recruit the best for this project, has resulted in the accomplishment of this book. They are a veteran in the field of academics and their pool of knowledge is as vast as their experience in printing. Their expertise and guidance has proved useful at every step. Their uncompromising quality standards have made this book an exceptional effort. Their encouragement from time to time has been an inspiration for everyone.

The publisher and the editorial board hope that this book will prove to be a valuable piece of knowledge for researchers, students, practitioners and scholars across the globe.

List of Contributors

X. A. Yin, Z. F. Yang and C. L. Liu
State Key Laboratory of Water Environmental Simulation, School of Environment, Beijing Normal University, Beijing, China

Christof Schneider and Martina Flörke
Center for Environmental Systems Research, University of Kassel, Kassel, Germany

Lucia De Stefano
Department of Geodynamics, Complutense University of Madrid, Madrid, Spain

Jacob D. Petersen-Perlman
Department of Geography, Western Oregon University, Monmouth, Oregon, USA

M. K. van der Molen, I. R. van der Velde and M. C. Krol
Meteorology and Air Quality group, Wageningen University, 6700 AA Wageningen, the Netherlands

R. A. M. de Jeu
VU University Amsterdam, Department of Earth Sciences, De Boelelaan 1085, 1081 HV Amsterdam, the Netherlands
Transmissivity B. V., Space Technology Centre, Huygensstraat 34, Noordwijk, the Netherlands

W. Wagner
Vienna University of Technology, Department of Geodesy and Geoinformation, Gusshausstrasse 27–29, 1040 Vienna, Austria

P. Kolari
Department of Physics, University of Helsinki, Erik Palménin aukio 1, 00014 Helsinke, Finland

J. Kurbatova and A. Varlagin
A. N. Severtsov Institute of Ecology and Evolution of the Russian Academy of Sciences, Leninsky Prospekt 33, 119071 Moscow, Russia

T. C. Maximov and A. V. Kononov
Institute for Biological Problems of the Cryolithozone, 677980, 41 Lenin Ave., Yakutsk, Republic of Sakha (Yakutia), Russia

T. Ohta and A. Kotani
Graduate School of Bioagricultural Sciences, Nagoya University, Furo-cho, Chikusa-ku, Nagoya, 464-8601, Japan

W. Peters
Meteorology and Air Quality group, Wageningen University, 6700 AA Wageningen, the Netherlands
Centre for Isotope Research, Energy and Sustainability Research Institute Groningen, Groningen University, Groningen, the Netherlands

Xuening Fang, Jingyi Ding, Yuanxin Liu and Xiao Zhang
State Key Laboratory of Earth Surface Processes and Resource Ecology, College of Resource and Technology, Beijing Normal University, Beijing 100875, P. R. China

Wenwu Zhao
State Key Laboratory of Earth Surface Processes and Resource Ecology, College of Resource and Technology, Beijing Normal University, Beijing 100875, P. R. China
Department of Earth Sciences, Indiana University–Purdue University Indianapolis (IUPUI), Indianapolis, Indiana 46202, USA

Lixin Wang
Department of Earth Sciences, Indiana University–Purdue University Indianapolis (IUPUI), Indianapolis, Indiana 46202, USA

Qiang Feng
State Key Laboratory of Earth Surface Processes and Resource Ecology, College of Resource and Technology, Beijing Normal University, Beijing 100875, P. R. China
College of Forestry, Shanxi Agricultural University, Taigu, Shanxi 030801, P. R. China

Chunwei Liu
Jiangsu Provincial Key Laboratory of Agricultural Meteorology, College of Applied Meteorology, Nanjing University of Information Science and Technology, Nanjing 210044, China

Ge Sun and Steven G. McNulty
Eastern Forest Environmental Threat Assessment Center, Southern Research Station, USDA Forest Service, Raleigh, NC 27606, USA

Asko Noormets and Yuan Fang
Department of Forestry and Environmental Resources, North Carolina State University, Raleigh, NC 27695, USA

D. Hawtree, F. Tavares-Wahren and K.-H. Feger
Technische Universität Dresden, Institute of Soil Science & Site Ecology, Pienner Str. 19, 01737 Tharandt, Germany

J. P. Nunes, J. J. Keizer, R. Jacinto, J. Santos, M. E. Rial-Rivas and A.-K. Boulet
University of Aveiro, Centre for Environmental and Marine Studies (CESAM), Department of Environment and Planning, 3810-193 Aveiro, Portugal

Gordon C. O'Brien, Retha Stassen and James MacKenzie
University of KwaZulu-Natal, College of Agriculture, Engineering and Science, School of Life Sciences, Scottsville, South Africa

Chris Dickens
International Water Management Institute, Silverton, 0127, South Africa

Eleanor Hines and Wayne G. Landis
Western Washington University, Institute of Environmental Toxicology, Bellingham, Washington, USA

Victor Wepener
Water Research Group (Ecology), Unit for Environmental Sciences and Management, North-West University, Potchefstroom, 2520, South Africa

Leo Quayle
Institute of Natural Resources NPC, Scottsville, 3209, South Africa

Kelly Fouchy
IHE Delft Institute for Water Education, 2601 DA Delft, the Netherlands

P. Mark Graham
University of KwaZulu-Natal, College of Agriculture, Engineering and Science, School of Hydrology, Centre for Water Resources Research, Scottsville, South Africa

Stephen T. Casey, Matthew J. Cohen and Subodh Acharya
School of Forest Resources and Conservation, University of Florida, Gainesville, FL, USA

David A. Kaplan
Engineering School of Sustainable Infrastructure and Environment, Environmental Engineering Sciences Department, University of Florida, Gainesville, FL, USA

James W. Jawitz
Soil and Water Science Department, University of Florida, Gainesville, FL, USA

Stephanie S. Day
Department of Geosciences, North Dakota State University, Fargo, ND 58102, USA

Karen B. Gran
Department of Earth and Environmental Sciences, University of Minnesota Duluth, Duluth, MN 55812, USA

Chris Paola
Department of Earth Sciences, University of Minnesota, Minneapolis, MN 55455, USA

Kai Duan and Yang Zhang
Department of Marine, Earth, and Atmospheric Sciences, North Carolina State University, Raleigh, NC, USA

Shanlei Sun
Department of Marine, Earth, and Atmospheric Sciences, North Carolina State University, Raleigh, NC, USA
Key Laboratory of Meteorological Disaster of Ministry of Education, Nanjing University of Information Science & Technology, Nanjing, Jiangsu Province, China

Ge Sun, Erika Cohen and Steven G. McNulty
Eastern Forest Environmental Threat Assessment Center, USDA Forest Service, Raleigh, NC, USA

Peter V. Caldwell
Coweeta Hydrologic Laboratory, USDA Forest Service, Otto, NC, USA

G. Cassiani, J. Boaga and M. T. Perri
University of Padua, Department of Geosciences, Padua, Italy

D. Vanella and S. Consoli
University of Catania, Department of Agriculture, Food and Environment, Catania, Italy

Martin A. Briggs and John W. Lane Jr
U.S. Geological Survey, Hydrogeophysics Branch, 11 Sherman Place, Unit 5015, Storrs, CT 06269, USA

Judson W. Harvey
U.S. Geological Survey, Water Cycle Branch, M.S. 430, Reston, VA 20192, USA

Stephen T. Hurley
Massachusetts Division of Fisheries and Wildlife, 195 Bournedale Road, Buzzards Bay, MA 02532, USA

Donald O. Rosenberry
U.S. Geological Survey, National Research Program, M.S. 406, Bldg. 25, DFC, Lakewood, CO 80225, USA

Timothy McCobb
U.S. Geological Survey, 10 Bearfoot Road, Northborough, MA 01532, USA

Dale Werkema
U.S. Environmental Protection Agency, Office of Research and Development, National Exposure Research Laboratory, Exposure Methods & Measurement Division, Environmental Chemistry Branch, Las Vegas, NV 89119 USA

Kimberly J. Van Meter
Department of Earth and Environmental Sciences, University of Waterloo, Waterloo, Ontario N2L 3G1, Canada

Michael Steiff
Department of Civil and Environmental Engineering, University of Waterloo, Waterloo, Ontario N2L 3G1, Canada

Daniel L. McLaughlin
Department of Forest Resources and Environmental Conservation, Virginia Tech, Blacksburg, Virginia 24061, USA

Nandita B. Basu
Department of Earth and Environmental Sciences, University of Waterloo, Waterloo, Ontario N2L 3G1, Canada
Department of Civil and Environmental Engineering, University of Waterloo, Waterloo, Ontario N2L 3G1, Canada

B. R. Voortman and R. P. Bartholomeus
KWR Watercycle Research Institute, 3430 BB Nieuwegein, the Netherlands

S. E. A. T. M. van der Zee
Soil Physics and Land Management, Environmental Sciences Group, Wageningen University, 6700 AA Wageningen, the Netherlands

M. F. P. Bierkens
Department of Physical Geography, Faculty of Geosciences, Utrecht University, 3508 TC Utrecht, the Netherlands
Deltares, 3508 AL Utrecht, the Netherlands

J. P. M. Witte
KWR Watercycle Research Institute, 3430 BB Nieuwegein, the Netherlands
VU University, Institute of Ecological Science, Department of Systems Ecology, de Boelelaan 1085, 1081 HV Amsterdam, the Netherlands

Jingyi Ding, Hao Fan and Yaping Wang
State Key Laboratory of Earth Surface Processes and Resource Ecology, Faculty of Geographical Science, Beijing Normal University, Beijing 100875, P. R. China

Wenwu Zhao
State Key Laboratory of Earth Surface Processes and Resource Ecology, Faculty of Geographical Science, Beijing Normal University, Beijing 100875, P. R. China
Department of Earth Sciences, Indiana University - Purdue University, Indianapolis (IUPUI), Indianapolis, Indiana 46202, USA

Stefani Daryanto and Lixin Wang
Department of Earth Sciences, Indiana University - Purdue University, Indianapolis (IUPUI), Indianapolis, Indiana 46202, USA

Qiang Feng
State Key Laboratory of Earth Surface Processes and Resource Ecology, Faculty of Geographical Science, Beijing Normal University, Beijing 100875, P. R. China
College of Forestry, Shanxi Agricultural University, Taigu, Shanxi 030801, P. R. China

Index

www.ingramcontent.com/pod-product-compliance
Lightning Source LLC
LaVergne TN
LVHW060529210926
848485LV00005B/1171